Coherence Phenomena in Atoms and Molecules in Laser Fields

NATO ASI Series

Advanced Science Institutes Series

A series presenting the results of activities sponsored by the NATO Science Committee, which aims at the dissemination of advanced scientific and technological knowledge, with a view to strengthening links between scientific communities.

The series is published by an international board of publishers in conjunction with the NATO Scientific Affairs Division

A	**Life Sciences**	Plenum Publishing Corporation
B	**Physics**	New York and London
C	**Mathematical and Physical Sciences**	Kluwer Academic Publishers
D	**Behavioral and Social Sciences**	Dordrecht, Boston, and London
E	**Applied Sciences**	
F	**Computer and Systems Sciences**	Springer-Verlag
G	**Ecological Sciences**	Berlin, Heidelberg, New York, London,
H	**Cell Biology**	Paris, Tokyo, Hong Kong, and Barcelona
I	**Global Environmental Change**	

Recent Volumes in this Series

Volume 280—Chaos, Order, and Patterns
edited by Roberto Artuso, Predrag Cvitanović, and Giulio Casati

Volume 281—Low-Dimensional Structures in Semiconductors: From Basic Physics to Applications
edited by A. R. Peaker and H. G. Grimmeiss

Volume 282—Quantum Measurements in Optics
edited by Paolo Tombesi and Daniel F. Walls

Volume 283—Cluster Models for Surface and Bulk Phenomena
edited by Gianfranco Pacchioni, Paul S. Bagus, and Fulvio Parmigiani

Volume 284—Asymptotics beyond All Orders
edited by Harvey Segur, Saleh Tanveer, and Herbert Levine

Volume 285—Highlights in Condensed Matter Physics and Future Prospects
edited by Leo Esaki

Volume 286—Frontiers of High-Pressure Research
edited by Hans D. Hochheimer and Richard D. Etters

Volume 287—Coherence Phenomena in Atoms and Molecules in Laser Fields
edited by André D. Bandrauk and Stephen C. Wallace

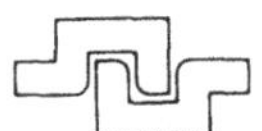

Series B: Physics

Coherence Phenomena in Atoms and Molecules in Laser Fields

Edited by

André D. Bandrauk

Université de Sherbrooke
Sherbrooke, Quebec, Canada

and

Stephen C. Wallace

University of Toronto
Toronto, Ontario, Canada

Springer Science+Business Media, LLC

Proceedings of a NATO Advanced Research Workshop on
Coherence Phenomena in Atoms and Molecules in Laser Fields,
held May 5-10, 1991,
in Hamilton, Ontario, Canada

Library of Congress Cataloging-in-Publication Data

Coherence phenomena in atoms and molecules in laser fields / edited by Andre D. Bandrauk and Stephen C. Wallace.
p. cm. -- (NATO ASI series. Series B, Physics ; v. 287)
"Proceedings of a NATO Advanced Research Workshop on Coherence Phenomena in Atoms and Molecules in Laser Fields, held May 5-10, 1991, in Hamilton, Ontario, Canada"--T.p. verso.
"Published in cooperation with NATO Scientific Affairs Division."
Includes bibliographical references and index.
ISBN 978-0-306-44190-5 ISBN 978-1-4615-3364-1 (eBook)
DOI 10.1007/978-1-4615-3364-1
1. Lasers--Congresses. 2. Coherence (Nuclear physics)--Congresses. 3. Multiphoton processes--Congresses. I. Bandrauk, André D. II. Wallace, Stephen C. III. North Atlantic Treaty Organization. Scientific Affairs Division. IV. NATO Advanced Research Workshop on Coherence Phenomena in Atoms and Molecules in Laser Fields (1991 : Hamilton, Ont.) V. Series.
QC685.C625 1992
621.36'6--dc20 92-3607
CIP

ISBN 978-0-306-44190-5

Originally published by Plenum Press, New York in 1992

PREFACE

This volume contains the lectures and communications presented at the NATO Advanced Research Workshop (NATO ARW 900857) which was held May 5-10, 1991 at McMaster University, Hamilton, Ontario, Canada. A scientific commitee made up of P.P. Lambropoulos (USC & Crete), P.B. Corkum (NRC, Ottawa), and H.B. vL. van den Heuvell (FOM, Amsterdam) guided the organizers, A.D. Bandrauk (Sherbrooke) and S.C. Wallace (Toronto) in preparing a programme which would cover the latest advances in the field of atom and molecule laser interactions.

Since the last meeting held in July 1987 on "Atomic and Molecular Processes with Short Intense Laser Pulses", NATO ASI vol 171B (Plenum Press 1988), considerable progress has been made in understanding high intensity effects on atoms and the concomitant coherence effects. After four years, the emphasis is now shifting more to molecules. The present volume represents therefore this trend with four sections covering the main interests of research endeavours in this area:

i) Atoms in Intense Laser-Fields
ii) Molecules in Intense Laser Fields
iii) Atomic Coherences
iv) Molecular Coherences

The experience developed over the years in multiphoton atomic processes has been very useful and is the main source of our understanding of similar processes in molecules. Thus ATI (above-threshold ionization) has been found to occur in molecules as well as a new phenomenon, ATD (above-threshold dissociation). Laser-induced avoided crossings of molecular electronic surfaces is also now entering the current language of high intensity molecular processes. Multiphoton processes also introduce strong coherences whenever excitation times are shorter than relaxation times. This enables one to create electronic or nuclear wavepackets, with time evolutions which enable one to unravel atomic and molecular dynamics. Ever increasing intensities lead to the observation of high order nonlinear harmonics in atoms and molecules, coulomb explosions in molecules. Coherences lead to the trapping and creation of new atomic and molecular states, control of linear and nonlinear photochemical and photophysical processes.

The clear message from the proceedings of the workshop presented here is that advancing laser technology is pushing forward the frontier of nonlinear photochemistry and photophysics, so that new nonlinear phenomena such as ATI, ATD, X-ray production, control of atomic and molecular processes, subpicosecond dynamical processes are being studied. The theoretical understanding of these new phenomena fosters the development of new nonperturbative numerical and modelling techniques. We are thus witnessing the emergence of a new discipline, nonlinear photochemistry and photophysics.

This timely workshop was generously supported by the NATO Scientific Affairs Division. It has also benefited from a Strategic Grant to A.D. Bandrauk by the Natural Sciences and Engineering Research Council of Canada. We believe that another workshop on the same subject in about four years time would be certainly appropriate, as the subjects covered in this book are expanding rapidly, both experimentally and theoretically.

André D. Bandrauk
Stephen C. Wallace

November 15, 1991

CONTENTS

ATOMS IN INTENSE LASER FIELDS

MOLECULES IN INTENSE LASER FIELDS

ATOMIC COHERENCES

MOLECULAR COHERENCES

SEARCH FOR THE "DIRECT" CHANNEL IN MULTIPHOTON DOUBLE IONIZATION OF MAGNESIUM WITH FEMTOSECOND PULSES

R. Trainham[(a)], N. J. van Druten[(b)], L. D. Noordam[(b)]
H. G. Muller[(b)], P. Breger[(a)], G. Petite[(a)], E. Mevel[(a)]
P Agostini[(a)] and A. Antonetti[(c)]

[(a)]DRECAM/SRSIM CEN Saclay, F-91191 Gif Sur Yvette Cedex
[(b)]FOM, Kruislaan 1098 SJ, Amsterdam, The Netherlands
[(c)]LOA, Ecole Polytechnique, ENSTA, F-91120 Palaiseau

INTRODUCTION

The removal of several electrons by multiphoton absorption from the outer shell of an atom is quite easy to observe with the currently available short pulse lasers. For instance, the outer shell of xenon atoms can be completely stripped out at 10^{16} Wcm^{-2}. After the absorption of the first few photons in the discrete part of the spectrum, the atom can be (say) doubly ionized, in principle, through two quantum paths: either all the intermediate states of the multiphoton transitions are single-electron states (in this case, the electron correlations reflect only weakly in the process, and this results in the formation of the singly-charged ion in its ground state which, in turn can be ionized through only single-electron states etc...), or the multiphoton transition goes mostly through doubly-excited states up to the double ionization threshold (and is therefore more sensitive to electron correlations[1]). In the literature the first case is usually referred to as "sequential" and the second one "direct". In the perturbation picture these names are clearly improper since, strictly speaking, multiphoton absorption is always a sequential process, only one photon being absorbed at each interaction. The more appropriate names of "*correlated*" and "*uncorrelated*" have been proposed.

The probability amplitude of a real multiphoton multiple ionization is the sum of the amplitudes over the two paths and all the open channels will contribute with various branching ratios. Historically, the so-called "direct" process was at first thought to be responsible for most of the multiply charged ions. The reason was double: on the one hand, experimentally, the double (and later on) multiple ionization appeared much easier than could be predicted on the basis of single-electron, non-resonant transitions alone[2]. On the other hand, the "slopes", i.e. the power laws followed by the rates were much smaller than predicted on the same basis. It was subsequently realized that many other explanations of this unexpected behaviour could be offered: first of all, Lambropoulos[3] showed, within the framework of standard perturbation theory, that the saturation intensities of n-photon processes were rapidly increasing and then reaching an asymptotic value as a function of n (typically for n>10). The immediate implication was that a process of order say 20 would appear (and saturate) at approximately the same intensity as a process of order 10. A corollary implication was that during the rise time of a real laser pulse, the low-order processes leading to the production of the

Coherence Phenomena in Atoms and Molecules in Laser Fields
Edited by A.D. Bandrauk and S.C. Wallace, Plenum Press, New York, 1992

first charge states would saturate first, thus enormously reducing the possibilities of "correlated" processes. This argument mostly applied to the multiple ionization of rare gases. Second, several experiments revealed the influence of ionic resonances in the case of alkaline-earth atoms that both increased the ionization rates and lowered the slopes thus removing most of the mysteries from the available data. For instance, in an alkaline earth atom, the absorption of the first few photons normally results in the excitation of only one of the outer electrons. Even though, for a suitably chosen wavelength, a multiphoton resonance through a doubly excited state may be forced, it will enhance essentially the production of ionic excited states through the absorption of a few "above-threshold" photons but, in general, will not lead to the double ionization threshold "directly". The second threshold is most likely to be reached through multiphoton (resonant) transitions which involve only (single-electron) ionic excited states. Actually, the "correlated" process, albeit possible in principle, is in practice competing with all the other channels and is normally responsible only for an extremely small part of the multiply charged ions. Although it has not been clearly demonstrated up to now, direct double ionization remains a fascinating goal for experiments and a potentially powerful tool to investigate multiply excited states (at least the doubly-excited states in alkaline earth atoms) and electron correlations.

In this paper, we first briefly review the methods to isolate this "correlated" process and then report the progress of the (so-far unsuccessful) quest in multiphoton double ionization of magnesium with femtosecond pulses.

FAVORING AND DETECTING THE MULTIPHOTON "CORRELATED" DOUBLE IONIZATION

Isolated Core Excitation (ICE) scheme

The doubly excited states (n_1l_1,n_2l_2), which normally lie above the first threshold, can be roughly classified in categories according to the expectation values of the radial coordinates $\langle r_1\rangle$ and $\langle r_2\rangle$. If $\langle r_1\rangle \gg \langle r_2\rangle$ and $\langle r_2\rangle \cong 1$ a.u. the outer electron is far away from the core and the corresponding energy region is between the singly-charged ion ground state and its first excited states. In this case n_1 is large and n_2 is small. These states form Rydberg series converging to the low-lying ionic excited states. If $\langle r_1\rangle \gg \langle r_2\rangle$ and $\langle r_2\rangle \gg 1$ a.u. they form Rydberg series converging to the highly excited ionic states. Finally if $\langle r_1\rangle \cong \langle r_2\rangle \gg 1$ a.u. the states are located just below the double ionization threshold and are sometimes referred to as Wannier states. In order to study the "correlated" process, one might be attempted to try the technique known as the ICE method[4] which has been extensively used to study the autoionizing region of the spectrum. A first laser (or possibly two lasers) resonantly excites one of the two electrons on a Rydberg bound state ns md for example. It is then far enough from the core to remain spectator while another laser excites the core electron to a p state resulting in, say, a np md autoionizing state. Although extremely efficient to produce cleanly low-lying autoionizing Rydberg series, this method, from its very principle, is not likely to allow driving the atom above the double ionization threshold or even to the Wannier region[5]. Actually, it will lead to mostly low lying states of the singly charged ion and, consequently to virtually no "correlated" double ionization.

Non-Resonant multiphoton transitions

Taking an opposite strategy, one may try to force a completely non-resonant multiphoton transition from the atom ground state up to the Wannier region and above. This works for instance in the XUV range, for single-photon double ionization[6]. However, this method fails too in the case of multiphoton transitions with low-energy (visible) photons. The experiments show that, because of the high density of autoionizing states just above the single ionization threshold, there is a very large branching ratio for channels leading again to the formation of low-lying excited ionic states

Symmetric resonant channels

An alternate route seems to be multiphoton transitions resonant only with strongly correlated states in which the two electrons have the same principal quantum numbers (although it is not clear in this case that n_1 and n_2 always remain good quantum numbers). This is not easy in practice however, since the energies of these states are largely unknown, and their ability to enhance multiphoton transitions is also unknown. Nevertheless, some observations[7] seem to indicate that this is the right way to go in the case of magnesium where a transition resonant with the $3p^2$ was found to lead to highly excited states of the ion, only one photon (2 eV) below the double ionization threshold.

Short pulses versus long pulses

The other experimental parameter is the pulse duration. There are at least two arguments in favor of short pulses. Since one wants to enhance the highest-order channel it is natural to take pulses as short as possible to increase the corresponding saturation intensity. Further, the shorter the pulse is the less "sequential" (in the sense of the two electrons being emitted independently of each other) the process can be, especially under the conditions where the first emitted electron has a low kinetic energy and the second a higher one. The electron-electron Coulomb interaction seems bound to play an important role in this case. However, the drawback of short pulses is that the threshold shifts are not fully compensated by the ponderomotive effect, and the photoelectron energies depend on the intensity at which they are released thereby shifting and broadening the electron peaks. This complicates, in the non-resonant case, the interpretation of the electron energy spectra which, as discussed in the next section remain potentially the best possible signature for the "correlated" process.

Electron energy spectra vs ionization rates measurements

One (and historically the first) way, to study multiphoton multiple ionization is to detect the ions produced and to analyze their charge state in a time-of-flight spectrometer. Then plotting the various rates as a function of intensity could in principle reveal the correlated channel, at least when it requires a different number of photons than the sum of the photon numbers required for the sequential steps. However, the difference is at most 1 and therefore the slope measurements must be extremely accurate, a difficult condition to meet, particularly for large slopes. Electron spectra, on the contrary, are easy to record and potentially contain a very clear signature since the "uncorrelated" process leads to discrete peaks while the "correlated" one produces, in principle, a continuous spectrum according to the condition:

$$E_1 + E_2 = Nh\nu - IP_1 - IP_2 \qquad (1)$$

where E_1 and E_2 are the kinetic energies of the photoelectrons released in the "correlated" process, IP_1, IP_2 the atom and ion ionization potentials respectively and $h\nu$ the photon energy. It is likely, however, that this continuous spectrum can easily be buried in the wings of the large discrete peaks corresponding to the various uncorrelated channels and the associated ATI spectra.

MULTIPHOTON IONIZATION OF MAGNESIUM WITH FEMTOSECOND PULSES

For magnesium, which was chosen for this study, $IP_1 = 7.55$ eV and $IP_2 = 15$ eV. Around 590 nm, the single ionization requires 4 - photon

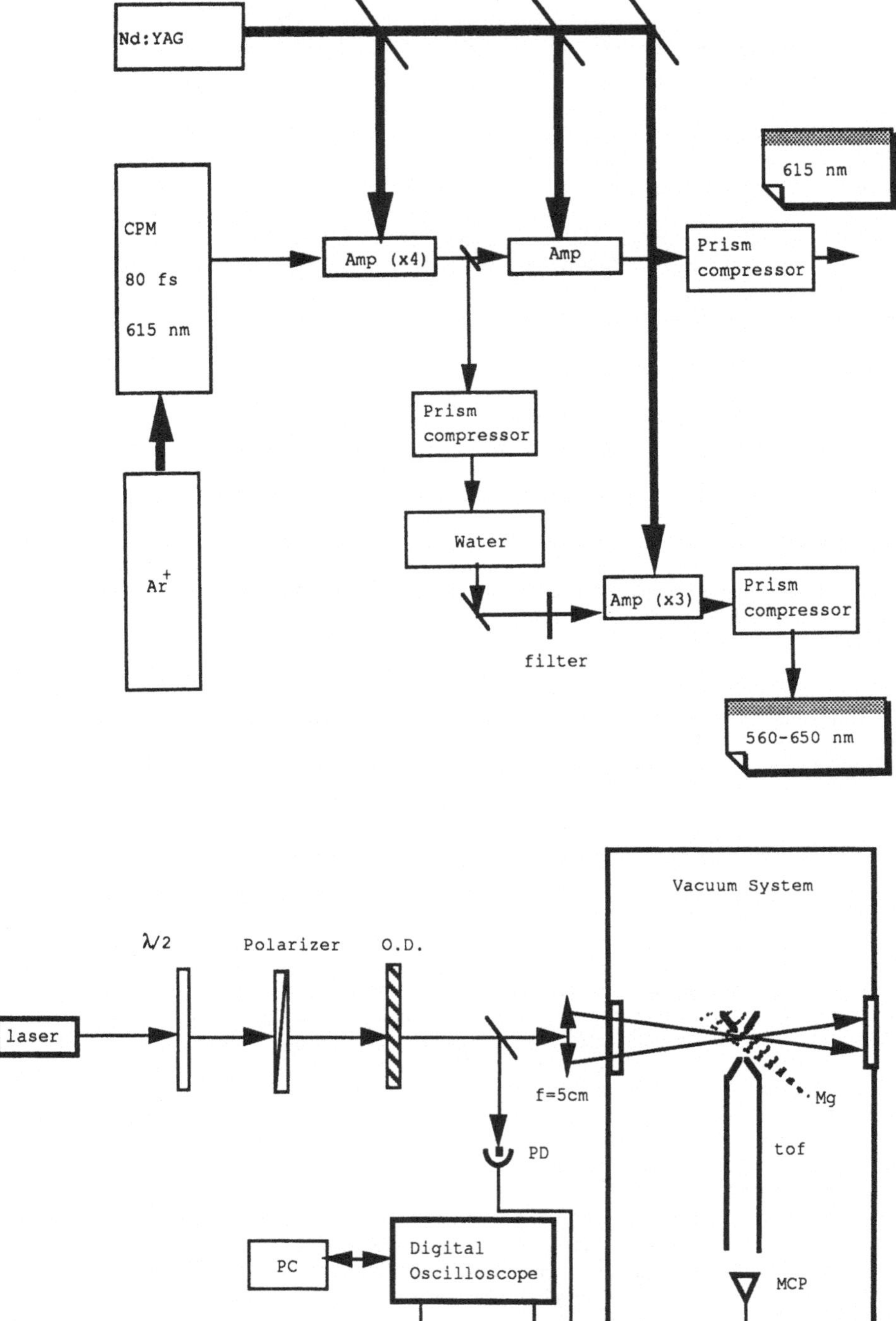

Fig. 1 Experimental setup

absorption while the "uncorrelated" and "correlated" double ionization require 12-photon absorption. Previous studies[7] in the picosecond regime have shown that it was possible to force (with a branching ratio of a few percent) 9- or 10-photon transitions to reach the 4p and 5p thresholds if the laser was tuned to the four-photon resonance with the $3p^2\ ^1S$ state.

The aim of the present work is to test the influence of the pulse duration in the subpicosecond regime and to try the covariant mapping technique to extract information from the energy spectra.

Experimental setup

The laser is an amplified 80 fs CPM dye laser and provides either 1 mJ, 615 nm pulses or, after continuum generation, wavelength selection, amplification and compression 200 to 400 μJ, 100 fs pulses in the range 560-650 nm. The beams are focused by a f=5 cm achromat between the poles of a magnetic-bottle spectrometer where it crosses a magnesium effusive thermal beam. Ions can be charge-analyzed and electrons energy-analyzed in a time-of-flight tube and detected on a twin channel-plates device. The signals are collected by a digital oscilloscope and processed by a computer. The laser energy is monitored by a fast photocell whose signal is used to bin the laser shots and the corresponding spectra into narrow laser intensity windows (Fig. 1). The pulse energy is varied either through natural fluctuations or through variations controlled by a polarizer-half wave plate device and optical density filters.

Ion detection: saturation and doubly to singly charged ions ratio

Fig. 2 shows the Mg^+ and Mg^{++} signals as a function of the laser pulse energy at 590 nm. Single ionization saturates around 4 mJ and the Mg^{++}/Mg^+ maximum ratio is about .1

Tuning the wavelength between 580 and 605 nm results in the Mg^+ and Mg^{++} lineshapes shown in Fig.3. The tuning range was here limited to the Rhodamine 6G & 610 range used in the amplifiers. It could be easily extended by using other dyes.

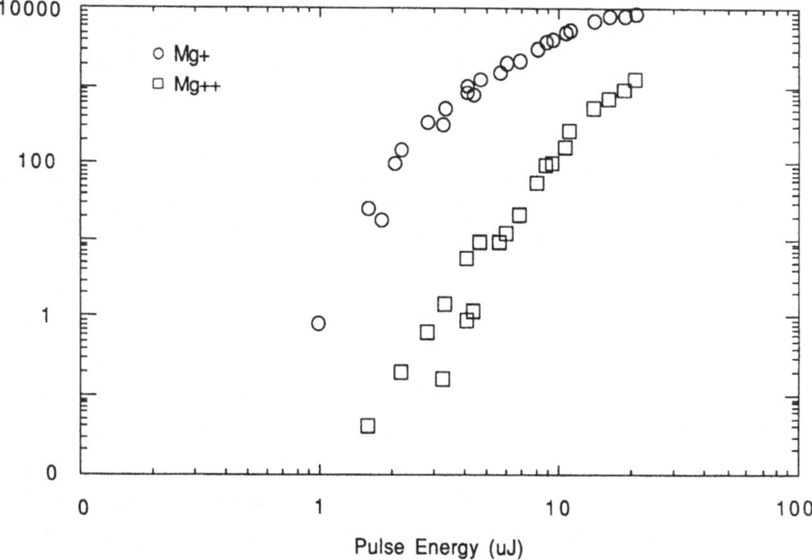

Fig. 2 Ions signals vs pulse energy

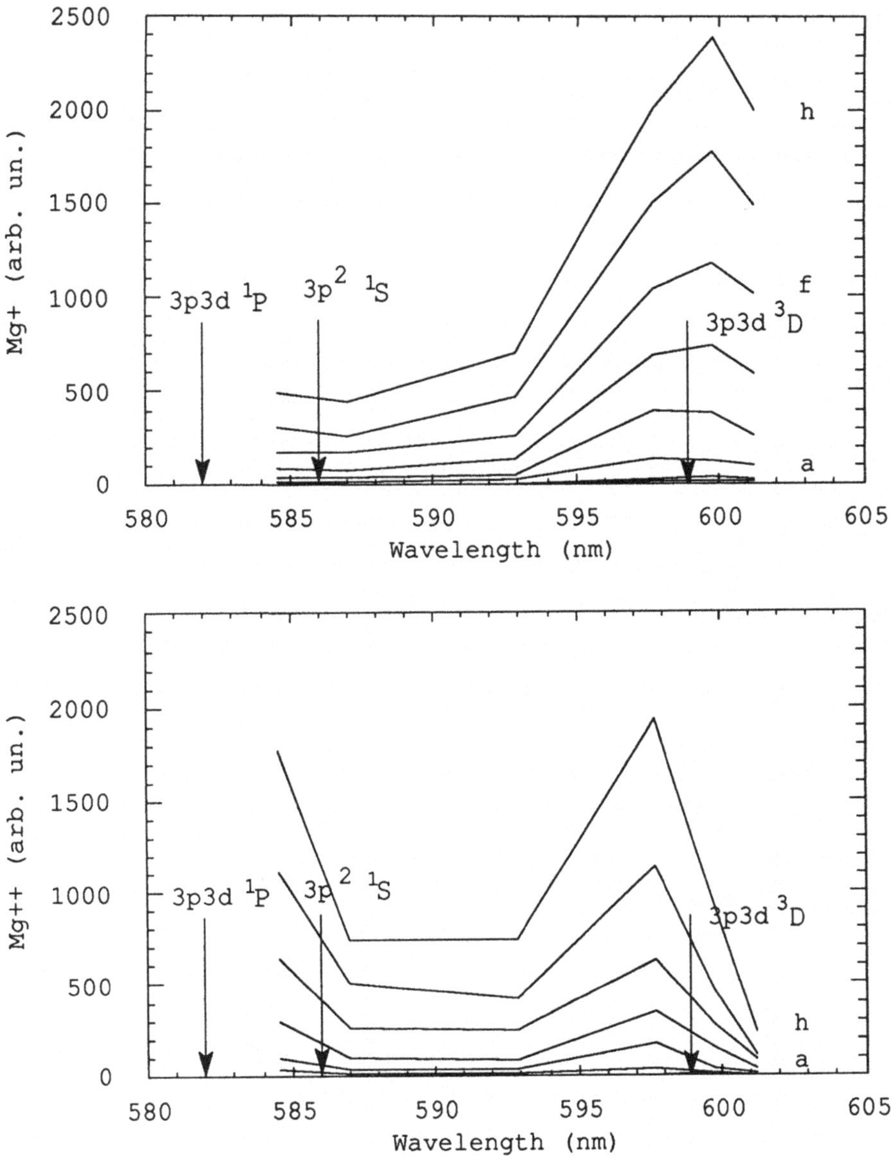

Fig. 3. Mg^+ (top) & Mg^{++} (bottom) ion signals vs wavelength. The curves a-f are for increasing laser pulse energy over a range of a few microjoules. The arrows indicate resonances for 4-photon absorption to the $3p^2\ {}^1S$ autoionizing state and 5-photon absorption to the two 3p3d states.

Electron energy spectra: shifts, covariance mapping

Electron spectra

Fig. 4 shows some electron energy spectra at 590 nm. The most prominent peaks correspond to the 4-photon (0.75 eV) ionization of Mg and the first ATI order. Several other transitions may be recognized (and will not be

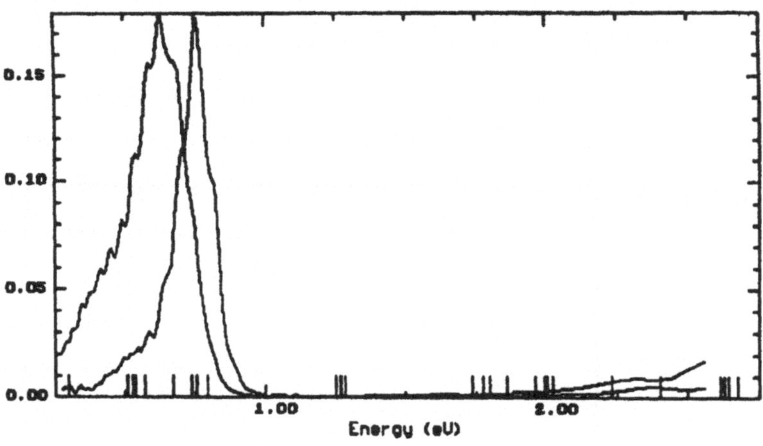

Fig. 4 Electron energy spectrum at 590 nm For two different intensities. The shift of the 0.75 eV peak is clearly visible.

commented upon here), but essentially no trace of the continuum due the correlated process can be identified. The 0.75 eV peak is shifting as a function of the pulse energy as expected for short pulses and nonresonant ionization.

Covariance mapping

To investigate the relationships between various electron peaks in a multiphoton ionization spectrum we employ the technique of covariance mapping. This technique has been applied to multiphoton molecular fragmentation[8] yielding insights into the photofragmentation process which were unavailable by previous experimental techniques. The idea is that the measurement of higher order moments of a physical process yields more information concerning the probability distribution function for that process than does a simple measurement of the first order moment (i.e., the average signal).

The electrons removed from different atoms are identically but independently distributed in energy, giving a distribution function for the ensemble of electrons which is the product of the individual probability functions. In such case there is no correlation in the distribution function, and the sample covariance tends to zero. If, however, two electrons derive from the same atom then the sum of their energies is constrained by the number of photons absorbed by the atom and by the final internal state of the ion. Moreover, there is a correlation between them. The probablility distribution function representing these two electrons includes a correlation term, and the computed sample covariance will be nonzero. It is the relationship between electrons emanating from a single atom that give rise to nonzero covariances. Even if several atoms are ionized at each laser shot the independent events will tend to a sample covariance of zero, and only correlated energies yield nonzero values. Of course, the noise level of the system will place a lower limit on a significant correlation, and detector efficiency will place an upper limit on the sample covariance given a particular value of the correlation.

If we consider two small ranges of energy then we can associate with each range an average rate, variance, and correlation. The distribution function for the pair is the bivariate normal distribution. And in general, for n variates (energy ranges) we have the multivariate normal distribution:

$$p(X) = \frac{1}{\sqrt{2\pi}^N \sqrt{|K|}} \exp\{-\frac{1}{2}(X^t K^{-1} X)\} \quad (2)$$

where $X = (x_1-m_1, x_2-m_2, \ldots, x_N-m_N)$ is a column vector of the rates minus their averages, X^t is its transpose row vector, and K^{-1} is the inverse of

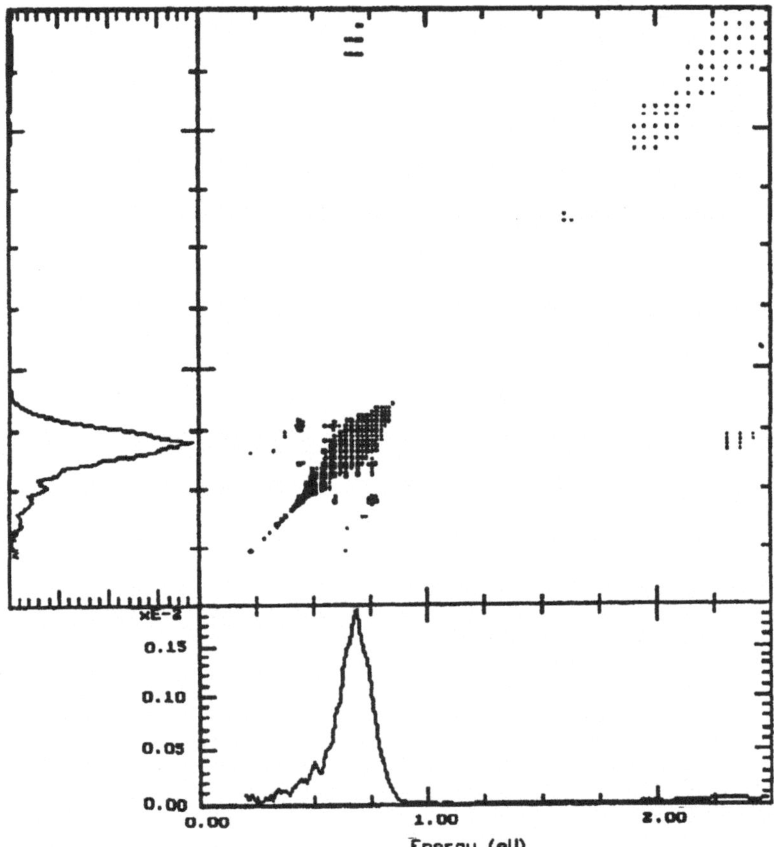

Fig. 5 Sample of covariance map for the electron energy spectrum at 590 nm.

the covariance matrix. For a collection of observations, the probability of observing that particular collection is the product of the individual probabilities. For M observations, the probability is

$$P = \frac{1}{(\sqrt{2\pi}\,)^{MN} \, (\sqrt{|K|})^M} \exp\{-\frac{1}{2} \sum X^t K^{-1} X\} \quad (3)$$

Since the argument of the exponential is a 1x1 matrix it is the same as the trace of the 1x1 matrix. A property of the trace of a product of matrices is that it is preserved under commutation of the product. Thus

$$\sum X^t K^{-1} X = tr(\sum X^t K^{-1} X) = tr(K^{-1} \sum X X^t) = tr (K^{-1} S) \quad (4)$$

where $S = \sum X X^t$ is a square matrix for which an individual element is $(S)_{ij} = \sum(x_i-m_i)(x_j-m_j)$ the sample covariance. If we require that the above probability be unconditionally maximized, i.e., we invoke the property of maximum likelyhood, we find that the distribution covariance which maximizes the probability is the one which is equal to the sample covariance.[9]

A serious problem arises when a third independent parameter correlates two electron energies which would otherwise be independent in the absence of the mutual correlation to the third parameter. The intensity of the laser light is just such a parameter. Laser fluctuations induce fluctuations in the electron production thereby inducing correlations among all of the electron peaks. This is a catastrophic effect because these artificial correlations can overwhelm the single-atom correlations. To circumvent this problem, one must take pains to reduce to a minimum the fluctuations of the laser intensity.

In the best case, our laser fluctuates by 10 percent from shot to shot. This level of fluctation is much too important and the consequent induced correlations can easily overwhelm the real correlations which we seek. One solution to this problem is to window the data collection by laser intensity. In this way, the laser fluctuations for a particular data set can be reduced to below the one percent level (the lower limit is set by the 8 bit resolution of the oscilloscope) and since the data is binned by laser intensity, one collects covariance data over a range of intensities. The residual correlation left in the data due to laser fluctuation can be treated by computing a partial covariance which mathematically freezes out the effect of the fluctuating laser. This technique is of limited utility since it removes only linear cross correlations[10].

An example covariance map is shown in figure 6. On the sides of the figure are shown the average electron spectra from which one can identify the peaks giving rise to covariance islands in the central part of the figure. Unfortunately the structure shown on the map is not readily explained by the energy level diagram for magnesium. The covariances are small and the correlations induced by laser fluctuations are not completely removed. This points to the fact that our detection is inefficient and that the range of allowed laser fluctuation is too large. We are currently improving the system and hope to present better maps soon.

CONCLUSION

In conclusion, the multiphoton double ionization of magnesium atoms by visible subpicosecond pulses still does not show substantial trace of the "correlated" process in the electron energy spectra. The more sophisticated technique of covariant mapping has been attempted but without conclusive result so far. More experimental effort is required, especially the spectroscopy of high-lying autoionizing states, to increase the chances to observe this rather elusive process.

ACKNOWLEDGMENTS: the support of the EC (contract SC1-0103C) is gratefully acknowledged.

REFERENCES

1. L.F. DiMauro, Kim Dalwoo, S. Fournier, and M. Saeed, "Magnesium atoms in an intense nonresonant laser field", *Phys. Rev. A* **41**, 4966 (1990); X. Tang, T. N. Chang, P. Lambropoulos, S. Fournier and L. F. DiMauro, "Multiphoton ionization of magnesium with configuration-interaction calculation", *Phys. Rev. A* **41**, 5265 (1990).
2. N. B. Delone, V. V. Suran and B. A. Zon, "Many-Electron Processes in Nonlinear Ionization of Atoms", in : *Multiphoton Ionization of Atoms,* S. L. Chin and P. Lambropoulos ed., Academic Press, New York (1984).

3. P. Lambropoulos, Phys. Rev. Lett. 55, 2141 (1985).
4. W. E. Cooke, T. F. Gallagher, S. A. Edelstein and R. M. Hill, *Phys. Rev. Lett.* **40**, 178 (1978).
5. P. Camus, P. Pillet and J. Boulmer, "Spectroscopic studies of 9dnd' doubly excited autoionising states of neutral barium", *J. Phys. B,* **18**, L481 (1985).
6. M. Ya. Amusia , *Atomic Photoeffect*, Plenum, New York (1990).
7. Hou M., P. Breger, G. Petite and P. Agostini, "Direct multiphoton transitions to highly excited two-electron states", *J. Phys. B* **23**, L583 (1990).
8. L.J. Frasinki, K. Codling, P.A. Hatherly, "Covariance Mapping: A Correlation Method Applied to Multiphoton Multiple Ionization", *Science* **246**, 1029 (1989).
9. John Parker Burg, David G. Luenberger, and Daniel L. Wenger, "Estimation of Structured Covariance Matrices", *Proc. IEEE* **70**, 963 (1982).
10. W.J. Krzanowski, *Principles of Multivariate Analysis*, (Clarendon Oxford, 1988).

INHIBITION OF ATOMIC IONIZATION IN STRONG LASER FIELDS

Bernard Piraux and Etienne Huens

Département de Physique, Université Catholique de Louvain
2 chemin du cyclotron
B1348 Louvain-la-Neuve (Belgium)

INTRODUCTION

When an atom interacts with an intense laser pulse which generates electric fields comparable to the atomic (Coulombic) field, high order processes become dominant as for example multiphoton ionization[1] and above threshold ionization[2]. In some particular circumstances however, coherence phenomena may be at the origin of a substantial inhibition of ionization and lead to the stabilization of the atom. In this contribution, we analyze in detail a physical mechanism which is responsible for strong suppression of ionization[3]. We show that the intense field excitation of the atom may generate a new kind of spatially extended wave packet through virtual transitions from the initial state via high-lying Rydberg and continuum states to a coherent superposition of Rydberg states with a small initial overlap with the nucleus. Conventional atomic wave packets[4] are created by direct short pulse excitation of overlapping Rydberg states from a compact initial source which ensures a large initial overlap with the nucleus and a substantial ionization. By contrast, our wave packet stems from extended high-lying states which are accessed virtually through Raman coupling rather than directly through short pulse excitation from compact low-lying states.

In order to illustrate this mechanism, we consider the one-photon resonant excitation two-photon ionization of atomic hydrogen initially in the metastable 2s-state. We show that a low-lying p-state is first populated through the resonant excitation; then, through the Raman coupling between p-intermediate states, the population in the low-lying state is gradually transferred to high-lying Rydberg states. Because this is not a direct process, the p-wave packet is formed relatively far away from the nucleus. As a result, its interaction with the field is expected to be weak. In addition, because it is made of many excited states, this wave packet is highly non-harmonic and spreads very rapidly; this rapid spreading prevents the overlap with the nucleus from becoming large at any time.

Our theoretical description of the excitation dynamics of atomic hydrogen is based on three different approaches : the numerical solution of the one- and three-dimensional time-dependent Schrödinger equation and the essential states method which has the merit to be analytical.
In the one-dimensional model[5], the atom is described by a "soft core"

Coherence Phenomena in Atoms and Molecules in Laser Fields
Edited by A.D. Bandrauk and S.C. Wallace, Plenum Press, New York, 1992

potential[6] which is Coulombic at large distances and the time-dependent Schrödinger equation is solved numerically in the Kramers-Henneberger accelerated reference frame[7]. The essential states approach[8] consists in expanding the total wave function in a reduced set of unperturbed states of atomic hydrogen; within the rotating wave approximation (R.W.A.) the perturbative expansion for the probability amplitudes is generated by means of a simple recursion formula and summed by using analytical continuation techniques[9]. In this case, the length form for the interaction Hamiltonian (evaluated in the Coulomb gauge and within the dipole approximation) is used. The three-dimensional time-dependent Schrödinger equation for atomic hydrogen with the interaction Hamiltonian in the velocity form is solved numerically after expanding the total wave function on a Sturmian basis[10].

Besides the present mechanism of suppression of ionization, two other processes which lead to atomic stabilization in the high frequency regime (photon energy larger than the unperturbed binding energy) have been studied in detail[11]. The first process which occurs when the ponderomotive energy is larger or equal to the photon energy may be explained in terms of pure kinematical considerations[12]. In the high frequency regime, ionization takes place through a hard collision during which the electron must acquire a drift momentum much larger than the characteristic atomic orbital momentum; when the ponderomotive energy becomes larger than the photon energy, one can show that energy is hardly conserved during the ionization process. The second process is discussed by Parker and Stroud in ref (13). They show through a series of numerical models that if the finite bandwidth of the laser pulse overlaps several, initially unpopulated states, these ones may be pumped into a coherent superposition that inhibits or prevents excitation of the atom, effectively trapping population in the initial state. This trapping of population in the initial state results from a transfer of population from states of the same parity as the initial state; the transfer occurs via the ionization continuum through a stimulated Raman transition. This mechanism is studied by Feodorov et al in ref (14) in the case where the initial state is a high lying Rydberg state. They show that atomic stabilization occurs when the ionization width of the Rydberg states becomes larger than the spacing between neighboring Rydberg levels.

This contribution is organized as follows : in the first section, we briefly describe the one-dimensional model and discuss the result for the ionization yield. The second section is devoted to the essential state approach; we first describe the theoretical method and then analyze in detail the results for the time evolution of various unperturbed state populations and of the p-wave packet, the ionization yield and the electron energy spectrum. In the last section, we give a brief description of the numerical method used to solve the three-dimensional time dependent Schrödinger equation and discuss preliminary results.

1. ONE DIMENSIONAL MODEL

1.1. DESCRIPTION OF THE MODEL AND METHOD OF SOLUTION

The potential associated to our model atom is given by the following expression (from now on unless stated, atomic units are used)

$$V(x) = \frac{-1}{\sqrt{1+x^2}} \tag{1.1}$$

This potential introduced by Eberly et al[6] presents several important properties : the electron motion is continuous around $x = 0$, parity is a good quantum number and it supports two series of Rydberg states.

The numerical solution of the corresponding stationary Schrödinger equation gives a ground state energy of -0.67 a.u. ($\sim$ -18 eV). We assume the following field time dependence :

$$E(t) = E_0 \sin^2(\pi t/T) \sin(\omega t) \tag{1.2}$$

where T is the pulse length and E_0, ω respectively the maximum amplitude and the frequency of the field. Instead of solving numerically the one dimensional time-dependent Schrödinger equation in the laboratory frame, the calculations are performed in the Kramers-Henneberger accelerated frame of reference[7] in which the electron is asymptotically free. This reduces the necessary computer time dramatically[5]. In this accelerated frame of reference, the time dependent Schrödinger equation reads :

$$i \frac{\partial}{\partial t} \Phi(x,t) = \left[-\frac{1}{2} \frac{\partial^2}{\partial x^2} + V(x+\alpha(t)) \right] \Phi(x,t) \tag{1.3}$$

where $\alpha(t)$ describes the oscillations of a free electron in the presence of the electric field (1.2). The maximum amplitude is given by $\alpha_0 = E_0/\omega^2$. In order to solve numerically eq(1.3), a large grid in space is defined with boundaries well removed from the nucleus. The time propagation of the solution is then performed by using the Crank-Nicholson algorithm[15].

1.2. RESULTS AND DISCUSSION

In the following, we consider a two-photon ionization process from the n=2 level of our model atom and analyze the time evolution of the initial state population and the ionization yield. We consider two different frequencies : in the first case, the laser is tuned just below resonance with the n=2 to n=37 transition (see figs 1a and 1b) and in the second case, just below resonance with the n=2 to n=5 transition (see fig. 1c and 1d). In both cases, the pulse duration is 8 cycles, and the peak intensity, 1.32 10^{13} Wcm^{-2} which corresponds to an electric field of the order of 10^8 Vcm^{-1}. This electric field should be compared to the atomic field. In the case of atomic hydrogen, the atomic field scales as n^{-4} where n is the principal quantum number; on the 2s-orbit, the Coulombic field is equal to $\sim$ 3 10^8 Vcm^{-1}. In the high frequency case, the short pulse duration leads to a transient excitation of many Rydberg states around n=37 and to the formation of a regular wave packet of the usual kind created by Fourier overlap which, as it prediodically traverses the nucleus, increases its ionization : at the end of the eight-cycle pulse, 2.32 % of the population is ionized (see fig. 1b). If the laser is tuned to lower lying levels, a different wave packet is generated which cannot be of Fourier overlap type because the energy gap between neighboring excited states is larger than the laser induced bandwidth. By contrast to the previous situation, the ionization is substantially suppressed as clear from fig. 1d : now, at the end of the eight-cycle pulse, only 0.18% of the population (i.e. a suppression of a factor 12) is ionized. An analysis[3] of the time evolution of the population in the continuum states clearly demonstrates the importance of Raman type transitions (via the continuum) which transfer population from the n=5 state to higher lying excited states. This numerical study also shows that the number of states which play a significant role in the ionization dynamics, is relatively small : a set of intermediate Rydberg levels, a single continuum, and essentially no continuum - continuum process. In these conditions, it is clear that an essential state type of approach is adequate.

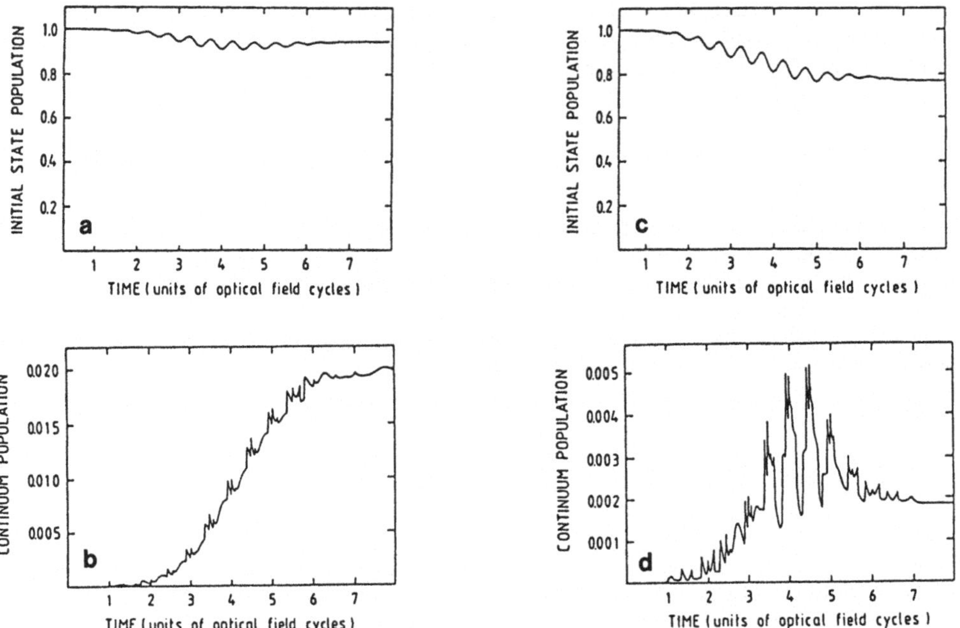

Fig. 1. Initial state and continuum population as a function of time for the case of a eight cycle pulse excitation with peak intensity of 1.32 10^{13} Wcm^{-2}. In (a) and (b) the frequency is very close to the n=2 to n=37 transition frequency and in (c) and (d), to the n=2 to n=5 transition frequency (from ref. 3).

2. ESSENTIAL STATE APPROACH

2.1. DESCRIPTION OF THE MODEL AND ESSENTIAL STATES METHOD

The radiation field is described classically (within the dipole approximation) as follows :

$$\vec{E} = \vec{\varepsilon}\, E_0 f(t) \cos(\omega t) \tag{2.1}$$

where $\vec{\varepsilon}$ is the polarization vector and as before, E_0 and ω represent respectively the maximum amplitude and the frequency of the electric field. Here, we assume that the pulse envelope f(t) is given by :

$$f(t) = \text{sech}\left(\frac{t}{\tau}\right) = \frac{2}{e^{t/\tau} + e^{-t/\tau}} \quad ; \tag{2.2}$$

the duration of the pulse is fixed by the parameter τ, the full width at half maximum (FWHM) of the envelope being given by 2.63 τ. We work in the Coulomb gauge and choose the length form for the interaction Hamiltonian $V(\vec{r},t)$:

$$V(\vec{r},t) = - \vec{r} \cdot \vec{E}(t) \ . \tag{2.3}$$

The essential states method consists in expanding the total wave function $\Psi(\vec{r},t)$ in a reduced basis of atomic hydrogen unperturbed state wave functions $\phi_\alpha(\vec{r})$; these unperturbed states are those which play a significant role in the physical process which is considered. As already indicated in section 1, just a few unperturbed states are expected to dominate the ionization dynamics : a set of p - Rydberg states, the initial 2s-state and the s- and d- continuum. This is also clearly indicated in the results obtained with the method described in section 3. In the following, and for clarity, we also neglect the contribution of the d-continuum. Therefore, we write the total wave function as follows :

$$\Psi(\vec{r},t) = \sum_{\alpha} a_{\alpha}(t)\ \phi_{\alpha}(\vec{r}) e^{-iE_{\alpha}t} \tag{2.4}$$

where α stands for s and p- states and the summation becomes an integration when α refers to a continuum state; E_α is the energy associated with the unperturbed state wave function $\phi_\alpha(\vec{r})$ and $a_\alpha(t)$ the corresponding probability amplitude. We introduce the transformations :

$$a_{np}(t) = e^{i(E_{np}-E_{2s}-\omega)t}\ b_{np}(t)\ , \tag{2.5}$$

$$a_{Es}(t) = e^{i(E_{s}-E_{2s}-2\omega)t}\ b_{Es}(t) \tag{2.6}$$

together with the change of variable

$$z = e^{t/\tau} \tag{2.7}$$

and obtain, within the RWA the following system of coupled equations for the amplitudes of probability :

$$\frac{z}{\tau}\frac{d}{dz}\, a_{2s}(z) = -\, i \left(\frac{z}{z^2+1}\right) \sum_n \Omega_{2s,np}\, b_{np}(z), \tag{2.8a}$$

$$\frac{z}{\tau}\frac{d}{dz}\, b_{Es}(z) = -\, i\, \Delta_{Es} b_{Es}(z) - i\left(\frac{z}{z^2+1}\right) \sum_n \Omega_{Es,np}\, b_{np}(z), \tag{2.8b}$$

$$\frac{z}{\tau}\frac{d}{dz}\, b_{np}(z) = -\, i\, \Delta_{np} b_{np}(z) - i\left(\frac{z}{z^2+1}\right)\left[\Omega_{np2s} a_{2s}(z) + \int_0^{\infty} dE_s\, \rho(E_s)\,\Omega_{np,Es} b_{Es}(z)\right] \tag{2.8c}$$

where we set :

$$\Delta_{Es} = E_s - E_{2s} - 2\omega, \tag{2.9}$$

$$\Delta_{np} = E_{np} - E_{2s} - \omega; \tag{2.10}$$

$\rho(E_s)$ is the density of state in the continuum which reduces to unity if the continuum state wave function is normalized in the energy space. The dipole matrix elements $\Omega_{\alpha,\beta}$ are given by :

$$\Omega_{\alpha,\beta} = E_0 < \alpha \mid \vec{r} \cdot \vec{\varepsilon} \mid \beta > . \tag{2.11}$$

Although the RWA is the weak point of the present analysis, the results obtained in section 3 clearly suggest that the counter-rotating wave contributions are negligible. A detailed analysis[8] of eqs (2.8) reveals that the probability amplitudes admit a taylor series around z=0. The convergence of this series is however very slow. In fact, it turns out that the coefficient of the nth power of z represents the nth order term of the standart perturbative expansion. In order to overcome the problem of the convergence, we introduce the following expansions :

$$a_{2s}(z) = \sum_{m=0}^{\infty} a_{2s}^{m} \left(\frac{z^2}{z^2+1}\right)^m , \tag{2.12a}$$

$$b_{Es}(z) = \sum_{m=0}^{\infty} b_{Es}^{m} \left(\frac{z^2}{z^2+1} \right)^m \tag{2.12b}$$

$$b_{mp}(z) = z \sum_{m=0}^{\infty} b_{np}^{m} \left(\frac{z^2}{z^2+1} \right)^m \tag{2.12c}$$

and use the method of the Padé approximants[9] to resum the series. The coefficients a_{2s}^{m}, b_{2s}^{m} and b_{np}^{m} are evaluated by means of a simple recursion formula obtained after substitution of expressions (2.12) into the system of eqs (2.8). It is easy to show[8] that the coefficients a_{2s}^{m}, b_{2s}^{m} and b_{np}^{m} may be expressed in terms of one quantity denoted by $\xi_{np,n'p}^{2m}$ which governs the ionization dynamics :

$$\xi_{np,n'p}^{2m} = \sum_{\nu} \frac{<np|V|\nu><\nu|V|n'p>}{E_{2s}+2\omega - E_{\nu} + \frac{2im}{\tau}} \tag{2.13}$$

where ν refers to the s-states, the summation becoming an integration when ν refers to a continuum state. When n=n', $\xi_{np,n'p}^{2m}$ contributes to both the ac-Stark shift and the induced width of each intermediate p-state; it also governs the Raman mixing between intermediate p-states. Let us remark at this stage that because of the 2s-2p degeneracy the Raman Coupling between the 2p and other p-states via the 2s-state is strongly overestimated in the RWA but this coupling is essentially cancelled when the counter-rotating component is included; we therefore deliberately exclude the 2p-state.

2.2. RESULTS AND DISCUSSION

In this section, we consider the two-photon ionization from the 2s-state by a hyperbolic secant pulse (see eqs (2.1) and (2.2)) with a photon energy around 2.8 eV i.e. quasi resonant with the 2s-5p transition.

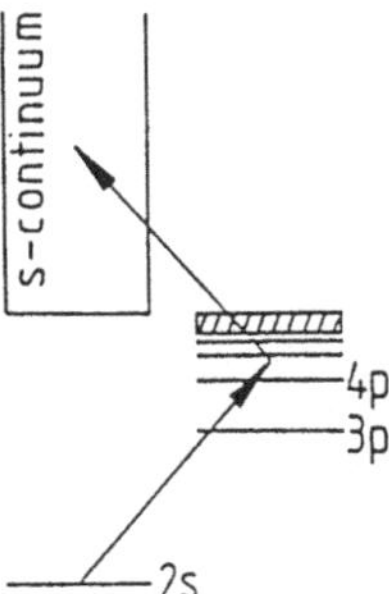

Fig. 2. Schematic representation of the excitation process studied in section 2.

All the results are obtained by using eqs (2.12) and the level scheme represented in fig. 2. We first analyze the ionization yield as a function of the peak intensity I_0 of the pulse for various times. Our non-perturbative results are compared to those obtained within second-order perturbation theory for a pulse duration of 32 fs. Our calculations

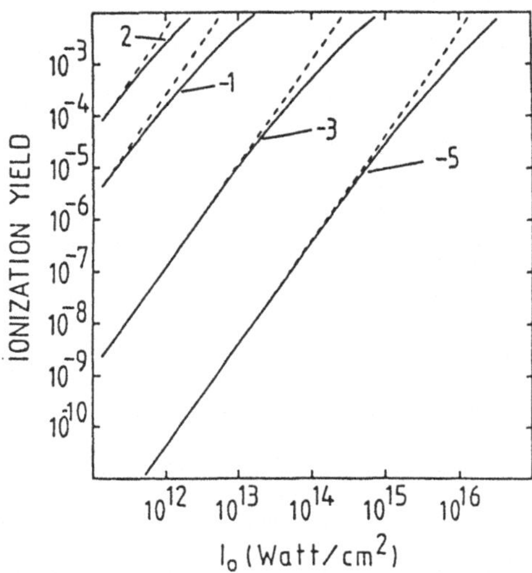

Fig. 3. Log-log plot of the ionization yield resulting from the excitation process represented schematically in fig. 2 as a function of I_0, the peak intensity of the hyperbolic secant pulse. The FWHM of the pulse equals 32 fs and the photon energy is equal to 2.8 eV. The label associated with each curve refers to the value of t/τ (τ=500 a.u.). The solid line corresponds to our non perturbative result and the dashed line to the perturbative result (from ref. 8).

include 27 p-states. On the log-log plot (see fig. 3), the perturbative results (dashed line) reduces to a straight line of slope equal to 2. We clearly see that the peak intensity at which the saturation of the ionization yield occurs depends strongly on the time, measured in units of τ and indicated by the label associated with each curve. For $t/\tau=2$, which corresponds essentially to the end of the pulse, non-perturbative effects become important already at peak intensities around 10^{12} Wcm^{-2}. As mentioned before, such a low peak intensity is expected because the internal Coulombic field on the 2s-orbit is much smaller than the atomic unit of electric field. In the following, we choose the peak value of the electric field strength equal to 10^8 Vcm^{-1} (i.e. 1/3 the Coulombic field on the 2s-orbit) which corresponds to a peak intensity of 1.3 10^{13} Wcm^{-2}. In figs 4,5, we analyze the time evolution of the (unperturbed) p-state population, the total population in the (unperturbed) p-states[16] and the ionization yield for two pulse durations : 6 fs and 32 fs; in both cases, we observe a very low ionization yield with 95% of the population trapped in a superposition of the initial 2s-state and the intermediate p- Rydberg states. More physical insight is gained by studying the time evolution of the ionization yield as a function of the number of intermediate p-states included in the calculations. The results are presented in fig. 6 for the same case as in fig. 5 (i.e. a pulse duration of 32 fs). The dashed line represents the results obtained by including the intermediate 5p state only, while the solid line corresponds to a calculation including 13 p-states. Clearly, there is a strong destructive interference effect among the contribution of many intermediate states. This destructive interference effect is also present in the electron energy spectrum shown in fig. 7 for the same parameters as in fig. 5 and for a time equal to 2τ. The solid line is the result obtained by including the intermediate 5p-state only, and the dashed line is the result obtained by including 13 p-states. The electron energy spectrum exhibits the usual n=5 Autler-Townes splitting which is substantially narrowed and reduced in peak heights when compared with a simple two-level model.

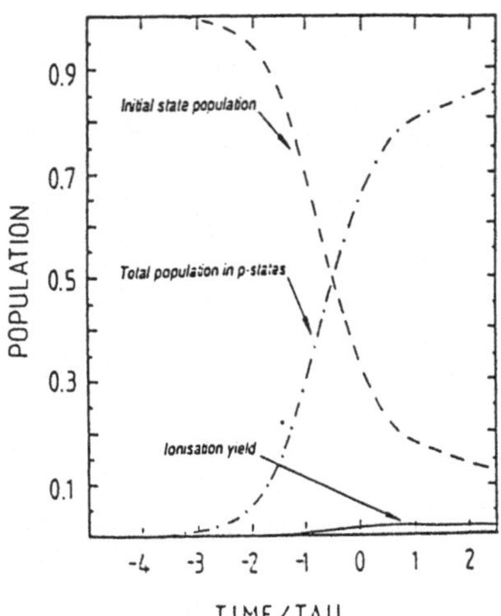

Fig. 4. Time evolution of various populations for the excitation process represented schematically in fig. 2. The hyperbolic secant pulse has a FWHM of 6 fs a peak intensity of 1.3 10^{13} Wcm^{-2} and the photon energy is equal to 2.8 eV (from ref. 3).

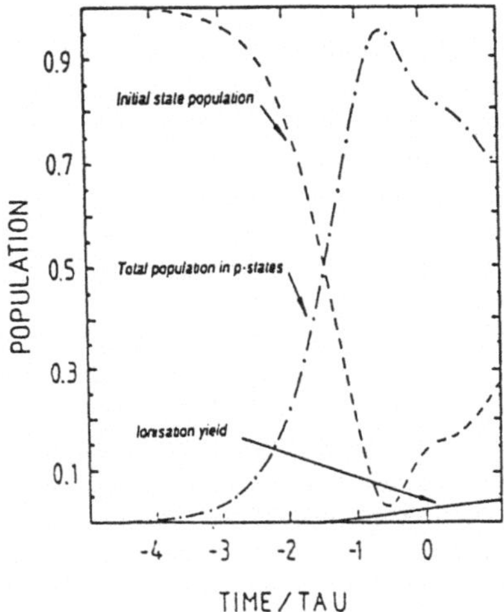

Fig. 5. Same as fig. 4 but for a pulse with a FWHM equal to 32 fs (from ref. 3).

Let us now study the formation of the radial p-wave packet w(r,t) defined as follows :

$$w(r,t) = r^2 |\Psi_p(r,t)|^2 = r^2 \left|\sum_n a_{np}(t)\phi_{np}(r)\right|^2 \tag{2.14}$$

where $\phi_{np}(r)$ is the unperturbed radial np-wave function. The time evolution of this wave packet is shown in fig 8 for a pulse duration of 6 fs

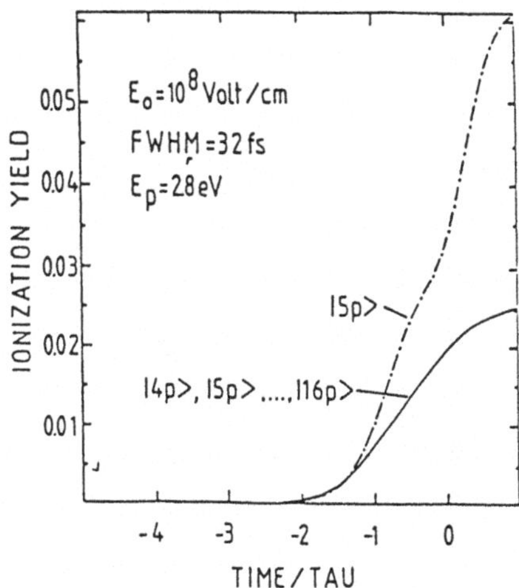

Fig. 6. Time evolution of the ionization yield for the same case as in fig. 5. The dashed line is obtained by including only the 5p state in the calculations and the solid line by including 13 p-states (from ref. 8).

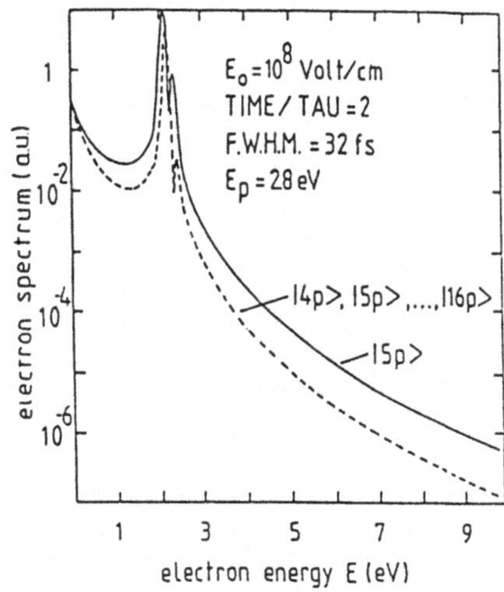

Fig. 7. Electron energy spectrum in a.u. resulting from the excitation process represented in fig. 2, for the same case as in fig. 5. The solid line is obtained by including only the 5p state in the calculations and the dashed line by including 13 p-states (from ref. 8).

and a photon energy of 2.8 eV and in fig 9 for a pulse duration of 32 fs and a frequency corresponding to the exact unperturbed resonance frequency of the 2s-5p transition. Both graphs clearly show the same features : when the system has just started to interact with the laser pulse, the wave packet exhibits a shape appropriate to a 5p-state. After a longer time, some population is ionized (essentially the population close to the nucleus), but more is transferred through the Raman coupling between intermediate states towards high-lying p-states so that the maximum amplitude extends out to roughly 50 a.u. and as time goes on, moves out to

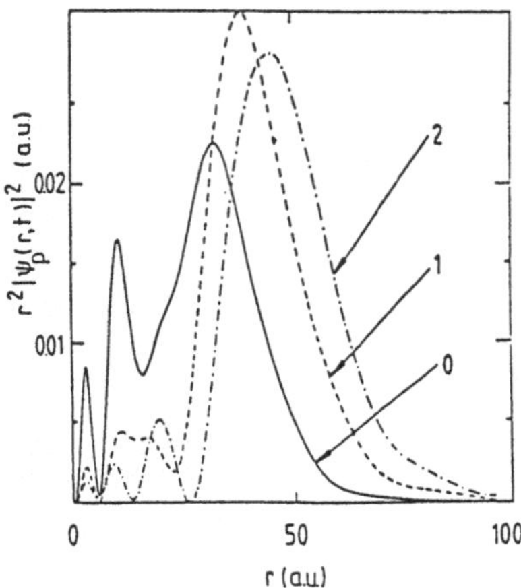

Fig. 8. Radial p-wave packet resulting from the excitation process represented in fig. 2 for the same case as in fig. 5. The labels 0,1,2 reflect the times in unit of τ (from ref. 8).

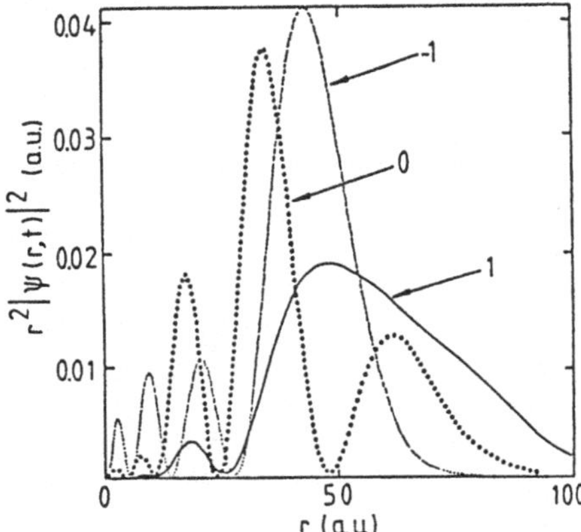

Fig. 9. The same as in fig. 8 but for a pulse with a FWHM equal to 32 fs and a frequency corresponding to the exact unperturbed resonance frequency of the 2s-5p transition (from ref. 8).

higher r values. This wave packet is so far from the nucleus that it weakly interacts with the field with only the inner "core" effectively ionized.

3. NUMERICAL SOLUTION OF THE THREE DIMENSIONAL TIME DEPENDENT SCHRODINGER EQUATION

3.1. THE METHOD

We work in the Coulomb gauge and use the "velocity" from for the interaction Hamiltonian :

$$V(\vec{r},t) = -\frac{1}{c}\vec{A}\cdot\vec{p} \tag{3.1}$$

where $\vec{A}$, the potential vector is given (in the dipole approximation) by :

$$\vec{A}(t) = \vec{A}_0 \; e^{-(t/\tau)^2} \sin(\omega t) \tag{3.2}$$

The full wave function $\Psi(\vec{r},t)$ which describes atomic hydrogen in presence of the laser pulse is expanded in a complete basis of Sturmian functions :

$$\Psi(\vec{r},t) = \sum_{n,l,m} a_{nl}(t) \frac{S_{nl}^{no}}{r} Y_{lm}(\hat{r}) \; ; \tag{3.3}$$

the Sturmian function S_{nl}^{no} is given by :

$$S_{nl}^{no}(r) = N_{n,l}^{no} \; r^{l+1} \; e^{-r/no} \; {}_1F_1(-(n-l-1);\; 2l+2;\; 2r/no) \tag{3.4}$$

where the normalization constant N_{nl}^{no} is calculated such that $\langle S_{nl}^{no} | S_{nl}^{no} \rangle = 1$. The Sturmian basis which has the advantage to be discrete is well adapted to the hydrogen problem; the matrices associated to the kinetic energy and the (Coulombic) potential energy operators are diagonal and the overlap matrix which results from the non-orthogonality of the Sturmian functions is tri-diagonal. These simple matrix structures allow large scale calculations. Details about the numerical procedure are given in ref (10).

3.2. RESULTS AND DISCUSSION

We consider the two-photon ionization of atomic hydrogen in the 2s-state by an intense laser pulse with a peak intensity of 10^{13} Wcm^{-2}. The potential vector associated to the laser electric field is given by eq 3.2 and the FWHM of the envelope is equal to 5 optical periods; the photon energy is equal to 2.8 eV. For these parameters, the ionization yield is only equal to 0.1. In fig. 10, we show the radial probability density in atomic units as a function of the radial distance r and for various times in unit of optical period. For short times, at the beginning of the interaction with the laser pulse, the radial probability density exhibits a 2s-shape. For $t \geq 0$, this density has two maxima; a peak around r=50 which corresponds to the p-wave packet and a peak around r=10 which results from "Rabi oscillations" of the population between the 2s-state and a group of p-states. The oscillatory behavior of the density at large distances reflects the asymptotic behavior of the radial Coulomb function; it represents the population which is in the s and d continuum. A detailed analysis of the numerical results also shows that the ns-states ($n > 2$) and the nd-states are hardly populated during the excitation process.

CONCLUSION

We have demonstrated that superintense-field excitation of atoms may generate spatially extended wave packets which substantially suppress the atomic ionization. We have identified the mechanism responsible for this wave-packet formation as Raman coupling via high-lying and continuum states and showed how it can be described analytically using a simple restricted essential-states analysis.

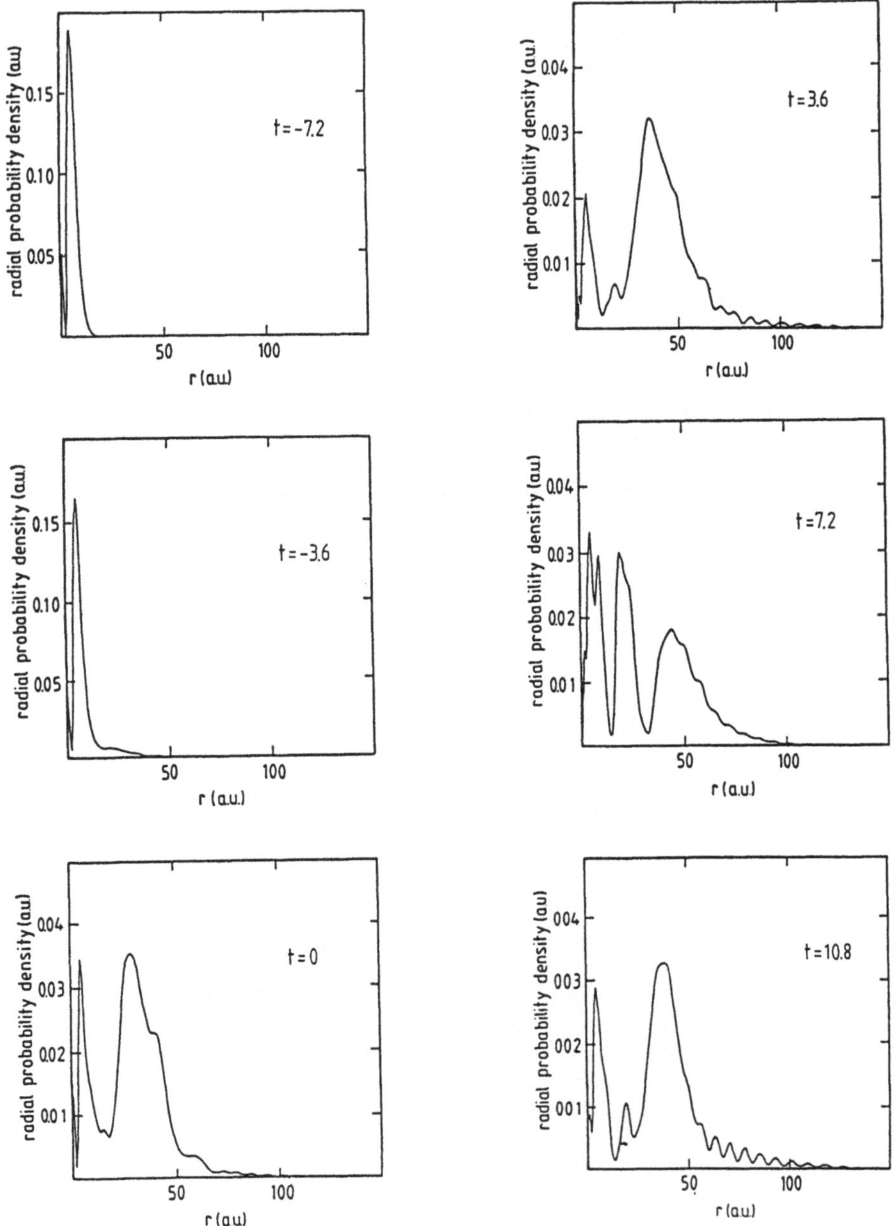

Fig. 10. Radial probability density in atomic unit resulting the two-photon ionization of atomic hydrogen in the 2s-state by a laser pulse. The potential vector has a Gaussian envelope of FWHM equal to 5 optical periods; the peak intensity is 10^{13} Wcm^{-2} and the photon energy equals 2.8 eV. Each graph corresponds to a different time t given in unit of optical period.

REFERENCES

1. P. Lambropoulos, Ad. At. Mol. Phys 12, 87 (1976); J. Morellec, D. Normand and G. Petite, ibid. 18, 97 (1982).
2. P. Agostini and G. Petite, Contemp. Phys. 29, 57 (1988).
3. K. Burnett, P.L. Knight, B. Piraux and V.C. Reed, Phys. Rev. Lett. 66, 301 (1991).
4. See for example : J. Parker and C.R. Stroud, Jr, Phys. Rev. Lett. 56, 716 (1986); G. Alber, H. Ritsch and P. Zoller, Phys. Rev. A34, 1058 (1986).
5. V.C. Reed and K. Burnett, Phys. Rev. A42, 3152 (1990).
6. See for example J. Javanainen, J.H. Eberly and Q. Su, Phys. Rev. A38, 3430 (1988).
7. H.A. Kramers, Collected Scientific papers (North-Holland, Amsterdam, 1956), p. 262; W.C. Henneberger, Phys. Rev. Lett. 21, 838 (1968).
8. B. Piraux, E. Huens and P.L. Knight, Phys. Rev. A44, 721 (1991).
9. C.M. Bender and S.A. Orszag in "Advanced Mathematical Methods for Scientists and Engineers" (McGraw-Hill, New York 1978) chap. 7.
10. M. Pont and R. Shakeshaft to be published; B. Piraux and E. Huens, submitted for publication.
11. M. Pont and M. Gavrila, Phys. Rev. Lett. 65, 2362 (1990); Q. Su, J.H. Eberly and J. Javanainen, Phys. Rev. Lett. 64, 862 (1990); K.C. Kulander, K.J. Schafer and J.L. Krause, Phys. Rev. Lett. 66, 2601 (1991).
12. M. Pont and R. Shakeshaft, submitted as a rapid communication in Phys. Rev. A.
13. J. Parker and C.R. Stroud, Jr, Phys. Rev. A40, 5651 (1989); A41, 1602 (1990).
14. M.V. Feodorov and A.M. Movsesian, J. Opt. Soc. Am. B5, 850 (1988); J. Phys. B21, L155 (1988); J. Opt. Soc. Am. B6, 928 (1989).
15. See for example W.H. Press, B.P. Flannery, S.A. Teukolsky and W.T. Vetterling in "Numerical Recipies" (Cambridge University press, 1989) chap. 17.
16. Although the s and p states are strongly mixed and lose their identity, it is instructive to analyze the time evolution of these populations in order to understand the excitation dynamics.

A HYDROGEN ATOM IN AN INTENSE, HIGH FREQUENCY, SHORT PULSE LASER

X. Tang S. Basile* and P. Lambropoulos

Dept. of Physics, University of Southern California
Los Angeles, CA 90089-0484
*Istituto di Fisica Teorica dell'Università
Casella Postale 50, 98166 Sant'Agata di Messina, Messina, Italy

The behavior of an atom under a superstrong, high-frequency field has been actracting quite a bit of attention in the community of theoretical multiphoton physics. This pertains to the question of possible stabilization with respect to ionization under such conditions as proposed by Pont *et al.*[1,2] A more recent paper by Pont and Gavrila[3] has raised considerable doubt about the possible direct experimental observation of such an effect. Having now evaluated the possibility for the atom to survive the lower (in relative terms) intensities during the rising and falling of the pulse, these authors obtain a negative answer. As has been shown in Ref. 4, the explicit time dependence of the laser intensity due to its pulsed nature plays a fundamental role in understanding the processes of interaction with strong fields and unless it is included in a calculation, one can not be sure about the validity of the prediction. The calculation of Pont and Gavrila[3] on the other hand, being inherently time-independent, provides only average lifetimes at fixed intensities and not the complete time evolution of the system in a realistic pulse.

There would be no doubt about the theoretical prediction of atomic stabilization with respect to ionization, if an atom could be exposed to a constant superstrong and high-frequency field. This has been proved by both time-independent[1,2,5] and time-dependent[6,7] calculation. Since high power lasers are inevitably pulsed, one of the major questions has to do with the dynamics of an atom in such a *realistic* laser *pulse*. The answer has been complicated further by the results of 1-dimensional atomic models which although easier to handle computationally, lack many of the fundamental properties of a real atom.

In this paper, we present results of 1- and 3-dimensional quantum- mechanical calculations for a hydrogen atom in an intense *pulsed* laser field. We use a pulse consisting of a Gaussian shape of the field which will obviously expose the atom to intensities lower than the peak one not only during the rise but also during the fall of the pulse.

The calculations reported here were performed in terms of square- integrable basis sets,[8] and in the framework of non-relativistic quantum mechanics and in the dipole approximation as has been the case with all related calculations. The time-dependent wave function of the electron is obviously going to spread out as a function of time. In this sense the length gauge form of the dipole interaction (the E gauge), which is proportional to $\vec{r}$, will increase without bound, and, as pointed out by Kulander *et al.*[7] recently, it will require much more effort to achieve numerical convergence in the results. Therefore we first convert the Hamiltonian to the velocity gauge form of the dipole interaction (the A gauge). As the A^2 term can be absorbed in a phase factor which does not affect the populations, the Schrödinger equation for the correctly transformed effective wave function is

$$i\hbar\frac{\partial}{\partial t}\Psi_A(\vec{r},t) = [H^0 + e\vec{A}(t)\cdot\vec{p}/mc]\Psi_A(\vec{r},t) \tag{1}$$

Coherence Phenomena in Atoms and Molecules in Laser Fields
Edited by A.D. Bandrauk and S.C. Wallace, Plenum Press, New York, 1992

Here H^0 is the Hamiltionian of the unperturbed atom and the vector potential $\vec{A}(t)$ is related to the electric field $\vec{E}(t)$ by $\vec{E}(t) = -(1/c)\frac{\partial}{\partial t}\vec{A}(t)$. Due to the assumed Gaussian pulse shape, the asymptotic values for both the field and the vector potential are $\vec{E}(\pm\infty) = 0$ and $\vec{A}(\pm\infty) = 0$. The relation between the wavefunction Ψ_E in the E gauge and the wavefunction Ψ_A in the A gauge is given by the well known Göppert-Mayer transformation:

$$\Psi_E(\vec{r}, t) = e^{ie\vec{r}\cdot\vec{A}(t)/\hbar c}\Psi_A(\vec{r}, t) \tag{2}$$

To obtain the final results in the E gauge, in principle one should first convert the initial state wave function of H^0 to the A gauge, perform the time integration, and then convert the solution back to the E gauge before calculating the population or probabilities.[8] The use of a Gaussian pulse shape will make the gauge transformation not necessary, since we will be interested only in the wavefunction at the end of the pulse.

We choose the vector potential to have the form:

$$\vec{A}(t) = \vec{\epsilon}\frac{cE_0}{\omega}\cos(\omega t)f(\frac{t}{\tau}) \tag{3}$$

where $\vec{\epsilon}$ is the unit polarization vector and $f(t/\tau)$ has the form $e^{-\frac{1}{2}(\frac{t}{\tau})^2}$. More details about the method of calculations can be found elsewhere.[8]

As a test for the method and the invariance under the gauge transformation, we first calculate single-photon ionization of hydrogen at different peak intensities (which defined as $I_0 = cE_0^2/8\pi$, is considered more as parameter of the numerical calculation than a quantity with a physical counterpart, since a rigorous definition of intensity for very short pulses is impossible), by using both the length and the velocity gauges. Because of the short pulse duration, the time derivative of the pulse envelope function f cannot be neglected in the definition of the electric field in terms of the vector potential. With the above definition of the vector potential, the electric field is then:

$$\vec{E}(t) = \vec{\epsilon}E_0[\sin(\omega t) + \frac{t}{\omega\tau^2}\cos(\omega t)]f(\frac{t}{\tau}) \tag{4}$$

The results listed in the Table 1 show a perfect agreement (when attention is paid to the correct definition of the electric field) between the length and the velocity forms of the interaction for the populations of the ground state and the total population in the continuum (positive energy) states, at the end of the pulse with τ=5 optical cycles. Through this test, we also found that increasing the peak intensity to higher and higher values, we have increasingly greater difficulty in obtaining numerically converged results when using the length gauge. This is because not only in the length gauge more CPU time is needed to perform the time integration (even with the same number of basis functions in both gauges), but also the length gauge requires more angular momentum functions to get the same degree of accuracy with respect to the velocity gauge. This result is consistent with the conclusion drawn in Ref. 7.

Table 1. Comparison between length and velocity gauge results for the populations of the ground state (g) and of the continuum states (c) at the end of the pulse with τ=31.4159 a.u. (5 optical cycles) for 3-dimensional hydrogen with photon energy $\hbar\omega$=1 a.u. The peak intensity I_0 is expressed in terms of the atomic unit $I_a = 3.51\times10^{16}$ W/cm^2.

Intensity I_0 (a.u.)	Length Gauge	Velocity Gauge
0.005	g=0.9507 c=0.0493	g=0.9507 c=0.0493
0.025	g=0.7773 c=0.2227	g=0.7772 c=0.2228

We have also numerically tested (for the sake of brevity the results are not presented here) that the relative importance of the second term in the definition of the electric field in Eq. (4) actually decreases with increasing the value of τ. Obviously, for $\tau \gg 2\pi/\omega$, a correct definition of the intensity in terms of the time average of the square of the electric field over one optical cycle becomes possible.

After having gained some confidence in the gauge invariance of our results, we have performed numerical calculations for both the very popular 1-dimensional model (the so-called soft core model) and the real hydrogen. We have investigated the surviving probability at the end of the pulse in a range of intensities for which the ionization rate is believed to be strongly suppressed. The binding potential[6] for this 1-dimensional model is:

$$V(x) = -\frac{1}{(1+x^2)^{\frac{1}{2}}} \tag{5}$$

The ionizing threshold for this model potential is 0.6698 atomic units. The behavior of this 1-dimensional model under a strong radiation field has been extensively investigated.[6,9,10] The stabilization of this model against ionization in intense, high-frequency radiation fields has been recently related to the properties of the x-p phase space.[11]

While there is no doubt about the usefulness of 1-dimensional calculations in very specific systems (*e.g.* solid state systems), their reliability in obtaining useful physical information for intrinsically 3-dimensional problems (like the general field of laser-atom interactions) is highly questionable. The conclusions, derived from the analysis of such a simplified 1-dimensional model, concerning the requirements upon intensity and frequency for the stabilization of real atoms in *superintense* laser fields, have been recently questioned on the basis of a purely classical analysis.[12]

Using a general non-perturbative method which was developed and presented most recently,[8] we are in the position to obtain answers to this question for a 1-dimensional model as well the real 3-dimensional atom, even if more than one electron is involved. Our purpose in this paper is to focus upon a precise evaluation of the difference between the 1D model and the real 3D atom, and to demonstrate the importance of the pulsed nature of the field.

In Table 2 we show numerical results for the ground and continuum state populations at the end of the pulse for the 1-dimensional model described above. The photon energy is such that the parameter $\alpha_0 = I_0^{1/2}/\omega^2$ a.u. ranges from 1 to 10. The intensity dependence of the continuum states' population shows a maximum which is particularly evident for the smallest value (50 a.u. $\approx$ 8 optical cycles) of the pulse shape parameter τ. The presence of this maximum is hardly seen for values of τ larger than 16 optical cycles.

Table 2. The populations of ground and continuum states at the end of the pulse for the 1-dimensional soft core model of hydrogen with $\hbar\omega$=1 a.u. The velocity gauge has been used. Symbol definition as in Table 1.

Intensity I_0 (a.u.)	τ=50 a.u.		τ=100 a.u.		τ=150 a.u.	
1	g=0.0574	c=0.9426	g=0.0033	c=0.9967	g=0.0002	c=0.9981
5	g=0.0293	c=0.9707	g=0.0008	c=0.9992	g=0.0000	c=0.9999
10	g=0.0686	c=0.9314	g=0.0046	c=0.9954	g=0.0008	c=0.9992
25	g=0.1039	c=0.8960	g=0.0103	c=0.9897	g=0.0006	c=0.9994
50	g=0.1283	c=0.8712	g=0.0164	c=0.9836	g=0.0028	c=0.9972
100	g=0.1356	c=0.8365	g=0.0245	c=0.9755	g=0.0039	c=0.9960

The same general behavior can be seen in Table 3. The photon energy is now such as to require a 2-photon ionization at low fields and the parameter α_0 is in the range from 1.7 to 5.36.

In Table 4 we show numerical results for 3D hydrogen. The parameter α_0 ranges from 0.45 to 10. We can see that a very small value of τ (in the first column τ is equal to 2 optical cycles which is almost four times smaller than the smallest one used in the 1-dimensional model) is required to show a significant evidence of the suppression of the ionization. We can also see from Table 4 that even with a pulse parameter τ as short as 5 optical cycles (τ=31.4159 a.u.= 0.76 fs), which is of course unrealistic, any significant suppression of the ionization at the end of the pulse is hardly seen. This is in agreement with the analysis of

Table 3. The populations of ground and continuum states at the end of the pulse for the 1-dimensional model of hydrogen with $\hbar\omega$=0.44522 a.u. (2-photon ionization at low field). The velocity gauge has been used. Symbol definition as in Table 1.

Intensity I_0 (W/cm^2)	τ=50 a.u.		τ=100 a.u.		τ=150 a.u.	
2×10^{16}	g=0.0856	c=0.8482	g=0.0048	c=0.9860	g=0.0002	c=0.9992
5×10^{16}	g=0.0428	c=0.8461	g=0.0086	c=0.9761	g=0.0006	c=0.9994
10^{17}	g=0.0672	c=0.7373	g=0.0155	c=0.9647	g=0.0017	c=0.9942
2×10^{17}	g=0.1072	c=0.7440	g=0.0244	c=0.9395	g=0.0025	c=0.9933

Table 4. The populations of ground and continuum states at the end of the pulse for 3-dimensional hydrogen with $\hbar\omega$=1 a.u. The velocity gauge has been used. Symbol definition as in Table 1.

Intensity I_0 (a.u.)	τ=12.56637 a.u.		τ=31.4159 a.u.		τ=62.8318 a.u.	
0.2	g=0.453	c=0.547	g=0.140	c=0.861		
1	g=0.037	c=0.963	g=0.001	c=0.999		
5	g=0.034	c=0.962	g=0.001	c=0.999		
25	g=0.054	c=0.872	g=0.001	c=0.998	g=0.000	c=1.000
50	g=0.043	c=0.823	g=0.004	c=0.992	g=0.000	c= 1.000
100	g=0.028	c=0.755	g=0.006	c=0.976	g=0.000	c=1.000

Ref. 12, stating that the requirements of high intensity and high frequency are less strict for 1-dimensional than for real 3-dimensional systems.

It should be noted that in some of the entries, the populations of the ground and continuum states do not add up to unity. The reason is that part of the population has remained in excited bound states which are populated because of the large bandwith of a short pulse. Such state tend to ionize slowly because their ionization cross sections are smaller than that of the ground state and the time is too short. Finally, not only the duration (width at half-maxmum) of the pulse but also the detailed shape, specifically how fast it rises, is equally critical. We have chosen to employ a realistic pulse shape. One can of course perform calculations with any desired pulse shape. it is not, however, obvious that one can shape to any arbitrary form a superintense.

For a pictorical appreciation of the interplay between pulse duration, intensity and ionization during the pulse, we have presented in Figs 1a-d the amount of "ionization" as function of time during the pulse. Clearly most of the ionization occurs during the rise of the pulse, well before it even reaches the maximum. it remains flat for a time interval of about 4τ and rises(much more slowly) during the fall of the pulse. The slow rise of ionization during the fall can be attributed to the depletion of the ground state as well as to the fact that part of the population is in excited states. We note in passing that 4 cycles at this frequency is 0.75 fs, and that the shape of the curve of ionization demonstrates that the process does not proceed with a constant rate. if one were to use an effective rate to obtain the correct final ionization, the resulting development in time would be misleading.

We undertook this investigation with a dual purpose. First we wanted to explore the possible danger in drawing conclusions about the dynamics of atoms in strong fields on the basis of one dimensional models. Our results have indeed demonstrated that under realistic pulse conditions the 1D model produces stabilization that can not be expected from the real atom.

Second, we wanted to produce realistic results for the real 3D hydrogen atom under realistic pulse shapes. Choosing a frequency sufficiently high and a range of intensities beyond present day possibilities (for that frequency) we have shown that suppression of ionization

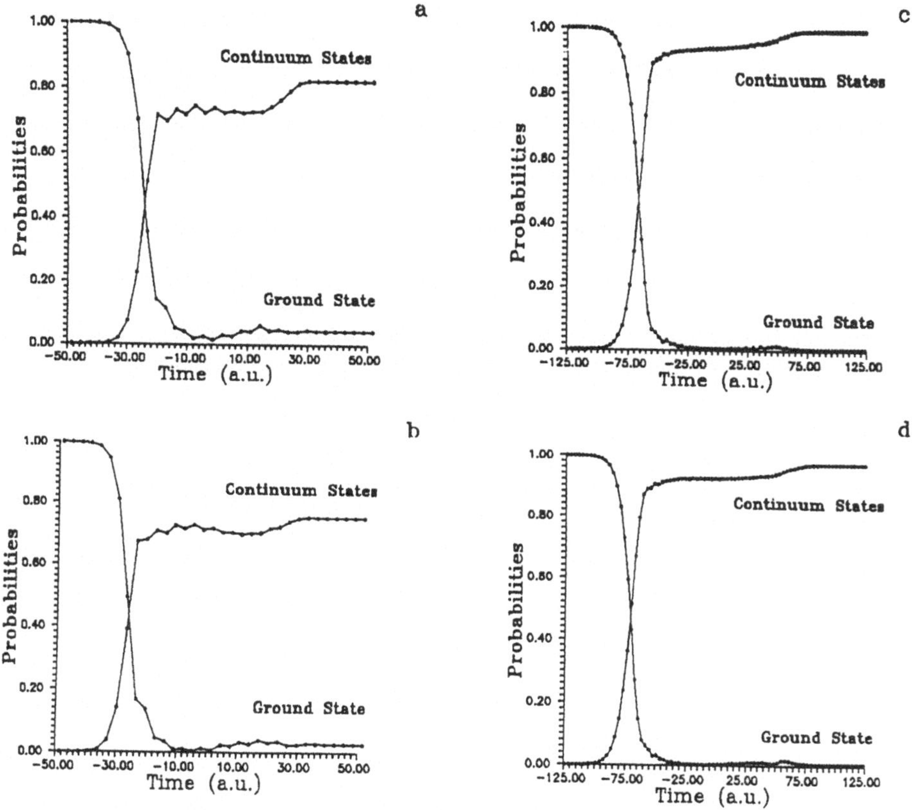

Figure 1. Probabilities in the unperbed ground and continuum states as a function of time for the atomic hydrogen with $\hbar\omega$=1 a.u. a) τ=2 optical cycles and intensity I_o=50 a.u. b) τ=2 optical cycles and intensity I_o=100 a.u. c) τ=5 optical cycles and intensity I_o=50 a.u. d) τ=5 optical cycles and intensity I_o=100 a.u.

(in the sense of decreasing ionization with increasing intensity) is obtained only for unrealistically short pulses ($\tau \approx 0.3$ fs). The effect disappears for slightly longer, but still unrealistically short ($\tau \approx 0.76$ fs) pulses. The effect found for the shorter pulses can be understood as a consequence of populating excited states - owing to the enormous bandwidth of the pulse (Fourier width) - which do not ionize easily during the time available to them. It should be noted that this aspect of the phenomenon is not (can not be) included in the predictions of time-independent calculations. We can also see from the table 4 that probabilities for trapping into higher exited states are quite dramaticl for $\tau \approx 0.3$ fs and $\alpha_0 \geq 5$ at end of the pulse, but that the trapping probabilities are decreasing as τ increases, which, of course, connected to suppression of ionization and can be also understood as an effect of the Fourier bandwidth.

We could go on with calculations at higher frequencies and intensities. We believe, however, that they would be nothing more than mathematical exercises, since our present combination of frequency and intensities already extends well beyond realistic expectations for the foreseeable future.

REFERENCES

1. M. Pont, N. R. Walet, M. Gavrila and C. W. McCurdy, Phys. Rev. Lett. 61, 939 (1988).

2. M. Pont, N. R. Walet and M. Gavrila, Phys. Rev. A 41, 477 (1990).

3. M. Pont and M. Gavrila, Phys. Rev. Lett. 65, 2362 (1990).

4. P. Lambropoulos, Phys. Rev. Lett. 55, 2141 (1985); P. Lambropoulos and X. Tang, J. Opt. Soc. Am. B 4, 821 (1987).

5. M. Dörr, R. M. Potvliege and R. Shakeshaft, Phys. Rev. Lett. 64, 2003 (1990).

6. Q. Su, J H. Eberly and J. Javanainen, Phys. Rev. Lett. 64, 862 (1990); J. H. Eberly and Q. Su, in *Multiphoton Processes, Proceedings of the 5th International Conference on Multiphoton Processes*, edited by G. Manfray and P. Agostini (Commissariat à l'Énergie Atomique, Paris, 1991), p. 81; K. Burnett, P.L. Knight, B.R.M. Piraux and V.C. Reed, Phys. Rev. Lett. 66, 301 (1991).

7. K. C. Kulander, K. J. Schafer and Jeffrey L. Krause, Phys. Rev. Lett. 66, 2200 (1991).

8. X. Tang, H. Rudolph and P. Lambropoulos, Phys. Rev. Lett. 65, 3269 (1990); P. Lambropoulos, X. Tang, H. Rudolph and H. Bachau, in *Multiphoton Processes, Proceedings of the 5th International Conference on Multiphoton Processes*, edited by G. Manfray and P. Agostini (Commissariat à l'Énergie Atomique, Paris, 1991), p. 133.

9. J. Javanainen and J.H. Eberly, Phys. Rev. A 39, 458 (1989).

10. M. Janjusevic and Y. Hahn, J. Opt. Soc. Am. B 7, 592 (1990).

11. R.V. Jensen and B. Sundaram, Phys. Rev. Lett. 65, 1964C (1990).

12. J. Grochmalicki, M. Lewenstein and K. Rzążewski, Phys. Rev. Lett. 66, 1038 (1991).

APPLICATION OF THE FINITE ELEMENT METHOD

TO THE 3-D HYDROGEN ATOM IN AN INTENSE LASER FIELD

Hengtai Yu, and André D. Bandrauk

Département de chimie, Faculté des sciences
Université de Sherbrooke
Sherbrooke, Qué, J1K 2R1, Canada

Vijay Sonnad

IBM Corporation
Kingston, N.Y. 12401, U.S.A.

INTRODUCTION

Present laser technology allows one to subject atoms and molecules to ever increasing radiation intensities so that eventually perturbation theory breaks down, i.e., renormalization of the electronic spectrum by the external field must be taken into account. Highly nonlinear effects such as above threshold ionization (ATI) readily occur in the case of atoms [1] and recently in molecules also [2-3]. Laser induced avoided crossings of molecular electronic potential curves is another result of such nonperturbative effects [2-6]. Clearly nonperturbative radiative effects predominate whenever laser field intensities approach atomic electric fields in an atom. The atomic unit (a.u.) of the electric field e^2/a_0^2 corresponds to a field intensity of 6×10^{16} W/cm^2 [7]. Such intensities are currently being attained, so that perturbative approaches to atom-laser interactions are no longer adequate.

Since with the present laser pulses considerable ionization is obtained on a short time scale, often less then a picosecond (10^{-12} sec), the total electronic wave function at any time must include simultaneously bound states and continuum electronic states. Furthermore, since the perturbation is time dependent, a nonperturbative time dependent description via the time dependent Schroedinger equation is in order. In the case of the hydrogen atom, successful numerical calculations have been carried out by solving the time dependent Schroedinger equation for the atom in an intense field. These calculations were based on implicit finite difference methods in view of the local nature of all potentials in the problem [8-11]. Extension of these numerical methods to multielectron problems have been unsuccessful due to the nonlocal nature of the exchange interaction [12-13]. Thus finite difference methods are limited so far to local potential problems.

Coherence Phenomena in Atoms and Molecules in Laser Fields
Edited by A.D. Bandrauk and S.C. Wallace, Plenum Press, New York, 1992

Another approach which could handle nonlocal potentials would be a basis set expansion, with sufficient flexibility to enable one to span localized bound state functions to highly delocalized free electron functions. Such flexible basis sets can be found in finite element (FE) methods. It is well known that simple finite difference and simple FE methods result often in identical approximation equations. Furthermore, one can show that with suitable local expansions all finite difference expressions can be cast as particular examples of trial function FE approximations [14-15].

Recent applications of FE methods in quantum chemistry have shown that these methods are well suited to treating non-local potentials, such as the exchange potential in the Hartree-Fock (HF) method for atoms [16-17] and recently molecules [18]. In the FE method, orbitals are expressed as a linear combination of basis functions with linear support thus avoiding the tendency towards linear dependence of global basis sets. In general, for bound state calculations of quantum chemistry, FE methods have been found to be as accurate but not necessarily as efficient as global basis set calculations (e.g. Gaussian bases). However in view of the great flexibility of FE basis sets, they would seem to be more appropriate for problems involving simultaneous localized (bound) and delocalized (continuum) states such as occurs in atom or molecule interactions with intense laser fields.

In the present case we study the 3-D hydrogen atom in a linearly polarized laser field using FE methods. We will show that solution of the time dependent Schroedinger equation by this method yields equivalent results to numerical finite difference methods published previously [8-11]. This therefore confirms our idea that FE methods should be very useful for time dependent HF problems where now the nonlocal exchange can be adequately treated. Furthermore, analysis of the wave function or state occupations is straightforward and illuminating concerning the excitation process.

TIME-DEPENDENT SYSTEMS

If wave functions vary with time, Schroedinger's time dependent equation applies,

$$H\Phi = i\frac{\partial\Phi}{\partial t} = i\dot{\Phi} \quad . \qquad (1)$$

H is the Hamiltonian of the system considered. In the case of the hydrogen atom, H can be written as follows in the polar coordinates for linearly polarized light,

$$H = H_o + H_i \quad , \qquad (2)$$

Where

$$H_o = -\frac{1}{2}\left(\frac{\partial^2}{\partial r^2} + \frac{2}{r}\frac{\partial}{\partial r}\right) + \frac{\ell(\ell+1)}{2r^2} - \frac{1}{r} \quad . \qquad (3)$$

and

$$H_i = r\varepsilon_o(t) \cos\theta \sin\omega t \quad , \tag{4}$$

where we define, ℓ, quantum number of angular momentum, $\varepsilon_o(t)$, amplitude of the electromagnetic field and ω, frequency of the laser pulse.

The wave function Φ in eq. 1 can be expanded in an angular momentum basis set,

$$\Phi(r,t) = \sum_{\ell=o}^{n} R_\ell(r,t)\; Y_{\ell,m}(\theta,\phi) \quad , \tag{5}$$

Substituting eq. 5 into eq. 1 for linear polarization (m=0) and the normalization condition,

$$\langle Y_{\ell_1} | Y_{\ell_2} \rangle = \delta_{\ell_1 \ell_2} \quad , \tag{6}$$

we get the time dependent coupled equations,

$$H_o(\ell)R_\ell \; - a \sum_{\ell'=o}^{n} \langle \ell(\cos\theta|\ell'\rangle\; R_{\ell'} = i\,\frac{\partial R_\ell}{\partial t} \quad , \tag{7}$$

where

$$a = r\varepsilon_o(t) \sin\omega t \quad , \tag{8}$$

and n is some maximum angular momentum in the basis set. One can write eq. 7 in matrix form for a limited n-basis set expansion

$$\begin{bmatrix} H_o(0) & C_o & 0 & 0 & 0 & .. \\ C_o & H_o(1) & C_1 & 0 & 0 & .. \\ 0 & C_1 & H_o(2) & C_2 & 0 & .. \\ 0 & 0 & C_2 & .. & .. & .. \\ 0 & 0 & 0 & .. & .. & C_{n-1} \\ 0 & 0 & 0 & .. & C_{n-1} & H_o(n) \end{bmatrix} \begin{bmatrix} R_o \\ R_1 \\ R_2 \\ \\ R_{n-1} \\ R_n \end{bmatrix} = i\partial/\partial t \begin{bmatrix} R_0 \\ R_1 \\ R_2 \\ \\ R_{n-1} \\ R_n \end{bmatrix} \tag{9}$$

where $C_\ell = a \times C_{\ell,\ell\pm1}$ and

$$C_{\ell\ell\pm1} = \langle \ell|\cos\theta|\ell'\rangle = \frac{\ell}{[(2\ell+1)\;(2\ell'+1)]^{1/2}}, \quad \ell'+1 = \ell \; ,$$

$$\frac{\ell + 1}{[(2\ell+1)\ (2\ell'+1)]^{1/2}},\ \ell'-1 = \ell \quad .$$

Note that $C_{\ell,\ell\pm1} \xrightarrow{\ell \to \infty} \frac{1}{2}$.

The solution of equation (1) can be propagated successively in exponential form by appropriate splitting of the total Hamiltonian [19],

$$\Phi^{n+1} = \exp\ (-\ i\Delta tH)\ \Phi^n \quad , \tag{10}$$

or in terms of a Taylor series expansion,

$$\Phi^{n+1} = (1 - i\Delta tH - \frac{1}{2}\Delta t^2H^2 + \frac{i}{6}\Delta t^3H^3 + \frac{1}{24}\Delta t^4H^4 + \ldots)\ \Phi^n \quad . \tag{11}$$

The latter is more practical for basis set methods where large expansions are required. It should be noted that H is generally a large matrix. For example, as will be shown below, if 150 basis sets are used and the maximum angular momentum number ℓ equals 10, one easily has to handle 1500 dimensional matrices.

Eq. 10 or 11 can be used to obtain the total wave functions at any time if the initial wave function is given. In the FE method used here, one need only to propagate the coefficients C_i in the expansion where N is the number of elements.

$$R_\ell = \sum_{i=1}^{N} C_{i\ell}\,\phi_i \quad , \tag{12}$$

Since a basis set of Legendre polynomials is used in the FE method which overlaps in different elements, such a basis set is not orthogonal. Thus radial functions R(r,t) are to be rewritten as

$$R(r,t) = \phi C(t) \quad , \tag{13}$$

where ϕ is the vector for the basis set and C is the matrix of coefficients for the different angular momentum states. Equation (7) now becomes

$$i\dot{R}(r,t) = i\phi\dot{C}(t) = iH\phi C(t) \quad , \tag{14}$$

or

$$i\phi^+\phi\dot{C} = \phi^+H\phi C \quad , \quad i\dot{C} = S^{-1}\phi^+H\phi C \quad , \tag{15}$$

where S is the overlap matrix $\phi^{+}\phi$. Propagating the coefficients allows the wave functions to be calculated from the numerical coefficients and the basis sets.

Normalization of the wave functions is used to check the results after propagation of eq. (11) since the total norm of wave functions should be conserved during the whole period of time propagation. It is interesting to note that the coefficients $C_{i\ell}$, eq. 12, are complex after propagation by use of eq. 11 and 15. It can be proved that

$$\int_{o}^{\infty} R_{\ell}^{*} R_{\ell} r^{2} dr = \sum_{\ell=0}^{n} [(\sum_{i=1}^{N} C_{i\ell}^{*} C_{i\ell} S_{ii} + 2 \sum_{j=2}^{N} \sum_{i=1}^{j-1} (ReC_{i\ell} * ReC_{j\ell} + ImC_{i\ell} * ImC_{j\ell}) S_{ij}] \quad (16)$$

where n is the maximum angular momentum included, and N is the number of elements.

INTERACTION TERM H_i

We reexpress the radiative interaction term in atomic units, (1 energy au = 27.2 ev),

$$H_i(au) = r(au)\ \varepsilon_o(t) \cos\theta \sin\omega t \quad ,$$

$$\varepsilon_o(t) = 0.53309178 * 10^{-8}\ I^{\frac{1}{2}} \left(\frac{w}{cm^2}\right) \quad (16)$$

where I is the peak intensity of the laser pulse, which is increased linearly from $1.75 * 10^{9}$ to 10^{14}, (w/cm^{-2}).

We shall use the same basis sets as those for previous time-independent calculations, i.e., in each element one has the following mathematical construction, [16-18],

$$I_a = \frac{1}{2}(1-t) \quad , \quad I_b = \frac{1}{2}(1+t) \quad , \quad (17)$$

$$B_i = (4i+2)^{-\frac{1}{2}} (P_{i+1} - P_{i-1}) \quad , \quad (18)$$

and

$$r = r_m + \delta t \quad . \quad (19)$$

I_a and I_b are the interpolation functions corresponding to the terminal, i.e., first and the last points in each element; B_i is a shape function in the element and P_i is the Legendre polynomial of order i and using the discretization coefficients,

$$c_{11} = \delta r_m^3 \quad , \quad c_{12} = 3\delta^2 r_m^2$$

and

$$c_{13} = 3r_m\delta^3 \quad , \quad c_{14} = \delta^4 \quad , \tag{20}$$

then it can be proved that the matrix elements between the two terminal functions I_a and I_b are

$$R_{aa} = \int_{-1}^{1} I_a r I_a r^2 dr = \frac{2}{3} c_{11} - \frac{1}{3} c_{12} + \frac{4}{15} c_{13} - \frac{1}{5} c_{14} \quad ,$$

$$R_{ab} = \int_{-1}^{1} I_a r I_b r^2 dr = \frac{2}{3} c_{11} + \frac{1}{15} c_{13} \quad ,$$

$$R_{bb} = \int_{-1}^{1} I_b r I_b r^2 \, dr = \frac{2}{3} c_{11} + \frac{1}{3} c_{12} \frac{4}{15} c_{13} + \frac{1}{5} c_{14} \quad . \tag{21}$$

The matrix elements between the polynomial B_i and B_j are

$$R_{ij} = \int_{-1}^{1} B_i r B_j r^2 dr = c_{11}B_{ij}(0) + c_{12}B_{ij}(1) + c_{13}B_{ij}(2) + c_{14}B_{ij}(3) \quad , \tag{22}$$

where B_{ij} is defined in appendix A.

The matrix elements between B_i and the terminal function I_a are

$$R_{ia(b)} = c_{11}B_{ia}(0) + c_{12}B_{ia}(1) + c_{13}B_{ia}(2) + c_{14}B_{ia}(3) \quad , \tag{23}$$

where now $B_{ia}(n)$ is defined in appendix B.

PROPAGATION OF WAVE FUNCTIONS

An external electric field acting on an atom gives rise to what is known as the Stark Effect [20]. If it is assumed that $t = \pi/2\omega$, i.e. $\sin \omega t = 1$, then the instantaneous Stark effect at that time can be calculated by diagonalizing the Hamiltonian matrix in equation (9). A preliminary calculation was therefore performed to gauge the appropriate grid and basis expansion for the nonperturbative problem at hand. Using 100 FE basis functions and a grid length r = 100 au (53 Å), under these numerical conditions energies of field free levels up to n = 20 are obtained with an absolute error of 10^{-7} au.

Diagonalizing the field dependent Hamiltonian (eq. 9) at $t = \pi/2\omega$ with the same numerical conditions gives the result reported in figure 1 for the lowest Stark shifted eigenvalue as a function of laser intensity. Since the transition moment of the 2S → 2p transition is 3 au [20], using the formula $\omega(\mathrm{cm}^{-1}) = 1.17 \times 10^{-3}\ u(\mathrm{au})\ I\ (\mathrm{W/cm}^1)^{1/2}$ for the radiative interaction (Rabi frequency) one would expect a Stark shift of about 80 eV (~ 3 au) at $I = 10^{12}$ W/cm². Clearly the lowest eigenvalue observed has a Stark shift of ~ 1000 au and therefore must come from highly lying Rydberg levels which have large transition moments. One observes from figure 1 that the basis function expansion, eq. 5, requires at least $n = \ell = 10$ angular momentum states for a proper description of intense field effects at intensities up to $I = 10^{16}$ W/cm².

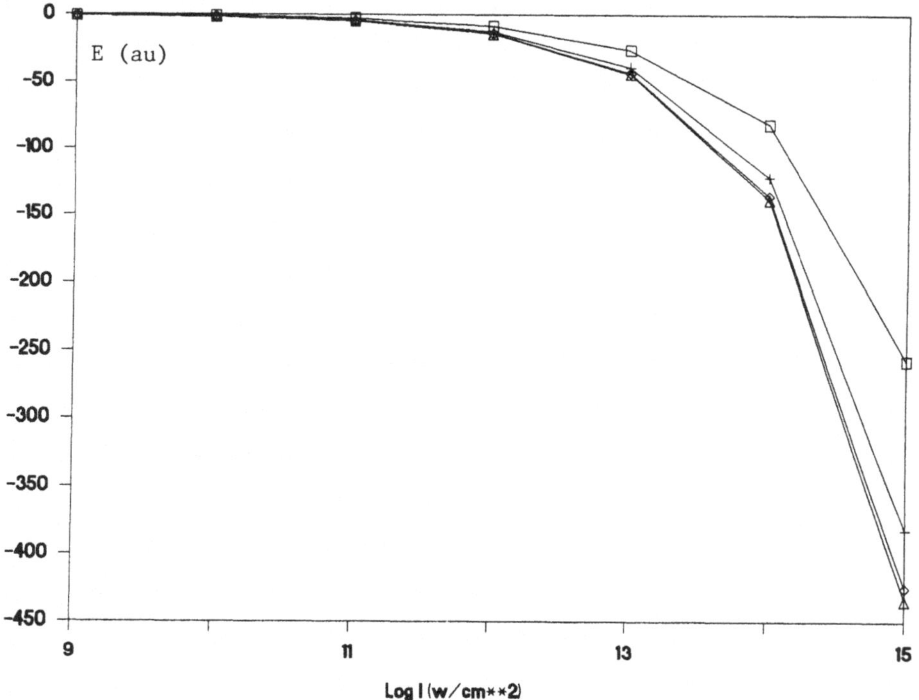

Fig. 1. Energy of lowest eigenstate at $t = \pi/2\omega$ as a function of basis expansion, eq. 5: □, n = 2; +, n = 4; ◇, n = 8; △, n = 12.

Figure 2 shows the initial 1S radial probability distribution $(R_{1S}(r))^2$, the final total radial probability distribution $|R(r)|^2$, after 30 cycles of an electromagnetic perturbation of frequency $\omega = 0.2$ au. The box size is 200 au, the integration time step is $\Delta t = 3 \times 10^{-3}$ au, and the intensity is 1.75×10^{14} W/cm². Figure 3 shows for the same conditions as fig. 2, the atomic energy increase as a function of time cycles (1 cycle = 31.4 au, 1 au = 2.42×10^{-17} s). The energy increase is non monotonic. Thus ionization begins to occur after 12 cycles with a different rate occuring for later times. The wave function (fig. 2) delocalizes rapidly with no discernible localized structure which can be associated to a distinct wave packet.

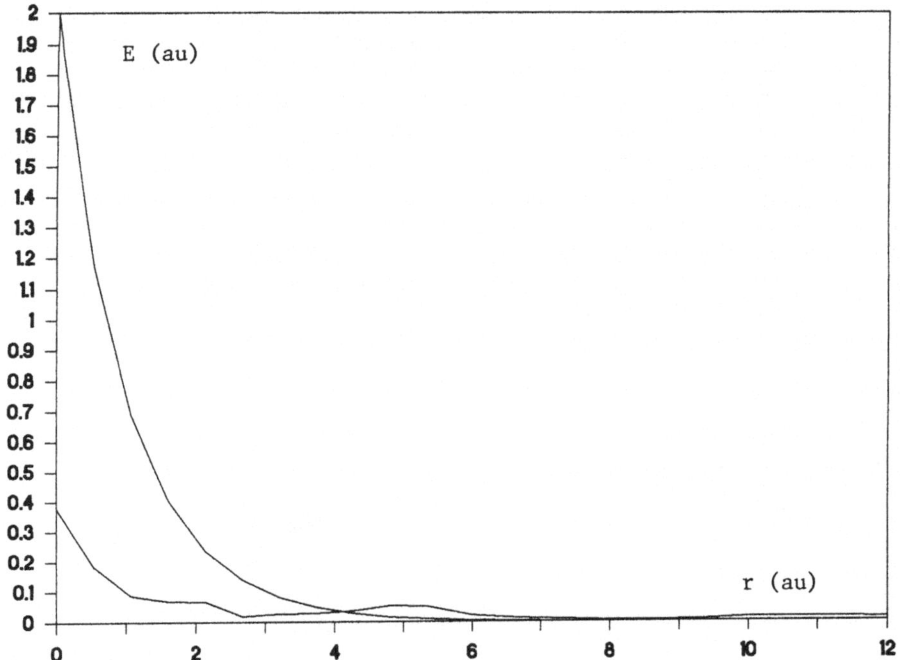

Fig. 2. Radial probability $|R(r,t)|^2$ as a function of electron distance r. I = 1.75 x 10^{14} W/cm^2, τ = 30 cycles, ω = 0.2 au.

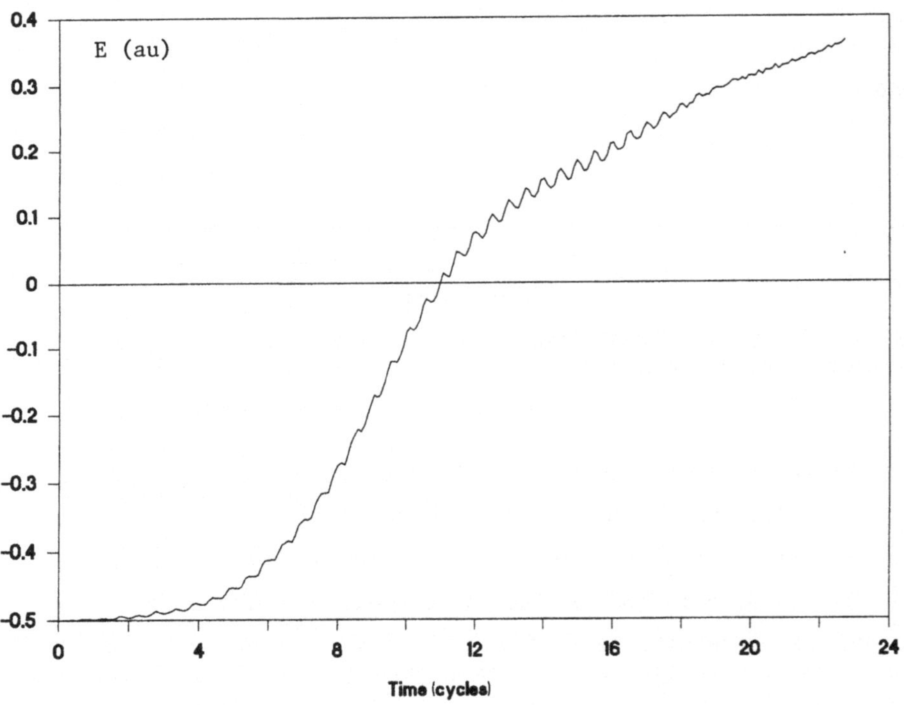

Fig. 3. Total atomic energy (au) as a function of time (1 cycle = 8 x 10^{-16} s). I = 1.75 x 10^{14} W/cm^2, ω = 0.2 au.

Figure 4 shows the atomic energy increase for the intensity I = 1.75×10^{15} W/cm^2. The box length L is now 2000 au, time increment Δt = 10^{-3} cycle, and ω = 0.2 au. The more refined mesh and box size was found necessary for accuracy. Ionization (defined as E > 0), now begins after 1/2 a cycle (~ 1/2 femtosecond). A slight decrease occurs between 0.5 and 0.8 cycles, indicative of persistent Rabi oscillations at these high intensities.

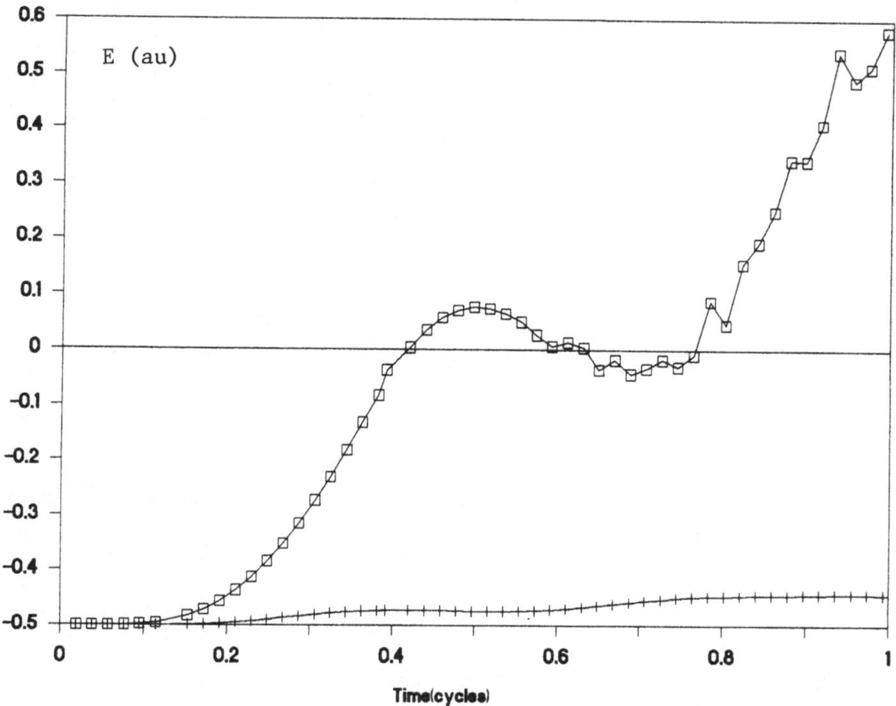

Fig. 4. Total atomic energy as a function of time. (1 cycle = 8 x 10^{-16} s). □,I = 1.75×10^{15} W/cm^2, ω = 0.2 au; +.. I = 1.75×10^{14} W/cm^2.

Finally figure 5 shows the populations of ℓ = 0(s), ℓ = 1(p) and ℓ = 9 states as a function of time for ω = 0.2 au, I = 1.75×10^{14} W/cm^2, corresponding to figures 2 and 3. One notices that after 10 cycles (7.5 femtoseconds), considerable population of higher angular moments occurs. Furthermore, distinct phase relations remain. Thus the p states are out of phase with respect to the S states, and the ℓ = 9 state is out of phase with both S and p states on the femtosecond time scale.

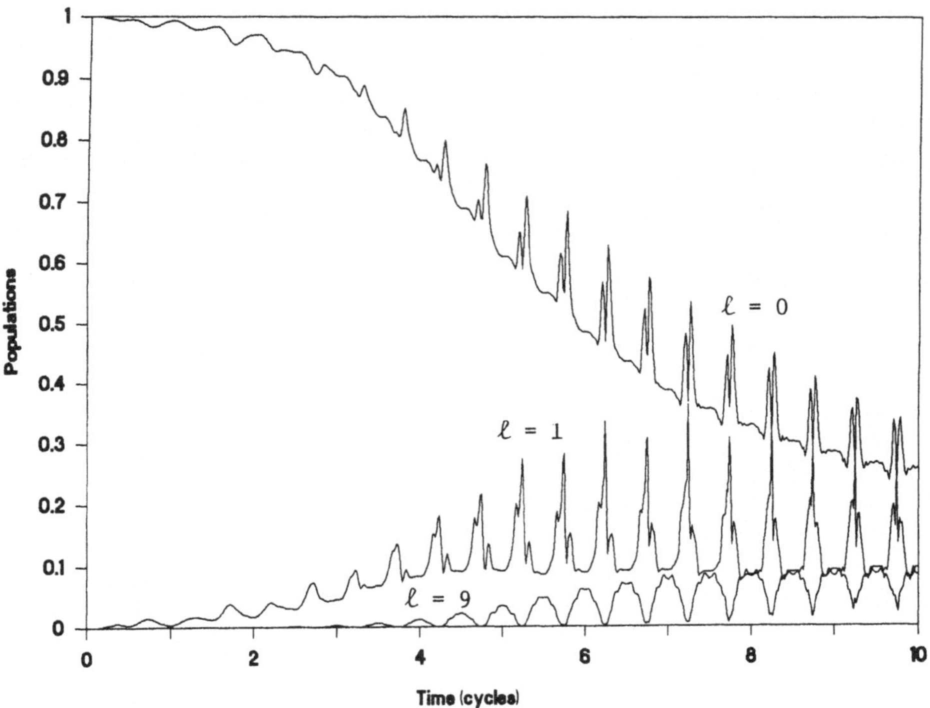

Fig. 5. Populations of ℓ = 0,1,9 states as a function of time (same conditions as fig. 4).

From calculations of the populations on a long time scale, up to 30 cycles, one observes a clear exponential decrease or linear decrease on a logarithmic scale in the population of the initial state. From this result one can estimate the ionization rate Γ if one assumes $P_{1s}(t) = e^{-\Gamma t}$. We tabulate these results below for various intensities (L = 500 au, $\Delta t = 10^{-3}$ cycle, ω = 0.2 au, N = # of FE function = 100, $n = \ell_{max}$ = 10).

TABLE 1

I (W/cm^2)	Γ (sec^{-1})
7 x 10^{13}	6.5 x 10^{13}
1.75 x 10^{14}	1.6 x 10^{14}
10^{15}	1.75 x 10^{15}

These numbers agree well with previously published numerical results (eg at 1.75×10^{14} W/cm^2, $\Gamma = 3.5 \times 10^{14}$ s^{-1} [10] and 2.9×10^{14} s^{-1} [8] by finite difference methods). Using a coarser grid, $\Delta t = 3 \times 10^{-3}$ cycle and a smaller box, L = 200 au, gave $\Gamma = 2.7 \times 10^{14}$ sec^{-1} by the FE method. We conclude therefore that high accuracy and transparent results are readily obtainable from a FE approach to treating ionization of the H atom in an intense, nonperturbative electromagnetic field. We are currently extending this method to multielectron atoms and molecules in a time dependent Hartree-Fock framework.

APPENDIX A

$$B_{ij}(n)=(4i+2)^{-1/2}(4j+2)^{-1/2}\int_{-1}^{1}[P_{i+1}P_{j+1}-P_{i+1}P_{j-1}-P_{i-1}P_{j+1}+P_{i-1}P_{j-1}]t^n dt. \tag{A1}$$

Define

$$C_2(i,j) = (4i+2)^{-1/2}(4j+2)^{-1/2} \quad , \tag{A2}$$

then (A1) can be transformed into

$$B_{ij}(n)=C_2(i,j)[P_n(i+1,J+1)-P_n(i-1,j+1)-P_n(i+1,j-1)+P_n(i-1,j-1)]. \tag{A3}$$

When n = 0,

$$P_o(i,j) = \frac{2}{2i+1} \; i = j \quad ; \quad 0, \; i \neq j \quad ; \tag{A4}$$

When n = 1,

$$P_1(i,j) = \frac{2i}{(2i+1)\,(2i-1)} \; , \; i = j + 1 \; ;$$

$$\frac{2(i+1)}{(2i+3)\,(2i+1)} \; , \; i = j - 1 \quad ; \tag{A5}$$

0 otherwise.

When n = 2

$$P_2(i,j) = \frac{2i(i-1)}{(2i+1)(2i-1)(2i-3)} - i = j + 2 \quad ;$$

(A6)

$$\frac{2i(4J^3+6j^2-1)}{(2i+1)^2(2j+3)(2j-1)}, \quad i = j \quad ;$$

$$\frac{2j(j-1)}{(2i+1)(2j+1)(2j-1)} - i = j - 2 \quad ;$$

0 otherwise.

When n = 3,

$$P_3(i,j) =$$

$$\frac{2(j+1)(j+2)(j+3)}{(2j+1)(2j+3)(2j+5)(2j+7)}, \quad i = j + 3 \quad ;$$

$$\frac{2}{2j+3}\left[\frac{(j+1)(j+2)^2}{(2j+1)(2j+3)(2j+5)} + \frac{(4j^3+6j^2-1)(j+1)}{(2j+1)^2(2j+3)(2j-1)}\right] \quad i = j + 1;$$

(A7)

$$\frac{2}{2j-1}\left[\frac{j(4j^3+6j^2-1)}{(2j+1)^2(2j+3)(2j-1)} + \frac{j(j-1)^2}{(2j+1)(2j-1)(2j-3)}\right], \quad i = j - 1;$$

$$= \left(\frac{2}{2j-5}\right)\frac{j(j-1)(j-2)}{(2j+1)(2j-1)(2j-3)} \quad i = j - 3 \;;$$

0 otherwise.

APPENDIX B

$$B_{ia}(n) = \frac{1}{2}(4i+2)^{-1/2}\int_{-1}^{1}(P_{i+1}t^n - P_{i-1}t^n \mp P_{i+1}t^{n+1} \pm P_{i-1}t^{n+1})dt \; . \qquad \text{(B1)}$$

The upper sign in ∓ or ± in (B1) indicates interaction between B_i and I_a; lower sign between B_i and I_b. Define

$$C_1 = \frac{1}{2} (4i+2)^{-1/2} \quad , \tag{B2}$$

then from properties of Legendre polynomials, we have,

$$P_1 = t \text{ and } P_o = t^o = 1$$

From this (B1) can be transformed into

$$B_{ia}(n) = C_1[P_n(i+1,0) - P_n(i-1,0) \mp P_n(i+1,1) \pm P_n(i-1,1)], \tag{B3}$$

where P_n is the same as P_n in $B_{ij}(n)$ of Appendix A.

REFERENCES

[1] "Atomic and Molecular Processes with Short Intense Laser Pulses", edit. A.D. Bandrauk, NATO ASI, Series B, Physics, vol. 171 (Plenum Press, NY 1988).

[2] P.H. Bucksbaum, A. Zavrijev, H.G. Muller, D.W. Schumacher. Phys. Rev. Lett. 64, 1883 (1990).

[3] J.W. Verschuur, L. Noordham, H.B. van Linden van den Heuvell, Phys. Rev. 40A, 4383 (1989).

[4] A.D. Bandrauk, M.L. Sink, Chem. Phys. Lett. 57, 569 (1978); J. Chem. Phys. 74, 1110 (1981).

[5] J.F. McCann, A.D. Bandrauk, Phys. Rev. 42A, 2806 (1990).

[6] A.D. Bandrauk, E. Constant, J.M. Gauthier, J. de Phys. II, 1, 1033 (1991).

[7] M. Mittleman, "Theory of Laser Atom Interactions", (Plenum Press, NY (1982)).

[8] K.C. Kulander, Phys. Rev, 35A, 445 (1987).

[9] M.S. Pindzola, G.J. Bottrell, C. Bottcher, J. Opt. Soc. Am, B7, 659 (1990).

[10] P. DeVries, J. Opt. Soc. Am, B7, 517 (1990).

[11] M.S. Pindzola, M. Dorr, Phys. Rev. 43A, 439 (1991).

[12] K.C. Kulander, Phys. Rev. 36A, 2726 (1987).

[13] M.S. Pindzola, D.C. Griffin, C. Bottcher, Phys. Rev. Lett. 66, 2305 (1991).

[14] B.A. Finlayson, "Method of Weighted Residuals",(Academic Press, N.Y. 1974).

[15] O.C. Zinkiewicz, K. Morgan, "Finite Elements and Approximation" (John Wiley, N.Y. 1983).

[16] J.R. Flores, E. Clementi, V. Sonnad, J. Chem. Phys. 91, 7030 (1939).

[17] J.R. Flores, E. Clementi, V. Sonnad, Chem. Phys. Lett. 163, 198 (1989).

[18] H. Murukami, V. Sonnad, E. Clementi - IBM prepublication, 1992.

[19] A.D. Bandrauk, H. Shen, Chem. Phys. Lett. 176, 428 (1991).

[20] H.A. Bethe, E.E. Salpeter, "Quantum Mechanics of One and Two Electron Atoms", (Springer-Verlag, Berlin (1957)), p. 233.

OPTICAL ANALOGS OF MODEL ATOMS IN FIELDS

P.W. Milonni

Theoretical Division, MS-B212
Los Alamos National Laboratory
Los Alamos, NM 87545

"Quantum mechanics is not a bad preparation for optics." - Dennis Gabor

INTRODUCTION

Everyone knows there are many analogies between optics and quantum mechanics. I would like to suggest here that some recent effects of interest for atoms in strong fields might possibly be realized in the propagation of light.

First recall that the time-dependent Schrödinger equation is formally the same as the paraxial wave equation of optics. This is well known to optical physicists. (See, for instance, Cook (1975) or Stoler (1981).) We will briefly derive this correspondence, with the slight generalization of allowing the refractive index to vary both axially and transversely. This will lead us to an optical analog of an atom in a monochromatic field.

Assume a linearly polarized monochromatic field with electric field amplitude $E(\mathbf{r})e^{-i\omega t}$. E satisfies the Helmholtz equation, $\nabla^2 E + k^2 n^2 E = 0$, where $k \equiv \omega/c$ and n is the refractive index. Write $E(\mathbf{r}) = E_o(\mathbf{r}_\perp, z)e^{ikz}$, where E_o is assumed to be slowly varying in z compared with e^{ikz}; $\mathbf{r}_\perp$ is the coordinate in the xy plane, perpendicular to the direction of propagation z. Then we can drop $\partial^2 E_o/\partial z^2$ compared with $k\partial E_o/\partial z$ in the Helmholtz equation and work with the paraxial wave equation

$$2ik\frac{\partial E_o}{\partial z} = -\nabla_\perp^2 E_o - k^2(n^2 - 1)E_o, \tag{1}$$

where $\nabla_\perp^2 \equiv (\partial^2/\partial x^2 + \partial^2/\partial y^2)$. Obviously the paraxial wave equation has the same form as the Schrödinger equation for a particle constrained to move in two spatial dimensions:

$$i\hbar\frac{\partial \psi}{\partial t} = -\frac{\hbar^2}{2m}\nabla_\perp^2 \psi + V\psi. \tag{2}$$

It is convenient to scale the time and space variables. Introduce in (2) the dimensionless time $\tau = (\hbar/ma^2)t \equiv \omega_o t$ and the dimensionless coordinate variables $X = x/a, Y = y/a$, where a is some convenient length. Then (2) becomes

$$i\frac{\partial \psi}{\partial \tau} = -\frac{1}{2}\overline{\nabla}_\perp^2 \psi + \frac{V}{\hbar\omega_o}\psi, \tag{3}$$

where $\overline{\nabla}_{\perp}^2 = a^2\nabla_{\perp}^2$ is the transverse Laplacian in the X, Y variables. Similarly introduce the dimensionless coordinate variables $Z = 2z/kb^2, X = x/b, Y = y/b$, where b is some convenient length scale for the optics problem, so that (1) becomes

$$i\frac{\partial E_o}{\partial Z} = -\frac{1}{2}\overline{\nabla}_{\perp}^2 E_o - \frac{1}{2}k^2b^2(n^2-1)E_o. \tag{4}$$

The effective potential in this "optical Schrödinger equation" is thus

$$V_{\text{opt}} \equiv -(\hbar\omega_o)\frac{1}{2}k^2b^2(n^2-1). \tag{5}$$

To keep things really simple we will consider a few one-dimensional examples, where equations (3) and (4) become respectively

$$i\frac{\partial\psi}{\partial\tau} = -\frac{1}{2}\frac{\partial^2\psi}{\partial X^2} + \frac{1}{\hbar\omega_o}V(Xa, \frac{\tau}{a})\psi \tag{6}$$

$$i\frac{\partial E_o}{\partial Z} = -\frac{1}{2}\frac{\partial^2 E_o}{\partial X^2} + \frac{1}{\hbar\omega_o}V_{\text{opt}}(Xb, \frac{1}{2}kb^2 Z)E_o. \tag{7}$$

Classical (Ray) Limit

The classical Newton equation of motion corresponding to (6) is $m\ddot{x} = -\partial V/\partial x$. If we use (7) to construct the corresponding "optical Newton equation," we obtain $d^2x/dz^2 \cong \partial n/\partial x$ for $n \cong 1$. This is the paraxial approximation to the ray equation $(d/ds)(nd\mathbf{r}/ds) = \nabla n$ for a position vector $\mathbf{r}$ of a point on a ray, with s a distance measured along the ray. (Born and Wolf, 1970)

This example brings out a simple but important point. In the approximation of geometrical optics one deals with *families* of rays, for a ray merely gives us some information about a point on the wavefront. In the same way the comparison of classical and quantum theories must involve an *ensemble* of trajectories. This is well known, of course, but sometimes it seems to be forgotten by those who express great surpise at the fact that classical systems can exhibit chaos (in the sense of a positive Lyapunov exponent), while the corresponding quantum systems do not. This can be understood in part from the simple consequence of Louiville's theorem that classical *distributions* of trajectories cannot exhibit the "very sensitive dependence on initial conditions" that is the hallmark of classical chaos. (Milonni, 1989)

The Harmonic Oscillator

According to equations (5)-(7) we can produce the optical analog of a harmonic oscillator by making $V_{\text{opt}}/\hbar\omega_o = -\frac{1}{2}k^2b^2(n^2-1) = CX^2b^2$, or $n(x) = 1 - Dx^2$ for $n \cong 1$, where C, D are positive constants. This defines a lenslike medium, (Yariv, 1976) also known as a *graded-index waveguide*. In this case the "optical Newton equation" above becomes $d^2x/dz^2 + Dx = 0$, so that the ray displacements oscillate harmonically with propagation in the lenslike medium. In the wave-optical description, as in the quantum-mechanical oscillator, the modes of the field in the lenslike medium are such that $E_o(x)$ is a Hermite polynomial, the lowest-order mode being the ubiquitous Gaussian beam. Since spherical mirrors affect rays in basically the same way as a thin lens, it should come as no surprise to anyone who knows elementary quantum mechanics that the modes of stable laser resonators are Gaussian fields.

Anharmonic Oscillator

Similarly we can construct the optical analog of an anharmonic oscillator with potential $V(x) \propto x^2 + Ax^4$, say, by choosing $n(x) = 1 - D(x^2 + Ax^4)$. By choosing the transverse index variation $n(x)$ appropriately we can "design" any anharmonic oscillator we like.

TIME-DEPENDENT POTENTIAL

Thus far we have taken n to be a function of x but not z. That is, we have not allowed the refractive index to vary along the direction of propagation. For the quantum problem, this means we have restricted ourselves to time-independent potentials. To develop optical analogs of driven quantum systems with time-dependent perturbations, we now allow n to be a function of both x and z.

A potential that has been used recently in numerical experiments on above-threshold ionization and high-order harmonic generation is (Javanainen and Eberly, 1988)

$$V(x,t) = \frac{-e^2}{\sqrt{x^2 + a_o^2}} - exA\cos\omega t, \tag{8}$$

or

$$\frac{1}{\hbar\omega_o} V(Xa, \frac{\tau}{a}) = -\frac{1}{\sqrt{X^2+1}} - XF\cos\mu\tau, \tag{9}$$

where we choose $\hbar\omega_o = e^2/a_o, a = a_o, \mu = \omega/\omega_o$, and $F = A(e/a_o^2)^{-1}$ is the field strength in atomic units. The optical analog of the Schrödinger equation with this potential is obtained when

$$n^2(x,z) - 1 = \frac{2}{k^2b^2}\left[\frac{1}{\sqrt{x^2/b^2+1}} + \frac{x}{b}F\cos\left(\frac{2\mu z}{kb^2}\right)\right], \tag{10}$$

in which case (7) becomes

$$i\frac{\partial E_o}{\partial Z} = -\frac{1}{2}\frac{\partial^2 E_o}{\partial X^2} + U(X)E_o - XF\cos(\mu Z)E_o. \tag{11}$$

with $U(X) \equiv -(X^2+1)^{-1/2}$. Numerical solutions of such a one-dimensional Schrödinger equation go back at least as far as Goldberg, Schey, and Schwartz (1967). More recently such numerical solutions have been used in studies of quantum chaos (Goggin and Milonni, 1988) and above-threshold ionization. (Javanainen and Eberly, 1988; Collins and Merts, 1988)

Since paraxial wave propagation with a periodically varying index like (10) is described by a Schrödinger equation for a particle in a time-independent potential plus a sinusoidal applied field, we can do all the usual things done to treat atoms in fields. For instance, we can use an expansion in basis states: Consider solutions of the "unperturbed" system defined by

$$i\frac{\partial E_o}{\partial Z} = -\frac{1}{2}\frac{\partial^2 E_o}{\partial X^2} + U(X)E_o. \tag{12}$$

(We will assume $U(-X) = U(X)$.) Write $E_o(X,Z) = g(X)e^{-iKZ}$, so that $g(X)$ satisfies the eigenvalue equation

$$-\frac{1}{2}\frac{d^2g}{dX^2} + U(X)g(X) = Kg(X). \tag{13}$$

The solutions $g_n(X)$ with eigenvalues K_n define the optical modes for the unperturbed paraxial wave equation (12). To solve (11) we can write

$$E_o(X,Z) = \sum_n a_n(Z) g_n(X) e^{-iK_n Z} , \tag{14}$$

and then (11) implies

$$i\frac{da_n}{dZ} = -F\cos(\mu Z)\sum_m X_{nm} e^{-i(K_m - K_n)Z} a_m(Z), \tag{15}$$

$$X_{nm} \equiv \int_{-\infty}^{\infty} dX g_n^*(X) X g_m(X). \tag{16}$$

Thus the mode amplitudes $a_m(Z)$ satisfy the same equations as the probability amplitudes of the corresponding quantum-mechanical problem.

Two-State Atom

Suppose the field is propagating in the lowest-order g_1 mode and encounters a sinusoidal perturbation along Z, such that $\mu \cong K_2 - K_1$ in equation (15). If $|K_n - K_1|$ and $|K_n - K_2|$ for $n > 2$ are sufficiently different from μ that all the $a_n(Z)$ obtained from (15) with $n > 2$ are very small, we can approximate (15) by

$$\begin{aligned} i\frac{da_1}{dZ} &= -F\cos(\mu Z) X_{12} e^{-i(K_2-K_1)Z} a_2(Z) \\ &\cong -\frac{1}{2} X_{12} F a_2(Z), \end{aligned} \tag{17}$$

$$i\frac{da_2}{dZ} \cong -\frac{1}{2} X_{21} F a_1(Z), \tag{18}$$

if we assume $\mu = K_2 - K_1$. (The approximation made in writing these equations will be recognized as the "rotating-wave approximation.") Thus

$$\frac{d^2 a_j}{dZ^2} + \frac{1}{4}|X_{12}|^2 F^2 a_j(Z) = 0, \quad j = 1,2. \tag{19}$$

This means that, if we have a sinusoidal index variation along the direction of propagation, such that (11) applies, and if the spatial frequency of this variation equals the difference $K_2 - K_1$ between the eigenvalues of the modes $g_2(X)e^{-iK_2 Z}$ and $g_1(X)e^{-iK_1 Z}$, then as the field propagates it will oscillate between these two modes at the "Rabi frequency" $\frac{1}{2}|X_{12}|F$. Obviously we can construct "optical Bloch equations" from (17) and (18).

One-Dimensional Hydrogen Atom

Now let us return to the optical analog (11) of a "one-dimensional hydrogen atom." Suppose that a wave corresponding to the lowest-order mode $g_1(X)$ is launched into a medium with refractive index (10). If $F = 0$ such a wave will, ideally, propagate as $g_1(X)e^{-iK_1 Z}$. (For the present discussion we can consider any superposition of eigenmodes, but for simplicity let us just assume the lowest-order mode corresponding to the "ground state" of the atom in the quantum analog. We will also ignore "turn-on" effects associated with the sinusoidal perturbation.) If $F \neq 0$ the sinusoidal perturbation can cause "transitions" among modes, as we have just seen for the optical analog of a two-state atom. But in addition to transitions caused by such resonances between the spatial frequency of the index variation and the difference between two mode wave numbers, we can induce transitions at *any* spatial frequency of the index variations if the amplitude of these variations is large enough. For sufficiently large F we can have "photoionization": the optical analog is a transverse spreading of the wave into the "continuum," i.e., a transition in which at least part of the field is no longer associated with a transversely confined propagating mode.

There are obviously optical analogs of above-threshold ionization, harmonic generation, and stabilization. Let us consider the latter phenomenon. To do this we will first derive the optical analog of the Kramers-Henneberger (KH) transformation.[1]

We can cast (11) in the language of bras, kets, and linear vector spaces of square-integrable functions:

$$i\frac{\partial}{\partial Z}|E_o\rangle = \left[\frac{1}{2}P^2 + U(X) - XF\cos(\mu Z)\right]|E_o\rangle, \tag{20}$$

with $[X, P] = i$. Write $F\cos(\mu Z) = -\partial A/\partial Z$ and define[2]

$$|E_o'\rangle = e^{-iP\int_0^Z dZ' A(Z') + (i/2)\int_0^Z dZ' A^2(Z')} e^{iXA(Z)}|E_o\rangle. \tag{21}$$

Then it follows from (20) that

$$i\frac{\partial}{\partial Z}|E_o'\rangle = \left[\frac{1}{2}P^2 + U(X + X_R(Z))\right]|E_o'\rangle, \tag{22}$$

or

$$i\frac{\partial E_o'}{\partial Z} = -\frac{1}{2}\frac{\partial^2 E_o'}{\partial X^2} + U(X + X_R(Z))E_o', \tag{23}$$

where $X_R(Z)$, satisfying $d^2X_R/dZ^2 = F\cos(\mu Z)$, is the ray displacement determined by geometrical optics with only the z-dependent part of the refractive index. The transformation from (20) to (22) is the "optical K-H transformation." Equation (23) is equivalent to

$$2ik\frac{\partial E_o'}{\partial z} = -\frac{\partial^2 E_o'}{\partial x^2} - k^2[n_o^2(x + x_R(z)) - 1]E_o', \tag{24}$$

where n_o is the z-independent part of the refractive index and $x_R(z)$ is the geometrical ray displacement *without* this part of the index.

Stabilization

By analogy to the atomic case, stabilization can occur if F is large and if a high-(spatial) frequency approximation is justified. In the high-frequency approximation we replace $n_o^2(x + x_R(z)) - 1$ by its average over z. Then in effect the z-dependent part of the index is removed, and there is nothing left to cause transitions among different modes. In particular, the field will remain confined transversely to a large degree.

---

[1] As I learned in a lecture by M.H. Mittleman (Los Alamos National Laboratory, April 3, 1991), this transformation may be traced at least as far back as W. Pauli and M. Fierz, Nuovo Cim. **15**, 167 (1938). In fact a very similar transformation appears in a primarily relativistic setting in F. Bloch and A. Nordsieck, Phys. Rev. **52**, 54 (1937).

[2] The reader may recognize the transformation from $|E_o\rangle$ to $|E_o'\rangle$ as a Power-Zienau transformation followed by the KH transformation.

CAN THESE OPTICAL ANALOGS BE CONSTRUCTED?

It is not difficult to imagine fabricating a medium with any transverse index variation we like, so let us assume this can be done. (It can pretty much be done with quantum wells, for instance.) The difficult part of producing an index variation like (10) is to get the part varying as $x\cos(Az)$.

Here is one way this might be done. Recall the paraxial ray equation $d^2x/dz^2 = \partial n/\partial z$ and suppose we bend the propagation path in such a way that the optical axis ($x = 0$) is displaced by $y(z)$. Then x is transformed to $x + y(z)$ and the paraxial ray equation is transformed to $d^2(x+y)/dz^2 = \partial n/\partial x$, or

$$\frac{d^2x}{dz^2} - \frac{\partial n}{\partial x} = -\frac{d^2y}{dz^2} \tag{25}$$

if $dx/dz, dy/dz$ are not too large. This is the equation one gets with index $n(x) - xd^2y/dz^2$. In other words, the index is changed from $n(x)$ to $n(x)+C(z)x$, where $C(z)$ is the curvature of the guiding optical axis. In particular,

$$n(x,z) = n(x) + A^2x\cos(Az) \tag{26}$$

if $y(z) = \cos(Az)$. This is exactly the type of index we need to realize the optical analog of a model atom in a monochromatic field.

Note that a simple sinusoidal variation with z of the index, as in a distributed feedback laser,[3] for instance, is not enough; we require a sinusoidal variation times x. This occurs when we bend the guiding axis as just described. Of course we should not bend the axis so strongly that the paraxial approximation breaks down, or that the boundaries of the medium intercept the beam.

REMARKS

The simplified discussion just given is sufficient to bring out a few important points. One of these is that phenomena such as stabilization are not distinctly quantum-mechanical, for we have seen that they can be realized with *classical* waves. It should also be clear that the same sorts of conclusions inferred from the one-dimensional model carry over to the real case of two-dimensional transverse beam variations.

As mentioned earlier, the optical analogy might be useful in studies of "quantum chaos." If the index variations are such that the ray trajectories are chaotic, then the corresponding classical-mechanical system will be chaotic. The question of how this chaos might manifest itself quantum mechanically is then mathematically identical to the question of how the chaos of geometrical rays might manisfest itself in wave propagation.

[3]Of course the distributed feedback case involves both backward- and forward-propagating waves, under conditions for which the paraxial approximation is not applicable.

ACKNOWLEDGEMENT

I would like to thank my friends Dick Cook and Dave Stoler for many discussions relating to this topic, especially relating to their papers cited herein.

REFERENCES

Born, M. and Wolf, E., *Principles of Optics* (Pergamon, Oxford, 1970), fourth edition, p. 122.

Collins, L.A. and Merts, A.L., "Model calculations for an atom interacting with an intense, time-dependent electric field," Phys. Rev. A**37**, 2415 (1988).

Cook, R.J., "Beam wander in a turbulent medium: An application of Ehrenfest's theorem," J. Opt. Soc. Am. **65**, 942 (1975).

Goggin, M. and Milonni, P.W., "Driven Morse oscillator: classical chaos, quantum theory, and photodisscociation," Phys. Rev. A**37**, 796 (1988); "Driven Morse oscillator: classical chaos and quantum theory for two-frequency excitation," Phys. Rev. A**38**, 458 (1988).

Goldberg, A., Schey, H.M., and Schwartz, J.L., "Computer-Generated Motion Pictures of One-Dimensional Quantum-Mechanical Transmission and Reflection Phenomena," Am J. Phys. **35**, 177 (1967).

Javanainen, J. and Eberly, J.H., "Comparison of Keldysh models with numerical experiments on above-threshold ionization," Phys. Rev. A**39**, 458 (1988).

Milonni, P.W., Letter to the Editor, Physics Today, April, 1989, p. 15.

Stoler, D., "Operator methods in physical optics," J. Opt. Soc. Am. **71**, 334 (1981).

Yariv, A., *Introduction to Optical Electronics* (Holt, Rinehart, and Winston, 1976).

ABOVE-THRESHOLD PHOTODISSOCIATION AND BOND SOFTENING AT HIGH INTENSITIES

P.H. Bucksbaum and A. Zavriyev[a]

Randall Laboratory of Physics
The University of Michigan
Ann Arbor, MI 48109

1. INTRODUCTION

Molecules subjected to intense optical fields can ionize and dissociate by the multiple absorption and scattering of photons. A new kind of dissociation and ionization has been seen in laser fields that approach or exceed the binding fields in the molecule. The electron spectra show many peaks, corresponding to the absorption of more photons than the minimum number necessary for ionization. This new behavior is called above-threshold ionization, or ATI.[1] The energies of dissociating molecular fragments also show absorption or emission of additional photons beyond the minimum order process; this has been called above-threshold dissociation, or ATD.[2] These two new phenomena will be reviewed in this paper. Recent experiments in molecular hydrogen and deuterium performed at the University of Michigan and at AT&T Bell Laboratories illustrate the main features.[3-6]

Two different kinds of lasers were used in the Bell Laboratories and Michigan investigations: a mode-locked Nd:Yag laser with a pulse duration of $\approx$100 psec;[3,4] and a passively modelocked dye or Ti:Sapphire laser, with duration of $\approx$160 fsec.[5,6] Electron and ion energy spectra show markedly different behavior in these two pulse length regimes. Although some of the differences are due to well-understood ponderomotive final state effects,[7] other spectral differences seem to have to do with the internal dynamics of the molecules. These effects include the orientation of molecular bonds with respect to laser polarization, and intensity-dependent dissociation channels.

Some general features of high intensity light are described in the next section. This lays the foundation for a description of above-threshold ionization in molecules in section 3. Here we describe bond softening and above-threshold dissociation. Finally, section 4 describes recent data on the difference between dissociation with short and with long laser pulses.

2. GENERAL FEATURES OF HIGH INTENSITY LIGHT

The new phenomena described in this paper occur when the laser intensity is above about 10^{13} W/cm^2, which corresponds to a peak electric field of $\approx$1 V/Å. The light-induced shifts in the energies of molecular states at this intensity become quite large, often larger than photon energy $\hbar\omega$. Ordinary perturbation theory is not very useful for calculating absorption rates in this regime, because dozens of photons may be absorbed

Coherence Phenomena in Atoms and Molecules in Laser Fields
Edited by A.D. Bandrauk and S.C. Wallace, Plenum Press, New York, 1992

during a single dissociation, and hundreds more scattered during the separation of the fragments. It is often more useful to treat the laser interaction as arising from a classical vector potential in the Schrodinger Hamiltonian.

The Ponderomotive Potential[8]

The extra energy induced by the laser on a loosely bound electron is comparable to a classical quantity, the "ponderomotive potential." This is just the cycle-averaged kinetic energy of a free electron wiggling in an oscillating electromagnetic field:

$$U_P = \frac{e^2 E_0^2}{4m\omega^2} \tag{1}$$

$$\approx 1 \text{ eV for } 10^{13}\text{W/cm}^2 \text{ and } \lambda = 1.06\mu\text{m}$$

where m is the mass of the classical particle, e is its charge, and E_0 is the peak electric field in the laser. For typical conditions of a Nd:YAG laser ($\lambda = 1.06\mu$m) focused to 10 TW/cm^2, U_P for a free electron is equal to about 1 eV. The photon energy in this case is also about 1 eV. the ponderomotive energy increases quadratically with wavelength, and is linear with intensity.

The light-induced shifts in the levels of a molecule can be comparable to U_P, and therefore should be directly observable in photoelectron spectra; however, the ponderomotive potential has an equally large effect on the free electrons as they leave the laser pulse. The time-averaged energy of wiggling may be transformed into translational energy, depending on the details of the electrons' exit from the laser focus. The energy transferred between the laser field and a wiggling particle in a spatially and temporally inhomogeneous field is generally very complicated; however, the problem simplifies in the "adiabatic" limit, where the laser turns on and off over many optical cycles, and the amplitude of the wiggle is much smaller than the waist of the focused laser beam. Our typical laser focus involves 50-1000 cycles focused to 10-20 μm, while the wiggle amplitude is generally less than 20Å. This is well within the adiabatic limit. The classical Hamiltonian for a nonrelativistic spinless electron in the laser field

$$H = \frac{\left[\mathbf{p} - \frac{e}{c}\mathbf{A}(\mathbf{x},t)\right]^2}{2m} \tag{2}$$

may be averaged over a cycle, yielding two terms:

$$\langle H \rangle = \frac{p^2}{2m} + U_P(\mathbf{x},t) \tag{3}$$

The first term is the "drift kinetic energy", i.e. the kinetic energy of the electron due to the time time-averaged drift of the particle. The second term is the ponderomotive energy, or the average energy due to wiggling. Although the source of this energy is kinetic, it acts like an effective potential energy. A charged particle in an inhomogeneous region of intensity, such as a laser focus, experiences a time-averaged force equal to the negative gradient of this ponderomotive potential. This leads to acceleration, and shifts in the electron energy spectra.

The total a.c. Stark energy shift of a bound state includes contributions from the ponderomotive potential and from second-order and higher-order couplings to other bound states. In the weakly bound Rydberg states, the ponderomotive contribution dominates. In the ground state, ponderomotive and bound state couplings tend to cancel. As a result, the ionization energy tends to increase with intensity, so that more photons are needed to ionize at high intensity than at low intensity. This extra energy goes into the ponderomotive potential of the photoelectrons.

Energy spectra may be sorted into a "long pulse" and a "short pulse" regime according to whether the laser pulse is longer or shorter than the time for the electron to drift out of the focus. In the long pulse regime, the electron leaves the focus before the light turns off, so that its initial ponderomotive potential energy is converted into translational kinetic energy, which is measured by the electron spectrometer. *In the long pulse regime, the measured photoelectron energy includes the ponderomotive potential.*

In the short pulse regime, the electron cannot move far enough during the laser pulse for any appreciable change in the drift velocity; the spectrometer therefore measures only the "drift energy," rather than the total energy, of the photoelectron. *In the short pulse regime the observed energy does not include the ponderomotive potential.*

Practically speaking, the long pulse regime includes laser pulses longer than about 50 psec, while the short pulse regime is observed for laser pulses shorter than 5 psec. For pulses of intermediate duration, ponderomotive acceleration is appreciable, but the electrons do not gain the full ponderomotive potential as they drift out of the focus. Short pulse and long pulse effects in atomic photoionization have been studied for several years.[7] In molecules, these effects have only begun to be investigated, as we show below.

3. EXPERIMENTS IN THE LONG PULSE REGIME

Above-Threshold Ionization in Molecules

The defining characteristic of ATI spectra, whether in the short pulse or the long pulse regime, is the appearance of multiple peaks or sets of peaks separated by $\hbar\omega$. Figure 1 shows two ATI spectra obtained using 532 nm light in H_2. In the top spectrum, obtained at a relatively low intensity of about 20 TW/cm^2, there is a series of peaks that can be identified with ionization to different final state vibrational levels of the ground state manifold of H_2^+. These experiments were carried out in the long pulse regime, described in the previous section. The close agreement between the peak energies and the energies of the ground state vibrational levels of H_2^+ indicate that the light-induced energy level shift between the ground state and the ionization limit in H_2 is very nearly equal to the ponderomotive potential, as it is in the rare gas atoms which have been studied extensively.

Franck-Condon integrals using the field-free internuclear potentials indicate that the favored final state should be $v^+ = 5$ of $H_2^+ \, 1s\sigma_g$. Furthermore, in the absence of Stark shifts, the hydrogen molecule is nearly resonant for five-photon excitation from the ground state into the B state ($v' = 3$).[9] This would then be followed by two or more photon ionization to the $v^+ = 5$ state. However, the large ponderomotive shifts and other ac Stark shifts evidently interfere with straightforward Franck-Condon analysis.[10] Vibrational states from $v^+ = 0$-7 appear in the spectrum.

Additional peaks in the spectrum can be matched to the ionization of atomic hydrogen, particularly hydrogen in the n=1 and n=2 state. The n=2 peak suggests molecule dissociation into excited states.[9,4]

The bottom spectrum shows the same experiment performed at higher intensity.[4] This shows a series of broader peaks similar to an atomic ATI spectrum. A major difference between the high and low intensity regimes is the amount of dissociation.[3] This can be measured in a simple charge-to-mass ratio spectrometer. The measurement shows little dissociation in the focus that produced the top spectrum in figure 1, but near total dissociation in the focus that produced the bottom spectrum.

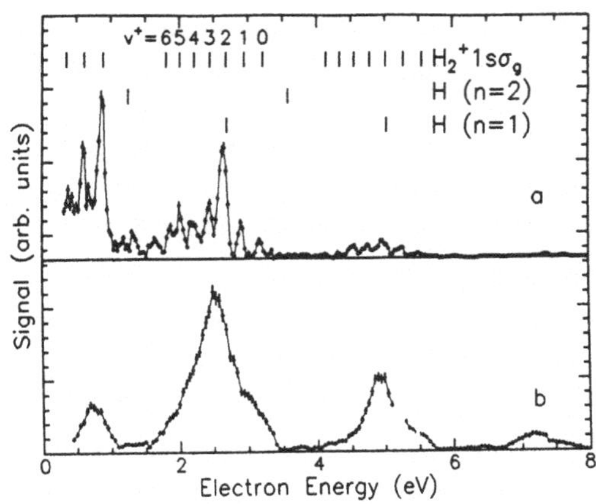

Figure 1. ATI spectra for H_2 ionized by intense 532 nm light. The electrons were detected along the laser polarization. (a): $I \approx 20 TW/cm^2$. This is a typical spectrum obtained in the "low intensity regime" and the "long pulse regime". The largest peaks correspond to 7 photon ionization into the lowest three vibrational levels of H_2^+. In the 8 photon ATI manifold between about 1.5 and 3.5 ev, the most prominent final state vibrational levels are $v^+ = 3$ and 5. The former may contain contributions from ATI of ground state atomic hydrogen. Also visible are peaks from 2-photon ionization from the n=2 state of atomic hydrogen. (b): $I \approx 100 TW/cm^2$. The spectrum has completely transformed into a series of broad ATI peaks centered near the position for ionization of ground state (n=1) atomic hydrogen. This is the "high intensity regime."

Dissociation Spectra

Still more information can be obtained by measuring the energies of the dissociation fragments. Figure 3 shows two energy spectra for the dissociation of hydrogen by 532 nm (2.33 eV) light at two intensities.[3] Most of the protons are concentrated in a broad peak around 0.4 eV. There are also peaks at higher energies, separated by 1.17 eV from each other. Figure 3 also shows angular distributions of H^+ fragments in the polarization plane. The distributions are narrowly peaked around the polarization direction, indicating that most of the ions dissociate in the direction of the laser electric field. Since the dissociation direction is along the internuclear axis, we conclude that most molecules dissociate when they are aligned with the field. This suggests dissociation via transitions to the repulsive $2p\sigma_u$ state, which are maximum for molecules oriented along the laser polarization.

We have also measured the relative fraction of molecules that dissociate, and find that this tends toward unity, even thought the angular distribution remains sharply peaked along $\hat{\varepsilon}$. It is as though the molecules are actually pulled into alignment by the field. This hint that the internuclear potentials may be strongly distorted by the intense laser field is corroborated by the proton energies. The most prominent spectral peak centered at 0.25 eV, if formed from dissociation along the $2p\sigma_u$ potential curve, would have come from transitions at R=2.3Å. This is outside the outer turning points of the first six vibrational levels of $1s\sigma_u$ which are populated by ATI from the ground state of

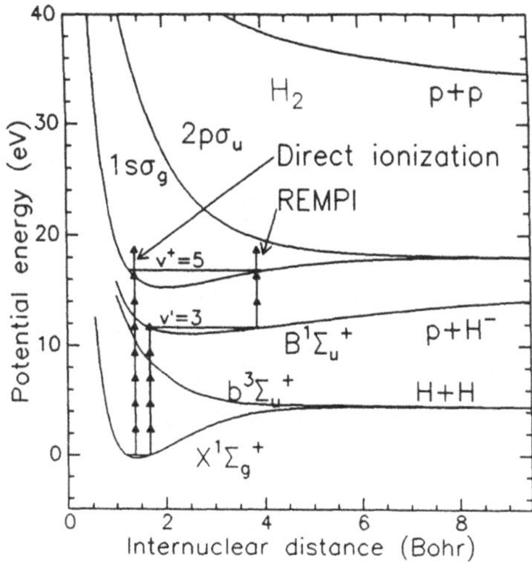

Figure 2. Potential diagram for H_2. The v = 0 ground state shown here is the initial state in this experiment. v′ is a resonant vibrational level of the B state, and $v^+ = 5$ is the final state vibrational level that is the largest peak in the low intensity regime shown in figure 1(a).

H_2, unless the ion potentials are distorted by several electron volts in the laser field.

This distortion of $1s\sigma_g$ and $2p\sigma_u$ can be calculated using Floquet theory.[2-4] The strong dipole coupling between these states causes the potential well to become more shallow, leading ultimately to the disappearance of nearly all of the ground state vibrational levels. This "bond softening" is a manifestation of the reduction in the screening of the ion repulsion in a strong laser field.

The results of an *ab initio* calculation of bond softening for 532 nm light is shown in Figure 4.[3,4]. Dressed state potentials are shown for several different intensities for polarization along the internuclear axis. The potentials deform where the two bare states are in multiphoton resonance and the diabatic dressed states cross. The dipole interaction opens gaps, or avoided crossings, between the dressed states at these points.

Higher order gaps representing multiphoton transitions produce molecular fragments moving faster, since more photons are absorbed. Competition between multiphoton and single photon dissociation leads to a spectrum with several kinetic energy peaks, shown in Figure 3.

The molecular ion dissociates into H^+ and a neutral hydrogen atom (Figure 4). The kinetic energy of the detected proton is given by:

$$E_{path[n]} = (E_{ion} + n\hbar\omega - E_D) / 2. \tag{4}$$

for paths 1,2, and 3, where E_D is the final potential energy of the system, and E_{ion} is the energy of the bound ion. The division by a factor of 2 shows that the available energy is divided equally between the two fragments.

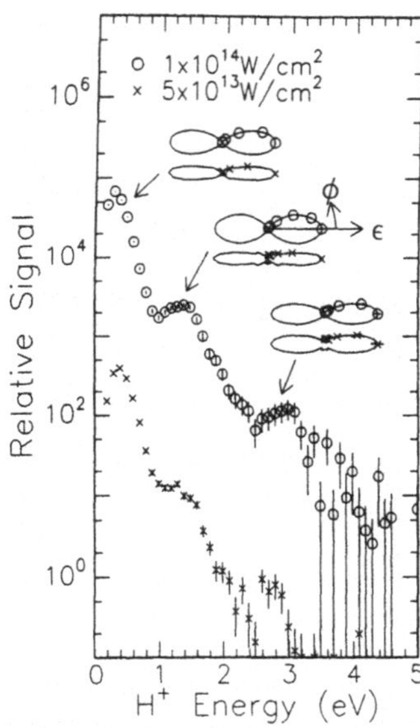

Figure 3. Energy and angular distributions from multiphoton dissociation of H_2 above 50 TW/cm^2. The peaks in the ion spectra are separated by $\approx \hbar\omega/2$. (From reference 3.)

Discussion

The softening of the molecular bond leads to dissociation; the fact that the dissociation is always along the laser polarization suggests that as the intensity increases there is also a tendency for the molecule to move into alignment with the field and dissociate. Thus the dissociation fraction tends toward unity at the same time that the angular distribution of dissociation fragments remains sharply peaked along the internuclear axis. The impetus to realign comes from a torque exerted by the field. We can estimate this by noting that an energy shift of up to $\hbar\omega$ is possible as the molecule moves from perpendicular to field, where the internuclear potential is not distorted by coupling between the $1s\sigma_g$ and $2p\sigma_u$ states, to parallel, where the coupling is maximal. Thus, an order of magnitude estimate for hydrogen gives a torque of 2 eV/rad, enough to overwhelm the rotational angular momentum in the ground state of the hydrogen ion; the molecule does not merely precess under this influence, it snaps into alignment in a few femtoseconds.

4. EXPERIMENTS IN THE SHORT PULSE REGIME

In order to investigate this issue of orientation, we can repeat the same experiments with much shorter pulses, where the rotational time is comparable to the time of the laser pulse. The notion that the laser exerts a torque on the molecule is a classical idea, derived in the Born Oppenheimer picture where the internuclear separation is a parameter rather than an operator in the molecular Hamiltonian. It is more proper to think of the laser field as an external perturbation which breaks the angular degeneracy of the molecules internal energy. Thus, rotation and vibration are no longer good quantum

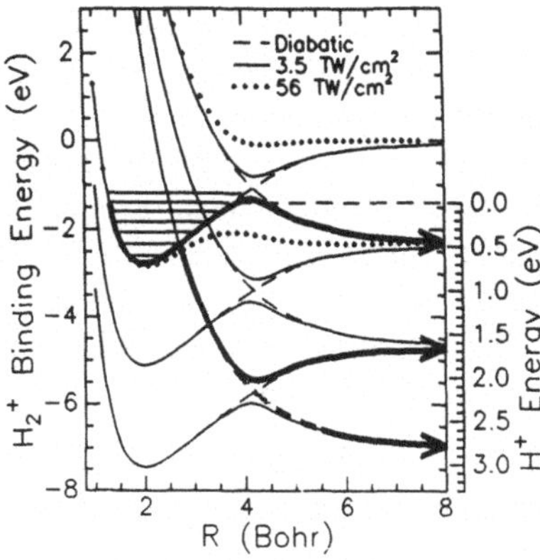

Figure 4. The H_2^+ molecular potential deforms in the presence of intense light polarized along the internuclear axis. The figure shows the results of a Floquet calculation, for two intensities. The bold lines trace the adiabatic and diabatic channels for dissociation of the molecule into the three peaks seen in figure 3. (From reference 4.)

numbers. However, if the molecule can make a transition in a time short compared to the rotational period, then it may form a coherent superposition state, or "wavepacket" with some classical properties. In the hydrogen molecular ion, the rotational period is around 0.5 psec, and twice that in the deuterium molecular ion. Thus, a laser pulse of about 100 fsec can virtually freeze the rotational motion.

Even vibrational motion may be affected on this time scale. The period of vibrational motion in the first vibrational state of H_2^+ is 20 fsec, and 30 fsec for deuterium. Although a 100 fsec pulse is considerably longer than this, some wave packet effects can still be observed.

To investigate this, H_2 and D_2 experiments were performed at the Center for Ultrafast Optical Science at the University of Michigan to ionize and dissociate the molecules by ultrashort pulses from a mode-locked and amplified Ti:Sapphire laser.[5,6] The focused (160 fs) pulses of 769 nm (1.612 eV) light had a peak intensity up to 2×10^{15} W/cm^2.

A typical ion kinetic energy spectrum for deuterium is shown in figure 5. The low energy regular peaks are separated by approximately one-half $\hbar\omega$, and are probably due to ATD, as with the long pulse dissociation data in figure 3. Now, however, the ratio of peaks is radically different. The second peak, corresponding to the net absorption of two photons, is much larger than the lower one-photon peak. (Compare this to figure 3, where the one-photon peak is more than ten times higher than the two-photon peak.) There is also a broad distribution of ions at higher energies, extending to nearly eight eV. This feature is totally absent in the long pulse data.

We investigated the effect of the short pulse on rotation, vibration, and dissociation, by comparing spectra from the dissociation of H_2 and D_2 under identical circumstances. Figures 6 and 7 show this comparison at two different laser intensities. The angular distributions of the different components in the spectrum are shown above the spectral peaks. The ions emerge along the laser polarization, but the angular spread of protons is somewhat broader for protons than for deuterons.

There is a classical physics explanation for the difference in the angular distributions. The hydrogen molecular ion is half the mass of the deuterium ion, with the same internuclear separation; therefore it rotates twice as fast as deuterium, and vibrates $\sqrt{2}$ times faster. The hydrogen ion will therefore rotate more during its dissociation than will the deuterium ion. The quantum mechanical explanation must involve the formation of a "rotational wavepacket." This is a transient state that is made when the laser bandwidth exceeds the spacing between rotational levels, which will always happen if the pulsewidth is less than the rotational orbit period. In this case, the transient state that is formed resembles a classical molecule.

The spectral differences between above-threshold dissociation in the long and short pulse regimes points to dynamics on the short pulse time scale in molecular dissociation. For an explanation, we return to the Floquet picture of the dissociation process, where the transitions between states occur at dressed state curve crossings (see figure 4). According to standard Landau-Zener theory, the ion will follow the adiabatic path (avoiding the crossings), so long as its rate of passage through the gap region is slower than the Rabi transition rate given by the adiabatic gap separation divided by $\hbar$.[11]

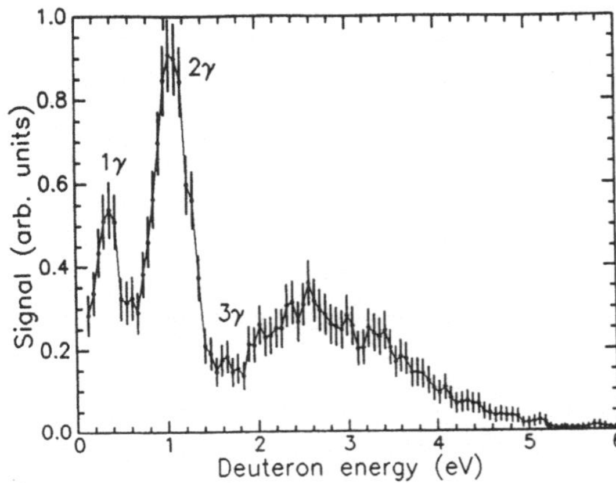

Figure 5. Energy spectrum of deuterons from dissociation of D_2 in a 100 fsec, 769 nm laser focus with a peak intensity of $\approx 10^{15}$W/cm^2.

For intensities above 10TW/cm^2, the one-photon gap is wide enough so that the adiabatic avoided crossing occurs with nearly unit probability, and the lowest order peak (path 1 in figure 4) has the largest signal. In the short pulse experiments, the pulse duration is still somewhat longer than a single molecular vibration, for either hydrogen (20 fsec for $v'=1$) or deuterium (30 fsec) ions; however there is a significant change in intensity over a single vibrational period, and this is enough to significantly change the branching ratio between the first and second ATD peaks.

In the extreme short pulse limit, ionization creates a vibrational wave packet in the molecular ion. Even for pulse lengths that are comparable to the vibration period, a wave packet may form due to the highly nonlinear increase in the ionization rate with intensity. For 769 nm light, at least 10 photons must be absorbed in order to ionize the molecular ground state. The time-integrated ionization probability during the rising edge of the pulse can easily exceed the vibrational period of the ion final state. In this

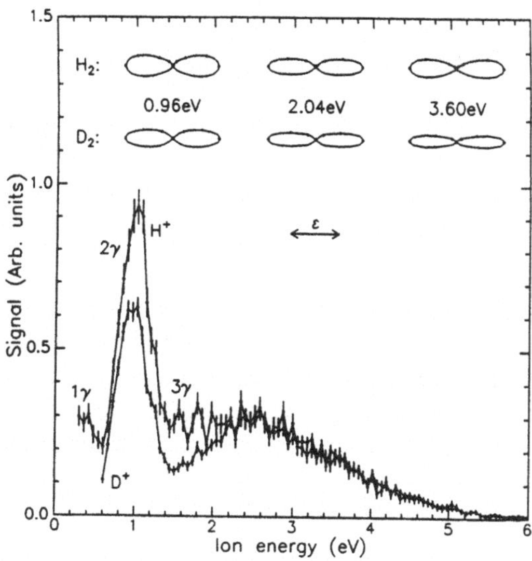

Figure 6. Spectrum and angular distribution of protons and deuterons detected following irradiation of H_2 and D_2 by 769 nm 160 fsec linearly polarized light. The spectrum shown is for ions emitted along the polarization direction. The peak intensity was $10^{15}W/cm^2$.

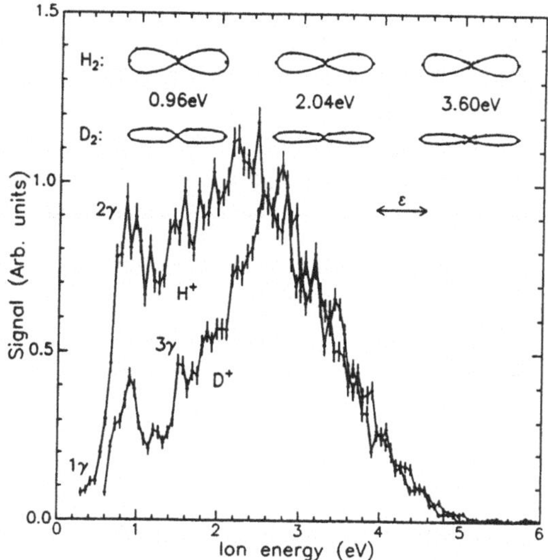

Figure 7. Same as Figure 6, but with intensity of $2\times10^{15}W/cm^2$ at the peak.

case a vibrational wave packet is created. This is a superposition state of two or more vibrational eigenstates that interfere constructively near the inner turning point of the $1s\sigma_g$ potential.

Consider as an example a D_2^+ ion in the 5th vibrational level of the $1s\sigma_g$ ground state. According to our data, this is a prominent channel for ionization at an intensity below $\approx 2\times10^{13}W/cm^2$ (figure 1). If the laser peak intensity is $10^{15}W/cm^2$, the ion is likely to be born on the rising edge of the pulse. The ion first reaches the outer turning point about 15 fsec later, still well below the peak of the laser pulse. For a 160 fsec pulse with the peak intensity in this example, the intensity will have risen to $\approx 3\times10^{13}W/cm^2$, and the potential will be slightly distorted; however, the one-photon gap is still small at this point, and the ion starts back toward the inner turning point. 40

fsec have elapsed since ionization when the ion returns to the three-photon gap. At this point the laser intensity is nearly 100 TW/cm^2, and the gap is quite wide. The ion has a high probability of absorbing three photons and shifting to the repulsive curve. One-photon is emitted during passage through the one-photon gap, and the so the ion fragments dissociate with a net two-photons absorbed. H_2^+ ions vibrate faster than D_2^+, so they are more likely to dissociate via one-photon absorption. Evidence for this can be seen in the higher ratio of 1γ peak to 2γ peak for hydrogen than for deuterium in the figure.

The broad distribution of faster ions with kinetic energies up to 6 eV is a new short pulse effect. These fast ions have a different intensity dependence than the slow bond-softening ions. There are several possible explanations for this new peak: It may be due to direct absorption between the ground state of the neutral molecule and the 2pσ_u state; or it may be due to excitation, or even ionization, of the second electron. A third possibility is double excitation of the hydrogen molecule, followed by dissociation, and ionization of either one or both hydrogen atoms.

Additional experiments are required to sort out these possibilities. However, the striking similarity in the high energy tail of the hydrogen and deuterium spectra suggests that the high energy distribution is caused by an electronic process, rather than a vibrational or a rotational one.

CONCLUSION

This paper has reviewed the new high field phenomena of Above-Threshold Ionization (ATI) and Dissociation (ATD) of H_2 and D_2 on with pulses from 0.1 nsec to 0.1 psec. Several new physical effects have been observed, including: the breakdown of the Franck-Condon effect; the reorientation of molecules in the intense field; bond softening; and the formation of vibrational wave-packets for sub-picosecond pulses. We also see broad distribution of fast (up to 6 eV) ions in spectra with sub-picosecond pulses.

Many people have helped us to understand these experiments. We thank M. Mittleman, F. Mies, A. Giusti, L. DiMauro, and B. van den Heuvell and his students. Parts of this work were supported by AT&T Bell Laboratories, by the University of Michigan, and by the National Science Foundation Center for Ultrafast Optical Science at Michigan.

REFERENCES

[a] Also at: Applied Physics Department, Columbia University, N.Y., N.Y.

1. P. Agostini, F. Fabre, G. Mainfray, G. Petite, and N. Rahman, Phys. Rev. Lett. **42**, 1127 (1979); P. Kruit, J. Kimman and M. Van der Wiel, Jour. Phys. B **14**, L597 (1981).

2. A. Giusti-Suzor, X. He, O. Atabek, and F.H. Mies, Phys. Rev. Lett. **64**, 515 (1990); A. Bandrauk and M. Sink, J. Chem. Phys. **74**, 1110 (1981).

3. P.H. Bucksbaum, A. Zavriyev, H.G. Muller, and D.W. Schumacher, Phys. Rev. Lett. **64**, 1883 (1990).

4. A. Zavriyev, P.H. Bucksbaum, H.G. Muller, and D.W. Schumacher, Phys. Rev. A **42**, 5500 (1990).

5. A. Zavriyev, P.H. Bucksbaum, J. Squier, F. Saline, and G. Mourou, to be published.

6. A. Zavriyev, D.W. Schumacher, F. Weihe, P.H. Bucksbaum, J. Squier, F. Saline, and G. Mourou, in *Short-Wavelength Coherent Radiation: Generation and Applications*, ed. by P.H. Bucksbaum and N. Ceglio, Proc. Opt. Soc. Am. **11**, 222 (1991).

7. R.R. Freeman, P.H. Bucksbaum, H. Milchberg, S. Darack, D. Schumacher and M.E. Geusic, Phys. Rev. Lett. 59, 1092 (1987).

8. L.S.Brown and T.W.B.Kibble, Phys.Rev.**133**, 3A (1964).

9. J.W.J. Verschuur, L. Noordam, and H. van Linden van den Heuvell, phys. Rev. A **40**, 4383 (1989); J.W.J. Verschuur, *Ionization and Dissociation of* NO *and* H_2 *in Strong Radiation Fields*, (Doctoral Dissertation), Amsterdam: FOM Institute for Atomic and Molecular Physics (1989).

10. B. Yang, M. Saeed, L.F. DiMauro, A. Zavriyev, and P.H. Bucksbaum, Phys. Rev. A **44**, R1458 (1991).

11. See, for example, L.D. Landau and E.M. Lifshitz, *Quantum Mechanics*, Course of Theoretical Physics **3**, Pergamon Press, New York (1965).

AN ARTIFICIAL CHANNEL PROCEDURE FOR MULTIPHOTON ABSORPTION LINESHAPE AND BRANCHING RATIOS IN INTENSE LASER FIELDS: APPLICATION TO H_2^+ PHOTODISSOCIATION

Salvador Miret-Artés
Instituto de Física Fundamental, Centro Mixto C.S.I.C.-U.C.M.
Serrano 123, 28006 Madrid, Spain

Osman Atabek
Laboratoire de Photophysique Moléculaire du CNRS, 91405 Orsay
France

André D. Bandrauk
Département de Chimie, Faculté des Sciences, Université de Sherbrooke
Sherbrooke, Québec, Canada J1K 2R1

Abstract

A coupled equation method combined with an artificial channel technique, properly describing the excitation step, is used to calculate absorption lineshapes and branching ratios of the Floquet states involved in the multiphoton absorption–emission processes of a dissociating molecular system. The formalism is applied to the photodissociation of $H_2^+(1s\sigma_g, v = 0, j = 1 \rightarrow 2p\sigma_u)$. Intense field behavior of laser induced resonance Fano profiles are analyzed in relation with the dissociation lineshapes. Strong mixings appear for intensities larger than $10^{13} W/cm^2$ and the distribution of higher energy peaks decreases with increasing intensity, due to stimulated emissions of the dissociating fragments.

1 Introduction

It is now a well known fact that multiphoton transitions in intense laser fields may radically change the dynamics of photodissociation. Nonlinear effects, above-threshold absorption and dissociation (ATD) leading to essential alterations in dynamical properties affecting optical spectra are expected for field intensities above $10^{11} W/cm^2$.Such powerful lasers are nowadays operational and open new areas of interest,both experimental and theoretical, in problems of half collision [1].

If in a weak field situation photodissociation cross section is given within a first order perturbation theory approach by Fermi's golden rule, this is no longer valid when increasing the field strength. A reliable model setting on an equal footing photon absorption, emission and molecular dissociation, should be of non perturbative nature and include the field in an appropiate way. The so-called dressed molecule picture, leading to laser-induced resonances through quasi-discrete states embedded in continua to which they are radiatively coupled, has recently been invoked as a possible way to reach intense field fragmentation dynamics [2,3]. It is however to be emphasized that these resonances cannot be directly related to physical observables for such intensity regimes. This is to be contrasted to the weak field case where a single quasi-energy lifetime gives the corresponding absorption cross section. It remains the possibility to reconstruct the absorption lineshape by combining the complete set of the resonances in a given energy region, but this may amount to the calculation of a very important number of quasienergies. A different, computationally attractive method, summarized in Section II, is a full collisional treatment [4] based on the introduction of two artificial channels as suggested by Bandrauk [5] and inspired from Shapiro's work [6].The first open artificial channel aims to transform the otherwise half collision situation into a full collision and the second closed artificial channel supports the true initial unperturbed (zero-field) molecular state, describing the preparation step. The power of the method relies on the fact that the summation over the resonances can be carried out in an indirect way, the information being extracted from the scattering amplitude between the artificial entrance channel and the final physical continuum. As a consequence of the coupling scheme , the manifestation of neighbouring resonances is through Fano's profiles. Section III is devoted to an application of the method to the case of H_2^+ photodissociation to which a great deal of effort is directed both experimentally and theoretically [7,2,3,4].

2 Theory

The total molecule plus radiation field Hamiltonian is taken as:

$$H = H_m + H_{rad} + V \tag{1}$$

where H_m and H_{rad} are the isolated molecule and free radiation parts respectively,V being the radiative coupling.For the gauge used here (i.e., radiation field RF or velocity form) one has [3b]:

$$V^{RF}(cm^{-1}) = 1.17 \times 10^{-3}\,\mu(a.u.)\left[I(W/cm^2)\right]^{1/2}((V_c - V_a)/\hbar\omega) \tag{2}$$

μ being the electronic transition moment and I the field intensity, V_a and V_c the bare potential energy surfaces of the initial, $\mid a >$, and final, $\mid c >$, electronic states, respectively, and ω the laser frequency. The unperturbed eigenstates of $H_m + H_{rad}$ are described by direct products of a molecular state $\mid a >$ (characterized by its quantum numbers a and energy E_a) and a field state (characterized by its photon occupation number $\mid n >$ and energy $n\hbar\omega$), and symbolized by $\mid a, n >=\mid a >\mid n >$. For simplicity, the index n will be usually dropped in this representation.

The temporal evolution of some initial state $\mid \psi(0) >$ is formally given by

$$\mid \psi(t) >= \exp(-iHt/\hbar) \mid \psi(0) > \tag{3}$$

which after expansion in terms of eigenstates of $H_m + H_{rad}$ results in:

$$| \psi(t) >= \sum_c \int dE_c \left(\sum_a I_{ca}(t) < a | \psi(0) > \right) | c > \quad (4)$$

$I_{ca}(t)$ is the time dependent transition amplitude between unperturbed bound $| a >$ and energy normalized continuum $| c >$ states,

$$I_{ca}(t) = (2i\pi)^{-1} \int dE \exp(-iEt/\hbar) G_{ca}(E^+) \quad (5)$$

and G the resolvent operator:

$$G(E^+) = \lim_{\varepsilon \to 0} (E + i\varepsilon - H)^{-1} \quad (6)$$

Using the well-known relation between resolvent (G) and transition (T) operators, $I_{ca}(t)$ can be expressed in terms of energy dependent transition amplitudes $T_{ca}(E^+)$. It can ultimately be shown, in the limit of radiative lifetimes shorter than pulse durations, that

$$\lim_{t \to \infty} | \psi(t) >= \sum_c \int dE_c \, T_{ca}(E_c) \frac{\exp[-i(E_c + n_c \hbar\omega)t/\hbar]}{[E_c + n_c \hbar\omega - E_a - n_a \hbar\omega]} | c > \quad (7)$$

provided the initial state$| \psi(0) >$ be $| a >$. The photodissociation probability P_{ca} for a transition from the initial bound state $| a >$ to a continuum state $| c >$ is thus

$$P_{ca} = \int dE_{c'} |< c' | \psi(t_\infty) >|^2 \quad (8)$$

which results in

$$P_{ca}(\omega) = \int dE_{c'} \frac{| T_{c'a}(E_{c'}) |^2}{(E_{c'} + n_{c'} \hbar\omega - E_a - n_a \hbar\omega)^2} \quad (9)$$

by the use of the orthogonality relation $< c' | c >= \delta(E_c - E_{c'})$. A different presentation of the results is obtained by referring to branching ratios or partial widths of given final continua,

$$\rho_{ca}(\omega) = \frac{P_{ca}(\omega)}{\sum_{c'} P_{c'a}(\omega)} \quad (10)$$

Intense field multiphoton processes are taken into account through the close coupling equations describing the time independent Floquet Hamiltonian:

$$[H_{aa} + (n+1)\hbar\omega - E] | a, n+1 > + V_{ac} [| c, n > + | c, n+2 >] = 0 \quad (11a)$$

$$[H_{cc} + n\hbar\omega - E] | c, n > + V_{ca} [| a, n-1 > + | a, n+1 >] = 0 \quad (11b)$$

where H_{aa} and H_{cc} stand for the Hamiltonian of the uncoupled $| a >$ and $| c >$ states. These equations describe a half-collision process and may lead to the determination of the field induced resonances as has been done recently [3]. If the widths of these resonances are proportional to the photodissociation cross section when their mutual overlapping are considered as negligible, such an assumption is not valid when strong fields are adressed to. In that case, a possibility remains to artificially transform the process to a full collisional one, by adding an open–entrance channel $| c_1 >$ as was first suggested by Shapiro [6]. T_{ca} is then amenable to an indirect numerical evaluation via the element S_{cc_1} of the scattering matrix.

In the weak field case, where the initial state $| a >$ is implicitely a well-defined, isolated and unperturbed state, Shapiro's treatment yields dissociation cross sections calculated within the first order Born approximation at the specific energy E_a of the state $| a >$, dressed by a single photon $\hbar\omega$. The treatment, which remains valid for the description of two step multiphoton excitation processes, provided the first radiative transition from the state $| a >$ be weak, has been extended to the calculation of the two photon dissociation cross section of IBr [8].

Intense incident fields produce however coherent mixing of the eigenstates of $H_m + H_{rad}$ via the radiative coupling V. Superposition of these states as well as the initially prepared state must be incorporated into the calculation. Although it is possible to extend Shapiro's treatment to calculate T_{ca}, exactly by summing up the Born perturbative expansion, the limitation to the only specific energies of the dressed unperturbed states $(E_a + \hbar\omega)$, at which T_{ca} can be obtained, remains. Contrary to the weak field limit, such energies have no more physical specificity since the field induced resonances resulting from states $| a >$ may be very much altered (shifted and broadened).The only meaningful information would be the total photodissociation probability given by Eq.(9) which necessitates the calculation of T_{ca} at any energy. Bandrauk has recently proposed the introduction of a second bound artificial channel $| d >$ which plays the role of the true initial unperturbed (zero field) molecular state weakly coupled to the total molecular–field manifolds[5]. This approach results in a simple numerical algorithm and has the merit to give more insight into the dynamics by focusing into field induced resonances. In practise, $| d >$ is some state $| a, n >$, a condition which is achieved using an electronic potential V_d identical to V_a ensuring $< a | d >= \delta_{ad}$ and $V_{ad} =< a | V | d >= 1$ (weak coupling) whereas $V_{da} = 0$ and $V_{ac} = V_{ca} \neq 0$ (symmetric strong coupling). The two artificial channels $| c_1 >$ and $| d >$ are supposed to be coupled in an asymmetric way, i.e. $V_{dc_1} \neq V_{c_1 d} = 0$ so as not to perturb the initial state $| d >$ (no shift or width). Use of the formalism of projection operators leads, after some algebra, to

$$S_{cc_1}(E) = -2i\pi \sum_d \frac{< c | T | a >< a | V | d >< d | V | c_1 >}{(E - E_a - n_a\hbar\omega)(E - E_d)} \tag{12}$$

where

$$T_{ca}(E) =< c | T | a >= (E - E_a - n_a\hbar\omega) \sum_L \frac{< c | t | L >< L | a >}{(E - E_L + i\Gamma_L/2)} \tag{13}$$

with $t = V + VP(E - PHP)^{-1}PV$, P being the projection operator over all continuum states. $| L >$ corresponds to field induced resonances with complex energies, $E_L - i\Gamma_L/2$. More precisely, they are eigenstates of $H_m + H_{rad} + QtQ$, Q being the projection operator over the discrete states. The overlaps $< L | a >$ describe the preparation of the initial state $| a >$ into the complex dressed state $| L >$, as would occur in sudden excitation. The transition operator t connects the dressed states $| L >$ to the physical final continuum state $| c >$. The coherence effects of the intense field is thus maintained through the $| L >$ states. Invoking the weak coupling limit, one has $| L > \rightarrow | a >$, $\Gamma_L \rightarrow 0$, and $E_L \rightarrow E_a + n_a \hbar\omega$ so that one obtains with $t = V$

$$T_{ca}(E) = < c \mid V \mid a > \tag{14}$$

This is for the formal definition of T_{ca}. It can, in principle, be obtained by the sum of the contributions of all the resonances as indicated by Eq.(13). Such a calculation is currently under investigation in our group. But it is also obtainable by the numerical evaluation of the S matrix element, S_{cc_1}, from Eq.(12). The sum over $| d >$ can be cancelled by the choice of an energy E close enough to E_d , but as this last energy can be varied at will, $T_{ca}(E)$ is obtained at any energy, such that Eq.(9) can now give the total photodissociation probability for a given final continuum $| c >$.

3 RESULTS

The theory which is presented in the previous section is illustrated by the example of the multiphoton photodissociation of H_2^+ bound in its ground electronic state

$$H_2^+(1s\sigma_g, v = 0, j = 1) + n\hbar\omega \rightarrow H^+ + H(1s) + \epsilon(n) \tag{15}$$

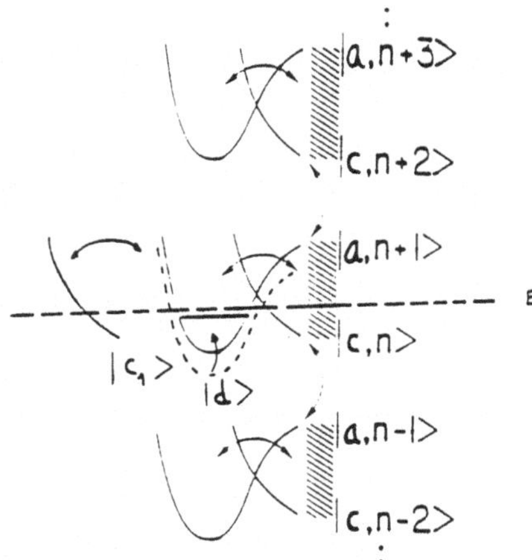

Fig.1. Schematic representation of the two-artificial channel method involving the dressed bound $| a >$ and dissociative $| c >$ electronic states. Each pair of crossing-curves (on the right) represents one Floquet block radiatively coupled to two adjacents ones as indicated by the arrows. They are open or close according to their asymptotic energies as compared to the collision energy E under consideration. $| c_1 >$ and $| d >$ are the two asymmetrically coupled artificial channels.

Fig. 1 displays the collection of dressed channels (close and open) described by the model. We only consider dissociation via the repulsive $H_2^+(2p\sigma_u)$ electronic state which is asymptotically degenerate with $H_2^+(1s\sigma_g)$. The corresponding potential energy curves are given by Morse–type representations, as for the transition dipole moment the Bunkin–Tugov form is taken following reference [9]. Any change in the rotation quantum number due to the interaction with the field is neglected. The calculations,done in the velocity gauge frame, are presented by considering three intensity regimes for the electromagnetic field (weak, if the Rabi frequency $\hbar^{-1}V_{ca}^{RF}$ is lesser than the vibrational frequency ω_v of H_2^+; intermediate if the two frequencies are comparable, strong if the Rabi frequency is larger).

3.1 Weak field regime

As has been pointed out previously [3], this regime (valid for intensity lower than $10^{11}W/cm^2$) corresponds to the linear behavior of the photodissociation rate $\mid T_{ca} \mid^2$ versus the intensity. This also corresponds to the region where Fermi golden rule is valid, such as a lowest order perturbation formula can successfully be applied. We have performed a series of calculations for a laser wavelength varying from 600 to 2000 $\mathring{A}$ and intensities ranging from $3.5\,10^6$ to $3.5\,10^{10}W/cm^2$ by using the three techniques we have at our disposal, i.e., Fermi golden rule, Shapiro's one artificial channel and the two artificial channel model, Eq.(12), by retaining only one Floquet block. At least three figures of accuracy is obtained and the results are in full agreement with those of Ref. [2,3] based on the calculation of resonance widths showing, in particular, that for such radiative couplings but one isolated laser induced resonance is responsible for the lineshape $\sigma(\omega) =\mid T_{ca}(E_a + \hbar\omega) \mid^2 /I$.

3.2 Intermediate field regime

For field strengths roughly extending from $I \approx 10^{11}W/cm^2$ to $I \approx 10^{13}W/cm^2$, a nonlinear behavior of the dissociation rates as functions of intensity is obtained. This can be related to off–the–energy shell contributions, as well as to neighboring resonance overlaps. A single–photon description involving one Floquet block but avoiding the use of the perturbative expansion , already contains the required information.In agreement with previous calculations, the increase of the intensity leads to a slight blue shift of the maximum of the lineshape as well as a flattening for large wavelengths. It is to be noticed that the unique Floquet block used in these calculations gives a convergence up to 98% and the difference between the two calculations (one or two–artificial channels) is less than 0.6%.

$\mid S_{kc_1}(E) \mid^2$ (k being an index for the kth open channel; i.e., on Fig.1,$k = 1$ for $\mid c, n >$; $k = 2$ for $\mid a, n-1 >$ and so on) as a function of the total energy, which results from Eq.(12) and enters the calculation of partial dissociation probabilities through $T_{ca}(E)$ (Eq.(9)), presents several interesting features. They may be related with the interferences arising from the interaction of higher vibrational ($v = 1, 2, 3, \cdots; j = 1$) field induced resonances which are supported by the $\mid a, n+1 >$ dressed state coupled to the $\mid c, n >$ continuum. The results, for $\lambda = 1200\mathring{A}$, $I = 8.8\,10^{11}W/cm^2$ and $I = 3.5\,10^{12}W/cm^2$, are displayed in Figs. 2a–b.

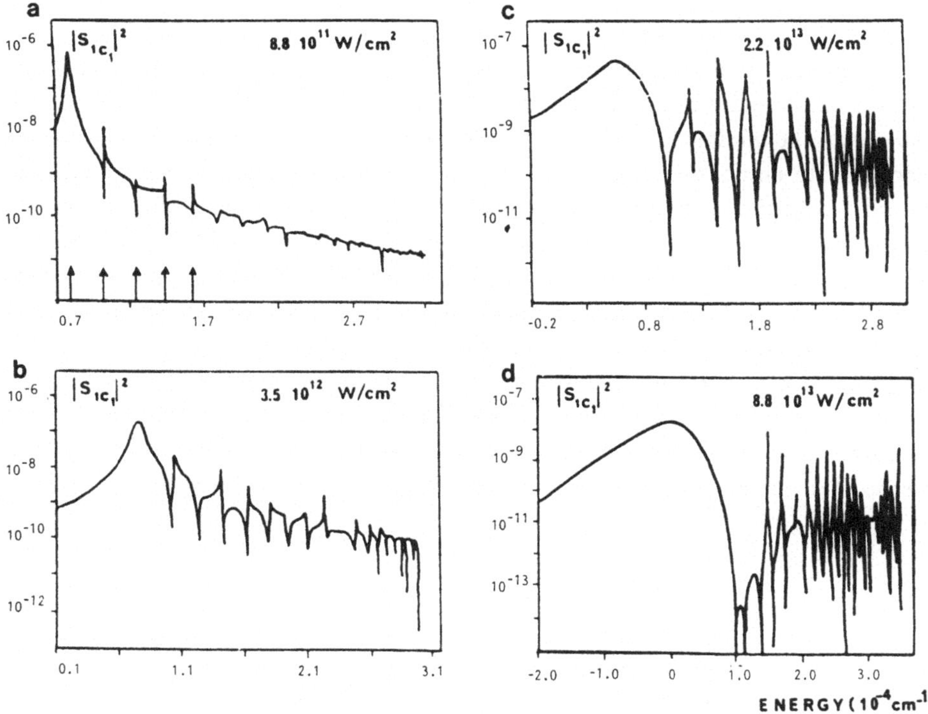

Fig.2. Square modulus of the scattering matrix element $| S_{11} |^2$ as a function of the collision energy for a laser wavelength $\lambda = 120nm$, in weak (a), intermediate (b) and strong field (c,d) regimes.The arrows in (a) indicate the position of unperturbed vibrational levels of $H_2^+(1s\sigma)$.

For the lowest intensity (Fig.2a), clearly five peaks can be identified which roughly correspond to the positions of $(v = 0,1,2,3,4; j = 1)$ resonances (i.e. the positions of the first five vibrational bound states of $| a >$, affected by a field induced shift). The successive peak amplitudes are very different (approximately one or two orders of magnitude of decrease between each of them). This is a direct consequence of the preparation step concerning the lowest vibrational level of $| a >$. Via the asymmetric coupling $V_{da} = 0, V_{ad} = 1, | d >$ is only coupled to the $(v = 0, j = 1)$ level of $| a >$, such that this level is picked out as the initial state which in turn is distributed over the complex dressed states $| L >$. The resulting lineshape presents an important maximum when the $(v = 0, j = 1)$ level of $| a, n+1 >$ is in doorway situation with respect to E (as is illustrated in Fig. 1). Upper resonances $(v = 1,2,3,4; j = 1)$ weakly interfer with $(v = 0, j = 1)$, only via their coupling through the continuum $| c, n >$, producing Fano profiles [10] with a peak and a dip structure. When the field strength is varied from $8.8\,10^{11} W/cm^2$ to $3.5\,10^{12} W/cm^2$ (Fig.2b) the position of the $(v = 0, j = 1)$ peak is shifted and its width is increased approximately as a linear function of the intensity. Fano profiles corresponding to $(v = 1,2,3,4,..; j = 1)$ levels, which are indirectly coupled among them via $| c, n >$ and $| a, n+1(v = 0, j = 1) >$, are also shifted and broadened, their amplitudes remaining however moderate as compared with the dominant peak.

3.3 Strong field regime

At higher intensities,depending on the photon energy $\hbar\omega$ and initial state, absorption of additional photons lead to multiphoton dissociation pathways that are much more favorable than the minimum–energy path. We refer to such processes as above–threshold dissociation to emphasize their similarity with the above–threshold ionization (ATI) process [11]. Typically nine Floquet blocks (four close and five open) plus the resonant one are necessary to achieve three figures accuracy . Figs.2c and 2d display converged results for $| S_{1c_1} |^2$ at the same wavelength and for intensities of $2.2\,10^{13} W/cm^2$ and $8.8\,10^{13} W/cm^2$. The preparation step again favors the resonance corresponding to the lowest vibrational level of $| a, n+1 >$ to which $| d >$ is directly coupled. The resulting peak presents a rather large bandwidth with a highly asymmetric behavior showing marked tendency to flatten at lower energies. A similar behavior is observed when the initial state is taken to be $(1s\sigma_g, v = 1, j = 1)$, as depicted in Fig.3. These patterns correspond to the absorption lineshapes ($v = 0$ and $v = 1$) displayed in Refs.[2,3a].

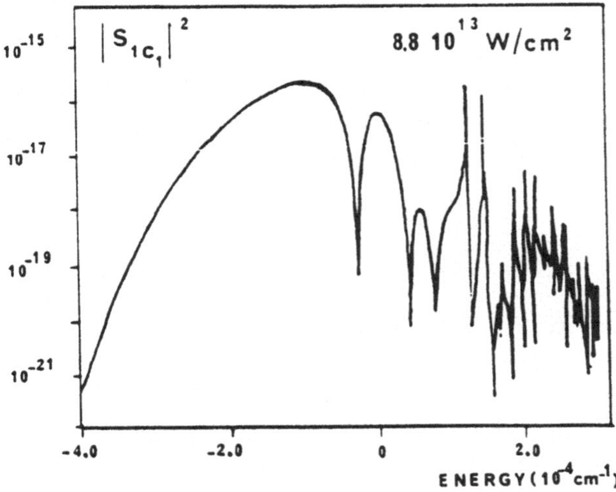

Fig.3. $| S_{11} |^2$ as a function of the collision energy for a laser wavelength $\lambda = 120nm$. and intensity $I = 8.8\,10^{13} W/cm^2$ with $(1s\sigma, v = 1, j = 1)$ as initial state.

With such high intensities the other complex dressed states are strongly interfering with each other. Important shifts and widths are affecting the resonances mediating the dissociation dynamics leading to a complicated peak and dip pattern which cannot be easily correlated with the set of vibrational levels of $| a, n+1 >$.

Branchig ratios, ρ_{ka}, are plotted on Fig.4 as a function of the laser field intensity ranging between $10^8 - 10^{14} W/cm^2$. They represent the relative probabilities to dissociate into channels $| c, n >, | a, n-1 >$ and $| c, n-2 >$. The wavelength which is chosen for the calculations, namely $\lambda = 3296.7\AA$, favors a three photon transition from $(v = 0, j = 1)$. Three different intensity regimes are obtained which can be interpreted in an adiabatic picture as following:

i) At field strengths below $3.5\,10^{12}W/cm^2$, due to a high and large tunneling barrier originating from an adiabatic decoupling of the surfaces associated with $\mid a, n+1 >$ and $\mid c, n >$, single–photon absorption resulting into fragments in channel $\mid c, n >$ has a very low probability. The absorption proceeds rather via a three photon transition to $\mid c, n-2 >$. The branching ratio ρ_{3a} largely dominates over ρ_{2a} and ρ_{1a}.

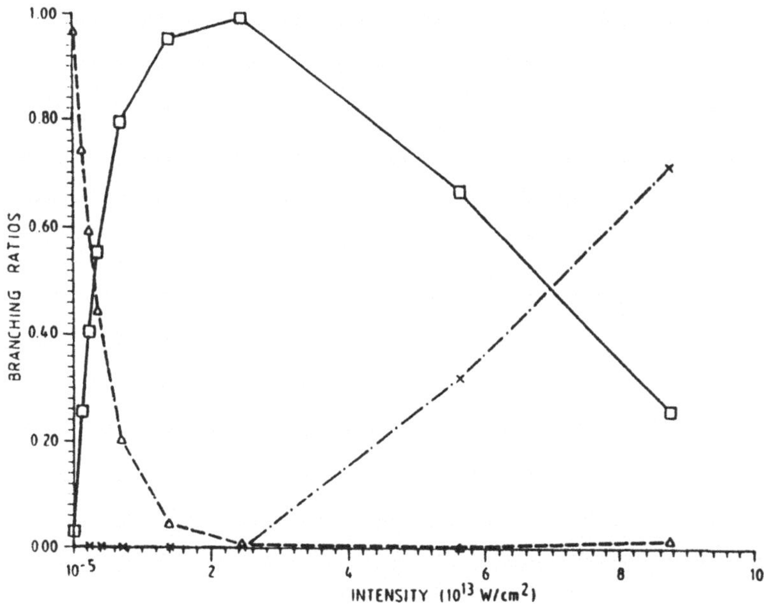

Fig.4. Branching ratios as a function of the field strength for a photon wavelength $\lambda = 329.7nm$. Crosses for the one-photon channel: ρ_1; squares for the two-photon channel: ρ_2; triangles for the three-photon channel: ρ_3.

ii) At higher intensities, the most striking observation is that, in spite of the I^3 dependence of the total photodissociation rate, the relative probability to find fragments in channel $\mid a, n-1 >$ dominates on $\mid c, n-2 >$ $(\rho_{2a} > \rho_{3a})$. Such a situation has already been discussed in the literature both from experimental [12] and theoretical [11] points of view. The avoided curve crossing between channels $\mid c, n-2 >$ and $\mid a, n-1 >$ taking part in the dynamics is invoked as an explanation . In the proximity of this point the three photon dissociating system returns one photon to the radiation field via stimulated emission, favoring thus the branching ratio of channel $\mid a, n-1 >$. The adiabatic path leading to $\rho_{2a} > \rho_{3a} \gg \rho_{1a}$ (three photon absorption , followed by an emission during dissociation resulting into velocity lowering of the protons) appears to be valid for increasing fields. This is precisely what one gets for intensities ranging between $3.5\,10^{12}W/cm^2$ and $2.5\,10^{13}W/cm^2$.

iii) Above $3\,10^{13}W/cm^2$ dissociation probability towards channel $\mid c, n >$ increases in such an extent that for $I = 7\,10^{13}W/cm^2$ it is of the same order than for channel $\mid a, n-1 >$ and for higher values of the field $(10^{14}W/cm^2)$ it describes the dominant process. A possible interpretation would be the lowering and flattening of the adiabatic potential barrier at the avoided curve crossing occurring between $| a, n+1 >$ and $| c, n >$. An estimation for changes that may affect dressed potential energy surfaces by intense radiation fields leading to important tunneling has very recently been given [12]. It appears that for the case of H_2^+, field intensities of the order of $5\,10^{13} W/cm^2$ are enough to completely flatten the potential barrier resulting from dressed molecular states with an electromagnetic field of wavelength $\lambda = 5320\AA$.

As a word of conclusion, this work is the first illustration of the ability of the double artificial channel procedure to reproduce partial dissociation probabilities in intense field regimes. The interpretation of the above–threshold photodissociation of H_2^+ is taken as an example motivated by the numerous experimental and theoretical investigations devoted to this system. The lowering of the recoiling proton kinetic energy as a consequence of a three-photon absorption followed by a one–photon emission at $\lambda = 329.7nm$ and $I > 3.5\,10^{12} W/cm^2$ is obtained as in previous calculations [11]. A new feature rises for intensities above $7\,10^{13} W/cm^2$ where two photons can be re–emitted during the fragmentation process, such that the asymptotic fragmentation channel corresponds to a net single photon absorption. This seems to be consistent with very recent experimental work on H_2, HD and D_2 photofragmentation . Interpretation of isotopic effects is undertaken in our group.

References

[1] M.H. Mittleman,"Introduction to the Theory of Laser. Atom Interactions", Plenum, New York, 1982.
[2] S.I. Chu, J. Chem. Phys 75, 2215 (1981).
[3] (a) X. He, O. Atabek and A. Giusti-Suzor, Phys. Rev. A38, 5586 (1988); (b) X. He, O. Atabek and A. Giusti-Suzor, Phys. Rev. A42, 1585 (1990).
[4] S. Miret-Artes, O. Atabek and A.D. Bandrauk, Phys. Rev. (submitted).
[5] (a) A.D. Bandrauk and G. Turcotte, J. Chem. Phys. 89 ,3039 (1985); (b) A.D.Bandrauk and O.Atabek, in "Lasers, Molecules and Methods", Edited by J.O. Hirschfelder, R.E Wyatt and R.D. Coalson (Wiley,NY,1989), Chap.19.
[6] M. Shapiro, J. Chem. Phys.56, 2582 (1972).
[7] C. Cornaggia, D. Normand, J. Morellec, G. Mainfray and C. Manus, Phys. Rev.A34, 207 (1986); T.S. Luk and C. Rhodes, Phys.Rev. A38, 6180 (1988); A. Zavriyev,P. H. Bucksbaum,H. G. Muller and D.W.Schumacher Phys. Rev. A42, 5500 (1990);B. Yang,M. Saeed,L. F. Di Mauro,A. Zavriyev and P. H. Bucksbaum,Phys. Rev. Lett. (submitted).
[8] A.D. Bandrauk, G. Turcotte and R. Lefebvre, J. Chem. Phys. 76, 225 (1982); M. Shapiro and H. Bony, J. Chem. Phys. 83 ,1588 (1985).
[9] F. V. Bunkin and I. I. Tugov, Phys. Rev. A8, 601 (1973).
[10] U. Fano, Phys. Rev. 124, 1866 (1961).
[11] A. Giusti–Suzor, X. He, O. Atabek and F.H. Mies, Phys. Rev. Lett. 64, 515 (1990).
[12] P. H. Bucksbaum, H. G. Muller, D. W. Schumacher and A. Zavriyev, Phys. Rev. Lett. 64, 1883 (1990).
[13] A.D. Bandrauk, E. Coutant, Canadian J. Phys. (in press).

INTENSE FIELD DYNAMICS OF DIATOMIC MOLECULES

L. F. DiMauro, B. Yang and M. Saeed

Department of Chemistry
Brookhaven National Laboratory
Upton, NY 11973

INTRODUCTION

Over the last decade studies examining the behavior of isolated atoms in intense radiation fields has resulted in a cumulative literature rich in new phenomena[1], unresolved issues, and exciting new predictions[2] for future studies. Very recently, similar investigations have begun to examine the general behavior of molecules in intense fields, especially in the studies of photodissociation dynamics. The additional challenge associated with studying molecules in strong fields result from the difficulty of sorting out the general behavior of field-induced effects from the details or specifics of the molecular structure. Previous studies on atoms[3] clearly demonstrate the importance of the atomic structure in describing the "non-resonant" ionization dynamics. Consequently, it is not surprising to expect enhanced structural effects in molecules due to the many internal degrees of freedom as compared to atoms. In addition, molecular strong-field studies raise some interesting and unique questions concerning the role and interplay of ionization and dissociation. Furthermore, the time-scales associated with the molecular nuclear motion, i.e. vibrations, are comparable to the duration of current ultra-short laser pulses, thus presenting the ability to freeze-out certain degrees of freedom. The attraction for such molecular studies derives its fundamental interest from the ability to control chemical dynamics by external variation of a laser field. The traditional chemical physics approach of using the laser frequency as the only external field parameter for achieving state-selective chemistry has met with little success. However, recent efforts, such as those represented in this book, have focussed on studying all aspects of the laser field for achieving control. Although premature in realization, an understanding of the influence of intensity, coherence, and pulse duration and shaping in concert with frequency may ultimately lead to our ability or inability to control chemical dynamics.

In this paper we will report on our progress in conducting a systematic investigation on the intense field behavior of homo- and hetero-nuclear diatomic molecules. These studies involve the "nonresonant" multiphoton excitation of diatomics in the intensity regime of $10^{11\text{-}14}$ W/cm^2 with the second and third harmonic radiation of a Nd:YAG or Nd:YLF laser system. The usage of the term "nonresonant" is used in a general context implying that the photon frequencies used in these experiments are not in resonance with any zeroth-order unperturbed molecular states. However, this does not exclude the possibility of dynamic Stark shifting of states into resonance by our laser field. In the situation where a nonlinear process proceeds without the influence of any specific intermediate states (truely nonresonant), we shall refer to this process as "direct". The specific diatomics used in these studies are hydrogen, deuterium, deuterium hydride, and oxygen. Measurements include energy resolved electron and mass spectroscopy the details of which will be described in the experimental section. The third section will

Coherence Phenomena in Atoms and Molecules in Laser Fields
Edited by A.D. Bandrauk and S.C. Wallace, Plenum Press, New York, 1992

describe the multiphoton ionization (MPI) of these molecules in an intense laser field. The role of dynamically Stark shifting bound state structure will be shown for H_2, D_2, and HD molecules. Furthermore, the study of these isotopic molecules provides a means of identifying the intermediate states involved in the ionization. Oxygen molecule provides a different scenario in which the interplay of multiple intermediate electronic state resonances produces a strong intensity dependent vibrational distribution in the molecular ion. A model is presented which explains most of the main features of the experiment. Finally, intense field dissociation of H_2^+ and D_2^+ will be dicussed in the last section. Both diatomic ions are formed via the nonlinear photoionization of the neutral precursor. This study provides the first observation of the dynamics associated with intense field dissociation by measuring the branching ratios of the different above-threshold (ATD) fragments. Our results support the dressed-molecule picture at least qualitatively, but differences do exist with the quantitative predictions.

EXPERIMENTAL

The photon sources used are well characterized Nd:YAG (1.06μm, 10 nsec, 10 Hz) and Nd:YLF (1.05μm, 50 psec, 1 kHz) lasers. Details of these laser systems have been described previously.[4] The harmonics are generated using standard nonlinear techniques either in BBO or KD*P crystals. The intensity ranges for 532 and 527 nm radiations were 8×10^{11} to 9.7×10^{12} W/cm^2 and 5×10^{12} to 4×10^{13} W/cm^2, respectively. The apparatus consists of ultra high vacuum chambers equipped with time of flight electron and mass spectrometers. The field-free photoelectron spectrometer has an energy resolution of about 50 meV for 1 eV electrons and an acceptance angle of 1×10^{-3} sr. It is calibrated by the well known ATI spectrum of xenon atoms. The mass spectrometer consists of a 42 cm long flight tube with a series of electric field acceleration plates. Fragment kinetic energy analysis is achieved by applying a uniform extraction field across the laser focus with the laser polarization parallel to the flight tube axis. This produces two nearly symmetric peaks in the time-of-flight spectrum corresponding to the two velocity components of the same fragment initially directed towards and away from the detector. The kinetic energy of the fragment is simply determined by measuring the arrival time difference between the peaks. The mass spectrometer calibration was performed by weak-field photodissociation of chlorine[5] molecules. Its resolution was estimated to be about 100 meV for 1 eV protons. The background pressure in the ultrahigh vacuum system was 1×10^{-9} torr. The photoelectron experiments are performed by adjusting the sample pressure to keep the number of ions produced in the ionization region to less than 50 per shot to minimize space charge effects. The data collection system is always operated in the pulse counting regime with counts rates much less than one count per laser shot.

MULTIPHOTON IONIZATION DYNAMICS

First, let us examine the intense-field ionization of hydrogen with second harmonic radiation. Molecular hydrogen represents the "model" molecular system for these studies because of its simple and well known molecular structure and as a result should provide the foundation for future field-molecule studies. Recently, effects of intense fields on H_2 have been studied by observing the above threshold ionization and dissociation (ATI and ATD) phenomena through resonant[6] and non-resonant[7-11] multiphoton ionization (MPI). In our experiment[10], H_2, D_2 and HD molecules were irradiated by the second harmonic of both Nd:YAG and Nd:YLF lasers. The first I.P. from the vibrationless H_2 ground state is 15.425 eV, requiring at least 7 photons for ionization in the current experiment. The well resolved intensity dependent electron and proton spectra along with the comparison of data taken at two wavelengths (527 and 532 nm) and for two different isotopes is very useful in elucidating the interaction dynamics predicted by existing high field models. Furthermore, our analysis provides an assignment of the intermediate states prevalent in the ionization process.

Figure 1 shows the photoelectron spectra of H_2 at 532 and 527 nm at different laser intensities. The 532 nm photoelectron spectra of D_2 as a function of laser intensity is shown in Fig. 2. Figure 3 shows the photoelectron spectrum of the three molecules recorded at the same low intensity. For the entire dynamic range of laser intensities, the vibrational structure of 7 and 8-photon processes is resolvable as well as

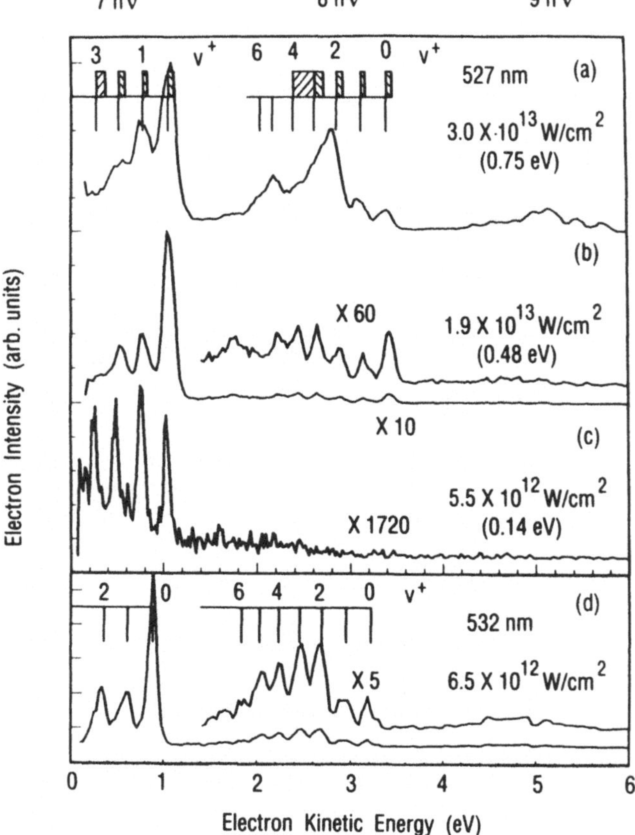

Fig. 1 Photoelectron spectrum of H_2 molecule taken at different laser intensities with 527 and 532 nm radiations. The data is normalized for gas density and number of laser shots. The tick marks indicate the unperturbed vibrational levels of the H_2^+ ground state. The hashed boxes are the shifts calculated using a Floquet method at the maximum 527 nm intensity of 3×10^{13} W/cm^2 (ref. 9). The numbers in parenthesis are the corresponding Ponderomotive energies at each intensity.

assignable to the unperturbed vibrational frequency of molecules within experimental error. Some important features of the spectra should be pointed out here. (1) The low intensity 527 nm photoelectron spectrum shown in Fig. 1(c) exhibits well resolved v^+=0-3 vibrational bands whose intensity distributions are similar to the Frank Condon factors expected from a direct 7-photon ionization between the $H_2\ {}^1\Sigma_g^+$ and $H_2^+\ {}^2\Sigma_g^+$ ground states. (2) As the 527 nm laser intensity is increased in Fig. 1 to 3×10^{13} W/cm^2, only the low energy electrons, specically the v^+=2 and 3 bands, are suppressed proportional to the expected Ponderomotive shifts in the ionization potential. No such suppression is observed with the 532 nm excitation due to the lower saturation intensity of H_2 in nanosecond pulses. (3) Over the entire intensity region for 532 nm excitation and those above 10^{13} W/cm^2 with 527 nm the photoelectron spectrum (PES) profiles are similar, i.e. for the seven photon process most ions are formed in the v^+=0 level, suggesting a probable intermediate state resonance. (4) At the lowest intensity which results in an observable rate, the three molecules show different vibrational distributions (see Fig. 3) which is also indicative of intermediate state effects. (5) The maximum in the photoelectron distribution of H_2 for the eight photon process (ATI) shifts gradually from v^+=5 to v^+=2 as the laser intensity is increased. (6) The most probable photoelectron kinetic energy at a given intensity for the

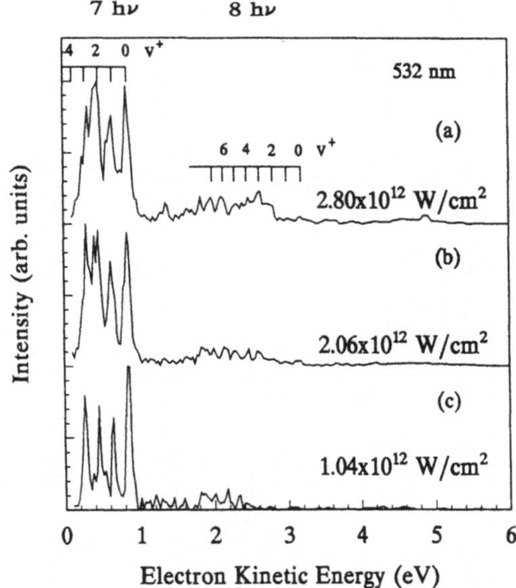

Fig. 2 Photoelectron spectrum of D_2 molecule taken at different laser intensities with 532 nm radiation. The tick marks indicate the unperturbed vibrational levels of the D_2^+ ground state.

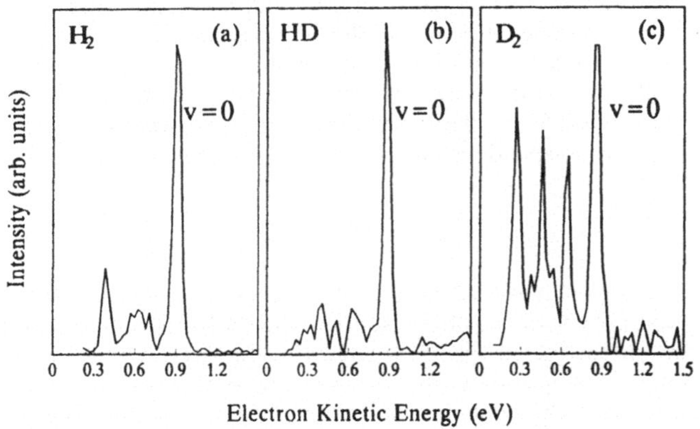

Fig. 3 Low intensity 532 nm PES of (a) H_2, (b) HD, and (c) D_2 at $1 \times 10^{12}\ W/cm^2$.

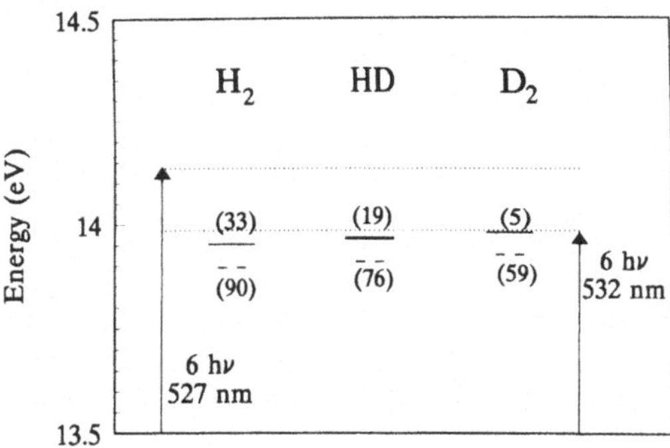

Fig. 4 The vibrationless levels of the J $^1\Delta_g$ (solid line), and I $^1\Pi_g$ (dashed line) state relative to the 6-photon energy of 532 and 527 nm radiations for H_2, HD, and D_2 molecules. The difference between the 6-photon energy of 532 nm and v=0 level is shown in parenthesis in meV. The energy of J $^1\Delta_g$ (v=0) for HD is estimated from energies of H_2.

eight photon process is the same for H_2, D_2, and HD and is thus independent of the molecule studied. (7) Finally, no dramatic shifts were observed in the positions of photoelectron kinetic energies as a function of laser intensities within our experimental resolution. However, the peaks are observed to become asymmetrically broadened towards lower electron energies at the highest intensities. This broadening is observed to be universal for all three isotopically substituted molecules.

The difficulty associated with assigning the intermediate states involved in the ionization is circumvented in our studies by the use of three isotopically substituted molecules and two slightly different laser wavelengths (532 and 527 nm). Speculations on the intermediate states involved in the 532 nm ionization of H_2 have been previously made by Verschuur et al.[8] and Zavriyev et al.[9] In Verschuur's[8] experiment, the ionization through the B $^1\Sigma_u^+$ (v'=3) with five photon resonance and GK $^1\Sigma_g^+$ (v'=2) with six photons were considered to be the major processes. Unfortunately, the photoelectron spectra in either experiment could not be resolved vibrationally. Figure 3 shows the 7-photon 532 nm photoelectron spectra of H_2, D_2, and HD taken at an intensity of 1 x 10^{12} W/cm^2. Each spectrum differs in the amount of contributions from higher vibrations (v^+=1-3) but all share the v^+=0 peak as the dominant feature. In fact, the HD spectrum is predominantly v^+=0 and indicative of a diagonal δv=0 transition from a Rydberg state to the ionic ground state. However, for 527 nm excitation of H_2 (Fig. 1) at low intensity, the lowest four vibrations have comparable amplitude but the distribution becomes similar to 532 nm PES at high intensity. These distributions can be described by dynamical resonances at the 6-photon energy and depending upon the intensity these states are shifted in and out of resonance. Consequently, the final state vibrational distribution of the ion is characterized by intensity dependent weighting factors for two paths involving direct and resonantly enhanced ionization. Our analysis implies that the major contributor to resonance enhancement occurs via Rydberg states, specifically the vibrationless level of the J $^1\Delta_g$, I $^1\Pi_g$, and G $^1\Sigma_g^+$ states. Figure 4 illustrates the location of the J and I state relative to the 6-photon energy for the three molecules studied. For 532 nm radiation, the energy difference between the 6 photons and unperturbed G $^1\Sigma_g^+$ (v'=0), I $^1\Pi_g$ (v'=0), and J $^1\Delta_g$ (v'=0) states of H_2 is only 121.9, 89.6, and 33.5 meV, respectively.[21] Ionization from these intermediate Rydberg states to the ionic ground states are diagonal transitions, i.e. δv=0. Consequently, there will be a wide range of laser intensities for which one can expect a resonant enhancement via dynamical shifts to occur for the v'=0 level, explaining its dominant population in our 532 nm PES. Moreover, for 527 nm excitation this difference is 271.9, 239.6, and 183.4 meV, respectively. This larger detuning implies that for low intensities the 527 nm process will be virtually nonresonant [see Fig. 1(c)]. However, for moderately high intensities a Ponderomotive shift ≥0.18 eV (7×10^{12} W/cm^2) will bring the various v'=0 states into resonance resulting in a PES spectrum similar to the 532 nm excitation. In addition, these states are consistent with the photoelectron spectra shown in Fig. 3 which are recorded at an intensity of 1×10^{12} W/cm^2 or ~25 meV Ponderomotive energy. A quick inspection of Fig. 4 immediately implies that the HD molecule would be more resonantly enhanced at this intensity resulting in a predominantly v^+=0 PES, as shown in Fig. 3. Thus, we conclude that the transition dynamics leading to the final state vibrational distribution via 532 and 527 nm excitation in the low intensity regime is dominated by the $J^1\Delta_g$(v'=0) intermediate level but at higher intensities may be influenced by the I state and to a lesser extent by the G state.

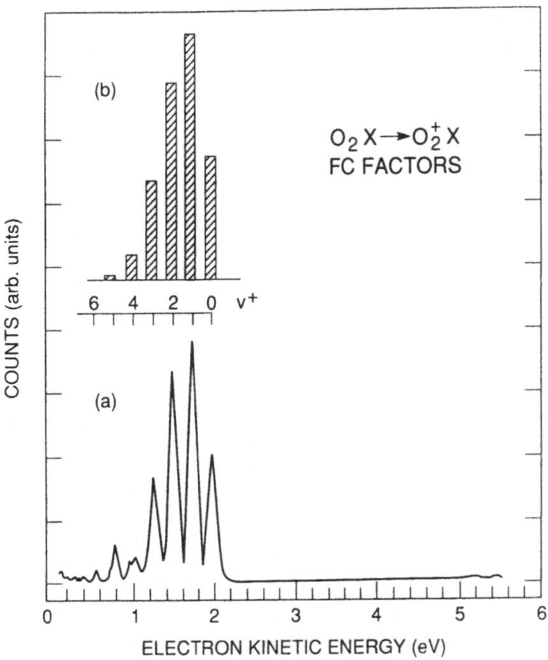

Fig. 5 Photoelectron spectrum of oxygen molecules via 355 nm ionization. The tick mark indicate the energies of the unperturbed vibrational levels of the O_2^+ ground state. The bar diagram (b) are the Franck-Condon factors resulting from a nonresonant $O_2\,X\,^3\Sigma_g^- \rightarrow O_2^+\,X\,^2\Pi_g$ transition.

As a second, more complex example, let us examine the MPI of oxygen molecules in an intense field which upon excitation with green photons (~2.3 eV) exhibits a final state ionic vibrational distribution which is strongly intensity dependent. Although the H_2 and O_2 studies share the same common goal of understanding the high field behavior of diatomic molecules, the O_2 study results in a very different physical interpretation than the MPI of H_2. As a result, we have acquired an insight into bound state structure of neutral oxygen and are able to derive from our analysis molecular constants for previously unobserved state. The adiabatic ionization potential of O_2 is 12.071 eV which corresponds to 4- and 6- photon ionization processes for 355 and 532 nm excitation, respectively. The neutral and ionic ground states have dissociation energies of 5.1 and 6.7 eV, respectively. An inspection of the potential curves reveals a large dissimilarity between the ground and excited states, as well as the ionic state. The variance between potential curves result in large Franck-Condon spreading in oscillator strength throughout the spectral region. This fact is clearly demonstrated in Fig. 5 which shows the oxygen photoelectron spectrum resulting from 355 nm, 4-photon ionization. The ionic vibrational structure is well resolved and assignable to the first seven vibrations, v^{+}=0-6, of the $O_2^+\ {}^2\Pi_g$ ground state. Weak 5-photon above threshold ionization (ATI) structure is observed which mimics the low order 4-photon vibrational distribution. Most notable, the relative amplitudes of the photoelectron peaks and energy distribution are unaltered over

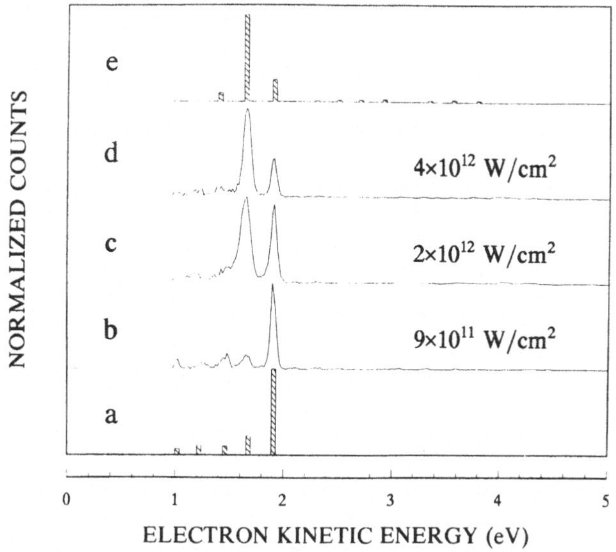

Fig. 6 Intensity dependent photoelectron spectra of oxygen molecules via 532 nm excitation taken at (b) 9 x 10^{11} W/cm^2, (c) 2 x 10^{12} W/cm^2, and (d) 4 x 10^{12} W/cm^2. The bar diagrams (a) and (e) are model simulations for 9 x 10^{11} W/cm^2 and 4 x 10^{12} W/cm^2 intensities, respectively.

the entire dynamic range of intensities used in the current experiment. Analysis of the intensity dependence for both the total O_2^+ yield curves and the individual electron peaks gives a slope consistent with 4-photon direct ionization. Likewise, analysis of the experimental photoelectron vibrational distributions in Fig. 5(a) is in excellent agreement with Franck-Condon factors calculated assuming a direct nonresonant O_2 X $\rightarrow O_2^+$ X transition, also shown in Fig. 5. The only anomalous feature in the photoelectron spectrum is for the v^{+}=5 peak but this could be understood by considering the near resonance at the 3-photon level with the v'=5 vibration of the D or $\beta\ {}^3\Sigma_u^+$ Rydberg state.[12]

Excitation with 532 nm radiation requires a minimum absorption of six photons to ionize oxygen, similar to xenon atoms. Figure 6 shows the photoelectron spectrum recorded at various intensities ranging from 0.9-4 x 10^{12} W/cm^2. All the major features in these spectra are assignable to vibrations of the O_2^+, X $^2\Pi_g$ ground state. However, a dramatic effect evident in Fig. 6 is the strong dependence of the amplitudes of the photoelectron distribution with varying laser intensity. This implies an energy-dependent change in the ion's vibrational content. The photoelectron spectrum in Fig 6(b) taken at an intensity of 9 x 10^{11} W/cm^2 shows a strong peak at 1.91 eV energy which is assignable to the $v^+=0$ vibration of the O_2^+ ground state. The weaker peaks fit to the $v^+=1$-4 vibrational progression of the ground state. As the intensity is increased in Fig. 6(c) and (d) the peak at 1.68 eV (tentatively assigned as $v^+=1$) increases in intensity and at saturation dominates over the $v^+=0$ by a factor of two. Likewise, the weaker structure assigned to $v^+=2$-4 becomes less pronounced. Also noticeable is the absence of appreciable ATI structure, electrons with kinetic energy in excess of a photon energy, at the highest intensities. An atomic xenon spectrum recorded under the same *unsaturated* intensities shows approximately 20% of the electrons ejected into the ATI channels. Closer examination of the intensity behavior of the electron spectrum reveals two important features. First, the electron peaks labeled, $v^+=0$ and 1, have an intensity dependence that scale as I^5 and I^7, respectively, and shows very little deviation from these power laws except at saturation. This result is verified by the total ion yield curves which gives a slope consistent with an average I^6 dependence. L'Huillier et al.[13] have measured the same I^6 total ion power depedence using 30 ps, 532 nm excitation. However, such an *averaged* result is obviously misleading since it would suggest an ionization process that is direct. Second, the 1.91 eV and 1.68 eV electron peaks show differing degrees of variation in linewidth as a function of laser intensity. The 1.91 eV ($v^+=0$) peak linewidth shows no intensity dependence within our spectrometer's resolution but 1.68 eV ($v^+=1$) peak changes by a factor of two in linewidth in the intensity regime where the rates of the two peaks become comparable. At higher intensities where the system starts to saturate the linewidth of the $v^+=1$ peak decreases to a limiting value which is about 50% larger than the $v^+=0$ peak.

The data resulting from third harmonic excitation are well described by a calculation using static Franck-Condon factors. Both the second and third harmonic excitations share the same final continuum state *gerade* symmetry (even number photon absorption) and total energy. Consequently, the strong vibrational intensity dependence observed with green excitation is not a final state effect but instead an intermediate state effect. The simplest intermediate state effect that could result in such an intensity dependence would involve dynamical Stark shifting of two successive vibrational levels (i.e. v'=0 and 1) belonging to the same *ungerade* Rydberg series. Here we have assumed that the measured power dependence implies that the resonances are occuring at the 5-photon level (~0.42 eV below threshold) and the Rydberg series is converging to the $^2\Pi_g$ ground state of O_2^+. At low intensity, the unperturbed v'=0 Rydberg level is resonant with 5-photons resulting in mostly $v^+=0$ ground state O_2^+ as a consequence of the $\delta v=0$ propensity rule. Assuming that most of the Stark shift comes from the Ponderomotive energy of electrons, we calculate that an intensity of ~9 x 10^{12} W/cm^2 is needed to shift the Rydberg level by one vibrational quanta ($\Delta\upsilon$ ~1900 cm^{-1}). However, this scenario is inconsistent with the experiment for a number of reasons. First, the observed intensities at which the distribution changes are an order of magnitude lower than the calculated intensities needed. Second, the predicted sign of the Stark shifts are opposite to those necessary to describe our results. Each Rydberg level is blue shifted similar to the ionization potential and thus the v'=0 level never shifts into resonance. Third, the slopes of the vibrational peaks fit well with an I^N scaling law. A dynamical Stark shift will give rise to intensity dependence more complex than this simple scaling, as demonstrated in cesium.[14,15] Even if one assumes that the Stark shifting occurs between two consecutive vibrations of Rydberg series with differing principal quantum numbers (using known quantum defects)[16], one finds similar inconsistencies with the data.

We propose a mechanism which is conceptually based on a model developed by Smith[17] and applied to describe the photoelectron spectra resulting from single photon ionization of oxygen.[18] Eland uses two continuum paths in Eq.(6) of Ref. 18 to describe the photoelectron data at various excitation wavelengths. Our model relies on three paths as illustrated in Fig. 7, (1) excitation through the $^3\Sigma_u^-$ Rydberg level (light shaded arrows), (2) excitation through $2^3\Pi_u$ adiabatic state (dark shaded arrows), and (3) direct X $\rightarrow$ X 6-photon nonresonant ionization (not shown). The final state vibrational distribution

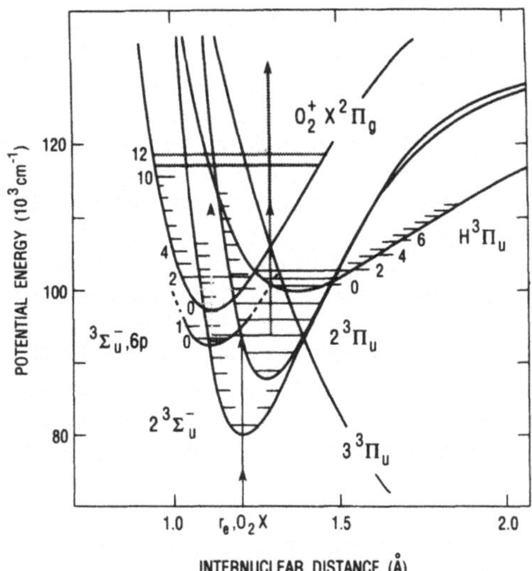

Fig. 7 An expanded view of the potential curves near the ionization potential. Five 532 nm photon absorption occurs adiabatically at r_e of the O_2 ground state while the one photon ionization proceeds via the two paths illustrated by the light and dark shaded arrows. See text for discussion.

results from an incoherent sum of these various channels determined by their Franck-Condon overlap with the O_2^+ ground state. The v'=0, 6p Rydberg series is estimated to be near resonant at the 5-photon level based on measured quantum defects.[19] The $2^3\Pi_u$, $3^3\Pi_u$ and $2^3\Sigma_u^-$ are adiabatic neutral potential curves calculated by Buenker et al.[20] The adiabatic potentials result from avoided crossings of Rydberg states with low lying valence states producing curves which are strongly mixed. Moreover, interaction among these states will be sensitive to internuclear distance since the balance between the valence-Rydberg character of these adiabatic potentials is strongly R dependent. These calculated potentials have resulted in assignments of three previously unidentified absorption bands[21] and have been ascribed to as perturbers to the v'=1 and 2 levels of the $H^3\Pi_u$ ($3s\sigma_g$) state.[22] The calculations do not explicity include interaction with (π_g,np) species which are expected to undergo significant mixing with the $2^3\Sigma_u^-$ and $2^3\Pi_u$ adiabatic states.

Our model assumes that a 5-photon absorption occurs vertically from the oxygen vibrationless ground state at an internuclear distance of 1.2075Å. Furthermore, this transition leads to near resonances at the 5-photon energy for both the v'=0, 6p Rydberg level and the v'=3, $2^3\Pi_u$ adiabatic state. These strongly mixed states are assumed to be long lived compared to the ionization step. The transition to the continuum can then be visualized as separate paths occuring from the various near resonant states and at very different internuclear distances. The analysis relies upon Franck-Condon projection of the individual zero-order excited vibrational wavefunctions onto the ionic ground state. The final state distribution can then be described as an incoherent sum of the individual Franck-Condon factors with intensity dependent electronic amplitudes. The essence of the model then relies upon absorption of an additional 532 nm photon to occur via two resonant paths (1) 6p Rydberg and (2) $2^3\Pi_u$ states. The direct ionization path (3) is also included in this formalism but with much less weight. The amount of excess energy above the I.P. resulting from 6-photon ionization can reach up to v^+=8 of the ionic ground state. The Franck-Condon factors via path (2) give poor overlap with the v^+=0-8 levels of the ionic ground state while path (1) proceeds via a δv=0 propensity rule. Consequently, at low intensity the photoelectron spectrum will be characterized by a strong v^+=0 (ε_0=1.91eV) plus weaker higher vibrational peaks corresponding to contributions from paths (2) and (3) illustrated as lightly shaded horizontal lines in Fig. 7. As the intensity is increased the transition rate favors absorption of an additional photon(s) in the continuum (ATI), i.e. 7-photons. Path (1) results in an electron peak that is characteristic of atomic ATI, that is with a total energy given by $(\varepsilon_o + \hbar\omega) = 4.24$ eV. However, due to favorable Franck-Condon factors path (2) results in a significant population in the $v^+=11$ and 12 levels of the $X^2\Pi_g$ ionic ground state (see dark shaded lines in Fig. 7). The electron energies via 7-photon ionization for $v^+=11$ and 12 are $\varepsilon_{11} = 1.91$ eV and $\varepsilon_{12} = 1.72$ eV, respectively. These electrons are near degenerate in energy with the $v^+ = 0$ and 1 electrons resulting from 6-photon ionization; in fact the differences are $\Delta\varepsilon_{0\text{-}11} = 0$ and $\Delta\varepsilon_{1\text{-}12} = 40$ meV and are not resovable with our electron spectrometer. However, $\Delta\varepsilon_{1\text{-}12}$ is discrenible as an intensity dependent broadening and shifting in the peak center at 1.68 eV, as seen in Fig. 6. The electron peak at 1.91 eV shows a constant 60 meV linewidth while the 1.68 eV peak undergoes a significant change in width with intensity. In both the low and high intensity extremes the linewidth is constant representing a peak characterized by low and high v^+, respectively, while the intermediate intensity regime where the linewidth is a maximum represents a blended $v^+=1$ and 12 peak with comparable amplitudes.

Figure 6 also shows a direct comparison at the two extreme laser intensities between the experimental photoelectron spectrum and a bar diagram simulation incorporating our model. The agreement is excellent and reproduces all the major features of the experiment at all intensities. The molecular constants for the adiabatic states are varied to give the best fit to a photoelectron spectrum at one intensity. All other molecular constants are taken from the literature.[21] Once the adiabatic constants are determined they are kept fixed in all subsequent simulations. Thus, the data is fit with three parameters corresponding to the electronic amplitudes for each path. The determined molecular constants for the $2^3\Pi_u$ adiabatic state that give consistent fits for our data are $\omega_e = 2500$ cm^{-1}, $\omega_e \chi_e = 36$ cm^{-1} and are $R_e = 1.32$ Å. The vibrational frequency and anharmonicity agree with those determined by Tonkyn et al.[22] and differ from the calculated[20] values by 1.5%. Our internuclear distance is 0.04 Å longer than that calculated by Beunker et al.[20] but this difference is well within numerical accuracy. Our determination of the electronic energy, T_e, is 0.3% higher than previously reported[20,22] with a value of ~87000 cm^{-1}. However, considering the inaccuracy associated with deriving a meaningful T_e in our "nonresonant" experiment our result should be viewed as consistent with the previous reports. It should be noted that our first approximation model although successful does not incorporate any light-induced state shifts which must be occurring to some extent. Dynamics should become more critical to our model as the pulse width is shortened (i.e. higher saturation intensity).

INTENSE FIELD DISSOCIATION DYNAMICS

A physically distinct motivation for studying molecules instead of atoms in an intense field is the presence of dissociative channels. Questions concerning the validity in extending our knowledge on intense field atomic ionization to describe molecular dissociation seem warranted. For example, does high field dissociation proceed in an analogous manner to above threshold ionization (ATI), that is via absorption of additional photons beyond the minimum needed to break a molecular bond. This process is appropriately called above threshold dissociation (ATD) or fragmentation (ATF). Furthermore, the atoms formed via the dissociation present the possibility to probe their behavior in a physically different environment as compared to the more common isolated atom experiments. In our experiment, we study the intense field dissociation of the H_2^+ molecular ion, a simple one-electron diatomic. As discussed in Section III, the ion is formed "nonresonantly" by absorption of seven or more green photons. The ion's vibrational state content is characterized via photoelectron spectroscopy. The H_2^+ molecule can subsequently dissociate by absorption of one or more photons into a proton and a neutral hydrogen atom. The proton's kinetic energies are then measured by our time-of-flight mass spectrometer. It should be noted that the nonlinear production of H_2^+ is the rate limiting step with the peak electric field determined by ionization saturation. Consequently, in the low order dissociation process following H_2^+ formation, the ion experiences an extremely intense laser field which differs from an experiment designed to produce an isolated beam of ions irradiated by an intense laser.

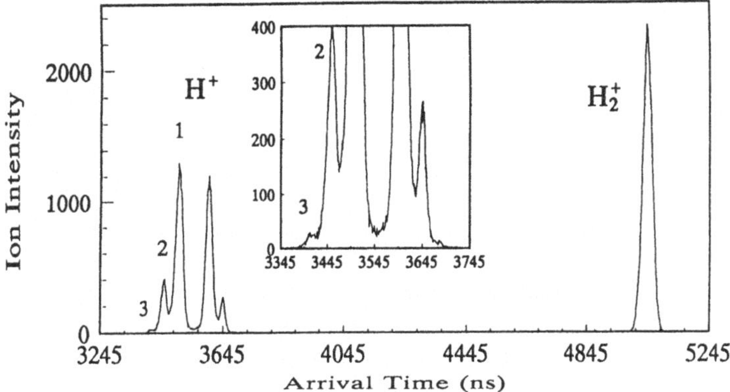

Fig. 8. Time-of-flight mass spectrum resulting from 532 nm MPI dissociation of H_2 at 9.7 x 10^{12} W/cm^2. Peaks labelled 1, 2, and 3 correspond the number of photons absorbed in dissociation.

In the latter case, the photodissociation of the ion will be best characterized by weak field physics regardless of the peak intensity asssociated with the laser.

The intense laser field photodissociation of H_2^+ has been explained by the mechanism of bond-softening[9,10,23] using a dressed molecular state formalism. In this model, the laser field is considered as a strong perturber to the molecular potential. The dressed potential curves couple and distort via the strong dipole interaction. The ground ionic state and the first repulsive state have symmetry $^2\Sigma_g^+$ $(1\sigma_g)$ and $^2\Sigma_u^+$ $(2p\sigma_u)$, respectively. The interacting potentials are governed by dipole selection rules (parity) and thus couple via odd photon absorption. The coupling strength or the width of the avoid-crossing is proportional to the field intensity. However, a *net even number* of photon dissociation can also occur due to a combination of absorption and stimulated emission of photons. Intense fields cause the laser induced avoided crossing gap to open up. Any ions with vibrational energy within the gap should become unstable. As the laser intensity is increased and the molecular potentials distort, one should observe a negative shift in the proton kinetic energy corresponding to dissociation from the lower vibrational states of H_2^+. Furthermore, the absorption of more photons ($N \geq 2$) from the same vibrational states results in a series of peaks separated by about a photon (total) kinetic energy (ATD). Likewise, the distorted potentials result in a change in the vibrational energy eigenvalues resulting in a positive shift in the photoelectron spectrum.

Figure 8 shows a typical time-of-flight mass spectrum resulting from the 532 nm nonresonant MPI of H_2 molecules at an intensity of 9.7 x 10^{12} W/ cm^2. The three peaks are labelled according to the number of photons being absorbed by H_2^+ ions leading to dissociation. According to our experimental results at low intensity most H_2^+ ions are dissociated from the $v^+(avg) = 5$ vibrational level following the absorption of a single photon (peak 1 in Fig. 8). The maximum of this peak is observed to shift from $v^+=5$ to $v^+=4$ at higer intensities. With increasing intensity another peak emerges corresponding to two photon dissciation via $v^+(avg) = 2$. Consequently, its separation is less than a photon total energy from the first peak due to differing initial vibrational states and is intensity dependent. At still higher intensities a third peak appears about one photon total energy away from the second peak corresponding to a three photon dissociation. The same general behavior is observed for D_2 and HD molecules. Similar kinetic energy shifts were observed in the 527 nm (psec) spectra over an intensity range of 5 x 10^{12} to 4 x 10^{13} W/cm^2. Shown in Fig. 9(b) is the dissociation fraction F^H and F^D as a function of laser intensity for H_2^+ and D_2^+, respectively and where $F^H = [H^+/(H^+ + H_2^+)]$ and $F^D = [D^+/(D^+ + D_2^+)]$. Note, that the dissociation fraction for each molecule increases with intensity and the F^D is larger than F^H at all intensities shown.

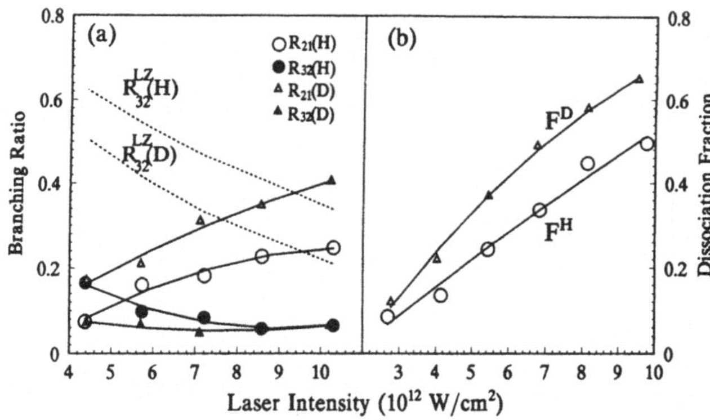

Fig. 9 A plot of the (a) branching ratios R, and (b) dissociation fraction F for H_2 and D_2 as a function of 532 nm laser intensity.

We have carefully examined the photodissociation of H_2^+ molecules by recording the proton kinetic energy spectra and the H_2 photoelectron spectra simultaneously as a function of laser intensity. The photoelectron spectrum not only depicts the final state of the ionization process but also acts as a sensitive probe of the initial distorted (dressed) potentials of the ionic dissociation. This assumes that the dissociation occurs immediately following ionization (at same peak field). Our results verify that the mechanism responsible for H^+ and D^+ formation is the photodissociation of the parent molecular ion. It appears that the amount of H^+ produced is porportional to that of photoelectrons due to (7+1) molecular ionization. In addition, energy conservation implies that seven second harmonic photons of YAG(YLF) can populate only up to $v^+=3(4)$ vibrational levels in the ionic ground state, whereas the lowest intensity data have proton energies peaked at $v^+=5$. Thus, we conclude that the H^+ peak labeled 1 in Fig. 8 (0.7 eV total energy) is due to the eight photon ionization of H_2 followed by one photon dissociation of H_2^+.

Our results are generally in qualitative agreement with the bond-softening model but there remains some quantitative differences. In the current experiment, we observed apparent negative shifts in proton kinetic energies with increasing laser intensity with both 532 nm (ns) and 527 nm (ps) pulses. The maximum population of the 1-photon proton kinetic energy peak shifts by ~200 eV over the entire intensity range. Likewise, the maximum of the ionic vibrational state distribution in the PES changes from high to low vibrational levels in a similar intensity range. Furthermore, the kinetic energy of the electrons with maximum amplitude in the 8-photon photoelectron distribution is independent of the molecule studied, which implies that the bond-softening mechanism to first order is predominantly an electronic coupling and not vibrational. However, we could not observe any commensurate positive energy shifts in the PES peak positions with increasing intensity although a broadening is clearly evident. An asymmetric broadening in our PES could result from spatial or temporal averaging in our experiment masking a "pure" energy shift. However, the sign of the observed broadening is opposite to that predicted by a bond-softening model.[9] The hashed boxes in Fig. 1 illustrates the range of shifting predicted by a Floquet calculation at a maximum intensity of 3×10^{13} W/cm^2. This inconsistency could imply a shortcoming in the numerical method since only the vibrational motion in the distorted potential are included in the Floquet calculation. Another contributing factor to the peak width could result via strong-field rotational pumping with increasing laser intensity, causing the photoelectron peaks to broaden towards lower energies.

The dynamics associated with high field dissociation are apparent in the plot of the branching ratios shown in Fig. 9(a). The ratio ($\mathbf{R_{21}}$) of dissociation via 2-photon versus the 1-photon channels for H_2^+ changes from 7% to 25% as the intensity is increased to 9.7×10^{12} W/cm^2. Furthermore, the ratio $\mathbf{R}_{32}$ for 3 versus 2-photon dissociation decreases from 16.5 to 6.5% in the same range. Interestingly, the behavior in D_2^+ is somewhat different, here the fraction $\mathbf{R}_{21}$ changes from 17 to 40% which is larger than H_2^+. However, the D_2^+ ratio $\mathbf{R}_{32}$ is smaller than H_2^+ and relatively constant (~6%) over the entire intensity range. These branching ratio effects can be all understood with the same model.[23] Physically, the increasingly laser intensity corresponds to a larger avoided-crossing gap resulting in a decrease in the 3-photon diabatic transition rate while favoring 2-photon adiabatic passage (3-photon absorption, 1-photon emission). Since H_2^+ and D_2^+ have a relatively large difference in vibrational frequencies due to their different masses, therefore at low intesities where the gap is not large enough to completely shut off the 3-photon channel, D_2^+ molecules branch more efficiently through the 2-photon adiabatic path as compared to H_2^+ due to the fact that a smaller vibrational frequency implies more adiabatic motion. Specifically, $\mathbf{R}_{32}(H) > \mathbf{R}_{32}(D)$ at low intensity. As the gap continues to widen, one should also observe the ratio $\mathbf{R}_{32}$ to decrease for both H_2^+ and D_2^+ molecules. Such behavior is clearly demonstrated in our experiments. Figure 9(a) also shows for comparison the results of the calculated fragment $\mathbf{R}_{32}(LZ)$ ratios predicted by a simple Landau-Zener (LZ) theory.[9] The theory predicts well the general behavior of the ratios as the light intensity changes. However, the calculated ratios are approximately three times larger than the experimental values which could imply that the degree of deformation of the calculated potential curves is beyond the limits of applicability of simple LZ theory.

ACKNOWLEDGMENTS

The authors would like to thank Joseph Dolce for his technical assistance. The authors gratefully acknowledge the contributions of Phillip Bucksbaum and Anton Zavriyev to the H_2 work. This research was carried out at Brookhaven National Laboratory under Contract No. DE-AC02-7600016 with the U.S. Department of Energy and supported, in part, by its Division of Chemical Sciences, Office of Basic Energy Sciences.

REFERENCES

1. See for example, "Atomic and Molecular Processes with Short Intense Laser Pulses", A. D. Bandrauk, ed., Plenum Press, NY (1988).
2. M. Pont and M. Gavrila, Phys. Rev. Lett. 65:2362 (1990); Q. Su, J. H. Eberly, and J. Javaninen, Phys. Rev. Lett. 64:862 (1990).
3. R. R. Freeman, P. H. Bucksbaum, H. Milchberg, S. Darak, D. Schumacher, and M. E. Geusic, Phys. Rev. Lett. 59:1092 (1987); Dalwoo Kim, S. Fournier, M. Saeed, and L. F. DiMauro, Phys. Rev. A 41:4966 (1990).
4. Dalwoo Kim, M. W. Courtney, M. Anselment, and L. F. DiMauro, Phys. Rev. A 38:2338 (1988); M. Saeed , Dalwoo Kim, and L. F. DiMauro, Appl. Opt. 29:1752 (1990).
5. L. Li, R. Lippert, J. Lobue, W. Chupka, and S. Colson, Chem. Phys. Lett. 151:335 (1988).
6. C. Cornaggia, D. Normand, J. Morellec, G. Mainfary, and C. Manus, Phys. Rev. A 34:207 (1986); S. W. Allendorf and A. Szoke, Phys. Rev. A 44:518 (1991).
7. T. S. Luk and C. Rhodes, Phys. Rev. A 38:6190 (1988).
8. J. Verschurr, L. Noordam, and H. van Linden van den Heuvell, Phys. Rev. A 40:4383 (1989).
9. A. Zavriyev, P. Bucksbaum, H. Muller, and D. Schumacker, Phys. Rev. A 42:5500 (1990).
10. B. Yang, M. Saeed, L. F. DiMauro, A. Zavriyev, and P. H. Bucksbaum, Phys. Rev. A 44:R1458 (1991).
11. Hanspeter Helms, private communication.
12. B. Walker, M. Saeed, T. Breeden, B. Yang, and L. F. DiMauro, Phys. Rev. A, in press.
13. A. L. L'Huillier, G. Mainfray, and P. M. Johnson, Chem. Phys. Lett. 103:447 (1984).
14. J. Morellec, D. Normand, and G. Petite, Phys. Rev. A 14:300 (1976).
15. Y. Gontier and M. Trahin, Phys. Rev. A 19:264 (1979).
16. P. H. Krupenie, J. Chem. Phys. Ref. Data 1:423 (1972).
17. A. L. Smith, J. Quant. Spectrosc. Radiat. Transfer 10:1129 (1970)
18. J. H. D. Eland, J. Chem. Phys. 72:6015 (1980).

19. M. Ogawa and K. R. Yamawaki, Can J. Phys. 47:1805 (1969).
20. R. J. Buenker and S. D. Peyerimhoff, Chem. Phys. Lett. 34:225 (1975); R. J. Buenker, S. D. Peyerimhoff, and M. Peric, *ibid*. 42:383 (1976).
21. K. P. Huber and G. Herzberg, "Constants of Diatomic Molecules", Van Nostrand Rheinhold Co., New York (1979).
22. R. G. Tonkyn, J. W. Winniczek, and M. G. White, J. Chem. Phys. 91:6632 (1989).
23. A. Bandrauk and M. Sink, J. Chem. Phys. 74:1110 (1981); A. Guisti-Suzor, X. He, O. Atabek, and F. Mies, Phys. Rev. Lett. 64:515 (1990).

CALCULATION OF POTENTIAL CURVES

FOR H_2^+ MOLECULES IN A STRONG LASER FIELD

H.G. Muller

FOM-Institute for Atomic and Molecular Physics
Kruislaan 407, 1098 SJ Amsterdam, the Netherlands

INTRODUCTION

Recently multiphoton ionisation experiments with molecular hydrogen have shown a large amount of dissociation into low-energy fragments (Bucksbaum *et al.*, 1990). The mechanism proposed for this process was *bond-softening* in H_2^+, i.e. a change of the potential curves under influence of the field that weakens the molecular bond and lowers the barrier to dissociation. This explanation has been very succesful, because it also predicts the occurrence of *above-threshold dissociation*, (the absorption of additional photons during the dissociation process leading to more energetic fragments) that was indeed observed in the same experiment.

The major part of the bond-softening is due to a strong dipole coupling that exists between the bonding ($1s\sigma_g$) and anti-bonding ($2p\sigma_u$) electronic states in H_2^+. In a dressed-states picture (figure 1) the potential curves associated with these states cross at the point where the electronic energies differ by exactly one photon energy. The coupling turns this crossing into an *avoided* crossing, leading to a set of new *adiabatic* potential curves. One of these very much resembles the bonding state at internuclear distances close to its potential minimum, but has a greatly reduced binding energy. The other also has a potential minimum, around the point of the original crossing, that in principle could support bound states as well. Some evidence for the existence of such *light-induced* vibrational states has been obtained (Verschuur *et al.,* 1989, Allendorf and Szöke, 1991).

In order to have a sound theoretical interpretation of these experiments, it has become desirable to accurately know the shape and size of these potential curves. This paper presents a novel method for calculating electronic eigenvalues in the presence of a periodic perturbation, as well as the results of applying this method to the H_2^+ molecule in the second harmonic of the Nd:YAG laser (λ=532 nm).

DETERMINATION OF QUASI-ENERGIES

In order to calculate the fate of a molecule subjected to an intense laser pulse, one can take advantage of the different timescales on which events take place. On the shortest timescale there is the motion of the electrons and, only slightly slower, the oscillation of the electro-magnetic field. The motion of atomic nuclei is much slower than that. Even with sub-picosecond laser pulses the change of the light intensity in time happens on the timescale of the nuclear motion. It makes sense to make use of the disparity in timescales by applying the Born-Oppenheimer approximation: One first treats the fast phenomena for each frozen value of the slow degrees of freedom, and then uses the results to calculate the behaviour of the time evolution of these slow variables. This second calculation is enormously simplified with respect to the full treatment, because it no longer has to bother with the details of the fast events.

In the practical case of intense-field photo-dissociation, the slow variables obviously are nuclear motion and the temporal shape of the light pulse. For each value of these parameters one can solve for the electronic motion under influence of the oscillating light wave, the amplitude of which is now fixed. It is worth a great deal to deal with a light wave of constant intensity, because it makes the Hamiltonian $H_{el}(t)$ that describes the electronic motion periodic in time:

$$\mathbf{H}_{el}(t+T) = \mathbf{H}_{el}(t) \qquad (1)$$

According to the Floquet theorem this means that the time evolution $\mathbf{U}(t)$ generated by $\mathbf{H}_{el}(t)$ through the time-dependent Schrödinger equation

$$d/dt\ \mathbf{U}(t) = -i\ \mathbf{H}_{el}(t)\ \mathbf{U}(t) \qquad (2)$$

can be factorised in a periodic and an exponential part :

$$\mathbf{U}(t) = \mathbf{\Phi}(t)\ \exp(-i\mathbf{F}t), \qquad \mathbf{\Phi}(t+T) = \mathbf{\Phi}(t). \qquad (3)$$

The operator F is called the Floquet-hamiltonian. Its eigenvalues are called quasi-energies, and are determined only upto a multiple of $2\pi/T$. This means that **F** is not unique, but, apart from the usual case of degenerate eigenvalues, its eigenfunctions are.

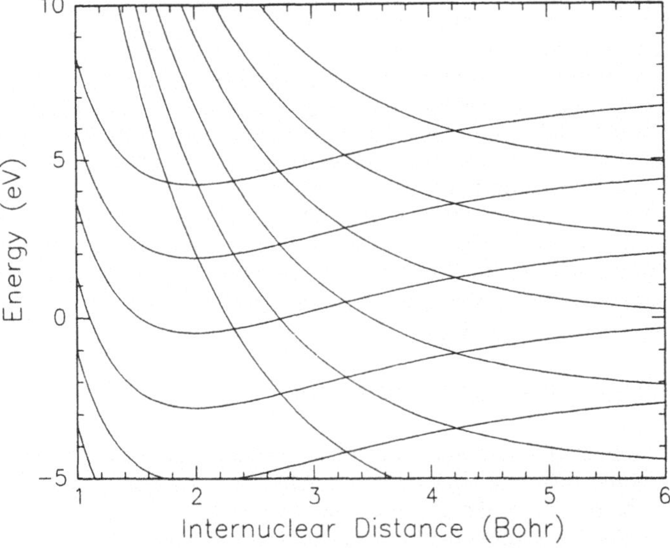

Figure 1. The dressed-molecule picture of the potential curves belonging to the $1s\sigma_g$ and $2p\sigma_u$ states of the H_2^+ molecule. The replicas of each curve are separated by the photon energy of 2.33 eV. At zero intensity the curves intersect at points where the n-photon transition between them is resonant. For n = 1,2,3 this happens at R = 4.2, 3.3 and 2.7 repectively.

The quasi-energies play the role that electronic energies usually play in the Born-Oppenheimer approximation: they define potential curves in which the heavy particles move. In this case the potential curves are also a function of the light intensity I(t) that might be time-dependent, but apart from necessitating a time-dependent treatment of the nuclear motion this is of no consequence. This paper only concerns itself with the solution of the first part of the problem, the calculation of the quasi-energies.

The most common way to exploit (3) is to develop $\Phi(t)$ into a Fourier series:

$$\Phi(t) = \Sigma\, \Phi_n \exp(-in\omega t). \tag{4}$$

On substitution in (2) this leads to time-independent relations between the Fourier coefficients. The Fourier coefficients can be given a physical meaning: they simply represent the contribution to the total state of the situation with that particular number of photons present in the field.

In quantum mechanics $\Phi(t)$ and $\mathbf{H}_{el}(t)$ represent operators. As long as one is dealing with a problem that only involves bound states, the most convenient way to handle these is to describe the problem in some finite basis set. In this case the operators simply become matrices, that are easy to handle by numerical methods. The Fourier components of $\Phi(t)$ are then also matrices. The relationship between the Φ_n has the form of an eigenvalue problem. If the time-dependence of the problem is such that M Fourier components are required, this requires diagonalisation of an MxM matrix the components of which are the Fourier components of $\mathbf{H}_{el}(t)$, which in itself are matrices of size NxN, say. Thus this requires diagonalisation of an MN x MN matrix, and thus an amount of work that in general increases as M^3N^3. This can be rather expensive.

This paper discusses another method for finding the quasi-energies, that retains the explicit time-dependence of all quantities. It is based on the fact that solving (2) for $\mathbf{U}(t)$ with initial condition

$$\mathbf{U}(0) = \mathbf{I} \tag{5}$$

leads to a solution of the form (3) that has the property that $\Phi(0) = \mathbf{I}$, the unit matrix. Since Φ is periodic this implies that also $\Phi(T) = \mathbf{I}$, and thus

$$\mathbf{U}(T) = \exp(-i\mathbf{F}T). \tag{6}$$

Since $\mathbf{U}(T)$ is a function of $\mathbf{F}$ these matrices have the same eigenfunctions, and their eigenvalues are related as (6). Thus once $\mathbf{U}(T)$ has been obtained by numerical integration of (2), one can diagonalise it to obtain the eigenfunctions of $\mathbf{F}$. The diagonal elements then are of the form exp(-iET), with E the sought-for quasi-energy.

The advantage of this approach is that although all matrix operations are still of order N^3, the amount of work is only linear in the number of time-steps M used in the numerical integration of (2). According to the Nyquist sampling theorem having M time-steps is roughly equivalent to having M Fourier components. It is not exactly the same, however, because the method that solves (2) is usually based on an integration rule that is derived on the assumption that between lattice points the function is a polynomial instead of a sum of sinc functions.

TIME-INTEGRATION METHOD

The main problem encountered in solving (2) is that it actually is a *stiff* equation. In order to get accurate results the number N of basis functions needed to get a faithful representation of the wavefunction is large. Unfortunately this means that $\mathbf{H}(t)$ has N eigenvalues. The lowest eigenvalues correspond to the physical states of interest, but the others are attempts to reproduce excited states. This means that they have a much higher eigenvalue (=energy) than the states we are after. Worse yet, the basis functions are probably not optimised for representing these excited states (they probably have the wrong exponential decay, for example). Thus a lot of meaningless states high up in the continuum are produced. Since these states will not be populated in practice, it would be very inconvenient if the integration time-step Δt had to be chosen small compared to the inverse energies of those states.

For accuracy the only requirement is that the time-step is small enough to be able to follow the changes in the physical states. Obviously this requires $\Delta t/T << 1$, but it also requires

$$|E_0(t)\ \Delta t| << 1, \tag{7}$$

with $E_0(t)$ the eigenvalues of $\mathbf{H}(t)$ for which the eigenvectors contain significant population. In the calculation of above-threshold dissociation the problem is dominated by the two lowest H_2^+-states, and by choosing a suitable origin of energy the eigenvalues corresponding to these states can be made reasonably small. In atomic units T is also fairly long for optical photons, which means that quite lare time-steps are acceptable.

The large Δt means that many of the non-populated, non-physical excited states do not fulfill (7), and in fact have $E_0(t)\ \Delta t > 1$. The integration method must not become unstable under these condition, because then any population that entered the non-physical states through rounding errors would blow-up exponentially and eventually couple back into the physical states. Thus an *absolutely-stable* integration method is required. In fact the method described in this section is more than that: Its area of stability is exactly equal the negative half-plane Re $z < 0$. This implies that for equations with a right-hand side that has purely imaginary eigenvalues, like (2), generates an evolution that is guaranteed to be unitary. This is a very desirable property for solving time-dependent Schrödinger equations, because it leads to automatic conservation of the norm of the wave-function.

If condition (7) is fulfilled, the results of the calculation start to converge to the exact answer if Δt is made still smaller. Usually the error is given as

$$\text{Error} = c\ |\max\{E_0\}\ \Delta t|^n. \tag{8}$$

The order n of the method determines how rapid this convergence occurs, and thus how large a Δt is still acceptable. The constant c is typically of the order 10^{-4} for most methods, and thus for a fourth-order method (n=4) $E_0\Delta t = 0.3$ (i.e. 20 steps per period) are required for convergence to the ppm level. A second-order method would require many more steps.

The requirement of high order is difficult to combine with that of absolute stability. Usually high-order methods require their accuracy by some form of extrapolation of the right-hand side of the differential equation based on the known values in a number of preceding points. Since (2) is a first-order differential equation, of the form

$$f'(t) = A(t)\ f(t), \tag{9}$$

its solution should be uniquely determined by one value from the past only, and thus using more than one previous point to boost accuracy is somewhat inconsistent and causes instabilities. Thus one wants to use a method that makes a time-step by looking only in two points: the old and the new one.

The most straightforward method would be to approximate the solution f(t) within the time-interval $[t_i,t_{i+1}]$ by a polynomial, and require that this polynomial obeys (9) in the end points t_i and t_{i+1}. This leads to the well-known trapezoid rule. Unfortunately this is only a second-order method. In order to do better the polynomial has to be of higher degree, which means that to determine it more conditions have to be imposed. To get a fourth-order method the extra conditions can be that the polynomial not only obeys (9), but also the derivative of (9):

$$f''(t) = A(t)\ f'(t) + A'(t)\ f(t). \tag{10}$$

Thus such a method would make explicit use of the time-derivative of the Hamiltonian, which of course can be easily obtained.

As it turns out some minor modifications of this idea are required to guarantee the unitarity and make it suitable for (non-commuting) operators. The integration rule that has all the desired properties is the following:

$$f(t_{i+1}) = (I + \Delta t/2\, B + \Delta t^2/12\, B^2)\cdot(I - \Delta t/2\, B + \Delta t^2/12\, B^2)^{-1} f(t_i) \qquad (11)$$

$$B = 1/2(A(t_i)+A(t_{i+1})) + \Delta t/12\, (A'(t_{i+1})-A'(t_i)) + [A(t_i),A(t_{i+1})] \qquad (12)$$

Since in (2) $A(t) = -i\mathbf{H}(t)$ the matrix B is actually anti-Hermitian, $B^\dagger = -B$. Thus the propagator in (11) is actually of the form $\Omega(\Omega^+)^{-1}$, where Ω and Ω^+ commute because they are both a function of the same matrix B. Therefore the propagator is unitary. The symmetric way in which (11,12) use $A(t_i)$ and $A(t_{i+1})$ further imply time-reversal symmetry: on integrating backwards in time the solution exactly retraces the forward solution.

The main disadvantage of (11) is that it requires a matrix inversion for each time-step, but this is the price one pays for absolute stability.

SYMMETRY CONSIDERATIONS

In the case of an H_2^+ molecule in a monochromatic field the problem actually consists of four similar quarter periods. This property can be used to cut the time needed for numerical integration of (2) in four. The hamiltonian appearing in (2) is in this case

$$\mathbf{H}_{el}(t) = \mathbf{H}_{mol} + \mathbf{D} \cos \omega t \qquad (13)$$

Thus $\mathbf{H}(t)$ is invariant under the time-reversal operator $\mathbf{T}$:

$$\mathbf{T}\, \mathbf{H}_{el}(t)\, \mathbf{T} = \mathbf{H}_{el}(-t) = \mathbf{H}_{el}(t). \qquad (14)$$

Therefore sandwiching (2) between $\mathbf{T}$s, using (14) and noting that $\mathbf{T}\, d/dt\, \mathbf{T} = -d/dt$ shows that $\mathbf{U}(-t)$ obeys the Schrödinger equation

$$d/dt\, \mathbf{U}(-t) = i\mathbf{H}_{el}(t)\, \mathbf{U}(-t), \qquad \mathbf{U}(0) = \mathbf{I}, \qquad (15)$$

which is just the complex conjugate of (2). Therefore the solution is also the complex conjugate, and we have

$$\mathbf{U}(-t) = \mathbf{U}^*(t). \qquad (16)$$

We will denote the evolution from t_0 to t_1 as $\mathbf{U}(t_1,t_0)$, where $\mathbf{U}(t)$ is just a short-hand notation for $\mathbf{U}(t,0)$. Then the evolution for a full period is obtained from that for a half period as

$$\begin{aligned} \mathbf{U}(T) &= \mathbf{U}(T,T/2)\, \mathbf{U}(T/2,0) = \mathbf{U}(0,-T/2)\, \mathbf{U}(T/2,0) = \\ &= \mathbf{U}^{-1}(-T/2,0)\, \mathbf{U}(T/2,0) = \mathbf{U}^\dagger(-T/2)\, \mathbf{U}(T/2) = \qquad (17) \\ &= \mathbf{U}^{*\dagger}(T/2)\, \mathbf{U}(T/2) = \mathbf{U}^t(T/2)\, \mathbf{U}\,(T/2). \end{aligned}$$

The evolution can be subdivided further by making use of the parity operator $\mathbf{P}$, because $\mathbf{H}_{mol}$ is even, but $\mathbf{D}$ is odd under $\mathbf{P}$. Therefore

$$\mathbf{P}\, \mathbf{H}_{el}(t)\, \mathbf{P} = \mathbf{P}\, \mathbf{H}_{mol}\, \mathbf{P} + \mathbf{P}\, \mathbf{D}\, \mathbf{P} \cos \omega t = \mathbf{H}_{mol} - \mathbf{D} \cos \omega t = \mathbf{H}_{el}(t-T/2). \qquad (18)$$

Sandwiching (2) between $\mathbf{P}$ and shifting the time by T/2 then gives the Schrödinger equation

$$d/dt\, (\mathbf{P}\, \mathbf{U}(t+T/2)\, \mathbf{P}) = -i\mathbf{H}_{el}(t)\, (\mathbf{P}\, \mathbf{U}(t+T/2)\, \mathbf{P}), \qquad (19)$$

which is identical to (2). Thus $(\mathbf{P}\, \mathbf{U}(t+T/2)\, \mathbf{P})$ is another solution of (2), and in order to have it meet initial condition (5) it is sufficient to right-multiply it by $\mathbf{U}^{-1}(T/2)$, after which it must be equal to $\mathbf{U}(t)$ since the initial condition uniquely specifies the solution:

$$\mathbf{U}(t) = \mathbf{P}\, \mathbf{U}(t+T/2)\, \mathbf{P}\, \mathbf{U}^\dagger(T/2). \qquad (20)$$

Writing $t = -T/4$ and applying (16) this gives after rearrangement:

$$\mathbf{U}(T/2) = \mathbf{U}^{\dagger}(-T/4)\, \mathbf{P}\, \mathbf{U}(T/4)\, \mathbf{P} = \mathbf{U}^{t}(T/4)\, \mathbf{P}\, \mathbf{U}(T/4)\, \mathbf{P}, \tag{21}$$

and thus expresses $\mathbf{U}(T/2)$ in $\mathbf{U}(T/4)$. All this means that the numerical integration of (2) only has to be performed upto T/4, after which (21) and (17) can be used to get $\mathbf{U}(T)$ with a few elementary matrix operations.

The symmetry that gives rise to (21) is simply parity conservation. This conservation law also shows up in a Floquet calculation, and makes it possible to delete half the Floquet states from the calculation (e.g. use only symmetric states with an even number of photons, and anti-symmetric states with an odd number of photons). Equation (17) reveals that there must be one more symmetry of the system that is somehow hidden in a Floquet calculation, but would enable one to cut the required number of Floquet blocks once more in half.

MATRIX DIAGONALISATION

In order to get the quasi-energies from $\mathbf{U}(T)$ it has to be diagonalised. Of course the evolution operator $\mathbf{U}(t)$ is unitary for any t, but (17) states that $\mathbf{U}(T)$ is symmetric as well as unitary, i.e.

$$\mathbf{U}^{\dagger}(T) = \mathbf{U}^{*}(T). \tag{22}$$

In general one prefers to diagonalise Hermitian matrices, because very efficient methods exist in that case. The matrices $(\mathbf{U}(T)+\mathbf{U}^{\dagger}(T))/2$ and $i(\mathbf{U}(T)-\mathbf{U}^{\dagger}(T))/2$ are Hermitian, and have the same eigenvectors as $\mathbf{U}(T)$. Their eigenvalues are just the real and imaginary parts of exp(-iET), so together they fix the quasi-energies E. Because of property (22) these two matrices are not only Hermitian, but also real, further simplifying the diagonalisation. Since they have the same eigenfunctions they can in fact be diagonalised in concert, leading to a more stable determination of the quasi-energy.

BASIS FUNCTIONS

The Schrödinger equation for the H_2^+ molecule is seperable, and the eigenvalues and eigenfunctions as a function of internuclear distance R are well known. Putting the molecule into a laser field destroys the separability, and of course perturbs the atom. One way to describe the molecule would be to express this perturbed wavefunction as a linear combination of unperturbed eigenfunction, that after all form a complete set. For a strong field the perturbation could be appreciable, and a lot of excited states would have to be mixed in. The disadvantage of this approach is that it requires knowledge (like dipole couplings) on all of these states.

In fact expanding the problem on a basis of eigenstates of H_2^+ converges rather slowly, because the perturbed wavefunctions, although perturbed, still have roughly their old energy, and thus the same exponential decay in the classically forbidden region. The eigenstates, on the other hand, all have very different energies and corresponding exponential decays. The right exponential decay has to be obtained at the expense of making linear combinations of many eigenstates to 'interfere away' their much to large exponential tails, and many of these eigenstates belong to the continuous spectrum. If one is only interested in a few states a much faster convergence can be reached by using a basisset of states that all have exponential decay of the same order as that of these states of interest.

The present calculation uses basis states that are well-adapted to solving H_2^+ type problems, but all with an exponential decay comparable to that of the unperturbed states. They make use of the *prolate-spheroidal* coordinate system in which the H2+ problem is separable. These coordinates (ξ,η,φ) are defined as

$$\begin{aligned} \xi &= (r_A + r_B)/R \\ \eta &= (r_A - r_B)/R \\ \varphi &= \arctan(y/x)\,, \end{aligned} \tag{23}$$

where r_A and r_B are the distances to the two protons positioned on the z-axis. For the present calculation only σ-states are required, and all functions are φ-independent. The coordinate ξ runs from 1 to ∞, while η runs from -1 to 1.For large distances they approach 2r/R and cos θ, respectively. The basisfunctions χ_{nm} for the calculation are all of the form

$$\chi_{nm}(\xi,\eta,\varphi) = \xi^n \eta^m e^{-\alpha\xi}, \tag{24}$$

with integer powers n and m, and real α. This makes the matrix elements of all operators involved trivial to evaluate. The value of α controls the exponential decay of the wavefunction at large radial distance. If a small number of basis functions is used, the eigenvalues of the unperturbed Hamiltonian $\mathbf{H}_{mol}$ depend rather critically on the value of α. For larger basis-sets there is enough freedom to compensate a slightly less desirable value of α with the polynomial pre-factors. In the light of this α was optimised in order to give accurate unperturbed energies with three basis functions (χ_{00},χ_{10},χ_{12}) for the $1s\sigma_g$, and four basis functions (χ_{01},χ_{11},χ_{03} and χ_{13}) for the $2p\sigma_u$ state, as a function of internuclear distance. Within one symmetry type, α was taken the same for all basis functions. The result of this optimisation was

$$\begin{aligned} \alpha_g(R) &= 0.745\ R - 0.045\ R^2 \\ \alpha_u(R) &= 0.625\ R - 0.05 \end{aligned} \tag{25}$$

These values were used throughout the rest of the calculation, irrespective of the number of basis functions that was actually used.

The disadvantage of basis-functions of the type (24) is that they are not able to give a faithfull representation of the wave-functions for large values of R, even in the case of no field. The reason is the dependence on η, which for two widely separated hydrogen atoms in their 1s state would be like cosh ηR/2. For any number of basis-functions there is a point where R is so large that the hyperbolic cosine can no longer be approximated by a polynomial in η, and the calculation becomes inaccurate. Since the main object was to get accurate potential curves in the region of the avoided crossing, this situation was remedied by simply taken enough basis functions to be able to do the calculation up to R=6 a.u.

The χ_{nm} of (24) are not orthogonal, and are made so before the calculation by means of Gram-Schmidt orthogonalisation. This method was preferred over the alternative of using the eigenfunctions of the overlap matrix, although the latter theoretically gives the larger numerical stability. This preference was motivated by the fact that the solutions of interest are very heavily dominated by the first few basis-functions of type (24). It thus seemed better to retain these functions in a pure form, rather than dilluting them by distribution over many eigenfunctions, some of them numerically ill-defined. Gram-Schmidt orthogonalisation offers this possibility, at the expense of having some of the orthogonalised basis-states even less well-defined. If these are states that the Hamiltonian doesn't like anyway, this is not a high price to pay.

CONVERGENCE OF THE CALCULATION

The accuracy of the results is dependent on the number of basis-functions N used to represent the wave-function, as well as on the number M of time-steps used in the time-integration to obtain $\mathbf{U}(T/4)$. The convergence with respect to N and M was checked for two cases, namely for the unperturbed molecule and for the largest intensity presented, with a peak electric field of 0.05 a.u. (corresponding to 87.5 TW/cm^2). Convergence with M was indeed fourth order, and N=20 was enough to approach the limit to within one wavenumber.

Table I. Error (meV) in the calculated energy of the $1s\sigma_g$ state for various combinations of basis-set size N and number of time steps per quarter period M as a function of internuclear distance R. All calculations were done for an intensity of 87.5 TW/cm^2.

N	20	20	20	20	20	20	12	07	30
R | M	**7**	**10**	**14**	**20**	**28**	**40**	**40**	**40**	**40**
1.00	31.3	8.15	2.45	0.89	0.52	0.42	1.98	3.65	0
2.00	2.62	0.68	0.22	0.10	0.07	0.06	0.65	5.65	0
3.00	0.38	0.19	0.15	0.14	0.13	0.13	1.14	20.87	0
4.00	0.85	0.85	0.84	0.84	0.84	0.84	5.22	69.82	0
5.00	-3.58	-2.19	-1.93	-1.86	-1.84	-1.84	10.25	139	0
6.00	-1.75	-0.99	-0.84	-0.81	-0.80	-0.79	9.66	266	0

The basis set was constructed by including all χ_{nm} with 2n+m up to some maximum value S. For S equal to 3, 5, 7 or 9 this gave rise to basis sets of 6, 12, 20 or 30 elements, respectively. The choice to have about twice as many different m values as n values was made because the η-dependence built into the basis functions was less natural the ξ-dependence. The functions (24) are not orthogonal, and as more and more are included their linear independence decreases. Even when calculating in double-precision (17 digits) not more than 30 basis functions could be tolerated before the calculation became unstable. Table I showns the residual error as a function of the number of basis functions.

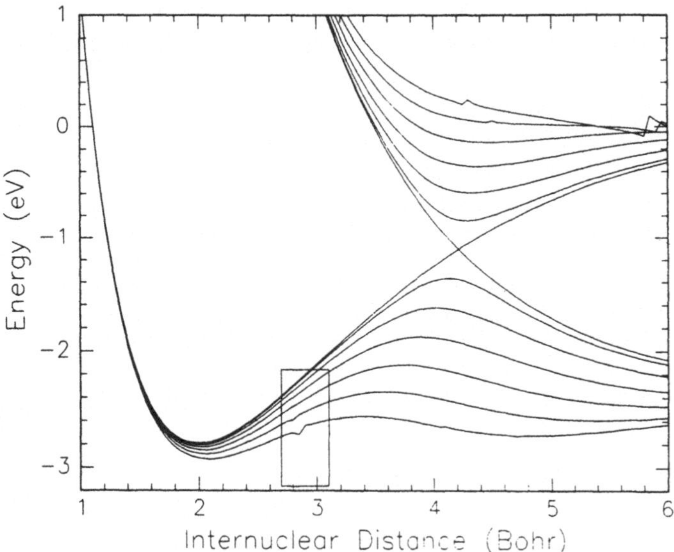

Figure 2. The 1-photon avoided crossing for intensities 0 (inner curves) 3.5, 14, 31.5, 56, 87.5 and 126 (outer curves) TW/cm^2. At the highest intensities irregularities appear due to opening-up of avoided crossings with other states. A close-up of the area within the box is shown in figure 3.

A good test for the convergence of the results is the gauge invariance. In the velocity gauge convergence with the number of basis functions is slower, but eventually leads to the same results, indicating that the results are indeed fully converged.

CALCULATED POTENTIAL CURVES

At a number of different values for the electric field the potential curves were calculated in the length gauge from R=1 to R=6, using 20 time-steps and 12 basis functions. From table I it can be seen that an accuracy of about 2 meV is expected over most of the internuclear distance range. The results are shown in figure 2.

At field strengths of 0.05 and higher, the curves start to show irregularities that degrade their initially smooth appearence. On closer inspection a calculation with a very small R-step reveals that these irregularities occur in the region where avoided crossings occur with other curves that are not shown in figure 2. The region in the neighbourhood of R=2.7 contains such an avoided crossing involving the bonding state, and this avoided crossing is shown in figure 3. The state that is avoided is actually the anti-bonding state dressed with three photons less. As such the width of the gap represents the three-photon coupling between those states.

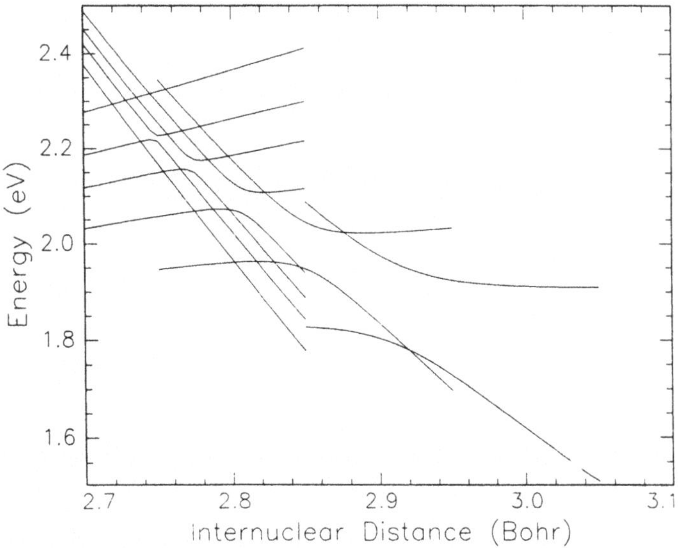

Figure 3. The potential curves in the vicinity of the 3-photon avoided crossing for intensities 0, 31.5, 56, 87.5, 126 and 171.5 TW/cm2. Due to ac-Starkshift of the curves the crossing point moves to larger internuclear distance.

The width of the gap is plotted as a function of intensity in figure 4, where its order of non-linearity confirms the three-photon character of the transition. In fact the order of non-linearity is slightly larger than three, so that at a field of 0.05 a.u. the width is about 20% larger than expected based on extrapolation from small intensities. Considering the fact that ac-Starkshift of the states makes the crossing move out to larger R, where the dipole moment is larger, this is not surprising. The predicted value for the three-photon transition-matrix element should be quite accurate, because both the initial and final states are represented well in the basis, and the virtual intermediate states should also have wave-functions that have roughly similar exponential decay.

Avoiding of crossings with other states than dressed versions of the two plotted in figure 2 also occurs. They represent transitions with six or seven photons to higher excited states of H_2^+. Since the basis-set was not really designed to represent these excited states very well, the position of these crossings is probably not obtained very accurately, except for maybe the lowest ones. Because of the large number of photons involved the associated gaps open up very aggressively if the intensity in increased still further, destroying the picture of coherent curves.

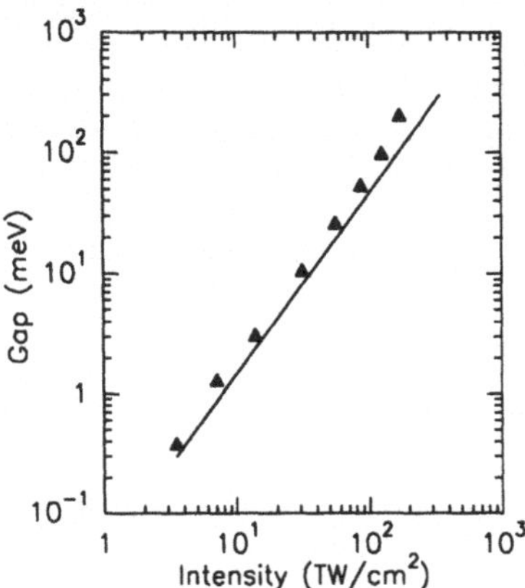

Figure 4. The width of the gap in the three-photon avoided crossing as a function of intensity. The line visualises the slope corresponding to a third-order process.

As transitions with increasingly large numbers of photons become possible, the calculation also starts to see transitions to the non-physical combinations of basis-functions high up in the continuum. At this point the calculated values lose all meaning. In fact as soon as transitions with the number of photons needed to reach the continuum become feasible, the result of the calculation becomes suspect, since no good representation of the continuum was included in the basis. Therefore the presented results only go up to I=88 TW/cm^2.

Acknowledgement - This work was supported by Fundamental Research on Matter (FOM) and the Netherlands Organization for the Advancement of Research (NWO). I wish to express my gratitude to D.W. Schumacher, who helped me making the computer plots.

References

Allendorf, S. W., and Szöke, A., 1991, High intensity multiphoton ionization of H2, **in:** "Short-Wavelength Coherent Radiation: Generation and Application," Opt. Soc. Am. 1991 Technical Digest Series

Bucksbaum, P. H., Zavriyev, A., Muller, H. G., and Schumacher, D. W., 1990, Softening of the H_2^+ molecular bond in intense laser fields, **Phys. Rev. Lett.** 64:1883

Verschuur, J. W. J., Noordam, L. D., and Van Linden van den Heuvell, H. B., 1989, Anomalies in above-threshold ionization observed in H_2 and its excited fragments, **Phys. Rev. A** 40:4383

INTENSE-FIELD PHOTOIONIZATION OF H_2

Sarah W. Allendorf

High Temperature Physics Division, L-438
Lawrence Livermore National Laboratory
Livermore, California 94550

Introduction

Studies of the interaction of intense lasers with molecules are no longer unusual. A number of workers are considering ionization and dissociation processes in these more complex systems. Several examples of the work in this field can be found elsewhere in this volume. Of particular interest is the importance of internal degrees of freedom in above threshold ionization (ATI). What are the roles played by vibration and rotation when a molecule is irradiated by an intense electric field? How are the molecular potentials affected by the laser? To date, most high intensity experiments on molecules have been performed with a limited range of laser wavelengths, either harmonic orders of a Nd:YAG laser, excimer wavelengths, or nontunable colliding pulse mode-locked dye lasers. The effects of the intense field on the internal structure of molecules have been probed only when molecular resonances were fortuitously encountered on the way to dissociation or ionization.

This paper concentrates on the truly molecular aspects of the molecule-laser interaction. We employ a tunable laser system, and we choose our wavelengths and intensities to probe regions of the molecular potential where a distribution of bond lengths exists. Systematically tuning our laser frequency allows us to examine specific intermediate state resonances in multiphoton ionization, and trace the behavior of the resonances in the presence of the intense laser field.

The experimental apparatus was described elsewhere. Briefly, we focus the frequency-doubled output of a pulse-amplified picosecond dye laser between the pole pieces of a "magnetic bottle" photoelectron spectrometer.[1] We record the photoelectron time-of-flight distributions for a number of different excitation wavelengths (288.5 nm $\leq \lambda \leq$ 304 nm) and peak laser intensities I_o, adjusting the H_2 pressure as I_o is increased in order to keep the number of electrons produced each laser shot constant. The spectra presented here have all been converted to photoelectron energy and normalized to an H_2 pressure of 1×10^{-8} torr. Because we use a short laser pulse (1.5 psec), the energy of a photoelectron produced at a given laser intensity reveals the energy of the corresponding photo-ion at the instant of ionization.

Coherence Phenomena in Atoms and Molecules in Laser Fields
Edited by A.D. Bandrauk and S.C. Wallace, Plenum Press, New York, 1992

The photoionization dynamics are greatly simplified by using ultraviolet photons (near 4 eV) because resonances can be encountered only at the three-photon level, and the fourth photon is sufficient to ionize the molecule except at very high laser intensity. The possible resonant states at are the $B\ ^1\Sigma_u^+$ and $C\ ^1\Pi_u$ electronic states; the photoionization dynamics of these two electronic states have been well characterized under relatively low field conditions.[2,3] Additionally, the use of 4 eV photons allows a large range of electron energies to be analyzed without the interference of overlapping ATI electron peaks. Finally, ponderomotive shifts are also smaller because of the smaller wavelength. Despite these simplifications, we find that the photoelectron spectra are quite rich and show a marked dependence on both laser wavelength and intensity.

Part of the difficulty in interpreting molecular ATI experiments is the increased number of possible intermediate resonances. In the case considered here, the C state in H_2 is the first Rydberg state in the series converging on the ground state of the molecular ion. The excited electron is far from the molecular core, and removal of this electron perturbs the ion core only a small amount. Ionization of Rydberg states often occurs with a relatively small geometry change, and can often be successfully predicted by use of the Franck-Condon (FC) formalism. We can also comfortably assume these Rydberg states will behave similarly to atomic Rydberg states in an intense field: if the photon energy is large enough to cause ionization, the state will most likely exhibit an a.c. Stark shift close in magnitude to the ponderomotive energy of a free electron in the same field.

It should be noted, however, that there are a number of very interesting examples of diatomic photoionization processes where FC calculations correctly predict only part of the observed photoionization dynamics of a Rydberg state.[4] The $C\ ^1\Pi_u$ state in H_2 has been found in nsec experiments to be such an case.[3,5] The temptation to apply a FC analysis blindly to the description of the ionization of any Rydberg state must be avoided. Nevertheless, our experimental resolution in these intense field ATI experiments is often insufficient to uncover the subtleties of non-FC ionization, so we will (carefully) employ FC computations as a guide in understanding our results.

The $B\ ^1\Sigma_u^+$ state in H_2 is a valence state. The behavior of valence states in an intense laser field is considerably less well known. Since the polarizability of a molecule depends on the internuclear distance, broad valence states may well distort, not just shift energetically, in the laser field. It is not clear whether a given valence state will a.c. Stark shift up or down, energetically. The shift depends on the geometry of the state, as well as what other states lie nearby that may interact with the valence state. In nsec experiments, there are many examples of valence states photoionizing to populate a broad range of vibrational levels in the molecular ion, in contrast to the usually sharp population distribution in Rydberg photoionization. At our wavelengths, the B state exhibits this valence state character in nsec experiments.[2] This behavior will probably extend to psec experiments, and so it may be possible to discern the influence of a Rydberg intermediate state from that of a valence intermediate state by analyzing the vibrational distribution produced in the ion, even in these intense field experiments.

In what follows we present the results of our experiments at a number of different wavelengths by discussing the different photoionization mechanisms observed. It is important to remember that our apparatus allows only for the detection of photoelectrons. We are unable to observe any dissociation processes that may occur after ionization unless further ionization of a dissociation product occurs.

Molecular Photoionization

Typical photoelectron energy spectra recorded at 288.5 nm are presented in Fig. 1 for electron energies ranging from 0.6 to 2.0 eV, and in Fig.2 for electron energies between 2.0 and 6.0 eV. Each spectrum was recorded at a different I_o (left vertical axis), and has been divided by the indicated scale factor (right vertical axis) then offset from the horizontal axis to improve the visibility of the graph. At 288.5 nm the energy of three photons is just above the field-free energy of the $C\ ^1\Pi_u$ (v'=2) vibrational state. The peaks in the lowest intensity spectrum in Fig. 1 are assigned to molecular photoionization of H_2, producing H_2^+ in several vibrational levels (v^+). At this I_o, the largest photoelectron peak is at 1.22 eV: this corresponds to the production of ions with two quanta of vibrational energy. Smaller peaks are seen at 0.98 eV and 1.45 eV, which indicate production of v^+=3 and v^+=1, respectively. This predominantly Δv=0 ionization is consistent with the FC overlap of the vibrational manifolds of the neutral $C\ ^1\Pi_u$ and the ionic $X\ ^2\Sigma_g^+$ electronic states. The photoelectron spectrum is similar to that recorded with a nsec laser.[5]

As I_o increases, the relative production of v^+=1 increases until the v^+=1 peak equals the v^+=2 peak. This happens because the resonant intermediate $C\ v'$=2 state is a.c. Stark shifted up and out of three-photon resonance by the electric field of the laser. Likewise, the C v´=1 state is a.c. Stark shifted into three-photon resonance as the peak laser intensity increases, then Δv=0 ionization produces H_2^+ ions in the v^+=1 vibrational level. The photoelectron energy of the v^+=1 peak initially decreases as I_o increases, due to the increased ponderomotive energy U_p in the portion of the laser focus where $C\ v'$=1 is in resonance. Further increases in I_o do not shift this peak further because it arises predominantly from Δv=0 resonant photoionization of the $C\ v'$=1 state, and at higher I_o this state is no longer in resonance. Similarly, the v^+=2 peak is initially shifted 20 meV which is consistent with a small U_p at $I_o \approx 5 \times 10^{12}$ W/cm^2. However, this peak does not shift to lower energies as I_o increases. This v^+=2 peak arises from the expanding volume of the laser focus with intensity where the $C\ v'$=2 state is in three-photon resonance. This behavior is the hallmark of resonances in short-pulse ATI, as has been discussed in great detail by others.[6] However, in molecules a new twist exists: ionic vibrational populations become intensity-dependent because the resonant intermediate state is intensity-dependent.

The shifting of ionic vibrational population with laser intensity was observed at all wavelengths studied, with the exception of the longest wavelengths. Figures 3 and 4 present the low (0.6 - 1.8 eV) and high (1.5 - 6 eV) energy photoelectron spectra recorded at 304 nm. At this wavelength, three photons are insufficient to reach even v'= 0 in the C state. Thus ionization proceeds only via vibrational levels in the B state. Unfortunately, the photoelectrons produced in four-photon ionization when λ = 304 nm have very low energy and so we can reliably detect only the v^+ =0 electrons. This peak is very broad and shifts to lower energy with increasing I_o. The signal level at this wavelength is much lower and the photoelectron spectrum is dominated by electrons with energy between 2 and 5 eV. When λ = 304 nm, only the B state is energetically accessible with three photons. Vibrational levels of this valence state will photoionize to produce a broad distribution of ion vibrational states, rather than the nearly Δv=0 ionization observed from the C Rydberg state. Thus, as the B state is a.c. Stark shifted by the intense field and new vibrational levels become resonant, ionization will proceed to a distribution of ionic vibrational levels. For example, more than one B (v') produces v^+=0 in the ion. So, the v^+=0 electron peak will continue to shift to lower energies, reflecting the increasing laser intensities at which the electrons are created. This is a clear example of how valence and Rydberg states behave differently in an intense electric field, which causes the resulting photoelectron spectra to differ.

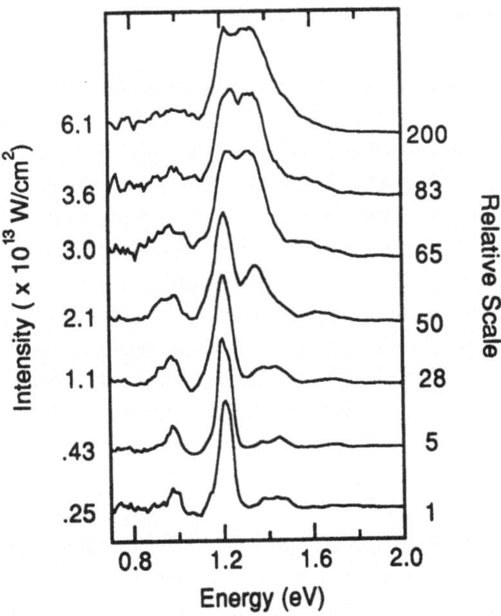

Figure 1. Photoelectron energy spectra between 0.7 and 2.0 eV recorded at 288.5 nm. Peak laser intensity for each spectrum is indicated on the left axis, and the amount by which each spectrum was divided for presentation purposes is indicated on the right axis.

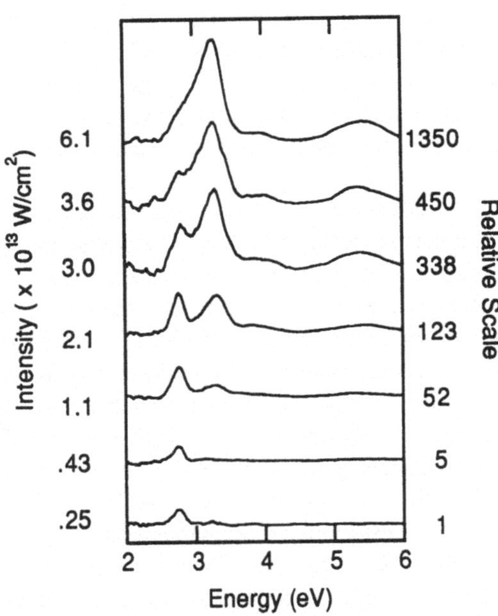

Figure 2. Photoelectron energy spectra between 2 and 6 eV recorded at 288.5 nm, with axes labeled as in Fig. 1.

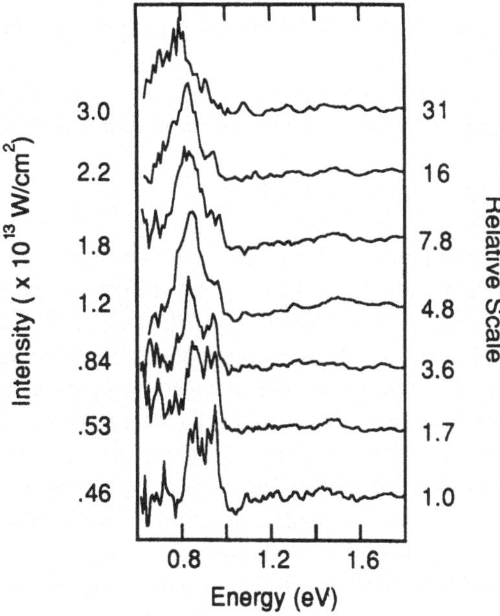

Figure 3. Same as Fig.1, at 304 nm.

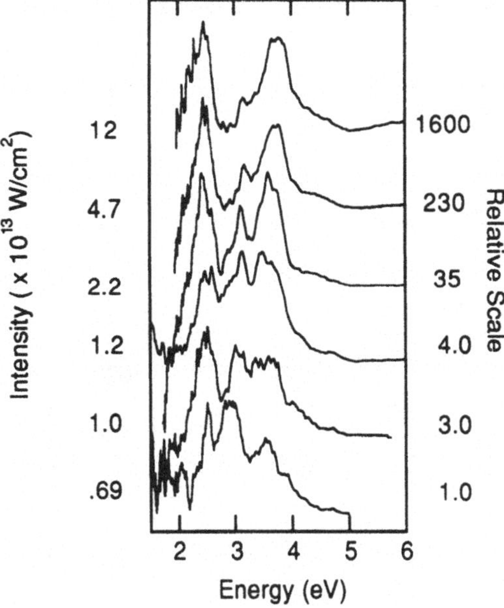

Figure 4. Same as Fig. 2, at 304 nm.

Finally, we assign the broad photoelectron peak at 5.4 eV seen in Fig. 2 as the first above threshold peak of this molecular photoionization. We lack sufficient energy resolution to observe the vibrational levels in this ATI electron peak. Figure 4 shows that ATI is not important at the very longest wavelength studied. At $\lambda = 304$ nm, an ATI peak near 5 eV is not discernable due to the much stronger peak seen near 3.8 eV which has a long high-energy tail. Evidently direct molecular photoionization via the valence *B* state is not an important process.

Dissociation Followed By Atomic Ionization

Electron energy spectra between 1.5 and 6.0 eV are presented in Fig.5 for $\lambda = 297.5$ nm. As the laser wavelength is tuned, the energies of the peaks shift in a manner that can be attributed to the change in the available total energy. At 297.5 nm, the energy of three photons is just below the field-free energy of the $C\ ^1\Pi_u$ (v′=1) vibrational state. At the lower laser intensities this energy region of the photoelectron spectra is dominated by a peak at 2.63 eV which is very sharp at most laser intensities. The peak also occurs at the energy expected for single-photon ionization of $H(3\ell)$ and does not shift in position as I_0 increases (the large shoulder toward lower photoelectron energy observed at higher intensities in Fig. 5 can be assigned to a different ionization mechanism because of its wavelength dependence). We assign the 2.63 eV peak in Fig. 5 to photoionization of $H(3\ell)$, resulting from direct dissociation of neutral H_2 followed by ionization. Photoionization of $H(3\ell)$ occurs readily since only one photon is required. However, we see no evidence for ionization of any other states of atomic H. At 297.5 nm, absorption of four photons is above the dissociation limit to produce $H(3\ell)$ but below the energy required to produce $H(4\ell)$. The absence of electrons resulting from ionization of $H(4\ell)$ suggests that dissociation occurs after the absorption of only four photons. The absence of any $H(2\ell)$ electrons suggest this dissociation occurs at internuclear distances R too large to allow curve-crossing to potentials that correlate to $H(2\ell)$+$H(1s)$ dissociation products.

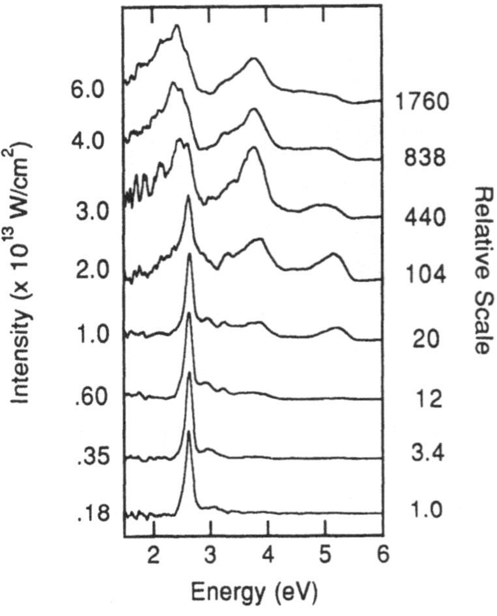

Figure 5. Same as Fig. 1, at 297.5 nm.

The final piece of evidence required to assign this peak arises from its wavelength dependence. At each wavelength, the peak was observed at the field-free energy for $H(3\ell)$ ionization, unless four photons is energetically below the dissociation limit to produce $H(3\ell)$. At these longer wavelengths ($\lambda > 299.5$ nm), a sharp peak was not observed. For example, in Fig. 4 there is no peak at 2.57 eV similar to the peaks at 2.79 and 2.63 eV seen in Figs. 2 and 5, respectively.

These observations are consistent with the population of high vibrational levels at the three-photon energy in the $B\ ^1\Sigma_u^+$ state which have rather large (R > 5 bohr) outer turning points. Absorption of a fourth photon at this R has a favorable FC overlap for transition into a neutral, doubly excited state similar to the $H_2^+\ |2p\sigma_u\rangle$ repulsive state, converging to the dissociation limit of $H(1s) + H(3\ell)$. The narrowness of the peak, even at high laser intensities, can be understood in the context of the relative a.c. Stark shift of the excited (n=3) state and the ionization potential. Recent calculations have shown that our use of >4 eV photons permits us to use the "high frequency approximation," because this excited state has an a.c. Stark shift nearly equal to that of the ionization potential.[7] Therefore, the excess energy available for the photoelectron is the same at every intensity for our 1.5 ps laser pulses.

Dissociative Ionization

The high-energy portion of the photoelectron spectra contains two peaks in addition to the peaks we have already discussed. In this section we describe the lower energy peak, and defer discussion of the higher energy peak to the next section. Figure 2 (λ=288.5 nm) has a peak near 3.3 eV which becomes important when $I_o > 1 \times 10^{13}$ W/cm^2. At 297.5 nm a strong peak grows in near 2.4 eV as I_o increases. These peaks are rather broad, and in general exhibit a low-energy tail. Spectra recorded at other wavelengths exhibit a similar peak. These high-energy peaks do not correspond to any low field processes, and are not observed in nsec experiments. In addition, the peaks do not shift significantly as the peak laser intensity is increased. To understand the mechanism responsible for these peaks, we examined their energy as a function of laser wavelength. We observe a marked decrease in photoelectron energy as the photon energy decreases, but the decrease is not a simple function of the photon energy.

We assign this peak to dissociative ionization occurring after a two-photon absorption from the outer turning point of a vibrational level in the B state. This process can be resonantly enhanced by any of a number of doubly-excited states at the four-photon level (the same states which dissociate to produce excited H atoms). If we estimate the turning point to be R=5.5 bohr, the energy of the $H_2^+\ |2p\sigma_u\rangle$ state is 18.45 eV. Five photons at λ=297.5 nm add to 20.84 eV. The excess energy available for partitioning between the electron and the dissociating $H(1s) + H^+$ is 20.84 - 18.08 = 2.76 eV, and the excess energy above the $H_2^+\ |2p\sigma_u\rangle$ state is 2.39 eV. The observed electron energy of 2.4 eV falls neatly between these two bounds. This bracketing of the electron energy by our model occurs at each wavelength studied, with the exception of the last wavelength (304 nm), where the observed electron energy is less than 5% larger than the upper bound calculated using our simple delta-function model for absorption followed by dissociative ionization. Dissociative ionization has been observed at other laser wavelengths by others, most notably by Luk and Rhodes.[8]

Note that the photoelectron can carry away either all of the excess energy in dissociative ionization, or none of it , or an energy in between these limits. The remaining energy then goes into the recoil energy of the dissociation, producing $H(1s) + H^+$. We observe electron energies indicating that at each wavelength, most of the energy goes into the electron kinetic energy. However, the $|2p\sigma_u\rangle$ repulsive curve is rather flat at large internuclear distance R, so a low recoil velocity for the dissociation products is not improbable. Unfortunately we are unable to record the proton kinetic energy in our experiments. Clearly such measurements at each wavelength would clinch this assignment.

Ionization into a transient laser-induced bound state

The evidence for a fourth photoionization mechanism in our study is as follows. With the possible exception of λ= 288.5 nm, the electron spectra display a prominent electron peak near the photon energy. For example, at 304 nm this peak is near 3.75 eV (Fig. 4), while at 297.5 nm this peak is near 3.8 eV (Fig. 5). Although the peaks are broad, they are fairly symmetric and we find a decrease in peak energy that is nearly 1:1 with decreasing photon energy. This rules out a conventional, multiple-photon ATI explanation for these high-energy electrons, and suggests that a single mechanism is the source of this peak at different wavelengths.

We assign these electrons to ionization into the adiabatic state formed by the avoided crossing of the n+1-photon dressed ionic bound state potential $|1s\sigma_g, n+1\rangle$ and the n-photon dressed ionic repulsive $|2p\sigma_u, n\rangle$ potential. This crossing creates two adiabatic states (upper and lower adiabatic potentials, UAP and LAP) from the diabatic states. These potentials are indicated in Fig. 6. The avoided crossing influences the molecular dynamics in two ways. First, molecular ions in the bound $|1s\sigma_g\rangle$ potential whose vibrations take them into the internuclear region near the crossing point become unbound in the presence of the laser as they travel along the LAP. This "molecular bond softening" does not affect the photoelectron energy distribution as much as it affects the proton kinetic energy spectrum. The second consequence of the avoided crossing of the two diabatic states is the possibility of transitions into the resulting adiabatic states. We find that our photoelectron spectra are quite sensitive to this second effect of the avoided crossing, and we propose the following mechanism for the high energy photoelectrons observed in our experiment. Three-photon excitation of the B state prepares an H_2 molecule with a large outer turning point, and then two or three further photons are absorbed at large R, with photoionization occurring into the UAP formed by the avoided crossing.

To test this mechanism we approximate the adiabatic potentials LAP and UAP by a simple two-state avoided crossing calculation involving only the $|1s\sigma_g, n+1\rangle$ and $|2p\sigma_u, n\rangle$ diabatic potentials. We diagonalize the adiabatic potentials to obtain their eigenvalues and eigenfunctions. For example, at λ=304 nm and I_o=1 x 10^{13} W/cm^2, the UAP is bound by 1 eV, and supports 14 vibrational states. Finally, we calculated the FC overlap between vibrations in the B state and vibrations in the UAP. For example, v_B=7 and v_{UAP} = 3 are indicated in Fig. 6. We find that the electron energies predicted by the best FC overlap with the B state agree qualitatively with the observed electron energies, but are approximately 0.4 eV lower than those observed. This agreement is reasonable, given the simple nature of our model. Further computations await a better understanding of the effect of the strong laser field on the valence B state.

This final mechanism is not inconsistent with the other mechanisms invoke to explain our photoelectron spectra. The branching ratio for photoionization into the UAP or LAP is determined by laser intensity, excess energy available, and the internuclear bond separation when photoionization occurs. At lower laser intensities, the molecule ionizes when sufficient photons (five) are absorbed to excite the molecule above the LAP. If this occurs when R > R_{cross} (the internuclear distance at which the diabatic potentials cross for a given laser wavelength and intensity), this mechanism is equivalent to dissociative ionization as discussed above. If absorption occurs when R < R_{cross}, molecular ions are formed in new vibrational states in the bound portion of the LAP. This results in molecular ionization, and can be distinguished in our photoelectron experiment only to the extent that the lower adiabatic potential is deformed relative

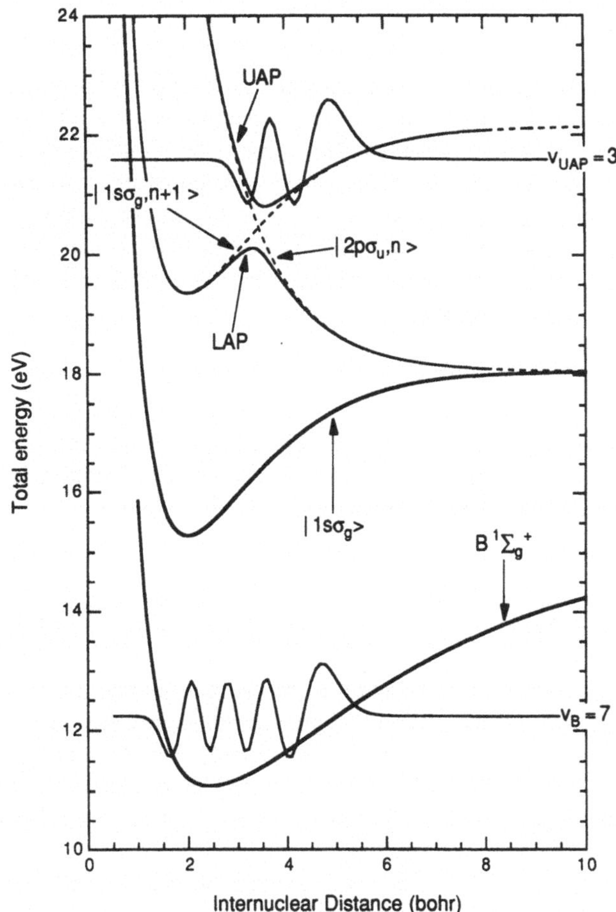

Figure 6. Enlarged potential-energy diagram, indicating with thick solid curves the neutral $B\ ^1\Sigma_u^+$ state and the ionic H_2^+ $1s\sigma_g$ state. The diabatic dressed states $|1s\sigma_g, n+1\rangle$ and $|2p\sigma_u, n\rangle$ for λ=304 nm at 1 x 10^{13} W/cm^2 are indicated by the dashed curves, while the UAP and LAP (see text) are drawn with thin solid curves. Vibrational wave functions in the $B(v'=7)$ and UAP($v_{UAP}=3$) states are also displayed.

to the H_2^+ $|1s\sigma_g>$ state. If this deformation changes the vibrational eigenvalues, the measured photoelectron energy will also change, and might be observable. When the laser intensity is sufficient to cause additional photon absorption to compete successfully with the above processes, sufficient energy is available for the ion core to be produced in the upper adiabatic state. Once again, the photoelectron energy is determined by the eigenvalue of the vibrational state produced. It is this final process, photoionization into the upper adiabatic potential, which we assign as responsible for the photoelectron peaks in our spectra near 4 eV.

Conclusions

We have presented results from an experimental study of photoionization in molecular H_2 at several laser wavelengths and intensities. Four different mechanisms were responsible for the photoelectrons observed: molecular photoionization, dissociation followed by atomic ionization, dissociative ionization, and ionization into a laser-induced bound state. The last two mechanisms are evidently more important at longer wavelengths. This this may be simply due to the decreased importance of competing pathways such as the first two mechanisms. We have explored the molecular aspects of the molecule-laser interaction, and have found that the resulting photoionization dynamics are rich indeed.

Acknowledgment

It is a pleasure to acknowledge my collaboration with A. Szöke. This work was performed under the auspices of the U. S. Department of Energy at Lawrence Livermore National Laboratory under contract number W-7405-Eng-48.

References

1. S. W. Allendorf and A. Szöke, High-intensity multiphoton ionization of H_2, Phys. Rev. A **44**, 518 (1991).
2. S. T. Pratt, P. M. Dehmer, and J. L. Dehmer, Resonant multiphoton ionization of H_2 with the $B^1\Sigma_u^+$, v=7, J=2 and 4 levels with photoelectron energy analysis, J. Chem. Phys. **78**, 4315 (1983); J.H.M. Bonnie, J.W.J. Verschuur, H.J. Hopman, and H. B. van Linden van den Heuvell, Photoelectron spectroscopy on resonantly enhanced multiphoton dissociative ionization of hydrogen molecules, Chem. Phys. Lett. **130**, 43 (1986).
3. S. T. Pratt, P. M. Dehmer, and J. L. Dehmer, Photoionization of excited molecular states. H_2 $C\ ^1\Pi_u$, Chem.Phys. Lett. **105**, 28 (1984).
4. S. T. Pratt, P. M. Dehmer, and J. L. Dehmer, Resonantly enhanced multiphoton ionization — Photoelectron spectroscopy as a probe of molecular photophysics and photochemistry, in: *Advances in Multi-photon Processes and Spectroscopy*, S. H. Lin, ed., World Scientific, Singapore, 1988.
5. S. N. Dixit, D. L. Lynch, B. V. McKoy, and A. U. Hazi, Electronic autoionization and vibrational-state distributions in resonant multiphoton ionization of H_2, Phys. Rev. A **40**, 1700 (1989).
6. R. R. Freeman and P. H. Bucksbaum, Investigation of above-threshold ionization using subpicosecond laser pulses, J. Phys. B: At. Mol. Phys. **24**, 325 (1991).
7. S. N. Dixit, S. W. Allendorf, and N. M. Khambatta, "Exact, frequency dependent a. c. Stark shifts in hydrogenic systems", Bull. Am. Phys. Soc. **36**, 1298 (1991).
8. T. S. Luk and C. K. Rhodes, "Multiphoton dissociative ionization of molecular deuterium", Phys. Rev. A **38**, 6180 (1988).

EXCITATION OF MOLECULAR HYDROGEN IN INTENSE LASER FIELDS

H. Helm, M. J. Dyer, H. Bissantz, and D. L. Huestis

Molecular Physics Laboratory
SRI International
Menlo Park, CA 94025

ABSTRACT

A high resolution photoelectron study of the response of molecular hydrogen to intense laser fields ($>10^{13}$ W/cm^2) at wavelengths between 310 and 330 nm shows that multiphoton ionization of the molecule competes with photodissociation of the neutral. The formation of excited atomic hydrogen photofragments which are subsequently photoionized and the ac-Stark shifting of molecular Rydberg states gives rise to similar photoelectron spectra at high field strength. The photoelectron spectra remain strongly dependent on wavelength at even the highest intensities investigated (8×10^{13} W/cm^2). By contrast, the total ionization rate does not change perceptibly in the wavelength range studied.

The molecular response is discussed in a framework of combined effects of AC-Stark shifting of molecular Rydberg states and predissociation of the molecular Rydberg states induced by the intense laser field.

INTRODUCTION

At intensities above 10 TW/cm^2 atoms exhibit above threshold ionization (ATI) phenomena, that is they absorb more photons than required for ionization.[1-3] The photoabsorption is generally followed by emission of photoelectrons, and the measurement of the photoelectron energy is a primary diagnostic tool for ATI. Molecules respond similarly as atoms do, however an additional degree of freedom appears because electrons have two positive coulomb centers to orient themselves, allowing the formation of antibonding orbitals and dissociation. Thus in multiphoton excitation the dissociation process may appear as competitor to ionization, or else lead to photodissociation of the molecular ion produced. Photodissociation of neutral H_2 has been observed in many stepwise photo-excitation experiments and has been interpreted as resulting from of sequential excitation, first to singly excited states, followed by excitation of the core electron.[4-6]

Here we report observations of intensity-dependent photodissociation of H_2 that competes with the direct ionization channel. This phenomenon is seen as a manifestation of (1) the deformation of molecular bonds that arises from bound-free absorption of the molecular ion core and (2) the AC-Stark shifting of molecular Rydberg states in the intense laser field.

Coherence Phenomena in Atoms and Molecules in Laser Fields
Edited by A.D. Bandrauk and S.C. Wallace, Plenum Press, New York, 1992

EXPERIMENTAL

The experiment is carried out in a magnetic bottle photoelectron spectrometer using light from a tunable dye-laser system with frequency doubling. The tunable oscillator is a synchronously pumped dye laser operating with DCM dye. The 3-500 mW average output of the dye laser is fed into 10m of single mode fiber to chirp and a double pass grating stage to compress the pulses down to 400 fs leading to an average output of typically 50 mW on the compressed light. Ten pulses of this train are amplified per second by a four stage amplifier system consisting of three Bethune prism cells with increasing bore diameters of 1, 3, and 6 mm. The beam is then expanded to a diameter of 20 mm and it double passes an open axicon amplifier.[7] The amplifiers are pumped by a 4-ns 450-mJ frequency-doubled YAG laser at 532 nm. The final output at the red wavelengths is between 2 and 4 mJ at pulse lengths between 400 fs and 2 ps. Frequency doubling in KDP leads to pulses up to 1 mJ in energy.

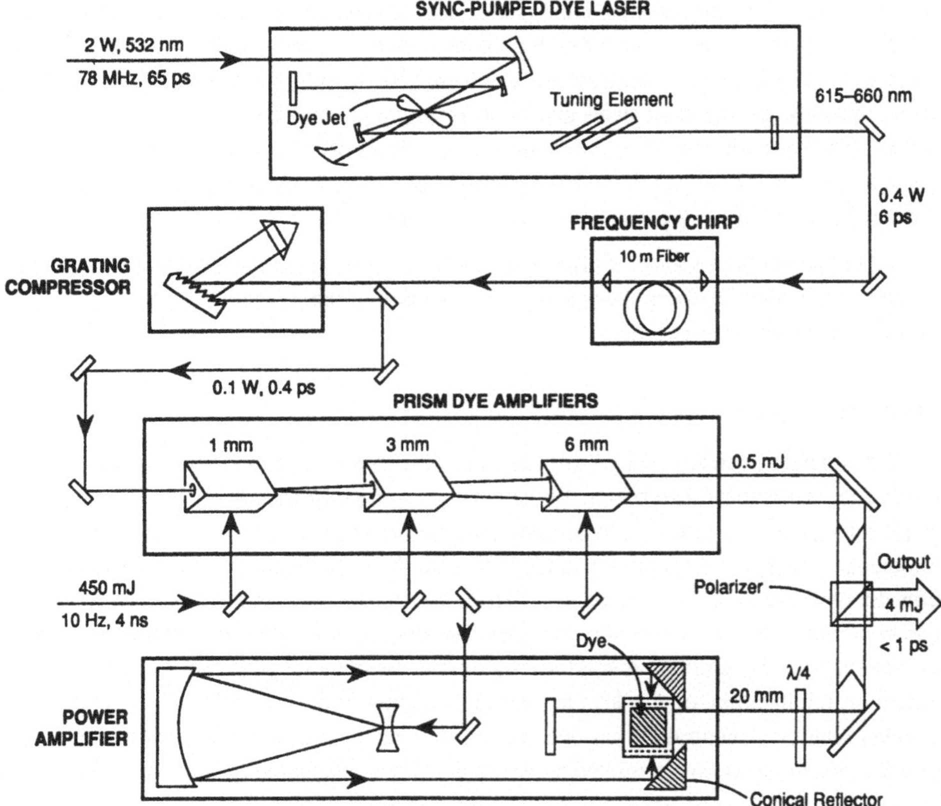

Figure 1. Schematic of dye laser system and amplifiers. The pulses from the synch pumped dye-laser are chirped and compressed and then amplified in three prism amplifiers. The final stage amplifier is a reflective axicon design for even amplification of the 20 mm diameter laser beam.

The photoelectron spectrometer is a magnetic bottle design[8] that uses permanent magnets and a 64 cm time-of-flight section. The photoelectron signal is monitored by a constant fraction discriminator whose output feeds a chain of 16 time-to-digital converters, each with 1 ns resolution. The pressure in the interaction region is adjusted such that several (typically less than 10) electrons are recorded per laser shot. In studies of resonant ionization of H_2 at low intensity we have demonstrated a resolution of 50 meV at 1eV electron energy using this device. The laser energy is encoded with the arrival time of each electron detected and stored on hard disk of an IBM personal computer. Final data are restricted to laser energy bins that fall within $\pm$ 20% of the quoted value. Experiments are carried out using linear laser polarization unless otherwise specified.

The base pressure in the photoionization region of the spectrometer is around 1×10^{-8} Torr, and the hydrogen pressure added reached as high as 5×10^{-6} Torr at the lowest intensities studied. The laser beam is focused using a 15 cm lens. All the hydrogen work described below was obtained with the dye laser output of ~1.5 ps, obtained from autocorrelation measurements and assuming a sech^2 pulse shape. The bandwidth of the laser falls typically in the range of 5 Å. Intensities quoted below are assumed lower limits, averaged over a conservatively chosen 28 μm spot diameter, twice the diffraction limit.

EXPERIMENTAL RESULTS

The wavelength range chosen for this investigation is such that the three-photon energy lies below the $C^1\Pi_u$ state (3-photon threshold: 300 nm) while the four-photon energy lies near the ionization limit. Thus at the wavelengths employed here (310 -330 nm) three-photon resonances are only expected to involve the $B^1\Sigma_u^+$ state while four-photon resonances may occur with a variety of vibrational levels of the Rydberg manifolds belonging to n = 3 and higher. We observed that scanning the dye laser in 1 Å intervals over a portion of the wavelength range studied (329 to 319 nm) showed no noticeable wavelength dependence in the total ionization yield at intensities around 3×10^{13} W/cm^2. The yield was comparable at the other wavelengths investigated where photoelectron spectra were recorded at intervals of 25 Å. Despite the insensitivity of the total ionization rate to wavelength we observe that the intensity distribution in the photoelectron energy spectra is strongly wavelength dependent.

In order to introduce the photoelectron features we first discuss relevant potential energy curves of molecular hydrogen. In Figure 2 we show a simplified diagram of molecular excited states, together with the lowest states of the molecular ion. Also indicated in this figure are the three, four, and five photon energies for photoexcitation at 326 nm, together with the vibrational wavefunction of ground state H_2, v = 0.

It is evident that in this case the Franck Condon overlap into the $B^1\Sigma_u^+$ state is poor and that the first "good" resonant states occur at the four-photon energy. This leads us to assume that an important ionization channel may involve four-photon excited states of the molecule:

$$H_2 + 4\,h\nu \rightarrow H_2^{**}(n\ell\lambda v) \tag{1}$$

that are subsequently ionized by a fifth photon forming molecular hydrogen ions in specific vibrational levels v^+:

$$H_2^{**}(n\ell\lambda v) + h\nu \rightarrow H_2^+(v^+) + e\,. \tag{2}$$

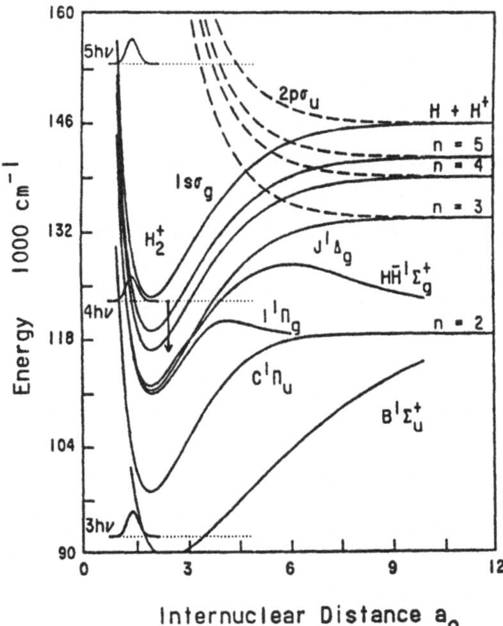

Figure 2. Selected potential energy curves of H_2 and the ground states of H_2^+. Also indicated are the resonant energies for 3, 4, and 5 photons at 326 nm together with the vibrational wave function of v = 0 of ground state H_2. The bound and repulsive Rydberg states are shown for zero quantum defect.

Alternatively a non-resonant five photon process might occur

$$H_2 + 5\,h\nu \rightarrow H_2^+(v^+) + e \tag{3}$$

which could also form molecular hydrogen ions in a variety of vibrational levels.

The observation is that indeed, at low intensities, the spectra show photoelectrons with energies that are consistent with the formation of H_2^+. As an example we give in Figure 3 spectra obtained at 3185 Å at various intensities.

The surprising observation in Figure 3 and in the spectra recorded at other wavelengths is that as the laser intensity is increased the photoelectron signal from the molecular hydrogen ion channel is apparently surpassed by the signal from one-photon ionization of excited atomic hydrogen

$$H(n\ell) + h\nu \rightarrow H^+ + e \tag{4}$$

where $n \geq 3$. This observation is interpreted as being due to a multiphoton dissociation process, but as we show below it is indistinguishable from the signal arising in one-photon ionization of ac-Stark shifted molecular Rydberg states. Plausible channels for the formation of excited atomic hydrogen are either the photodissociation of molecular Rydberg states:

$$H_2(n\ell\lambda,v) + h\nu \rightarrow H(1s) + H(n\ell). \tag{5}$$

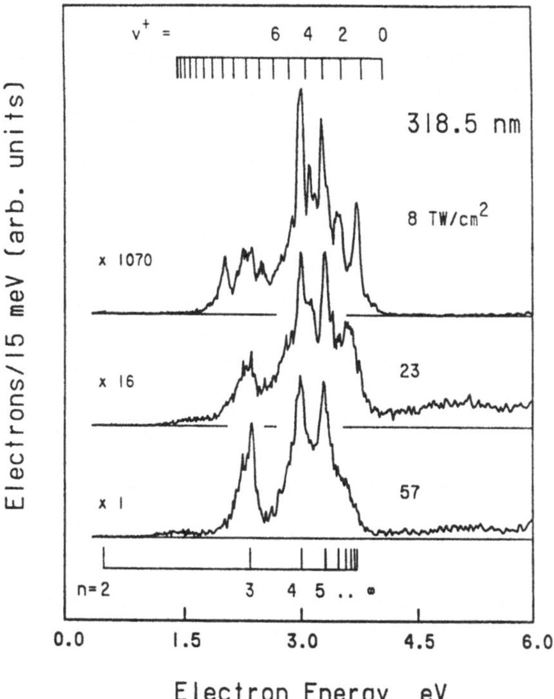

Figure 3. Photoelectron spectra of H_2 at 318.5 nm as a function of intensity. The tickmarks at the top refer to photoelectron energies as they should appear from five-photon nonresonant ionization forming H_2^+. The tickmarks at the bottom refer to photoelectrons produced in one-photon ionization of excited atomic hydrogen fragments $H(n\ell)$.

or also dissociative ionization processes involving at least eight photons:

$$H_2 + nh\nu \rightarrow H^+ + H(n\ell) + e\,. \qquad (6)$$

In Figure 3 we indicate at the top of the figure the electron energy peak location expected for non resonant five-photon ionization (process (3)) when the molecular ion is formed in the lowest rotational level. The one-photon ionization process of atomic hydrogen gives rise to a unique energy position that unambiguously identifies the excited atomic hydrogen fragment $H(n\ell)$ and the energies expected from process (4) are marked at the bottom of the figure.

As mentioned above we find that the intensity distribution in the photoelectron spectra is strongly wavelength dependent. This dependence exists at low intensities where the presence of many individual peaks indicates that processes (1-3) lead to the formation of H_2^+ ions in a variety of vibrational levels but also at higher intensities where the spectrum simplifies to atomic-like photoelectron features.

As an example we give in Figure 4 photoelectron spectra obtained at 3260 Å for a range of intensities. The spectra appear quite different from those obtained at 3185 Å. An additional surprising feature in the spectra in Fig. 4 is that the ionization process tends to favor states with lower principal quantum number as the intensity grows. While the signal corresponding to (n = 3) is only a weak channel at the lowest intensity (see the lowest trace in Figure 4) this channel dominates the spectrum the highest intensity value. We also note in this figure the appearance of the first ATI peaks of the photoelectron spectra at the higher intensities. ATI peaks of similar magnitude are present at all other wavelengths studied.

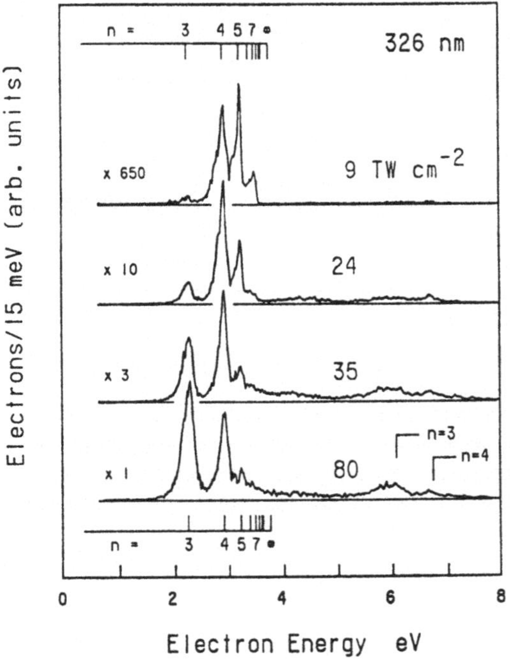

Figure 4. Photoelectron spectra of H_2 at 326 nm as a function of intensity. An intensity dependent shifting of population into lower dissociation limits is apparent.

DISCUSSION

Numerous investigations on multiphoton ionization of molecular hydrogen in intense fields have been reported in recent years.[9-12] A common feature than can be extracted from these data is that the ionization process remains wavelength specific at even high intensities. The phenomenon of strong wavelength dependence is not surprising in view of the various electronic, vibrational and rotational degrees of freedom of the intermediate excited states that can participate in the multiphoton excitation process H_2.

The data presented here add another dimension to the ongoing discussion on molecular hydrogen in strong fields. Their interpretation must be viewed as applicable to the specific wavelength range investigated. We have carried out similar studies at wavelengths from 610 to 660 nm and find photoelectron spectra that are quite different from those reported here in that individual peak structures are lost at high intensity at the expense of broad features that appear up to electron energies of ~16 eV.

Before we discuss the mechanism of photoionization and photodissociation via four-photon excited intermediate states to explain the observed photoelectron features we first discuss specific properties of the excited states of molecular hydrogen as well as the effect of ponderomotive forces on the appearance of the photoelectron spectra.

Potential Energy Diagram

The potential energy diagram for molecular hydrogen given in Figure 2 is highly schematic in its representation of the bound and dissociative Rydberg states. The states shown are drawn assuming zero quantum defect. The full diagram known to date for the excited singlet manifold of states is considerably more dense owing to the fine structure arising from the non zero quantum defects of the various ℓ-components in each principal quantum number. Furthermore, the avoided crossings[13] between states of the same symmetry that are either the singly excited Rydbergs (orbital configuration $(1s\sigma_g)(n\ell\lambda)$ or the doubly excited states $(2p\sigma_u)(n\ell\lambda)$ modify the appearance of the adiabatic states at large ($R>4a_o$) internuclear distances. In addition, the $^1\Sigma$ state potential energy curves are modified at large separation by the strongly attractive ion-pair character. This is apparent for the $B^1\Sigma_u^+$ and the $H\bar{H}^1\Sigma_g^+$ curves which are included in Figure 2.

The simplified potential energy diagram in Figure 2 nevertheless serves to indicate the strongest transition expected for reaching excited states at the wavelengths considered in this paper: At the internuclear distance of the vibrational wavefunction of the v = 0 level of the ground state, $H_2X^1\Sigma_g^+$, a vertical multiphoton excitation may find a first energy resonance with either the $B^1\Sigma_u^+$ state (3-photon transition) or else with the gerade states of the manifolds n = 3 or higher at the four photon energy. At the short internuclear separation where vertical excitation can be efficient the latter are nearly purely the singly excited states $(1s\sigma_g)(n\ell\lambda_g)$ with ℓ = s, d, and g. In the vicinity of the lowest vibrational level these gerade states have small quantum defects (typically < 0.1). At the wavelengths of this experiment the Franck-Condon factors to reach the bound B-state levels is always at least an order of magnitude smaller than the best overlaps with vibrational levels of the four-photon excited states. We will therefore initially concentrate on ionization processes that involve four photon excitation of H_2 to the bound Rydberg states and only later discuss the role of three-photon resonances with the B-state.

Ponderomotive Effects

When photoelectron energy spectra are recorded with intense lasers and short pulse lengths we need to consider the effect that the ponderomotive forces can have on the measured photoelectron energy. We note that the pulse lengths utilized here are sufficiently small, and the interaction volume sufficiently large such that the ponderomotive energy is not recovered by slow photoelectrons. This occurs because electrons do not exit the laser focus during the time of the laser pulse (a 3 eV electron takes 1 ps to travel 1 μm). To prove this point we used our laser in a photoionization experiment in xenon for which numerous previous data exist for comparison. We show a typical photoionization spectrum of xenon recorded at 620 nm (~2 ps) in Figure 5. At this wavelength seven photons are required to reach above the zero-field ionization threshold. We have marked in the figure where photoelectrons were expected from non-resonant seven photon ionization to form $Xe^+(^2P_{3/2})$. We note that the observed electron energy peaks are shifted to lower energies by an amount that is consistent with the process[1-3] where a six-photon resonance of bound neutral xenon is shifted into resonance as the intensity of the field grows and is subsequently ionized by an additional photon. To the extent that the excited states that form intermediate resonances experience an AC-Stark shift that is similar to the ponderomotive shift of the free electron

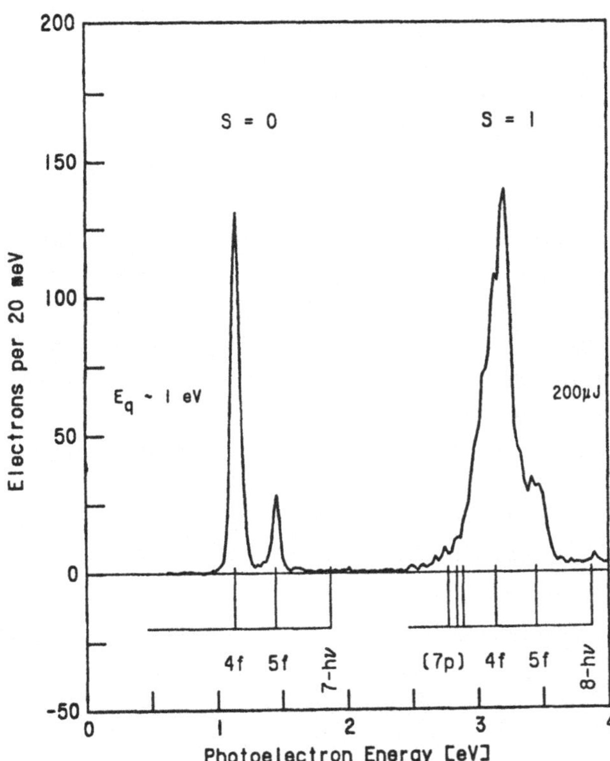

Figure 5. Photoelectron energy spectrum obtained in xenon at ~$3 \cdot 10^{13}$ W/cm^2. The pulse length is 2 ps and the wavelength is 620 nm. The photoelectron peak position expected from non-resonant 7 photon ionization to form $Xe^+(^2P_{3/2})$ is marked at the bottom of the figure. The peak energies in the 7 photon peaks as well as the higher ATI peaks are consistent with resonant excitation via the 4f and 5f resonance levels.

$$\varepsilon_p = 750\, I\, \lambda^2\ (\mathrm{cm}^{-1}) \tag{7}$$

the photoelectron's kinetic energy in the field is reduced from that of the non-resonant process

$$E_e = 7\ h\nu - IP \tag{8}$$

by an amount equivalent to ε_p. In equation (7) I is the laser intensity in units of 10^{12} W/cm^2 and λ is the wavelength in μm. The ponderomotive energy of equation (7) is converted into kinetic energy of the electron if the laser is on over a time span sufficient for the electron to "surf" off the focal volume[15] thereby converting the gradient of the ponderomotive potential into kinetic energy. If on the other hand the laser pulse is short compared to the time required to exit from the focal region, the ponderomotive energy remains with the electromagnetic field and the electron's energy is given by

$$E_e = 7\ h\nu - IP - \varepsilon_p\ . \tag{9}$$

The energy positions in the photoelectron spectra in Figure 5 indicate that the intermediate states 4f and 5f are the dominant contributors to the ionization process in xenon at this wavelength and that the AC-Stark shift of these states is quite similar to that given by equation (7) for the free electron. We note that no contributions from intermediates of the 7p and 8p manifolds of xenon appear and that this shifted resonance structure is reproduced in the higher order ATI peaks although the experimental resolution diminishes somewhat there.

Therefore we expect that under the laser pulse and focal properties employed in our experiment on molecular hydrogen the ponderomotive shifts of the free electron are not recovered from the electromagnetic field.

Wavelength Dependence

The absence of any marked wavelength dependence in the total photoionization yield indicates that either a non-resonant ionization process dominates over one that utilizes resonant intermediates or else that the manifold of excited states that participate in resonant intermediate formation is sufficiently dense as well as sufficiently broadened by the presence of the intense laser field such that excited state ro-vibrational resonance structures merge into a quasi-continuum. At least three processes can be identified that contribute to this broadening, and most likely all three participate with varying degree in our experiment.

For one, an energy broadening of the excited state will arise from the short lifetime of the excited state against one-photon ionization. Typical values for these lifetimes can be calculated using the generalized cross section formula given by Huestis.[16] A value of 200 fs is obtained for $n = 4$ at 10^{13} W/cm^2. It is quite evident that one-photon ionization of excited states should be quite efficient.

A second process responsible for broadening is related to the intensity dependent lifetimes of the molecular Rydberg states against photodissociation. We note here that this process involves the core electron of the Rydberg and we will discuss this process in detail in the next section.

A third factor is that even if no resonance appears at the four-photon energy at zero field, the AC-Stark shift of the excited states can shift rovibrational excited states into resonance at sufficient intensity, giving the appearance of continuous, wavelength-independent ionization. While the exact magnitude of the AC Stark shifts are not known, an order of magnitude estimate may be obtained from the ponderomotive shift (equation 7). At our wavelengths ε_p is about 100 mV per 10^{13} W/cm^2. The process of shifting intermediates into resonance is known to dominate atomic photoionization[1-3] at high fields and the example of xenon given in Figure 5 falls under this category. It it tempting to suspect a similar process in the molecular case but we must also consider the possibility of purely non resonant ionization.

If non-resonant ionization were an important contribution at our wavelengths then we would expect that the wavelength dependence of the electron energies formed in process (1) follows a five-photon dependence. This is contrary to what we find in the electrons' spectra. We show in Figure 6 the location of the dominant electron energies observed at the various wavelengths. It appears from the guiding lines drawn at a slope of one that most peaks fall onto segments that vary with the one-photon energy. Also - if the five-photon process occurs in a purely non-resonant fashion at the intensities utilized in this work we would expect that the variation in the ponderomotive shift of the photoelectrons that are created at widely different intensities should lead to a distribution of electron energies for each final state v^+ generated.

We conclude that the most likely ionization channels involve four-photon excited intermediates that are shifted into resonance at a specific intensity. The photoionization of the shifted state will lead to a photoelectron energy that shifts linear with the one-photon energy over a wavelength range over which the Stark shift can be accomplished by the intensities employed.

Four-Photon Resonance Formation

The existence of four-photon resonant intermediates can be used to develop a picture that can explain the observed shifting from ionization of excited H_2 to dissociation into neutral fragments and it can also convey the origin for the switching to lower energy photoelectron peaks as the intensity grows (see Figure 4).

One aspect of this picture involves the AC Stark shift of the Rydberg electron. At 80 TW/cm^2 the ponderomotive energy of a free electron amounts to ~6400 cm^{-1}, approximately three vibrational spacings of H_2^+. The magnitude of this shift is indicated in Figure 2 by the downward arrow. If the highly excited states of H_2 experience AC-Stark shifts like that of a free electron, then the excited state manifold will shift upwards with respect to the four-photon dressed ground state as the intensity grows.

It would thus appear that bound states that lie for zero field strength at energies below the four-photon energy can be shifted into resonance as the intensity grows. Ionization may then be envisioned to occur whenever the four-photon dressed ground state encounters a resonance with good transition strength. This picture similar to that generally accepted in the atomic ionization case.

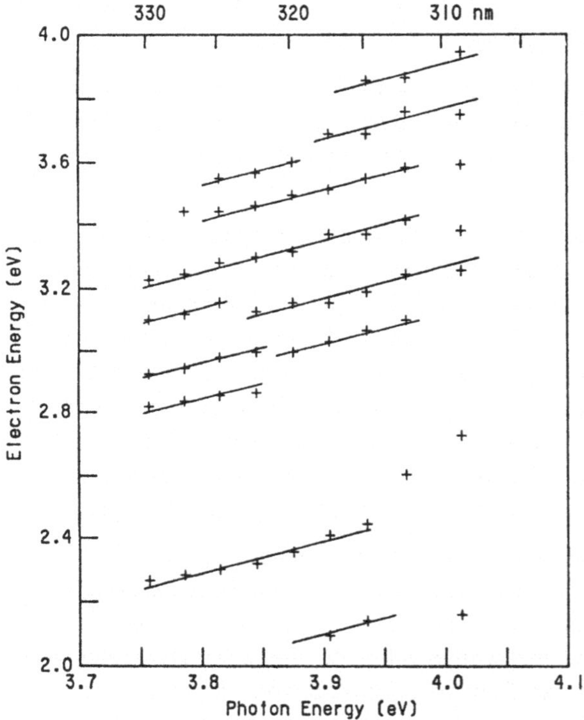

Figure 6. Energy of dominant features of the photoelectron spectra at wavelengths between 310 and 330 nm. Most peak positions follow short segments of slope of one, indicating that one-photon ionization of bound states that are AC-Stark shifted into resonance is the origin of these photoelectrons.

The open question is of course what is the actual Stark shift of the four-photon excited states at high field strength. At the wavelengths used here no bound transitions can be driven with one photon from Rydberg states with $n \geq 3$. A good first-order assumption may therefore be that the AC-stark shift of the higher Rydberg states is comparable to that of the free electron. However how will the core respond to the field? In the classical picture of a Rydberg state we may consider the ionic core as being separate from the Rydberg orbital. If we neglect electrostatic shielding of the core by the Rydberg orbital we should consider the response of the isolated core ion, H_2^+ to the field. This response has been the subject of several recent papers and we use the picture developed by Bandrauck et al[17] and Gusti et al[18] in the following discussion.

Dressed State Picture

We now combine the predicted response of the isolate core ion[18] with the picture of a Rydberg electron that experiences an AC-Stark shift as if it were a free electron. This leads to a potential energy diagram of adiabatic potential energy curves such as shown in Figure 7. The full curves give a simplified representation of the Rydberg states that originate from the dissociation limits H(1s)+H(n = 3) and H(1s)+H(n = 4), calculated for a field strength of 50 TW/cm^2. The unperturbed ground state of H_2^+ is indicated by the dashed line and the 4-photon energy at 326 nm is indicated at the left. The potential energy curves are the results of a Floquet calculation[18] that we performed for the diabatic singly and doubly excited states, $1s\sigma_g n\ell\lambda_g$ and $2p\sigma_u n\ell\lambda_g$, for 50 TW/cm^2. In this calculation the Rydberg states with n = 3 and n = 4 are considered separately, with no matrix element connecting them. In the dressed state picture the coupling between the singly excited gerade states and the doubly excited ungerade states leads to avoided crossings between the states {(n = 3)gerade+even photon number (N = 0)} and {(n = e)ungerade+odd photon number (N = 1)}, and likewise for 4g and 4u. The continuae 3,4g N ≥ 2 and 3,4u N ≥ 3, which interact with these states in higher order are omitted for reasons of clarity.

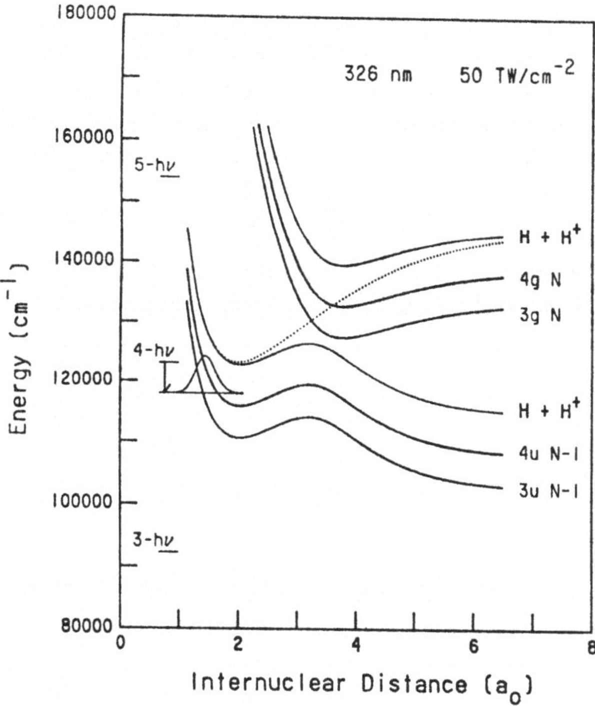

Figure 7. Results of a Floquet calculation for the n=3 and n=4 states of H_2 with the orbital configurations $1s\sigma_g n\ell\lambda_g$ and $2p\sigma_u n\ell\lambda_u$ at 50 TW/cm^2 and 325 nm. The ponderomotive shift for the location of the four photon energy relative to the unperturbed potential energy curve of H_2^+ (dashed curve) is also indicated.

As the intensity of the laser field increases a reduction of the number of bound levels occurs in the excited state manifolds and concurrent with this effect the four-photon resonant energy continually sweeps through the excited state manifold. In this fashion an intensity dependent competition between neutral dissociation and ionization occurs because with increasing intensity the higher vibrational levels in the Rydberg states will appear as continuum states that dissociate rapidly. The classical times for dissociation of molecular hydrogen is extremely short, it takes typically 10 to 20 fs to accelerate to distances of 10 Å. Consequently dissociation in competition with direct ionization of the neutral excited molecule is rather plausible in this adiabatic picture. Concurrent with opening the molecular bonds of the Rydberg, the shifting of states in energy at short internuclear distance leads to a progressively better overlap of the 4-photon dressed ground state with states from lower principal quantum number - thereby explaining the switching of the photoelectron features, to peaks corresponding to lower values of the principal quantum number n, as the intensity increases.

Model Calculation

Since specific vibrational levels of the molecular ion are rapidly photodissociated at the wavelengths used here, we should also consider the fate of the final ion state, and for that matter the fate of the molecular Rydberg core when subjected to intense fields. Photodissociation of the molecular ion proceeds through a transition from the $1s\sigma_g$ orbital to the repulsive $2p\sigma_u$ orbital.[19] The molecular Rydberg states share this very same core. To estimate the importance of photodissociation of the Rydberg (reaction 5), we approximate this cross section, σ_{diss}, by the photodissociation cross section calculated[12] for H_2^+. In Table I we compare these with the photoionization cross section for reaction (2), σ_{ion}, calculated using the generalized equation given by Huestis[16] with zero quantum defect.

The results in Table I show that at our wavelengths dissociation dominates over ionization for the higher vibrational levels. We conclude that Rydberg states with high vibrational excitation will follow path (5) rather than path (2). However, owing to the small quantum defects, the electron signature from ionizing excited atoms formed in process (5) using a sixth photon is indistinguishable from that from molecular ionization. Hence both dissociation followed by ionization of the excited atom (4), as well as direct ionization into H_2^+ are consistent with the energy spectra observed at high intensities.

Table 1. One-Photon Ionization[16] and Dissociation[19] Cross Sections for Molecular Rydberg States of H_2 at 326 nm in Units of 10^{-18} cm^2.

n	3	4	5	6	7
σ_{ion}	1.5	.36	.12	.047	.022
v	5	4	3	2	1
σ_{diss}	13	4.1	.23	.003	.000006

To simulate the intensity dependence of the photoelectron spectra we use a simple model that neglects contributions from a three-photon intermediate. The model assumes as bound states the ground state of $H_2(v = 0)$ and the four-photon excited Rydberg $H_2(n\ell\lambda,v)$

and as continuum states the products from reaction (2) and (5). Following Holt et al.[20] the loss of ground state hydrogen via a specific intermediate $|n\ell\lambda,v\rangle$ can, under our experimental conditions, be approximated by the time-independent rate:

$$\Gamma_{n\ell\lambda v} = \gamma_2\Omega^2/4\ [(4h\nu - E_{H_2(n\ell\lambda,v)} - \delta_2)^2 + \gamma_2^2/4\ ]^{-1} \tag{10}$$

Here γ_2 is the rate of loss due to ionization and dissociation:

$$\gamma_2 = (\sigma_{ion} + \sigma_{diss})\ \Phi\ , \tag{11}$$

where Φ is the photon flux. The ac-Stark shift of the Rydberg state, δ_2, is approximated by relation (7). We express the four-photon Rabi frequency, Ω, as

$$\Omega_{n\ell\lambda v} \propto \langle\chi_{v=0}|\chi_{v+}\rangle|^2\ n^{-3/2} \tag{12}$$

where the χ's represent the vibrational wavefunctions of the ground state of H_2 and of the core state of the Rydberg.

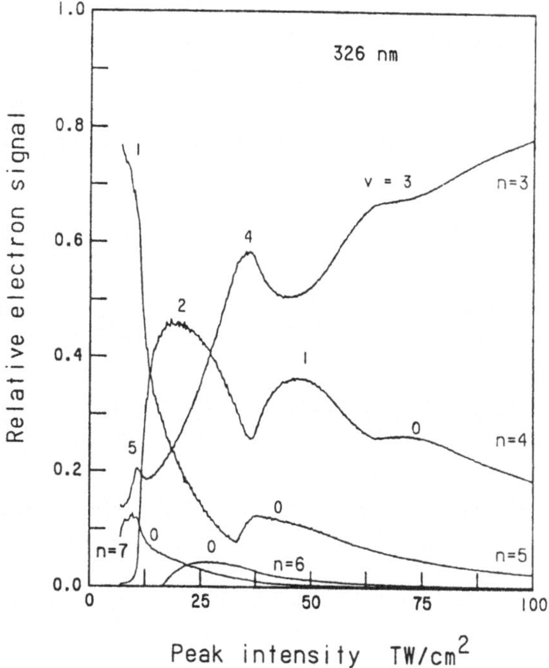

Figure 8. Calculated relative intensity of photoelectron peaks from ionization of the molecular Rydberg states and from photodissociation of the Rydbergs followed by ionization of the excited atom as a function of laser peak intensity. The n numbers refer to the principal quantum number of the Rydberg intermediate as well as to the electrons' energy group in Figure 2. The v labels along each curve indicate the intensity at which specific vibrational levels of each Rydberg are tuned through resonance. The data scatter is due to the finite stepsize (10^{11} W/cm^2) for the integration over the laser intensity profile.

We have used expressions (10-12) to calculate rates for the four-photon intermediates with core states $v^+ = 0 - 10$ for the manifolds $n = 3 - 10$. The quantum defects of the molecular hydrogen states were taken as zero. The rates have been folded with a spatially and temporally gaussian laser pulse profile. To simulate the electron energy spectra we evaluate the contributions from each Rydberg state (quantum number n) to the photoelectron signal at the energy $E_n = h\nu - R/n^2$. This energy is expected from reaction (2) for states with zero quantum defect and $v = v^+$. Relative intensities of the calculated signal at energy E_n are shown in Figure 8 for excitation at 326 nm.

The model predicts that at high intensity the photoelectron peak for n = 3 dominates over the contribution from n = 4 and n = 5 just as is observed in Figure 4. As the peak intensity is lowered the contribution from n = 4 first overtakes that from n = 3 and at yet lower intensity the contribution from n = 5 dominates, again consistent with what is found experimentally. Also predicted is a contribution from n = 7 at the lowest intensities, and this is indeed found in the top trace in Figure 4 where the signal at n = 7 exceeds that from n = 6. The prediction of the basic features of the observed spectra is taken as indication that the underlying model for explaining the photoelectron spectra is sound.

Why do more complex spectra appear at shorter wavelengths and at low intensities (see Figure 3)? At shorter wavelength we access higher vibrational levels in the B-state with better Frank-Condon factors. The non-Rydberg character of the B-state well allows transitions that are <u>not</u> diagonal in vibration to four-photon states and to H_2^+. Photoelectron energies will then reflect the change in vibrational quantum number <u>and</u> the (unknown) ac-Stark shift of the B state. The photoelectron spectra (as in Figure 3) can simplify as the intensity is raised for two reasons. First, the ac-Stark shift of the B state, if positive in energy, will shift vibrational levels with good Franck-Condon factors out of three-photon resonance - leaving open only the four-photon access to the Rydberg states. Second, the rate of four-photon excitation of the Rydbergs will eventually exceed the rate of three-photon excitation to the B-state as the intensity rises.

CONCLUSIONS

The wavelength and intensity dependence of photoelectron energies observed from molecular hydrogen at intensities above 10^{13} W/cm^2 and at wavelengths where five photons are required to reach the ionization continuum show that the most likely ionization channels involve four-photon excited intermediate states from the Rydberg manifolds with n = 3 - 7. While specific rovibrational assignment of these states cannot be given at this point (in part because little experimental information exists about hydrogen states at these high energies) this identification allows to develop a framework of the behavior of molecular Rydberg states in intense fields. This framework should be generally applicable to molecules in strong fields.

In molecular Rydberg states the effect of the AC-Stark shift on the Rydberg electron as well as on the core ion need to be taken into account. If the core ion possesses an electronic transition that is strongly driven by the applied electromagnetic field then the core's response to the field must play a major role in the response. In the framework of dressed states that are adiabatic in the electromagnetic interaction the core's response results in both a level shift (that is generally opposite to the direction of the ponderomotive shift of a free electron) and in a level broadening owing to predissociation and dissociation. The combination of this effect with the AC-Stark shifting of the Rydberg orbital provides a qualitative explanation of the observed switching from ionization to form H_2^+ to dissociation into neutral excited atomic hydrogen, as well as the observed switching to progressively lower quantum number states as the intensity grows. This simple discussion of a Rydberg electron and a core ion that respond independently of each other to the electromagnetic field remains to be tested in detail by theory and further experiments.

ACKNOWLEDGMENTS

This research was supported by the National Science Foundation under grant number PHY-90-24710 and by the Air Force Office of Scientific Research under Contract No F49620-88-K-0006.

REFERENCES

1. R. R. Freeman, P. H. Bucksbaum, H. Milchberg, S. Darack, D. Schumacher, and M. E. Geusic, Phys. Rev. Lett. **59**, 1092 (1987).
2. T. J. McIlrath, R. R. Freeman, W. E. Cooke, L. D. van Woerkom, and T. J. McIlrath, Phys. Rev. A **40**, 2770 (1989).
3. P. Agostini, P. Breger, A. L'Huillier, H. G. Muller, G. Petite, A. Antonetti and A. Migus. Phys. Rev. Lett. **63**, 2208 (1989).
4. J.H.M. Bonnie, J.W.J. Verschur, H. J. Hopman and H. B. van den Linden van den Heuvell, Chem. Phys. Lett. **130**, 43 (1986).
5. M. A. O'Halloran, S. T. Pratt, P. M. Dehmer, and J. L. Dehmer, J. Chem. Phys. **87**, 3288 (1987).
6. E. Y. Xu, T. Tsuboi, R. Kachru and H. Helm, Phys. Rev. A **36**, 5645 (1987).
7. M. J. Dyer and H. Helm, Optics Letters, submitted for publication.
8. T. Tsuboi, E. Y. Xu, Y. K. Bae, and K. T. Gillen, Rev. Sci. Instrum. **59**, 1357 (1988).
9. T. S. Luk and C. K. Rhodes, Phys. Rev. A **38**, 6180 (1988).
10. J.W.J. Verschuur, L. D. Noordam, and H. B. van Linden van den Heuvell, Phys. Rev. A **40**, 4383 (1989).
11. A. Zariyev, P. H. Bucksbaum, H. G. Muller, and D. W. Schumacher, Phys. Rev. A **42**, 5500 (1990).
12. S. Allendorf and A. Szöke, Phys. Rev. A **44**, 518 (1991).
13. L. J. Lembo, N. Bjerre, D. L. Huestis, and H. Helm, J. Chem. Phys. **92**, 2219 (1990).
14. P. Avan, C. Cohen-Tannoudji, J. Dupont-Roc, and C. Fabre, Le Journal de Physique **37**, 993 (1976).
15. P. H. Bucksbaum, M. Bashkansky, and T. J. McIlrath, Phys. Rev. Lett. **58**, 349 (1987).
16. D. C. Lorents, D. J. Eckstrom, and D. Huestis, "Excimer formation and decay processes in rare gases," Report No. MP73-2, Stanford Research Institute, Menlo Park, CA (September 1973).
17. A. D. Bandrauk and N. Gélinas, J. Chem. Phys. **86**, 5257 (1987).
18. A. Guisti-Suzor, X. He, O. Atabek, and F. H. Mies, Phys. Rev. Lett. **64**, 515 (1990).
19. G. H. Dunn, Phys. Rev. **172**, 1 (1968).
20. C. R. Holt, M. G. Raymer, and W. P. Reinhardt, Phys. Rev. A**27**, 2971 (1983).

HIGH INTENSITY MOLECULAR MULTIPHOTON IONIZATION

G. N. Gibson*, R. R. Freeman, and T. J. McIlrath*

AT&T Bell Laboratories
Holmdel, NJ 07733

INTRODUCTION

Multiphoton ionization (MPI) of molecules has been studied over a wide range of conditions which can be broadly divided into three general categories. First, resonantly-enhanced multiphoton ionization (REMPI) is accessible with moderate intensities ($10^9 - 10^{11}$ W/cm^2).[1] REMPI generally employs one or more narrow band lasers to excite various zero-field resonances in the molecule. Much detailed spectroscopic information can be obtained with this technique. Second, REMPI can be extended to higher intensities ($10^{11} - 10^{13}$ W/cm^2) with high-power short-pulse lasers. In this situation the potential curves responsible for the zero-field resonances begin to be affected by the intense laser field.[2] There is currently much interest in exactly how these potential curves are distorted by the strong laser field. Third, if there are no zero-field resonances in a molecule, MPI at high intensities still proceeds through resonance enhancement, although the resonances are entirely induced by the laser field, itself.[3]

In this paper, we investigate the third category using photoelectron spectroscopy in the short-pulse regime. This techinque is particularily well suited to studying laser induced resonances as was realized in atoms.[4] We have studied the dynamics of the MPI of the diatomic molecules nitrogen, oxygen and carbon monoxide.

The multiphoton electron energy spectra of atoms in the short-pulse regime up to intensities of 10^{15} W/cm^2 are now well understood in terms of multiple level crossings induced by the spatially and temporally varying laser field.[5] Applying this understanding of atoms to molecules would lead one to expect molecules to yield a very complex electron energy spectrum due to the multitude of potential curves and vibrational and rotational structure in the molecule. On the contrary, recently obtained ATI electron spectra of diatomic molecules show relatively simple features, some of which are surprisingly similar to the spectra of atoms and are straightforward to interpret.[3] The simplification from what one might expect occurs because of a $\Delta v = 0$ propensity rule associated with excitation out of Rydberg states and the fact that the valence states play a rather small, though observable role.

EXPERIMENTAL APPARATUS

The experiments discussed here were performed with 308 nm radiation produced with a CPM dye laser running at 616 nm and amplified to 1 mJ, 150 fs pulses at 10 hz. These pulses were doubled to 308 nm and again amplified to 1 mJ in an XeCl excimer laser. The electron time-of-flight (TOF) spectrometer[6] consists of a 1 m drift tube and a parabolic electron mirror biased at -5 V or -10 V allowing greater than 2π collection solid angle and an energy resolution of 30 meV at an electron energy of 2 eV. Spectra were binned according to laser energy into a bin width of ± 10% in energy. Sample pressures ranged from 5×10^{-7} to 7×10^{-5} torr to avoid the effects of space charge.[7] The spectra are all normalized to electrons/ev/shot/mtorr.

SHORT-PULSE PHOTOELECTRON SPECTROSCOPY

As with any photoelectron spectroscopy, the ionized electrons have a kinetic energy of $E_{elec} = mE_\gamma - E_{IP}$, where E_γ is the photon energy, m is an integer greater than or equal to the minimum number of photons required to ionize, and E_{IP} is the ionization potential of the atom in the laser field. The ionization potential in the field is simply given by the ionization potential at zero field plus the ponderomotive potential, $U_p = e^2E^2/4m\omega^2$ ($U_p = 9.33 \times 10^{-14} I[W/cm^2]\lambda^2[\mu m^2]$). The difference in a.c. Stark shift between the ground states of the atom and the ion is neglected. In the short-pulse regime the electron is not accelerated as it leaves the laser focus. As a result, the detected electrons will have an energy

$$E_{elec} = mE_\gamma - E_{IP}(\text{zerofield}) - U_P(I) \qquad (1)$$

where I is the laser intensity. Since U_p is a linear function of I, the electron energy gives a direct measurement of the absolute intensity when the electron was ionized provided the final state of the ion is known.

Although, for an arbitrary laser frequency, the atom has no resonances, the energy levels can be tuned into resonance by the a.c. Stark shift. A particular level will be in resonance at a well defined intensity and consequently will show up in the electron spectrum as a narrow feature at an energy given by Eq.(1).

In both atoms and molecules we can place states into two categories, Rydberg and valence, which behave rather differently in a strong optical field. In the limiting case of large principle quantum number, n, Rydberg states are completely decoupled from the ion core. As a consequence, the a.c. Stark shift is very well approximated by the ponderomotive energy, U_p. All the levels in a Rydberg state will stay together as they are shifted in energy by the laser field. In this special case, where the a.c. Stark shift is equal to the ponderomotive potential, the condition for resonance is that $U_p = m'E_\gamma - E$, where E is the zero field energy of the state above the ground state and m' is an integer. Eq.(1) then becomes, for Rydberg states in particular:

$$E_{elec}(\text{Rydberg}) = (m - m')E_\gamma - E_b \qquad (2)$$

where $E_b = E_{IP} - E$ is the binding energy of the state.

There is, however, a complication due to the molecular vibrational structure, which would prevent the application of high-intensity short-pulse photoelectron spectroscopy to molecules were it not for an important simplification. This vibrational structure would appear to produce a far more complex ATI spectrum in molecules compared to atoms, for there are many possible ionization paths for each molecular Rydberg state which is shifted into resonance. Each ionization path has a different E_{IP} and thus a different E_b. Figure 1 shows a diagram of several potential curves of N_2,[8] although the following discussion is perfectly general. Consider the vibrational level v^+ of the N_2^+ B final state and the vibrational level v' of a Rydberg state built on the B state. If α^+ and α' are the vibrational constants of these states the effective binding energy from the v' to v^+ state will be $E_b(v' \rightarrow v^+) = E_b(0 \rightarrow 0) + \alpha^+ v^+ - \alpha' v'$. This results in $E_{elec} = m\hbar\omega - E_b(0 \rightarrow 0) - \alpha^+ v^+ + \alpha' v'$. There are two simplifications which result in a single value for this energy: first, for high lying Rydberg states, α' converges to the ionic limit α^+ and typically differs by only a few tens of wavenumbers,[8] and, second, because the shape of the ionic and Rydberg potential curves are nearly identical,[9] the Frank-Condon overlap integral ensures that there will be a vibrational selection rule of $\Delta v = 0$ between the resonant and final states. The only possible electron energy is $m\hbar\omega - E_b(0 \rightarrow 0)$ independent of v', exactly as in the atomic case. This propensity rule of $v^+ - v' = 0$ is also seen in REMPI. However, generally in REMPI only one intermediate vibrational state is involved. In the strong field case many intermediate vibrational states can be brought into resonance, but each one contributes to the same photoelectron peak. This implies a difference from the atomic case where, as stated above, the energy of the photoelectron reveals the laser intensity at the time of excitation. In the molecular case, the multiple final states result in an ambiguity in the laser intensity producing photoelectrons. In addition to the vibrational degree of freedom, there are also rotations in a molecule. However, we neglect this because the energy scales characteristic of rotational motion is too small to be seen in these experiments.

Although in the above discuss we considered the limit of very high lying Rydberg states, it appears that the approximations used are still valid for the n=4 Rydberg states.

The behavior of the valence states in the strong field will be rather more complicated. The a.c. Stark shift will not have any simple relationship to the ponderomotive potential. Furthermore, the different angular momentum sublevels may not even shift together. The Frank-Condon overlap between the valence state and the final ionic state will not lead to

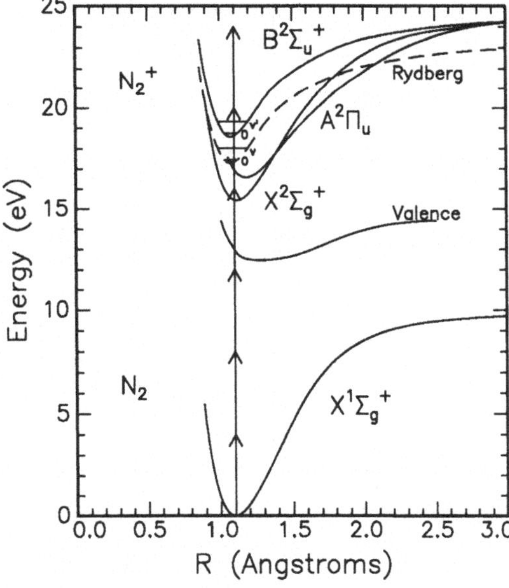

Fig. 1. Various potential curves of N_2 from Ref. 4. The arrows show the photon energy at 308 nm.

the simple selection rule $\Delta v = 0$, and, thus, a valence resonance may show vibrational structure. Finally, the a.c. Stark shift of a valence state may depend on the molecule's spatial orientation. All of these effects will lead to a smearing out of the valence state resonances into the photoelectron background.

NITROGEN DATA

Of the three gases, nitrogen, oxygen and carbon monoxide, nitrogen shows most clearly the behavior of a molecule in a strong 'non-resonant' laser field. Fig. 2 shows electron spectra of nitrogen using the 308 nm beam at four different laser energies. Because of the MPI nature of the interaction, the electron spectrum is duplicated every photon energy (see Eqs. (1) and (2)). In the lowest order MPI the electrons would come out with an energy between 0 and 4 eV. However, the spectrometer has a low energy cut-off at about 1 eV and, thus, we display the section from 2 to 6 eV, instead. In the low intensity limit U_p in Eq.(1) will be small and so the electrons will come out with energies slightly less than $mE_\gamma - E_{IP}$. In nitrogen there are three low lying ionization potentials corresponding to the removal of the $3\sigma_g$, $1\pi_u$, and $2\sigma_u$ electrons (see Fig.1), resulting in the ionic states $X^2\Sigma_g^+$, $A^2\Pi_u$, and $B^2\Sigma_u^+$, respectively. The value of $mE_\gamma - E_{IP}$ is marked in Fig.2 by the vertical line at the base of the arrows for each of these electrons. It is evident, in Fig.2a, that the three peaks in the spectrum are the result of ionizing to the three final states X, A, and B of the ion. This shows that any of the three outer electrons of nitrogen can be ionized with a strong field.

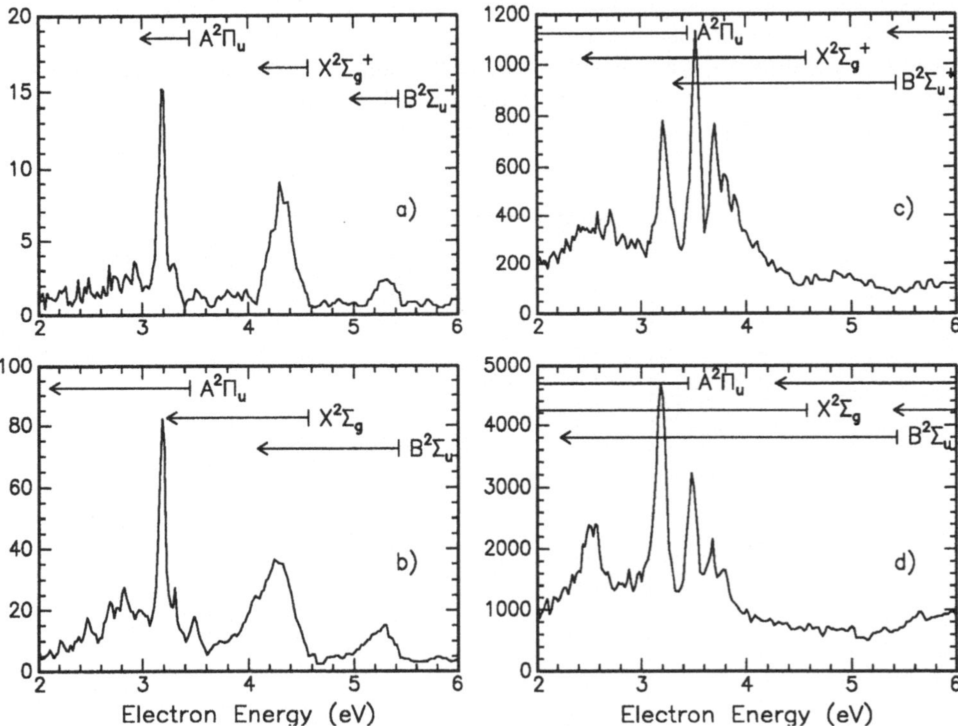

Fig. 2. Photoelectron spectrum of N_2 using 308 nm radiation at a) 5.2×10^{13} W/cm^2, b) 1.5×10^{14} W/cm^2, c) 2.4×10^{14} W/cm^2, d) 3.6×10^{14} W/cm^2, corresponding to ponderomotive shifts of 0.46 eV, 1.34 eV, 2.12 eV, 3.20 eV, respectively.

In order to make sure that all of the electron energy peaks came from the MPI of the neutral molecule and not the ion or perhaps a dissociated fragment, we measured the ion spectrum at the highest intensity used in these experiments. At this intensity there was no evidence for any dissociation as no N^+ was detected. Although some N_2^{2+} was detected, it was only about one tenth of the N_2^+ signal indicating the the electron spectrum really results from the MPI of the neutral molecule.

The spectrum in Fig.2a also shows two types of resonances: Rydberg and valence. The peak at 3.18 eV is due to the n=4 Rydberg state built on the N_2^+ $A^2\Sigma_g$ state. This resonance will be with four photons ($E_\gamma = 4.03$ eV) and, thus, the allowed states will have either np or nf character. The np states have a know quantum defect[9] of about 0.7 and would occur in the spectrum at 2.78 eV. The nf electrons will have essentially zero quantum defect, placing the n=4 peak at 3.18 eV. The n=5 peak of the np series can be seen, however, at 3.79 eV. The strength of the nf series relative to the np series supports the propensity rule for high angular momentum states previously observed in atomic ATI.

The a.c. Stark shift required to bring this state into resonance is, $4E_\gamma - (E_{IP}(1\pi_u) - E_b(4f)) = 0.27$ eV where $E_{IP}(1\pi_u)$ is the ionization potential of the $1\pi_u$ electron and $E_b(4f)$ is the binding energy of the 4f state. The laser energy required to just bring this peak into resonance was determined to be 13 μJ. This gives an absolute measurement of the ponderomotive potential vs. laser energy for our focusing conditions: 20 meV/ μJ. (The arrows on the figures show the calculated value of the peak ponderomotive shift for the specified laser energy. Resonant features arising from a given threshold must occur within the the electron energy space delineated by the arrow.)

The positions of the two other peaks in Fig.2a do not correspond to Rydberg resonances, and, thus must be due to valence states. Although the two peaks indicate ionization to different final states, they may be due to the same intermediate resonance state. It is difficult from just this data to determine the intermediate state, as there is some Stark shift. However, it may be a two-photon resonance with the shifted $a^1\Pi_g$ state.

The low intensity spectrum discussed above reveals the role of different types of resonances, both Rydberg and valence, in the MPI process. At high intensities the contribution of the Rydberg states begins to dominate the spectrum, as shown below. As the intensity is increased above that used in Fig.2a the Rydberg series based on the N_2^+ $X^2\Sigma_g^+$ state will be brought into resonance. The binding energy of the 5d state is 0.57 and photoelectrons will appear in the data at E_γ - 0.57 = 3.46 eV. The a.c. Stark shift required to bring this into resonance is 1.11 eV. Using the relationship between laser energy and a.c. Stark shift determined above, the 5d peak should come in at 55 μJ. Indeed, it is seen in Fig.2b taken at 67 μJ.

As the laser energy is further increased the next resonance to come in will be the Rydberg series built on the N_2^+ $B^2\Sigma_u^+$ with five photons at roughly 100 μJ, and Fig.2c shows the fully developed Rydberg series, although the intensity in Fig.2c was not quite high enough to reach the 4d resonance.

An important observation from a more complete model of electron spectra in the short-pulse regime[7] is that when a Rydberg series is brought through resonance with a given number of photons the relative height of the different members of the series will decrease rapidly with principle quantum number, n. In argon it was predicted that the peak heights would decrease almost an order of magnitude between successive states. In Fig.2c it is seen that this is not the case for the B series. This suggests that the laser intensity is sufficiently high to begin to saturate those resonances.

At the highest intensity, Fig.2d, the 3d peak of the B Rydberg series has come in at around 2.45 eV. Finally, we note that at this intensity the A state has still not had a chance to go through the full Rydberg series of resonances, and, thus, the spectrum is dominated by resonant MPI to the N_2^+ $B^2\Sigma_u^+$ final state.

OXYGEN AND CARBON MONOXIDE DATA

Like nitrogen, oxygen has no zero-field resonances with 308 nm radiation, and thus, is expected to show the same general behavior as nitrogen in a strong laser field. This is indeed the case, although the details of the photoelectron spectrum of oxygen are affected by its particular molecular structure. Fig.3a shows the photoelectron spectrum of oxygen taken at a relatively low intensity of 10^{13}W/cm^2. While there are many more peaks than in the corresponding N_2 spectrum (Fig.2a) all of the peaks can be identified with the different ionization potentials of O_2, just as with the N_2 spectrum. The electronic configuration of O_2 is $(1\sigma_g)^2(1\sigma_u)^2(2\sigma_g)^2(2\sigma_u)^2(1\pi_u)^4(3\sigma_g)^2(1\pi_g)^2$. Many low lying levels of the molecular ion are formed by the removal of the different outer electrons.

Again, although the low intensity spectrum is rather complicated and dependent of the specific molecular structure, the high-intensity spectrum is remarkably different, as seen in Fig.3b. Here, the photoelectron spectrum is dominated by the characteristic Rydberg series. This Rydberg spectrum is presumably composed of many overlapping series built on the different low-lying levels of the ion. Unlike N_2 it is not possible to sort out the contributions from the individual Rydberg series in O_2. However, the overall behavior of O_2 shows the same pattern as N_2: at low intensity a strong dependence on the

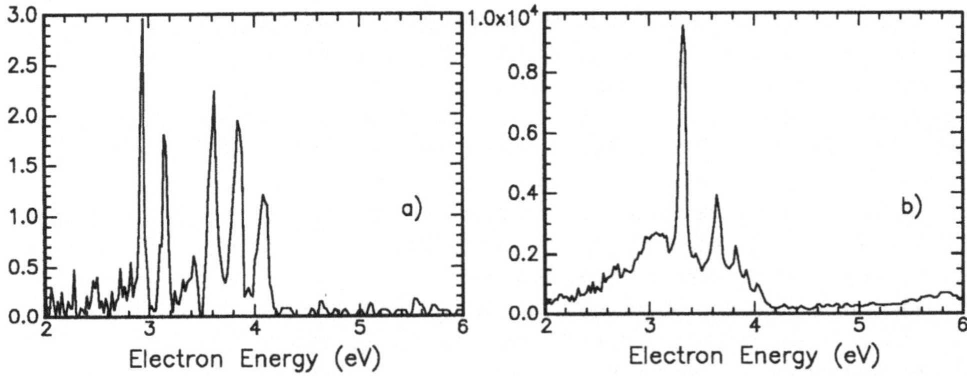

Fig. 3. Photoelectron spectrum of O_2 using 308 nm radiation at a) $1.8x10^{13}$W/cm^2, b) $2.8x10^{14}$W/cm^2, corresponding to ponderomotive shifts of 0.16 eV, 2.52 eV, respectively.

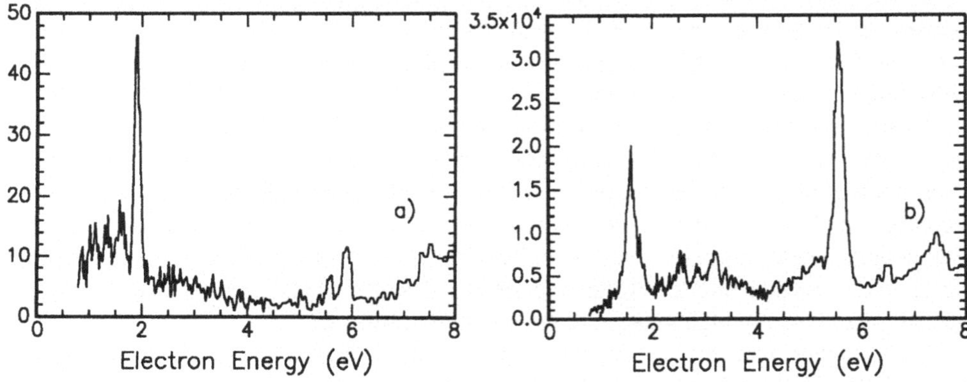

Fig. 4. Photoelectron spectrum of CO using 308 nm radiation at a) $3.2x10^{13}$W/cm^2, b) $2.9x10^{14}$W/cm^2, corresponding to ponderomotive shifts of 0.28 eV, 2.60 eV, respectively.

molecular structure is seen, while at high intensity the spectrum is dominated by a simple Rydberg series. Again, the possibility of atomic oxygen was ruled by measuring the ion spectrum.

Finally, we contrast the general behavior of N_2 and O_2 with CO. CO has two states ($A^1\Pi$, $I^1\Sigma^-$) which have a zero-field two-photon resonance with 308 nm radiation, within the bandwidth of the laser (3 nm). These conditions fall within the second category discussed in the introduction. Figs.4a and 4b show the photoelectron spectrum at two intensities in the range used for N_2 and O_2. For CO the spectrum consists of a single photoelectron peak corresponding to four photon ionization through the zero-field two-photon resonant intermediate state. As the intensity is increased no additional lines appear in the spectrum. This implies that the resonant ionization is so strong that essentially all the molecules are ionized at low intensity and none survive to the higher intesities. Thus, the Rydberg series characteristic of the high intensity spectrum of N_2 and O_2 never has a chance to be seen in CO.

CONCLUSION

In this paper we explored a regime of molecular multiphoton ionization where the molecule has no zero-field resonances with the laser. Under these conditions, we find relatively simple and general behavior in which the low intensity spectrum is sensitive to the molecular structure while the high intensity spectrum results in a simple atomic-like Rydberg series. We showed that this results from a cancellation of the vibrational structure in the photoelectron spectrum and a dominance of Rydberg resonances over valence state resonances at high intensities.

ACKNOWLEDGMENTS

This work was partially supported by AFOSR grant 88-0018.

*Also at University of Maryland.

REFERENCES

1. For some recent examples, see E. F. McCormack, S. T. Pratt, J. L. Dehmer, and P. M. Dehmer, Phys. Rev. A **42**, 5445 (1990); A. Fujii, T. Ebata, and M. Ito, J. Chem. Phys. **88**, 5307 (1990).
2. A. Zavriyev, P. H. Bucksbaum, H. G. Muller, and D. W. Schumacher, Phys. Rev. A **42**, 5500 (1990).
3. G. N. Gibson, R. R. Freeman, and T. J. McIlrath, to appear in PRL, 2 Sept 1991.
4. R. R. Freeman, P. H. Bucksbaum, H. Milchberg, S. Darack, D. Schumacher, and M. E. Geusic, Phys. Rev. Lett. **59**, 1092 (1987); P. Agostini, P. Breger, A. L'Huillier, H. G. Muller, and G. Petite, Phys. Rev. Lett **63**, 2208 (1989).
5. For a recent review, see R. R. Freeman and P. H. Bucksbaum, J. Phys. B: At. Mol. Opt. Phys. **24**, 325 (1991).
6. D. J. Trevor, L. D. Van Woerkom, R. R. Freeman, Rev. Sci. Inst., **60**, 1051 (1989);
7. T. J. McIlrath, P. H. Bucksbaum, R. R. Freeman and M. Bashkansky, Phys. Rev. A, **35**, 4611 (1987); L. D. van Woerkom, R. R. Freeman, W. E. Cooke, T. J. McIlrath, J. Modern Optics, **36**, 817 (1989).
8. A. Loftus and P. H. Krupenie, J. Phys. Chem. Ref. Data **6**, 113 (1977).
9. E. Lindholm, Arkiv for Fysik **40**, 97 (1968).

COMPETITION BETWEEN MULTIPLE IONIZATION AND DISSOCIATION OF DIATOMIC MOLECULES

D. Normand, C. Cornaggia and J. Morellec

Centre d'études de Saclay
Service des Photons, des Atomes et des Molécules, Bat.522
91191 Gif/Yvette France

INTRODUCTION

The molecular response to intense radiation field is a subject of growing interest. For diatomic molecules, the Multielectronic Dissociative Ionization (MEDI) has recently attracted a lot of attention[1-15]. Experimental works have been devoted to N_2 (ref.1-4), HI (ref.5), H_2 (ref.5), CO (ref.7-10) and O_2 (ref.11). Except for H_2, the interaction leads to the formation of transient multicharged molecular ions which dissociate into energetic fragments.To understand the molecular response to the strong laser field excitation, we have to identify the transient molecular ions, we have to know at what time of the interaction they are created and what is their potential energy.

PRINCIPLE OF THE EXPERIMENT

The experimental procedure consists in two steps: the ion energy measurements and the identification of the dissociation channels.

1) Ion Energy Measurements

Firstly we have to determine the mass, charge and energy of the ionic fragments issued from the MEDI processes. The ion energy measurements are based on the observation that the molecules whose internuclear axis is aligned with the laser polarization vector ε are preferentially ionized (Fig.1).

If ε is perpendicular to the detection axis (Fig.1a), then up and down ions have symmetric trajectories: they arrive at the same time at the detector. The width of the ion peak then reflects the ion angular distribution.

Coherence Phenomena in Atoms and Molecules in Laser Fields
Edited by A.D. Bandrauk and S.C. Wallace, Plenum Press, New York, 1992

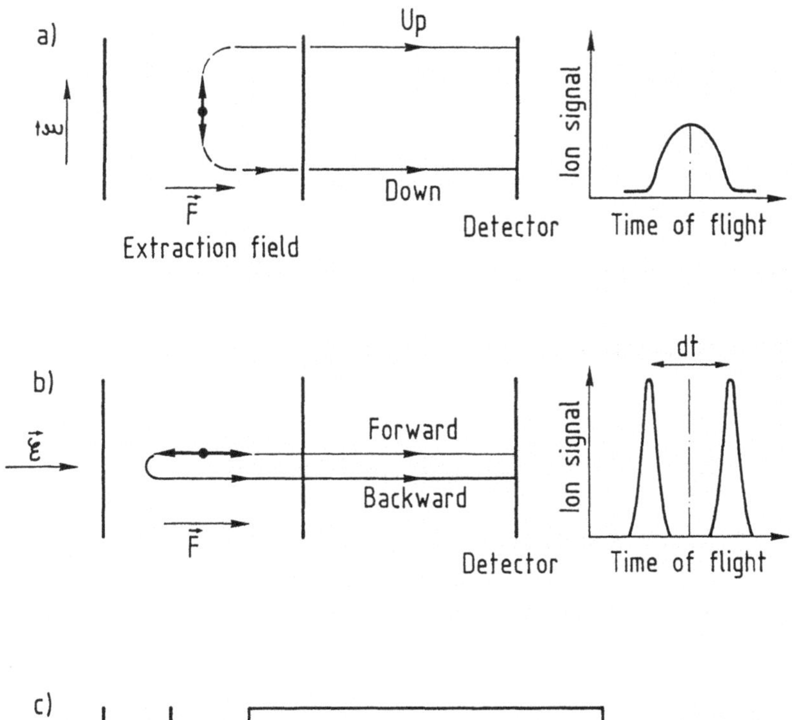

Fig.1. Schematic ion collection set up showing the ion trajectories and the corresponding TOF ion spectra when the laser polarization vector ε is: a) perpendicular b) parallel to electric extraction field F. Fig.1c shows the real arrangement with the double chamber ion spectrometer.

If ε is then set parallel to the detection axis (Fig.1b), each ion fragment energy class will result into two ion peaks on the TOF spectrum: the "forward" component in the TOF scale is related to the creation of fragments with an initial momentum directed towards the detector. The "backward" one represents fragments initially headed in the opposite direction and whose momentum has been reversed by the extraction electric field. The time interval δt separating two associated peaks allows us to determine the initial kinetic energy E_0 of the fragments according to the relation:

$$E_0 = e Z^2 F^2 \delta t^2 / 8M \qquad (1)$$

in which e denotes the electron charge, Z and M the charge and the mass of the fragment, and F the collection electric field in the interaction chamber.

2) Determination of the Dissociation Channels

Secondly, we have to associate by pairs the ionic fragments which proceed from the same (transient) parent ion. The "marriage" of the fragments A^{p+} and B^{q+} is based on different criteria:

a) First of all, due to energy and momentum conservations, the product of the mass by the energy must be the same for the associated fragments, that is:

$$\text{mass}(A)\ \text{energy}(A^{p+}) = \text{mass}(B)\ \text{energy}(B^{q+}) \qquad (2)$$

This condition is absolutely necessary but it is not sufficient, since it may happen that different dissociation channels yield similar ion energies (within the experimental uncertainty of +-10%).

b) We also want the abundances of the associated fragments to be the same (accounting for the detector sensivity which may be different for different ion species[16]). Obviously, each time an $AB^{(p+q)+}$ molecular ion dissociates, one A^{p+} and one B^{q+} fragments are formed.

c) Moreover we must check that the associated fragments present the same dependences on the laser intensity which should be indeed the intensity dependence of the parent ion.

d) Finally, it can be checked that both ionic fragments are formed during the same laser shot: that is the smart idea of the "covariance mapping" technique developped by the English group of Reading [7].

In the present experiments, we only use the first three criteria. Indeed the covariance mapping technique demands a very high ion collection efficiency only attainable using very short TOF tubes (25mm) . With such short tubes the ion TOF spectra show an overlap of the different ion species leading to a poorer ion energy resolution. The ideal experiment should be equipped with two TOF tubes: a short one adequate for the covariance mapping and a long one suitable for an accurate determination of the different ionic fragment energies.

EXPERIMENTAL SET UP

1) The Laser System

The experimental arrangement, including laser system, has already been described in detail elsewhere [4]. Briefly, the basic components of the experimental set up are the followings. The pulses delivered by a synchro-pumped dye laser are amplified through a four stage dye amplifier pumped by a Nd-YAG laser operating at 10 Hz. The measurements of the laser pulse characteristics give a pulse energy of a few mJ and a pulse duration of 2 ps at 610 nm. The 305 nm laser radiation is obtained after frequency doubling in a BBO cristal (conversion efficiency 15%) which shortens the laser pulse duration to 1.4 ps at this wavelength. The laser light is usually focused by a spheroparabolic lens of 60 mm focal length leading to maximum peak powers of 10^{15} W/cm^2 at 305 nm and 5 10^{15} W/cm^2 at 610 nm. Alternatively a 150 mm lens can be used to record data with a larger interaction volume (allowing lower laser intensities), in order to determine the laser intensities for which the different ion species appear (appearance intensities).

The 100 fs experiment has been performed at the Laboratoire d' Optique Appliquée (Palaiseau). There, the laser is based on a colliding pulse mode-locker oscillator amplified in a 5-stage dye amplifier pumped by a single-mode Nd-YAG laser[17].

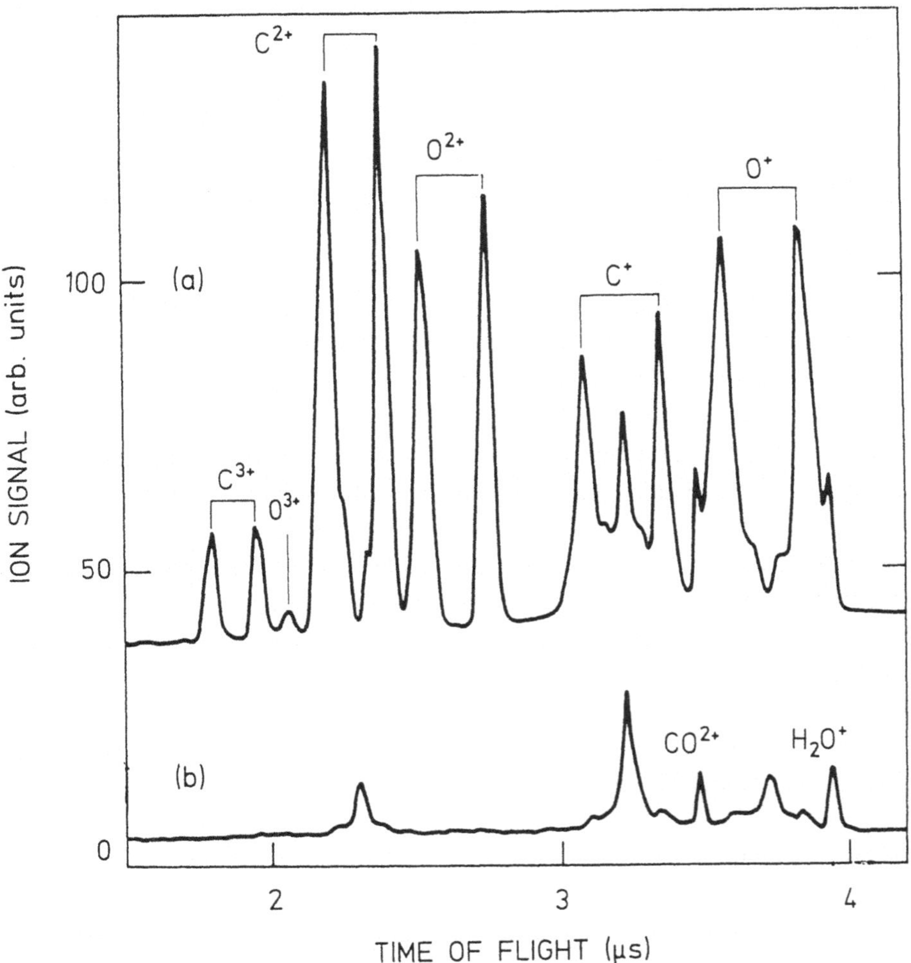

Fig. 2. Two ion TOF spectra of CO, recorded at 615 nm, for a laser intensity of 2.5 10^{15} W/cm^2. The laser polarization vector is a) parallel b) perpendicular to the detection axis.

2) The Double Chamber Ion Spectrometer

The detection system consists in a double chamber TOF ion mass spectrometer pumped at a background pressure of 10^{-9} Torr [4] (Fig.1c). The application of a weak collection electric field in the interaction chamber makes possible to determine the fragment kinetic energies with a good precision (10%) whereas the second chamber strongly accelerates the ions to ensure a good collection efficiency. Finally, the ions are mass separated through a 18 cm long drift tube and give a signal which feeds a Lecroy model 9400 (or 9450) transient digitizer with a 100 MHz (respectively 400 MHz) sampling frequency.

Two typical ion spectra are shown on Fig.2. They have been obtained by submitting the CO molecules to a laser intensity of 2.5 10^{15} W/cm^2 at 615 nm, with 100 fs pulse duration[15]. In Fig.2a, the laser polarization vector ε is set parallel to the detection axis; the forward-backward splittings are nicely resolved for all fragments. The only missing peak is the O^{3+} backward component which lies under the strong C^{2+} forward peak. In contrast, when ε is turned at 90 degrees (Fig.2b), most fast ions are lost: only ions with almost zero kinetic energies (C^{++}, C^{+}, CO^{++}, O^{+} and H_2O^{+}) can hit the detector. They show up as single peaks located right in the middle of the corresponding forward-backward peaks of Fig.2a.

ESSENTIAL EXPERIMENTAL FINDINGS

With this experimental technique, we have investigated the MEDI processes in N_2, CO and O_2 at two laser wavelengths (305 and 610 nm), over a wide range of laser intensities (10^{13} W/cm^2 to 5 10^{15} W/cm^2) and for different laser pulse durations (100 fs to 2 ps) [15]. The amount of data now available makes possible to bring out the essential features of the MEDI processes.

1- Selectivity

When compared with other excitation techniques such as electron bombardement or synchrotron radiation absorption, the laser excitation proves to be very selective: in most cases, a given dissociative channel such as

$$AB^{(p+q)+} \longrightarrow A^{p+} + B^{q+} + E_0 \qquad (3)$$

yields a single value of kinetic energy release E_0. For example, the laser induced double ionization of CO at 610 nm yields only one energy (5.7 eV) whereas the excitation of CO by synchrotron radiation[16] or by electron impact[17] open 5 and 3 channels, respectively. This selectivity must be underlined because the laser wavelengths investigated have not been chosen to be resonant.

2- Ionization is sequential

Ion spectra recorded for increasing laser intensities show that the multiple ionization processes are sequential. The first step is the formation of the molecular ion AB^+ whose ion yield saturates around 10^{13} W/cm^2. The singly charged fragments appear at an intensity of a few times 10^{13} W/cm^2: those fragments can only proceed from the dissociation of the molecular ions since the interaction volume contains no more neutral molecules at this intensity level.

The appearance laser intensities for the different ionization stages at 610 nm are indicated below. They are nearly identical for N_2 (ref.4), CO (ref.10) and O_2 (ref.11).

		appearance laser intensity
AB^+	$\dashrightarrow A^+ + B$	10^{13} W/cm^2
AB^{2+}	$\dashrightarrow A^+ + B^+$	$3\ 10^{13}$ W/cm^2
AB^{3+}	$\dashrightarrow A^{2+} + B^+$	$5\ 10^{13}$ W/cm^2
AB^{4+}	$\dashrightarrow A^{2+} + B^{2+}$	$8\ 10^{13}$ W/cm^2
	$\dashrightarrow A^{3+} + B^+$	10^{14} W/cm^2
AB^{5+}	$\dashrightarrow A^{3+} + B^{2+}$	10^{15} W/cm^2

The multicharged molecular ions are transient ions: except for some doubly charged ions, they are not detected in the TOF spectra. The multiple ionization stages occur successively, at increasing laser intensity. Surprisingly, the intensity range over which one passes from two to four electron removal is rather narrow (only a factor of 3). Conversely, the fifth electron removal is observed only at much higher laser intensity (10^{15} W/cm^2).

3- Fragment Energies are Sharp Peaks

During the fragmentation, the potential energies of the multicharged molecular ions are converted into the kinetic energies of the fragments. Measuring the fragment energies thus gives access to the potential energy of the parent molecular ion, provided the dissociation channels are identified.

Astonishingly enough, the forward-backward ionic fragments appear as sharp peaks in the TOF spectra , implying that the transient ions have well defined energies. As an illustration, Fig.3b shows the ion TOF spectrum of nitrogen, obtained at 615 nm for an intensity of 10^{14} W/cm^2 and 100 fs pulse duration: we observe essentially the backward-forward splittings associated with two ion classes, N^+ ions with an energy of

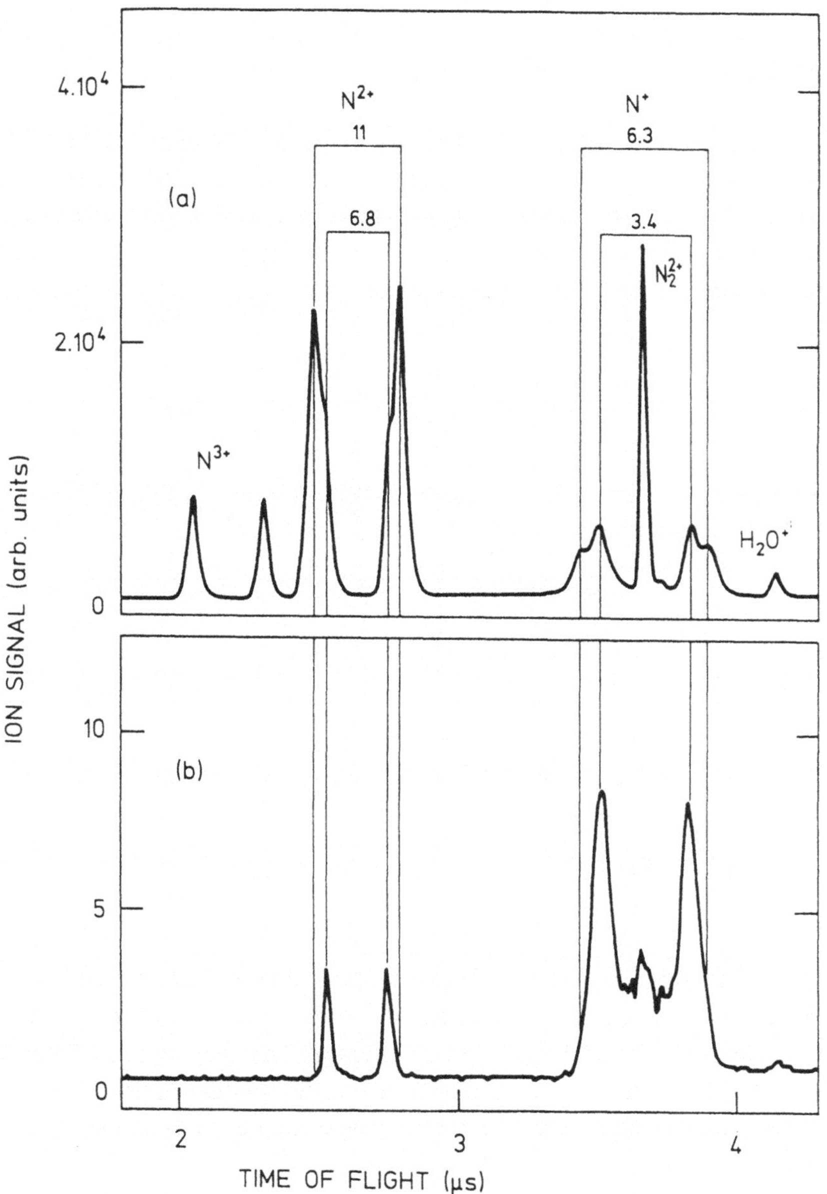

Fig. 3. Two ion TOF spectra of N_2, recorded with the laser polarisation vector parallel to the detection axis, at 615 nm, for a laser intensity of a) 2.5 10^{15} W/cm^2 b) 10^{14} W/cm^2. The figures are the kinetic energies (eV) of the fragments deduced from the backward-forward splittings.

3.4 eV and N^{++} ions with an energy of 6.8 eV. All peaks are indeed very sharp. This point deserves to be emphasized: it suggests that the successive ionization steps cannot proceed only via coulombic repulsive states, otherwise the molecular ion energies would consist of broad bands rather than of sharp peaks.

4- Fragment Energies are Independent of the Laser Intensity

Another important feature of the MEDI processes is the independence of the fragment energies on the laser intensity. There is a minimum intensity required for a given dissociation channel to be observed, but when the intensity is increased beyond this appearance intensity, even far beyond, the kinetic energy release remains unchanged. In other words, the fragments energies appear to be "robust" against the laser intensity.

Table 1. Comparison of the kinetic energies released by the dissociation of the doubly charged molecular ions O_2^{++}, $N_2^{++}$and CO^{++}, at 305 and 610 nm.

Laser wavelen. \ Molecule	O_2	N_2	CO
305nm	4.4	5.4	4.2
610nm	4.7	7	5.7

The comparison of Figs.3a and 3b shows clearly that the slow N^+ and N^{++} ions observed at 10^{14} W/cm^2 are still present at 2 10^{15} W/cm^2. At the higher intensity, fast N^+ and N^{++} ions, as well as N^{3+} ions appear, revealing the opening of additional dissociation channels. The branching ratios between the different dissociation channels depend on the laser intensity: as the intensity is increased, ions with higher charge states and higher kinetic energies are favored.

5- Influence of the Molecular Structure

Table 1 gives evidence of the role of the molecular structure in the MEDI processes investigated in the 5 10^{13} - 5 10^{15} W/cm^2 laser intensity range. There, we list the kinetic energies released in the dissociation of the molecular ion AB^{2+} for the six situations experimented, that is CO, N_2 and O_2, at 305 and 610 nm, with 2 ps laser pulse duration. The energies differ from one molecule to the other, thus proving that the molecular structure still plays a role at that intensity level. In addition, we observe that, except for oxygen, the energy releases in the dissociation of a given molecule, are different at 305 and at 610 nm, indicating that different states of AB^{2+} are populated.

6- Influence of the Laser Wavelength

The influence of the laser wavelength on the MEDI processes[4,10,11] is very important. Indeed, comparison of data recorded at 305 and 610 nm at the same intensity, show that, at 305 nm:

i) the fragmentation rate is much larger (for all three molecules studied),

ii) the stage of ionization reached is lower (for N_2 and CO),

iii) the energy released by a given dissociative channel is lower (for N_2,CO), (Table 1).

7- Influence of the Laser Pulse Duration

The MEDI of N_2, CO and O_2 has been investigated around 600 nm, with different pulse durations ranging from 100 fs to 2 ps. Comparison of data shows that the kinetic energy released in a given dissociative channel, does not depend on the laser pulse duration (see for example Table I, II and III of ref.15). This observation proves that the molecular dynamics is, to a large extent, uncorrelated with the laser pulse risetime. This is a surprising result if we think of the successive ionization steps as transitions from one repulsive state (such as CO^{3+} for instance) to another repulsive state (CO^{4+} in that case). Indeed, in the absence of bound states to impose the transition internuclear distances, R, (to go from CO^{3+} to CO^{4+}), we expect R to depend on how fast the laser pulse can provide the intensity required to ionize CO^{3+}. This remark shows the limits of the Coulomb explosion model.

DISCUSSION

1- What MEDI is Not

We know that the MEDI processes do not proceed via Coulombic states, at the equilibrium internuclear distance R_0 of the neutral molecule. If it were so, the potential energy E_{pq} of the molecular ion $AB^{(p+q)+}$ dissociating into A^{p+} and B^{q+} would be given by

$$E_{pq}\,(eV) = 14.4\ p.q\,/\,R_0 \qquad (4)$$

where R_0 is expressed in Angstroems. Thus, for instance, for the CO^{3+} (with $p=1$, $q=2$) and CO^{4+} (with $p=2$, $q=2$) molecular ions, the energies released would be 26 and 52 eV respectively whereas we measure only 11 and 19 eV for these dissociative channels.

We also know that the multicharged atomic ions detected cannot proceed from the subsequent ionization of neutral fragments nor of singly charged fragments. To illustrate that point, let us suppose for instance that the molecular ion AB^{2+} fragments into

A^+ and B^+ , these atomic ions being subsequently ionized into A^{2+} and B^{2+}. Would such a process occur, we would detect different charge states of the same atom (A^+, A^{2+},A^{3+}) with identical energies, which is generally denied by the experimental data.

2- What MEDI Could Be

a) First approach: the TFD model.

The Thomas-Fermi-Dirac (TFD) model developed by Brewczyk and Frasinski [13,14] is presented in the present volume. For the sake of consistency, let us briefly mention that it is a statistical model in which the molecule is treated as an electron gas and the laser field as a static electric field. The essential conclusions of this model are:

i) the molecular ions can lose additional electrons during their dissociation, that is at increasing internuclear distances;

ii) the ultimate charge state of the molecule is function of the laser intensity. The predicted appearance intensities are found to be larger than the measured values by about a factor of 5 for the first stages of ionization. For higher charged states, the agreement might be better, but the identification of the highly charged molecular states in the experiment is uncertain due to the weakness of the signals. A more significant test of the model would require additional experiment performed at laser intensities in the range of 10^{16} W/cm2 and over.

iii) the charge symmetric fragmentation channels are strongly favored. This prediction is in good agreement with the experiments performed both at 305 nm [20] and 610 nm. Only the experiment performed at 248 nm is interpreted in terms of asymmetric channels [3].

In conclusion, in view of the complexity of the physics involved, this simple model gives quite promising results. Unfortunately, the TFD model is unable to predict the kinetic energies of the various fragment ions and the considerable differences observed for each channel at 305 and 610 nm. Moreover, the molecular specificity of the MEDI process is clearly ignored.

b) Suggestions for a more realistic model.

We now tentatively define the frame of a model that could reproduce the essential features of the MEDI processes.

First of all, the independence of the fragment energies on the laser pulse risetime suggests that the ionization occurs at the equilibrium distance of the neutral molecule (Fig.4).

Secondly, the selectivity of the MEDI processes, as well as the robustness of fragment energies against the laser intensity indicate that the ionization proceeds via resonance states; but since the molecular levels are unavoidably shifted by the laser field, these resonances must be laser tuned resonances. In this case, since the resonance signals stand out over the non resonant background, the experiment selects the resonant events which occur at those intensities which adjust the resonances, whatever the peak intensity. Such a situation is met in the multiphoton ionization of atoms under strong laser fields conditions[21,22]. The role of the molecular Rydberg states which are ac Stark shifted into resonances has also been reported very recently on N_2 (ref.23) and H_2 (ref.24).

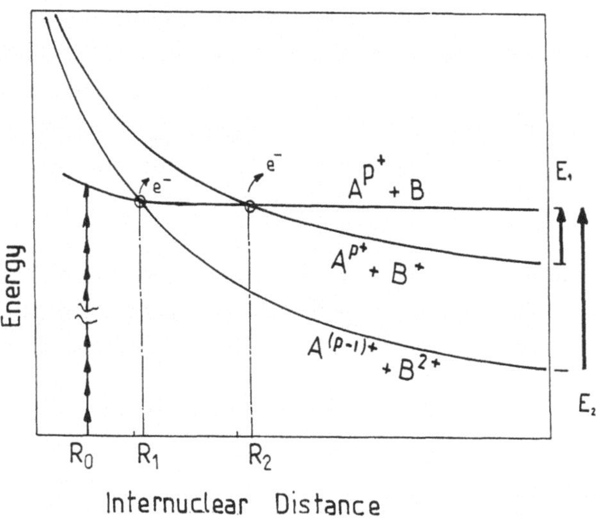

Fig. 4. Schematic energy diagram showing the different steps proposed to reproduce the essential features of the MEDI processes. 1) formation of charge asymmetric molecular states, through ac Stark shifted resonances, at the equilibrium internuclear distance R_0 of the neutral molecule. 2) charge transfers and electron emissions at the crossing points with Coulombic curves. E_1 and E_2 are the kinetic energy releases accessible to experiment.

Thirdly, the fact that the fragment energies are small (compared with those expected from a vertical Coulomb explosion) implies that those resonant molecular states are "flat" states, such as [A^{p+} + B] (the brackets mean that it is a molecular ion that dissociates into the fragments A^{p+} and B at infinity). The ionic fragments detected being predominantly charge symmetric, we must consider that, during the fragmentation processes, charge transfers such as:

$$[A^{p+} + B] \dashrightarrow [A^{p+} + B^{+}] + e^{-} \quad (5)$$

$$\dashrightarrow [A^{(p-1)+} + B^{2+}] + e^{-} \quad (6)$$

occur, due to curve crossings of the "flat" states with the Coulombic states. Interestingly, this hypothesis implies that the energy releases (E_1, E_2 in Fig.4) will take precise values, since the crossing points (occurring at R_1, R_2) fix accurately the potential energy of the transient molecular ions. Note that it is necessary to suppose an electron emission at the crossing points, otherwise curves with the same total number of charges do not cross.

CONCLUSION

The amount of data presently available on the interaction of the N_2, CO and O_2 molecules with intense laser fields, has made possible to bring out the essential features of the MEDI processes. Amazingly, the data show a challenging simplicity: i) very few dissociation channels are opened, ii) the kinetic energy releases are sharply peaked; iii) the fragment energies are independent of the laser pulse risetime, independent on the peak laser intensity, but they strongly dependent on the laser wavelength and on the molecule.

As far as the interpretation of the experiments is concerned, there is obviously a strong need for theoretical help. We hope that the very naive description that we have suggested, based upon our experimental results, will help defining a more realistic approach.

ACKNOWLEDGMENTS

The authors wish to express their gratitude to Pr A. Giusti and F. Mies for very stimulating discussions.

REFERENCES

1-Frasinski L.J., Codling K., Hatherly P.A., Barr J.R.M., Ross I.N. and Toner W.T., Phys. Rev. Lett. **58**, 2424 (1987)

2-Frasinski L.J., Codling K. and Hatherly P.A., Phys. Lett. A **142**, 499 (1989)

3-Boyer K., Luk T.S., Solem J.C. and Rhodes C.K., Phys. Rev. A **39**, 1186 (1989)

4-Cornaggia C., Lavancier J., Normand D., Morellec J. and Liu H.X., Phys. Rev. A**42**, 5464 (1990)

5-Codling K., Frasinski L.J., Hatherly P. and Barr J.R.M., J. Phys. B **20**, L525(1987)

6-Codling K., Frasinski L.J. and Hatherly P.A., J. Phys. B **21**, L433 (1988)

7-Frasinski L.J., Codling K., Hatherly P.A., Science **246**, 1029 (1989)

8-Codling K., Frasinski L.J.,Hatherly P.A. and Stankiewicz M., Physica Scripta **41**, 433 (1990)

9-Hatherly P.A., Frasinski L.J., Codling K., Langley A.J. and Shaikh W., J.Phys.B **23**, L291 (1990)
10-Lavancier J., Normand D., Cornaggia C.,Morellec J. and Liu H.X., Phys. Rev. A **43**, 1461 (1991)
11-Normand D., Cornaggia C., Lavancier J., Morellec J. and Liu H.X., Phys. Rev.A**44**, 475 (1991).
12-Normand D., Cornaggia C., Lavancier J., Morellec J. Proceedings of the 5th conference on Multiphoton Processes, Paris (1990)
13-Codling K., Frasinski L.J. and Hatherly P.A., J. Phys. B **22**, L321 (1989)
14-Brewczyk M. and Frasinski L.J., J. Phys. B**24** L307 (1991)
15-Cornaggia C., Lavancier J., Normand D., Morellec J., Agostini P., Chambaret J.P. and Antonetti A., Phys. Rev. A (1991 accepted for publication).
16-Schram B.L., Boerboom A.J.H., Kleine W. and Kistemaker J., Proceedings of the 7th Int. Conf. on Phenomena in Ionized Gases, Belgrade (1965)
17-Agostini P., Breger P., L'Huillier A., Muller H.G., Petite G., Antonetti A. and Migus A., Phys. Rev. Lett. **63**, 2208 (1989)
18-Lablanquie P., Delwiche J., Hubin-Franskin M.-J., Nenner I., Morin P., Ito K., Eland J.H.D., Robbe J.-M., Gandara G., Fournier J. et Fournier P.G. Phys. Rev. A **40** 5673 (1989)
19-Dujardin G., Hellner L., Hamdan M., Brenton A.G., Olsson B.J. et Besnard-Ramage M.J., J. Phys. B **23** 1165 (1990)
20-Codling K., Cornaggia C., Frasinski L.J., Hatherly P.A., Morellec J. and Normand D. to be published
21-Freeman R.R., Bucksbaum P.H., Milchberg H., Darak S., Schumacher D. and Geusic M.E., Phys. Rev. Lett. **59**, 1092 (1987)
22-Agostini P., Antonetti A., Breger P., Crance M., Migus A., Muller H.G. and Petite G., J. Phys B **22**, 1971 (1989)
23-Gibson G.N., Freeman R.R. and McIlrath T.J. Phys. Rev. Lett. **67**,1230 (1991)
24-Helm H., Dyer M.J. and H. Bissantz, Phys. Rev. Lett. **67**,1234 (1991)

CHARGE-SYMMETRIC FRAGMENTATION OF DIATOMIC MOLECULES IN INTENSE PICOSECOND LASER FIELDS

L.J.Frasinski, K.Codling and P.A.Hatherly

J J Thomson Physical Laboratory, University of Reading
Whiteknights, P O Box 220
Reading RG6 2AF, UK

ABSTRACT

Analysis of time-of-flight (TOF) spectra of N_2 and CO at 600 nm using covariance mapping shows that fragmentation is predominantly charge-symmetric. However, experiments at 305 nm and 248 nm using conventional TOF mass spectroscopy suggest that fragmentation is charge-asymmetric. Covariance mapping at 305 nm has led to a partial re-interpretation of the data and the conclusion that the fragmentation is, indeed, charge-symmetric. It is anticipated that the same will be true at 248 nm.

INTRODUCTION

The detailed dynamics of multiphoton multiple ionisation and fragmentation of diatomic molecules exposed to intense picosecond laser fields can be studied conveniently by time-of-flight (TOF) mass spectroscopy (Frasinski et al 1987). For example, Boyer et al (1989) used a laser of wavelength 248 nm, pulse length 0.6 ps and focused intensity of about 10^{15} W/cm^2 to ionise N_2. Their analysis led them to conclude that the molecule tends to fragment asymmetrically (for example $[N_2^{2+}] \rightarrow N^{2+} + N$ or $[N_2^{4+}] \rightarrow N^{3+} + N^+$, where the square brackets denote transient molecular ions). The first observation of fluorescence emission in N_2 at 55.8 nm, thought to be associated with a specific excited state of the N_2 ion, was felt by Gibson et al (1989) to provide confirmation of this asymmetric behaviour. More recently Normand et al (1991a) have studied the fragmentation of N_2, CO and O_2 at both 610 nm and 305 nm. They found that charge-symmetric fragmentation was dominant in all three gases at 610 nm, whereas charge-asymmetric fragmentation was the favoured process at 305 nm in the case of N_2 (Cornaggia et al 1990) and CO (Lavancier et al 1991).

However, analysis of one-dimensional TOF spectra can be at best ambiguous and at worst in error. For example, in a homonuclear molecule such as N_2, TOF spectra cannot differentiate the asymmetric process $[N_2^{2+}] \rightarrow N^{2+} + N$ from the symmetric process $[N_2^{4+}] \rightarrow N^{2+} + N^{2+}$, since the mass spectrometer is insensitive to neutral particles and the kinetic energies of the two fragments are bound to be the same. In a heteronuclear molecule such as CO, one finds it difficult to differentiate, for example, the fragmentation channels $[CO^{2+}] \rightarrow C^+ + O^+$ and $[CO^{3+}] \rightarrow C^{2+} + O^+$ if the kinetic energy ranges of the O^+ ions produced by the two processes overlap substantially.

Coherence Phenomena in Atoms and Molecules in Laser Fields
Edited by A.D. Bandrauk and S.C. Wallace, Plenum Press, New York, 1992

Recently Frasinski et al (1989a) devised a technique, covariance mapping, that allows unequivocal correlation of the various fragment ion pairs with their parent molecular ion precursors, determination of partial cross sections for fragmentation and the kinetic energies and angular distributions of the atomic ions (Hatherly et al 1990). We hope to show that application of the technique can significantly improve and indeed modify the analysis of TOF spectra. In particular we wish to suggest that molecular fragmentation is predominantly charge-symmetric when lasers of visible and ultraviolet wavelengths are used.

EXPERIMENTAL PROCEDURE

Two lasers were used in this work. The first is housed at the Central Laser Facility of the Rutherford Appleton Laboratory (RAL) in the UK and provides an output at 600 nm with a pulse length of about 0.6 ps at a repetition rate of 10 Hz. Focusing with an f/3 system produces an intensity of about $3x10^{15}$ W/cm^2 in the focal region. The second laser is housed at Saclay in France (Cornaggia et al 1990) and provides an output at 610 nm with a pulse length of about 2 ps and a focused intensity up to $5x10^{15}$ W/cm^2. Alternatively the laser frequency is doubled using a BBO crystal with a conversion efficiency of 15%. The pulse duration is then somewhat shorter(~1.4 ps) and a focused intensity up to 10^{15} W/cm^2 is achievable.

The time-of-flight mass spectrometer used in these experiments is described briefly in the publication of Hatherly et al (1990). The short drift tube length (~25 mm) ensures that all ions are collected, regardless of their kinetic energy or charge state, or whether they are emitted towards (forward ions) or away from (backward ions) the microchannel plate detectors. This high collection efficiency ensures that the covariance mapping technique will work. Of course this short drift tube compromises the energy resolution of the system but a second, longer drift tube is incorporated to obtain accurate ion energies, once the various ion-ion correlations have been determined.

RESULTS AND DISCUSSIONS

Frasinski et al (1989b) have used covariance mapping to analyse the TOF spectrum of N_2 using the RAL 600 nm picosecond laser. Figure 2 of Frasinski et al (1989b) publication shows the result for a laser intensity of about $3x10^{15}$ W/cm^2. One notes that the symmetric channel $[N_2^{4+}] \rightarrow N^{2+} + N^{2+}$ is by far the strongest, whereas the asymmetric channel $[N_2^{4+}] \rightarrow N^{3+} + N^+$, is barely visible; see also the data of Cornaggia et al (1990). A more recent experiment at 610 nm using the Saclay picosecond laser and covariance mapping confirms this general behaviour. Thus the analysis of the two experiments at a photon energy of about 2 eV disagree with those of Boyer et al (1989) at 248 nm and Cornaggia et al (1990) at 305 nm. However, one might argue that this is not surprising when one considers the large difference in photon energies employed in these experiments (5 eV and 4 eV).

Therefore covariance mapping has been used recently at Saclay to study N_2 at a laser wavelength of 305 nm and a focused intensity of about 10^{15} W/cm^2. On the covariance map obtained (Codling at al 1991) the symmetric channel $[N_2^{4+}] \rightarrow N^{2+} + N^{2+}$ + ~10 eV is clearly seen but the asymmetric channel $[N_2^{4+}] \rightarrow N^{3+} + N^+$ is missing entirely. This is consistent with the work of Cornaggia et al (1990); they saw no sign of the N^{3+} ion at 305 nm. However, they rejected symmetric fragmentation as a possible explanation for their 5.15 eV N^{2+} ions.

Also observed on the covariance map at 305 nm is the channel $[N_2^{3+}] \rightarrow N^{2+} + N^+$, a channel not recognised by Cornaggia et al (1990). They explained the observation of 5.15 eV N^{2+} ions in terms of the asymmetric channel $[N_2^{2+}] \rightarrow N^{2+} + N + 10.3$ eV. We, on the other hand, suggest that the process $[N_2^{3+}] \rightarrow N^{2+} + N^+ + 10.3$ eV can explain some of these ions, but one must now locate N^+ ions in the TOF spectrum of Cornaggia et al (1990) with an energy of 5.15 eV. These are thought to be hidden in the wings of the broad feature associated with N^+ ions, see figure 9 of that publication. Thus, the observation of two new channels, namely $N^{2+} + N^2$ and $N^{2+} + N^{2+}$, unrecognisable using conventional one-dimensional TOF spectroscopy, indicates the power of covariance mapping.

The provisional conclusion from the covariance mapping experiments at 600 nm and 305 nm is that symmetric fragmentation is generally favoured at both wavelengths, in agreement with the Thomas-Fermi-Dirac (TFD) field-ionisation model (Codling et al 1989; Brewczyk and Frasinski 1991).

In order to confirm this tendency towards charge-symmetric fragmentation, the isoelectronic molecule CO has also been studied. Codling et al (1990) used covariance mapping to determine the various channels at 600 nm and the symmetric channel $[CO^{4+}] \rightarrow C^{2+} + O^{2+}$ was found to be much stronger than the $[CO^{4+}] \rightarrow C^{3+} + O^+$ channel. In a more recent publication Lavancier et al (1991) obtained conventional one-dimensional TOF spectra at 610 nm which agreed well with the earlier data. This led them to similar conclusions - that sequential ionisation occurs in a "non-vertical" multielectron dissociative ionisation (MEDI) process. That is, the molecule continues to ionise on a time-scale of femtoseconds, even as it dissociates. Indeed, the molecules CO, N_2 and O_2 all appeared to ionise at quite specific internuclear separations with the ions having kinetic energies given by a simple "Coulomb explosion" model (Normand et al 1991a).

However, at 305 nm the picture suggested by conventional, one-dimensional TOF spectra is quite different. Figure 8 of the Lavancier et al (1991) publication shows TOF spectra taken at three laser intensities. At $3x10^{14}$ W/cm^2, a group of O^{2+} ions appear with an energy of 3.75 eV; these could not be associated with a group of C^{2+} ions of the same (3.75 eV) energy also present in the TOF spectrum because conservation of momentum must be obeyed (the energies must be in the ratio 12:16). Moreover, there were no C^+ or O^+ ions of appropriate energy with which to correlate these double-ions in processes $[CO^{3+}] \rightarrow C^{2+} + O^+$ or $[CO^{3+}] \rightarrow C^+ + O^{2+}$. This led Lavancier et al to conclude that the atomic double-ions must result from two asymmetric processes involving neutral atoms, namely $[CO^{2+}] \rightarrow C^{2+} + O +$ 6.5 eV and $[CO^{2+}] \rightarrow C + O^{2+} + 8.8$ eV.

Since this behaviour disagrees with the results of the TFD field-ionisation model, it seemed worthwhile to perform covariance mapping on CO with the Saclay laser at 305 nm in order to shed more light on the situation. On the map obtained (Codling at al 1991) one can clearly observe the channels $C^+ + O^+$ and $C^{2+} + O^+$. Moreover, by comparing the covariance map (which contains only ion-ion correlations) with the concurrently accumulated one-dimensional TOF spectrum (which contains, in addition, ions from ion-neutral fragmentations), one finds that the $C^{2+} + O$ channel is no more than 10% of the $C^{2+} + O^+$ channel. The total kinetic energy release in the $C^{2+} + O^+$ channel ranges from 3.5 to 8.8 eV. Therefore, we conclude that the process designated as $[CO^{2+}] \rightarrow C^{2+} + O + 6.5$ eV is, instead, $[CO^{3+}] \rightarrow C^{2+} + O^+ + 6.5$ eV.

One must now locate a group of O^+ ions at an energy of about 2.8 eV. In fact the width of the structure associated with the O^+ ions (figure 8 of Lavancier et al) is sufficiently large to allow an unresolved 2.8 eV O^+ ion to be hidden in the wings. The O^{2+} feature at 3.75 eV is weak, consistent with the lack of observation of the channel $C^+ + O^{2+}$ on the 305 nm map. However, the features in figure 8 of Lavancier et al, associated with C^+ and C^{2+} ions are also broad and could accommodate the possibility of the channels $[CO^{3+}] \rightarrow C^+ + O^{2+} + 8.8$ eV and $[CO^{4+}] \rightarrow C^{2+} + O^{2+} + 8.8$ eV. These channels fragment with much greater kinetic energy (10 and 19 eV) at 610 nm, according to Lavancier et al. The conclusion is, once again, that "symmetric" fragmentations (i.e. where fragment ion pairs differ by no more than one unit of charge) are favoured.

Let us now consider why Boyer et al (1989) suggested charge-asymmetric fragmentation processes. In the soft X-ray, "single photon" ionisation of a diatomic molecule, core hole production followed by rapid Auger rearrangement is the dominant route to double (or multiple) ionisation. For example, in CO the ejection of a C(1s) electron at 305 eV leads to channels $C^+ + O^+$, $C^{2+} + O^+$, $C^+ + O^{2+}$ and $C^{2+} + O^{2+}$ (Codling et al 1990). Note that no charge-asymmetric fragmentation is observed in this case. In N_2, where the shell structure is $KK(2\sigma_g)^2(2\sigma_u)^2(1\pi_u)^4(3\sigma_g)^2$, creation of a N(1s) core hole may be followed by the production of two holes in the valence shells. Auger electron-ion coincidence studies of Eberhardt et al (1987) showed that two-hole states at about 70 eV (for example $2\sigma_g^{-1}$, $1\pi_u^{-1}$) fragment asymmetrically: $[N_2^{2+}] \rightarrow N^{2+} + N$. Boyer et al suggested that the similarities in ion TOF spectra between the single photon and multiphoton cases, and in particular a large N^{2+} peak at 6-7 eV, could be explained in terms of an identical fragmentation process.

If this model of asymmetric fragmentation is pursued, the energies of the N^+, N^{2+} and N^{3+} ions are explained by Boyer et al in a straightforward manner. The laser acts as a δ-function and all molecular ions are created instantaneously at the normal internuclear separation of 1.10 Å. In a simple Coulomb explosion model, the dissociation energy is given in eV by $U(r) = 14.4\, q_1q_2/r$, where r is the internuclear distance in Å. This energy can be kept low and in agreement with experimentally determined kinetic energies only if $q_1 = 1$ for all fragmentation channels; hence the suggestion of asymmetric fragmentation. The alternative explanation for low ion energies - symmetric fragmentation involving a sequential ionisation process with r increasing - was not considered.

The asymmetry in the Saclay data at 305 nm was explained somewhat differently. Here it was suggested that the lifetime of the excited $[CO^{2+}]$ state relative to the laser pulse duration is of crucial importance. If the lifetime is longer than the laser pulse duration, the sequence of Coulomb explosions cannot start. Only "vertical" transitions are considered and it would require a high energy (~ 50 eV) to make a transition from $[CO^{2+}]$ to $[CO^{3+}]$ at the equilibrium internuclear separation (see figure 6 of Lavancier et al 1991). Hence there is a low probability of further ionisation and consequently the only states accessed are those of CO^{2+}; these fragment either into $C^+ + O^+$ with low kinetic energies or into $C^{2+} + O$ and $C + O^{2+}$. In the case of the 610 nm radiation, states accessed are thought to have lifetimes much shorter than the laser pulse duration; the doubly-charged ions dissociate during the laser pulse. Consequently the more highly ionised species can be accessed with the expenditure of less energy - they are created at larger values of r. Such explanations for the differences in TOF spectra at 305 and 610 nm may not be valid, however, since covariance mapping has now shown that the analysis of the 305 nm data must be partially revised.

Considerable progress has been made in the theoretical description of the process of multiphoton ionisation of diatomic molecules in moderately high fields (Zavriyev et al 1990; Guisti-Suzor and Mies 1991) but one of the simplest approaches, and one that should be applicable at high laser intensities, is the TFD field-ionisation model. Using this approach, Brewczyk and Frasinski (1991) have determined the maximum electronic charge a diatomic molecule can retain when exposed to an axial, static electric field, as a function of field strength and internuclear separation. They consider a simple dynamic situation of a molecule exposed to a laser field of finite risetime at a frequency sufficiently low that the electron cloud can react adiabatically. They track the evolution of the system as a function of increasing E-field and ion separation and find that, once the ions are so far apart that the inner potential barrier inhibits electron charge transfer between the two potential wells, the remaining charge will inevitably have been shared equally between the two ions through the action of the oscillating laser E-field. In other words, charge-symmetric fragmentation will be favoured.

Regardless of the temporal pulse shape, the TFD model predicts a small number of dissociation channels for a particular peak field. The lower ionisation stages seen experimentally are produced in regions surrounding the focal volume, where the laser field is less intense. The model should predict the correct behaviour of ultimate charge state reached versus laser intensity, for intensities in excess of 10^{14} W/cm^2. Unfortunately, the model is unable to account for the wavelength dependence of the appearance of the various atomic ions as a function of laser intensity, as exemplified by the recent experimental data on N_2 (Cornaggia et al 1990), CO (Lavancier et al 1991) and O_2 (Normand et al 1991b). Nor is it capable of predicting the kinetic energies of the various fragment ions and the considerable differences observed for each channel at 610 and 305 nm. Moreover, the molecular specificity of the MEDI process is clearly not incorporated in the TFD model.

ACKNOWLEDGEMENTS

We are pleased to acknowledge the Science and Engineering Council (UK) for their financial support and the use of the Laser Support Facility of Rutherford Appleton Laboratory. We are extremely indebted to the Saclay (France) group of Professor J Morellec for allowing us access to their picosecond laser system.

REFERENCES

Brewczyk M and Frasinski L J (1991) J. Phys. B 24 L307

Boyer K, Luk T S, Solem J C and Rhodes C K (1989) Phys. Rev. A 39 1186

Codling K Frasinski L J and Hatherly P A (1989) J. Phys. B 22 L321

Codling K, Frasinski L J, Hatherly P A and Stankiewicz M (1990) Physica Scripta 41 433

Codling K, Cornaggia C, Frasinski L J, Hatherly P A, Morellec J and Normand D, to be published (1991)

Cornaggia C, Lavancier J, Normand D, Morellec J and Liu H X (1990) Phys. Rev. A 42 5464

Eberhardt W, Plummer E W, Lyo I-W, Carr R and Ford W K (1987) Phys. Rev. Lett. 58 207

Frasinski L J, Codling K, Hatherly P A, Barr J, Ross I N and Toner W T (1987) Phys. Rev. Lett. 58 2424

Frasinski L J, Codling K, and Hatherly P A (1989a) Science 246 1029

Frasinski L J, Codling K, and Hatherly P A (1989b) Phys. Lett. A 142 499

Gibson G, Luk T S, McPherson A, Boyer K and Rhodes C K (1989) Phys. Rev. A 40 2378

Guisti-Suzor A and Mies F H (1991) in "Multiphoton Processes" editors G Mainfray and P Agostini (CEA) p. 169

Hatherly P A, Frasinski L J, Codling K, Langley A J and Shaikh W (1990) J. Phys. B 22 L291

Lavancier J, Normand D, Cornaggia C, Morellec J and Liu H X (1991) Phys. Rev. A 43 1461

Normand D, Cornaggia C, Lavancier J and Morellec J (1991) in "Multiphoton Processes" editors G Mainfray and P Agostini (CEA) p. 191

Normand D, Cornaggia C, Lavancier J, Morellec J and Liu Hx (1991b) Phys. Rev. A 44 475

Zavriyev A, Bucksbaum P, Muller H G and Schumacher D W (1990) Phys. Rev. A 42 5500

INTERMEDIATE STATES IN MULTIPHOTON FRAGMENTATION OF SMALL MOLECULES

W.T. Hill III, S. Yang, D.L. Hatten, Y. Cui,* J. Goldhar* and J. Zhu

University of Maryland, Institute for Physical Science and Technology
College Park, MD 20742 USA

INTRODUCTION

Contemporary studies have shown that competition between ionization and dissociation channels is fundamental to the decay dynamics in highly excited diatomic molecules.[1,2,3] Such competition is possible because the nuclear and electronic motion can be comparable in high-lying Rydberg states embedded in ionization or dissociation continua. Competitive decay comes in two flavors -- spontaneous and induced. Spontaneous competition occurs between two or more decay channels, all of which are open at the same energy of the system. Induced competition occurs between one decay channel at one energy of the system and another channel at a different energy; the two channels, however, are coupled by an applied radiation field. The simplest example of the latter, which occurs even at moderate field intensities, would be competition between photoionization and (pre)dissociation. As the intensity of the radiation field is increased, a variety of highly nonlinear, higher order fragmentation processes comes into play such as above-threshold ionization, above-threshold dissociation and Coulomb explosions. In general, these intense field processes can all compete amongst themselves and with the lower order processes.

In this paper we will look at how intermediate states can influence the competition between various multiphoton fragmentation channels. In particular, we will review recent fragmentation studies that we have made of H_2 and CO induced by 193 nm radiation at intensity levels of 10^{10} to 10^{12} W/cm^2.[1,2,4]

*Permanent Address: University of Maryland, Department of Electrical Engineering, College Park, MD 20742 USA.

Coherence Phenomena in Atoms and Molecules in Laser Fields
Edited by A.D. Bandrauk and S.C. Wallace, Plenum Press, New York, 1992

APPARATUS

We induce fragmentation with pulsed 193 nm radiation of either 12 ns or 10 ps duration. The 12 ns pulses are generated by an ArF* excimer laser (Lambda Physik 150 ET) with a spot size of about 4 mm × 22 mm and a beam divergence of about 0.3 mrad. The beam is about 70% polarized and delivers a power density of 10^{10} W/cm^2 when focused with a 500 mm lens. The details of this laser have been discussed elsewhere.[4]

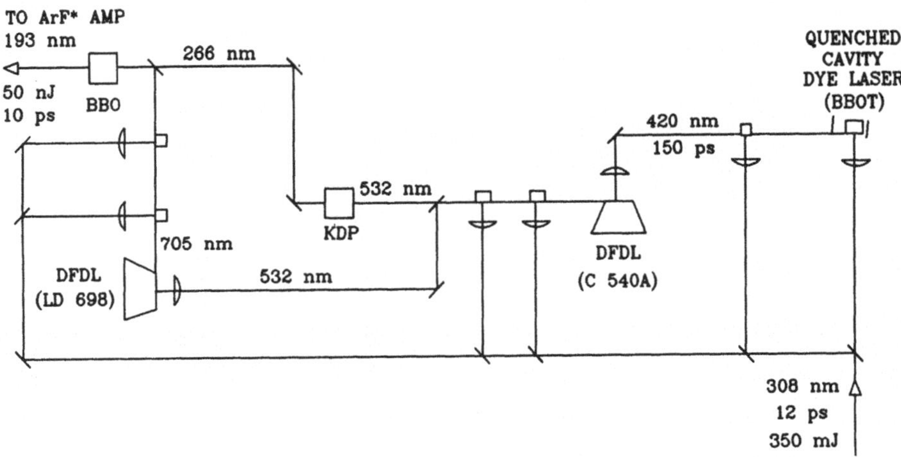

Figure 1. Schematic for the generation of picosecond 193 nm radiation.

Picosecond 193 nm pulses are generated by a scheme involving two distributed feedback dye lasers (DFDL) and two ArF* excimer amplifiers. A sketch of the design, shown in Fig. 1, is based on the technique first described by Szátmari and Schäfer[5] Initial pulse shortening by a factor of about 10 is accomplished with a quenched cavity dye laser. This laser pumps one DFDL,[6] which produces pulses about 10 ps in duration at 532 nm. Part of the output of the first DFDL is used to pump a second DFDL, which produces pulses of duration less than 10 ps at 705 nm. The remaining 532 nm radiation is doubled (to give 266 nm) in KDP and then mixed with the 705 nm radiation in BBO to generate about 50 nJ of the desired 193 nm radiation. The picosecond 193 nm radiation is amplified in our Lambda laser to give 5 to 10 mJ of nearly diffraction limited energy. In the studies presented here, this radiation was focused to give about 5×10^{12} W/cm^2. Additional details regarding the generation of intense picosecond 193 nm radiation will be presented in a future communication.[7]

The goal of our investigation is to study fragmentation processes quantitatively. This requires that we identify the species of each fragment and determine both its charge state and energy state (internal and kinetic). The apparatus used to achieve this is built around a time-of-flight (TOF) positive ion detector shown in Fig. 2. This detector is optimized to discriminate not only ions with different charge-to-mass ratios (q/m) but ions with the same q/m and different initial kinetic energies. Optimization is accomplished with a three-region arrangement (after Wiley and McLaren)[7] in which the electric field between G1 and G2 is kept small (10 - 40 V/cm) to allow the ions to separate in time according to their initial speeds regardless of their value of q/m.[4] The ions are then accelerated between G2 and G3 (100 - 200 V/cm) and allowed to drift field free between G3 and G4, where they are further separated in time according to q/m. To interrogate neutral fragments with this detector, we employ multiphoton ionization (MPI) to label atoms in specific internal energy states and the TOF spectrum to measure their kinetic energy. As discussed by Turner *et al.*[4] and Hill *et al.*[1] the ion collection efficiency changes radically as a function of ion energy because of the small field of view of the detector ($\approx 10^{-3}$ sr). Furthermore, the strength of the electrostatic potential and, hence, the electric field perpendicular to the TOF axis varies significantly within the first two regions. Consequently, it is necessary to calibrate the detector for both collection efficiency and flight time in order to extract initial kinetic energies and yields for specific fragments from their TOF spectra. Calibration involves modeling the potentials, particularly in the first two regions, and calculating the ion trajectories as a function of initial kinetic energy and ejection angle. We are able to achieve about 10% energy resolution when (1) the length of each of the first two regions is about 1 cm, (2) the length of the third region is 20 to 50 cm and (3) the TOF signal is digitized at a rate of 0.5 to 1 gigasamples per second. The reader is referred to Ref. 4 for further details on the TOF apparatus.

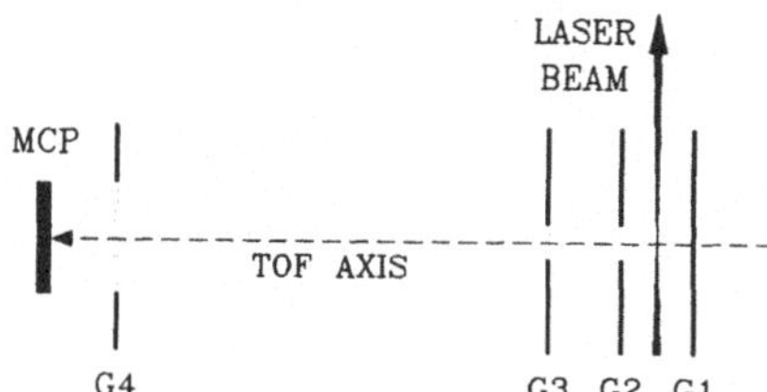

Figure 2. The optimized TOF detector arrangement.

SPONTANEOUS COMPETITION

Molecular hydrogen provides an example of spontaneous competition when it is excited about 19.2 eV above the ground state, which requires three 193 nm (6.4 eV) photons. At this energy H_2 can undergo ionization ($H_2 \rightarrow H_2^+ + e^-$), dissociation ($H_2 \rightarrow H + H^*$) and dissociative ionization ($H_2 \rightarrow H + H^+ + e^-$).[1] The ionization channels open at an internuclear separation (R_s) of about 0.5 Å at 19.2 eV while the dissociation and dissociative ionization channels open respectively at about 1.5 Å and 2.5 Å.[b] It is possible to alter the balance between the three channels by selecting the route of excitation. In our case of three-photon excitation, this means that the final state at 19.2 eV will depend on which intermediate state is resonant with the absorption of one or two photons. The double minimum $E,F\ {}^1\Sigma_g^+ 1s\sigma_g 2s\sigma_g + (2p\sigma_u)^2$ electronic state in H_2 (see Fig. 3) provides two-photon resonant intermediate states at 193 nm (an energy of 12.8 eV). The equilibrium internuclear separation for states sustaining partial waves localized in the inner well is about 1 Å while that of states in the outer well is more than 2 Å. Consequently, when the wavepacket describing an intermediate state of the system is localized in the inner well, the final states reached via photoexcitation must be described by partial waves that can exist at small R_s. The range of R_s over which the wavepacket is appreciable when localized in the inner well is indicated by the forward-hatched region at the three-photon level in Fig. 3. This region corresponds predominantly to ionization. On the other hand, dissociation and dissociative ionization can be excited more readily from a wavepacket that is localized in the outer well. The range over which the wavepacket is appreciable is indicated by the backward-hatched region. Transitions from the outer well will be able to excite all three channels.

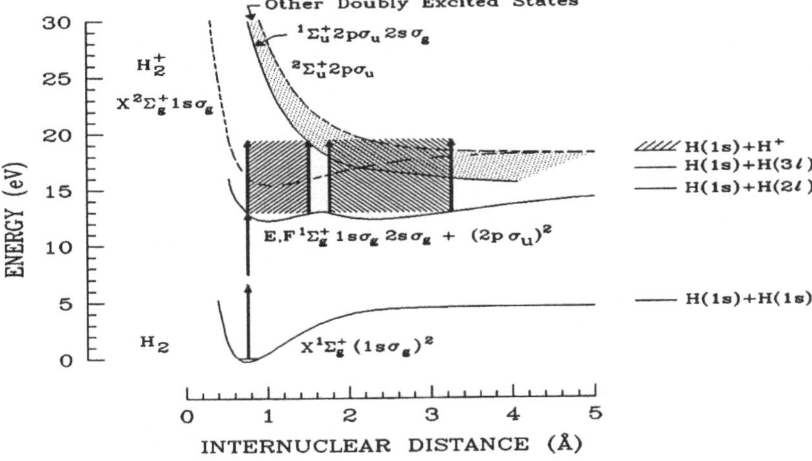

Figure 3. Selected potential curves of H_2 (solid) and H_2^+. Several of the dissociation limits are indicated at the right.

The upper trace of Fig. 4 shows a typical TOF spectrum of the charged fragments following photoexcitation of H_2 in which one observes a single H_2^+ peak and three H^+

[b]There is also a very weak contribution to the dissociative ionization channel from the inner wall of the ground state ion potential curve that also opens at about 0.5 Å.

peaks. A similar spectrum for HD is shown in the lower trace of the figure. Both the H_2 and HD spectra have nonzero and near zero kinetic energy components. The nonzero kinetic energy ions appear in two of the peaks, one corresponds to ions initially directed toward the detector and the other to ions initially directed away from the detector; the backward-going ions arrive at the detector at a later time. The ions with near zero kinetic energy also form a pair of peaks. However, for the conditions under which the displayed spectrum was taken the two peaks are unresolved. We have plotted the relative ion yields (the signal corrected for the change in collection efficiency) vs initial kinetic energies of H^+ and D^+ in Fig. 5 for the TOF spectra of Fig. 4. The scale above each trace indicates the allowed ranges of kinetic energies for ionization, dissociation and dissociative ionization.

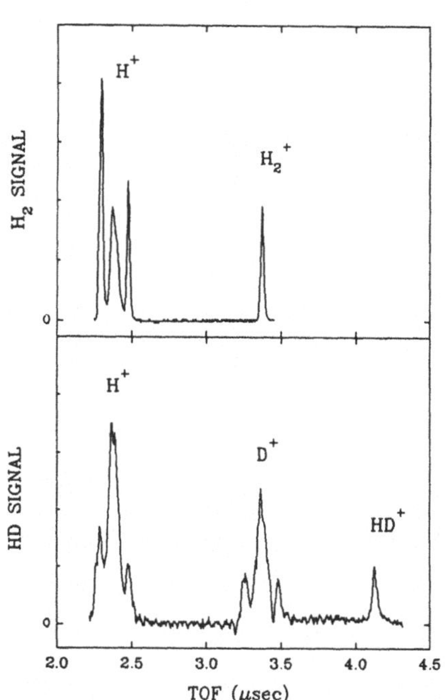

Figure 4. Typical TOF spectra of H_2 and HD taken with 12 ns pulses of intensity 5×10^{10} W/cm^2.

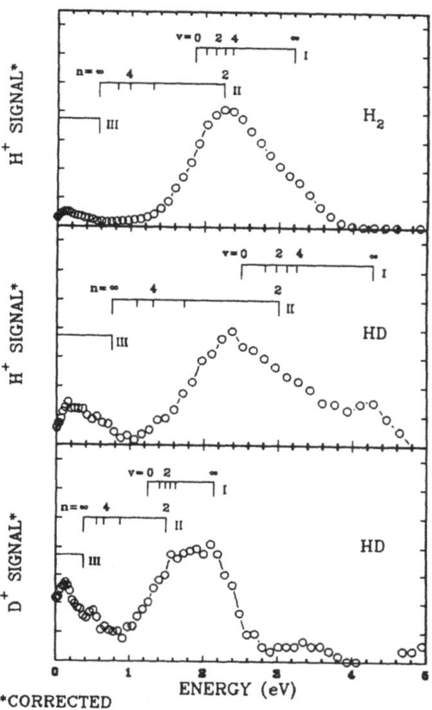

Figure 5. Relative ion yield vs initial kinetic energy corrected for collection efficiency.

From the above discussion, we can explain the general difference between the H_2 spectrum and the HD spectrum by the fact that excitation in H_2 takes place over smaller R_s than excitation in HD. There is more access to final states that are restricted to large R_s, those responsible for dissociation and dissociative ionization, in HD than there is in H_2.

More quantitatively, the ratio between the integrated proton signal of the slow peak to that of the fast peak in Fig. 5 increases by a factor of six to ten between H_2 and HD.[1] Thus, it is possible to alter the balance between the different channels significantly by selecting different intermediate states.[c]

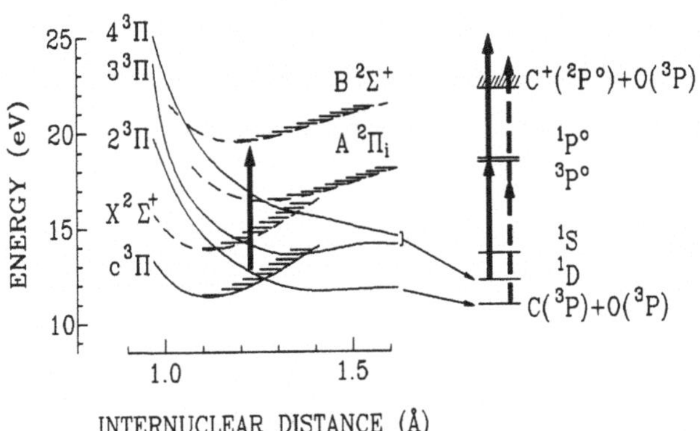

Figure 6. Diabatic potential curves for CO in the 10 - 25 eV range with several dissociation thresholds indicated to the right.

INDUCED COMPETITION

In this section we will look at competition between dissociation of CO, which occurs after absorption of two 193 nm photons (an energy of 12.8 eV), and ionization of CO, which requires three photons (an energy of 19.2 eV). In this experiment CO was excited to the $c^3\Pi$ Rydberg state. The *c* state is strongly perturbed by a family of $^3\Pi$ valence states (see Fig. 6) causing predissociation. Thus, there will be competition between spontaneous dissociation and photo-induced ionization.

A typical TOF spectrum obtained with 12 ns pulses ($I \approx 5 \times 10^{10}$ W/cm^2) is shown in Fig. 7. It consists of a double C^+ peak, which corresponds to forward and backward going ions with about 0.2 eV of initial kinetic energy, and a very small CO^+ peak.[d] The energy

---

[c]Since the electronic structure of H_2 and HD are essentially identical, we expect the photofragmentation spectra to be very similar. Thus, changing from H_2 to HD should have the same effect as tuning from an intermediate state that is represented by a wavepacket localized in the inner well to one localized in the outer well.

[d]Nearly all of the peak marked CO^+ is due to the background Si diffusion pump oil.

of C^+ corresponds to what one would expect if the *c* state were to dissociate into $O[^3P]$ + $C[^1D]$. As discussed by Hill *et al.*[2] and Turner *et al.*,[4] C^+ is generated via resonant MPI of $C[^1D]$ through the $^1P^0$ state (see Fig. 6). One might also expect another pair of peaks corresponding to faster C^+ ions due to dissociation into $O[^3P]$ + $C[^3P]$ followed by MPI of $C[^3P]$. At the same time, two complementary pairs of O^+ peaks corresponding to MPI of $O[^3P]$ following dissociation would be expected. Neither the faster C^+ nor either of the O^+ peaks were observed. Evidently, the fact that MPI of $C[^3P]$ and $O[^3P]$ are not resonantly enhanced at 193 nm renders their yield too small to detect in our system at these power densities. Following the analysis of Hill *et al.*[2] we conclude from Fig. 7 that the *c* state dissociates so efficiently that photoionization with 5×10^{10} W/cm^2 (12 ns pulses) essentially does not occur This is a case in which the intermediate states turn off higher order multiphoton fragmentation channels.

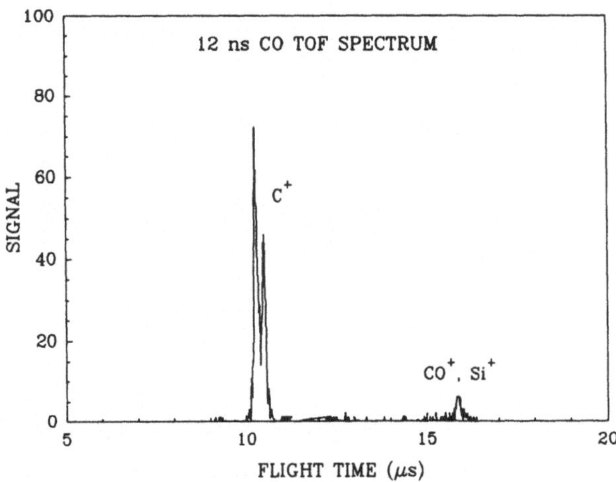

Figure 7. TOF spectra of CO showing a pair of C^+ peaks near 10 μs and a small peak near where CO^+ should be at 16 μs.

We can turn the higher order processes on by increasing the power density and shortening the pulse length. Figure 8 shows the spectrum of CO taken with 10 ps pulses and an intensity of about 5×10^{12} W/cm^2. We again see the 0.2 eV C^+ peaks. When the time scale is expanded we see additional C^+ peaks.[e] One pair of the new peaks has tentatively been assigned as the forward and backward going ion peaks due to MPI of fast $C[^3P]$ atoms that we could not detect in the long pulse experiment. We also observe peaks corresponding to CO^+, O^+ and C^{++}. Two of the O^+ pairs (four peaks in all) have been assigned tenta-

[e]The C^+ and CO^+ (Si^+) peaks do not occur at the same times in Figs. 7 and 8 because neither the potentials on G1 - G3 nor the field free path lengths between G3 and G4 are the same for the two experiments.

tively as originating from MPI of fast and slow O[^{3}P], which are created with the neutral carbon atoms discussed above. The other O^+ and C^+ peaks, which have energy up to about 5 eV, are probably due to dissociation of CO^{++}. Finally, the C^{++} peaks, which range from about 0.5 eV to nearly 5 eV, are most likely due to dissociation of CO^{+++} into C^{++} and O^+. These higher energy peaks and multiply-ionized atomic fragments are currently being investigated further in this lab and will be the subject of a future paper.[9]

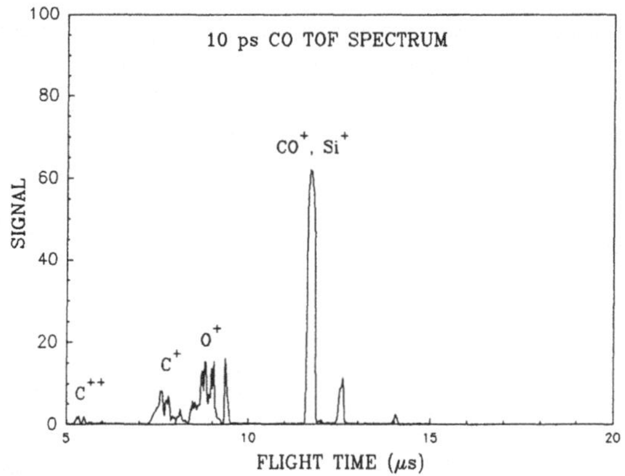

Figure 8. TOF spectrum of CO showing several C^+ and O^+ ions between 7.5 and 9.0 μs, and a large CO^+ peak near 11.75 μs.

CONCLUSION

In conclusion, it is clear that intermediate states play an important role in multiphoton fragmentation processes. We demonstrated that these states give the investigator influence over the competition between fragmentation channels and, thus, a knob to tweak for controlling, to some degree, the partitioning of strength between various fragmentation channels. Further control is possible by changing the pulse length and intensity of the light. In CO, we saw that we can turn on ionization at high intensity and short pulse lengths that are off under the conditions of long pulse, moderate intensities. Thus, we were able to observe evidence for multiple ionization at 193 nm. We are currently performing similar experiments on H_2 and its isotopes as well as other light diatomics.

ACKNOWLEDGMENTS

This work was sponsored by the US National Science Foundation under Grants Nos. PHY-84 06192 and PHY-84 51284 and the US Air Force Office of Scientific Research under Grant No. AFOSR-89-0451.

REFERENCES

1. W.T. Hill III, B.P. Turner, S. Yang, J. Zhu and D.L. Hatten, Phys Rev A43:3668 (1991).

2. W.T. Hill III, B.P. Turner, H. Lefebvre-Brion, S. Yang and J. Zhu, J. Chem. Phys. 92:4272 (1990).

3. Also see, H. Rottke and H. Zacharias, Opts. Commun. 55:87 (1985); D. Normand and J. Morellec, J. Phys. B21:L625 (1988); R.J. Miller, L. Li, W.A. Chupka and S.D. Colson, J. Chem. Phys. 90:754 (1989); P.J.H. Tjossem and K.C. Smyth, J. Chem. Phys. 90:2041 (1989).

4. B.P. Turner, W.T. Hill III, S. Yang, J. Zhu, A. Pinkas and L. Bao, Rev. Sci. Instrum. 61:1182 (1990).

5. S. Szatmári and F.P. Schäfer, Appl. Phys. B33:95 (1984).

6. A.N. Rubinov and T.Sh. Efendiev, Opt. Acta 32:1291 (1985).

7. Y. Cui, D.L. Hatten, J. Goldhar and W.T. Hill III, To be published.

8. W.C. Wiley and I.H. McLaren, Rev. Sci. Instrum. 26:1150 (1955).

9. S. Yang, Y. Cui, D.L. Hatten, J. Goldhar and W.T. Hill, To be published.

INTENSE CO_2 LASER IONIZATION OF D_2O VAPOUR

J.E. Decker, G. Xu, S.L. Chin

Centre D'Optique, Photonique et Laser
Département de Physique, Université Laval
Québec (Québec) Canada G1K 7P4

Experimental results of the ionization of heavy water vapour by a CO_2 laser are presented. Experimental conditions are such that ionization in the tunnelling regime is expected. A qualitative model is proposed explaining the processes leading to D_2O^+ and $(D_2O)_2^+$ ions.

1. INTRODUCTION

The interaction of very intense laser pulses with atoms has revealed interesting new phenomena such as above threshold ionization (ATI), and ponderomotive effects on electrons produced in the process. A review of this subject has recently been published by Freeman and Bucksbaum (1991). Naturally, investigations are moving towards effects of strong fields on molecules. Multiphoton absorption and dissociation of molecules was a popular subject several years ago for the purposes of laser isotope separation. More recently, attention has turned to the area of intense laser interaction with molecules, including experimental interrogations of multiple ionization and Coulomb explosion phenomenae in small molecules (Codling et al., 1987; Luk and Rhodes 1988; Lavancier et al., 1990 and references therein), and effects associated with intense field interactions such as above-threshold dissociation and bond-softening (Bucksbaum et al., 1990; Zavriyev et al., 1990). Theoretical efforts involve non-perturbative calculations based on the dressed state model (Chu 1981; George 1982; Bandrauk and Turcotte 1985; He et al., 1988; Bandrauk and Atabek 1989; Guisti-Suzor et al., 1990). The common goal of these studies is to further understanding of the fundamental processes involved in the interaction of intense radiation with matter.

Absorption processes of CO_2 laser radiation in molecules have been studied in some detail (Evans et al., 1989 and references therein), although, not much is known about the ionization processes occurring in molecules subjected to strong, low frequency fields. The first CO_2 laser ionization experiments were performed by Chin (1977) and Chin and Faubert (1978), however, the process is still not well understood. The ionization mechanism likely to occur in a given atom by a strong laser pulse depends upon the lasers

Coherence Phenomena in Atoms and Molecules in Laser Fields
Edited by A.D. Bandrauk and S.C. Wallace, Plenum Press, New York, 1992

frequency and intensity. The adiabaticity parameter (or gamma parameter) introduced by Keldysh in 1964, makes use of these quantities in comparing the field frequency, ω, with the tunnelling frequency, ω_t, to determine which ionization process is more likely occurring in an experiment: multiphoton ionization (MPI) or tunnel ionization (TI) corresponding to the limit case $\omega \ll \omega_t$. The gamma parameter is often expressed as $\gamma = \omega \sqrt{2E_0}/F$, where ω and F are the laser's frequency and electric field strength, E_0 the ionization potential of the atom. Therefore, if $\gamma \gg 1$, the ionization process would be best described by multiphoton ionization, whereas if $\gamma \ll 1$ the ionization process is better described by tunnelling formulae. Still, these criteria discerning the processes of multiphoton ionization and tunnelling are somewhat vague. Precicely how much less than one should γ be in order for tunnel ionization approximations to be valid? Recently Ilkov et al., (1991) have shown that the the practical limiting value of γ need only be ≤ 0.5 for approximate tunnel ionization formulae to adequately model the ionization rate.

In a recent paper (Decker et al., 1991a) we reported the observation of tunnel ionization of D_2O molecules. In this article, we discuss the processes leading to D_2O^+ and $(D_2O)_2^+$ observed in experiments performed in our laboratory on the ionization of heavy water vapour by intense CO_2 laser pulses. In the present experiment, $0.36 \leq \gamma \leq 0.78$ corresponding to a laser intensity range of $5 \times 10^{12} > I > 10^{12}$ W/cm^2. The intensity for which $\gamma = 0.5$ is 2.4×10^{12} W/cm^2. Thus, the conditions are such that the experiment falls into the tunnel ionization regime as defined by Ilkov et al., (1991) for at least the higher end of our intensity range. A qualitative model is proposed explaining this ionization process. Our experimental data is comapared with a theoretical calculation of ion number in the tunnelling regime. An estimation of the probability corresponding to multiphoton excitation and strong field dressing of the molecule is also discussed.

2. Experiment

Details of the CO_2 laser chain used in these experiments are described in Decker et al., (1991b). Briefly, the chain begins with a hybrid CO_2 laser oscillator producing linearly polarized, single-longitudinal, single-transverse mode laser pulses. The oscillator is followed by a Pockel's cell, which allows the option of selecting a cut-out from the oscillator output. The beam is then amplified in two stages separated by spatial filtering. The laser pulse duration at full-width-half-maximum (FWHM), after amplification, is 2.1 ± 0.2 ns. Amplified pulses are focused within a high vacuum chamber by a NaCl lens of 10 cm focal length to a radius of 35 ± 2 μm. Experimental uncertainty in the intensity measurement is about 35%. The base pressure of the vacuum chamber is 2×10^{-7} Torr, while experiments are performed with a sample pressure of 2×10^{-5} Torr heavy water vapour (MSD Isotopes, 99.7 Atom % D). Ions are detected by the method of time-of-flight. A static electric field directs the ions into a field-free flight region terminating in an electron multiplier tube. A typical single shot, ion time-of-flight spectrum is shown in fig. 1. Signals are displayed on a LeCroy 9400 Transient Digitiser.

The experiment was performed using the 10P(20) line of the CO_2 laser at 10.6 μm. This frequency is near three photon-resonant with the ν_3 asymmetric stretching mode of the D_2O molecule. Ion number is plotted against the laser intensity on a log-log scale, an example of which is shown in fig. 2. Each point represents the average of 100 laser shots. Results have been corrected for the detector sensitivity dependence on ion mass,

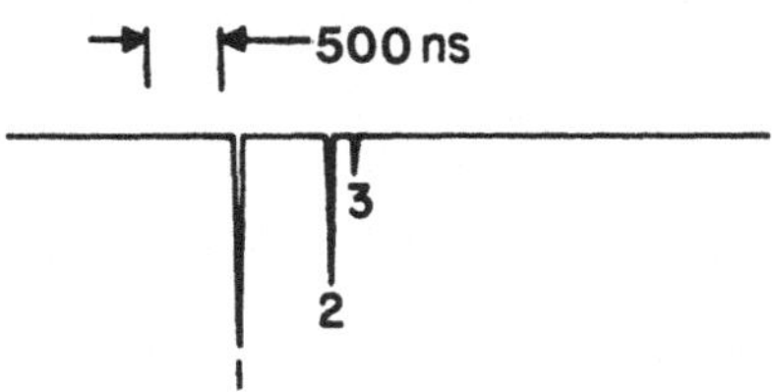

Figure 1. Typical oscilloscope trace of the ion time-of-flight for heavy water vapour ionized by intense CO_2 laser pulses. Labels correspond to the following ions: 1. D_2O^+, 2 $(D_2O)_2^+$, 3. CD_3CO^+ (impurity; deuterated acetone fragment).

M, which approximately follows $M^{-1/2}$ (Schram et al., 1965).

3. Discussion

3.1. Comparison with Tunnel Ionization in Atoms

Let us compare the facility of ionization by strong CO_2 laser pulses known for xenon (Chin et al., 1985; Yergeau et al., 1987) and potassium (Xiong et al., 1988, 1991a,b) atoms with the facility of ionization of the D_2O molecule. To do this, the ionization potential of the species E_0, is compared with the laser appearance intensity, I_{ap}. The appearance intensity is customarily defined as the intensity at which a few ions are detected at a fixed pressure of neutral species, and is read from a log-log plot of ion signal vs. laser intensity such as that of fig. 2. In all the above mentioned experiments, the pressures of the neutral species are similar, $\sim 10^{-5}$ Torr. Due to the highly non-linear nature of intense laser ionization processes, a small change in laser intensity commonly corresponds to a change of one or more orders of magnitude in the ionization rate. As a result, the intensity representing the threshold for ionization is not very accurate (Gibson et al., 1990). It is, however, useful in the context of a rough gauge of the ease of ionization with respect to the atoms ionization potential. For xenon where $E_0 = 12.1$ eV, $I_{ap} = 1.2 \times 10^{13}$ W/cm^2 and for potassium, $E_0 = 4.34$ eV, and $I_{ap} = 5 \times 10^{11}$ W/cm^2. In view of the appearance intensities for xenon and potassium, we would not expect to ionize the D_2O molecule ($E_0 = 12.6$ eV) even at the peak laser intensity used in the present experiments. In fact, the appearance intensity for D_2O is $I_{ap} = 1.2 \times 10^{12}$ W/cm^2. The monomer behaves as if its ionization potential was lower from the point of view of the minimum intensity required for the onset of ionization. The ionization potential for $(D_2O)_2$ is about 11 eV, based on the value for the $(H_2O)_2$ cluster ion appearance potential measured by Shiromaru et al., (1987). For the singly charged dimer ion, $I_{ap} = 1.7 \times 10^{12}$ W/cm^2. These appearance intensities are calibrated using a single measurement; hence, there is no ambiguity in their difference.

The difference in the appearance intensities of the monomer and the dimer may be a result of the vastly different initial sample densities of monomers and dimers expected for our water vapour sample at room temperature. For similar ionization potentials, a lower appearance intensity would be expected for the species whose sample density is greater. However, the dimer ionization potential is lower than that of the monomer. The higher dimer appearance intensity may be a result of the low neutral sample density relative to that for the monomer compensating for the lower ionization potential (see below). Nevertheless, the appearance intensities are, in general, lower than expected, based on

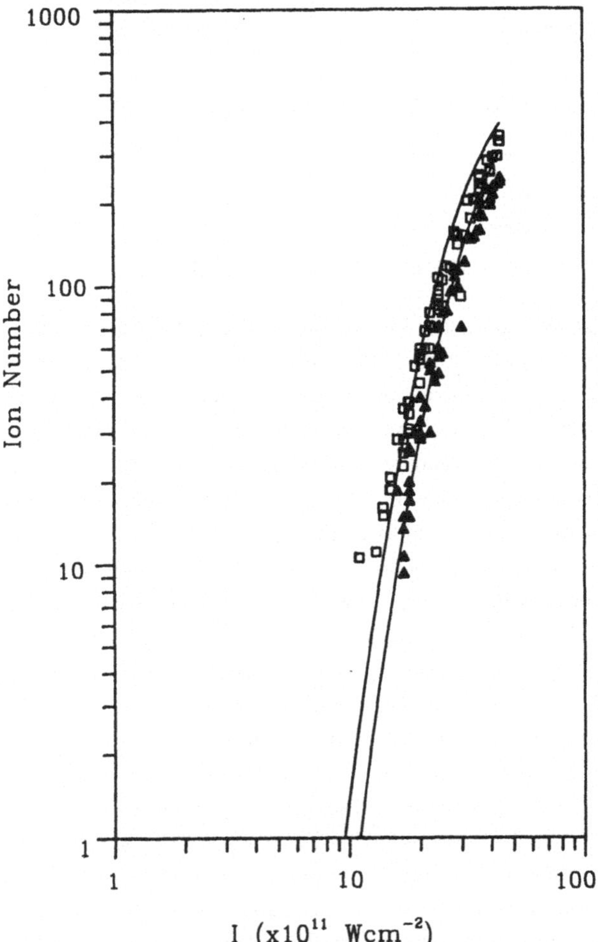

Figure 2. Measured ion number vs. 10.6 μm laser intensity plotted on a log-log scale. Each point represents the average of 100 laser shots. Open squares represent data for D_2O^+, filled triangles for $(D_2O)_2^+$.

similar data known for atoms. Details of the ionization process and an explanation for why the molecules behave as if their ionization potentials were lower, are hypothesised below.

3.2. Tunnel Ionization in Molecules

Potential energy surfaces for the ground state and the first excited state of D_2O are shown in fig. 3a). Consider the D_2O molecule in the presence of a strong, linearly polarized, laser field. During the rise-time of the laser pulse, the molecule is exposed to a sufficiently intense laser field that multiphoton vibrational excitation is easily attained. After a certain threshold intensity, the field strength may be sufficient so as to induce strong field dressing of the ground electronic state.

As the intensity increases to a value for which strong field dressing is significant, it is the highly vibrationally excited D_2O molecule that sees the intense field and is subject to distortion of the electronic states. A "field-dressed molecular channel state" corresponds to the coherent mixing of the electronic states with the laser photons (see He et al., 1988; Guisti-Suzor et al., 1990 and references therein). The ground electronic state $\tilde{X}\,^1A_1$ may be dressed by 1,3,5,... photons resulting in the curve crossing with the dissociative $\tilde{A}^1B_1$ state, shown schematically by the dashed lines of fig. 3b). In the case under discussion, dressing by a number of photons would be necessary for significant crossing of the $\tilde{X}\,^1A_1$ and $\tilde{A}^1B_1$ potential surfaces since the CO_2 laser photons are small (0.117 eV). As the field strength increases, the potential surfaces repel each other at the crossings giving rise to the laser-induced resonances indicated by the solid lines of fig. 3b). The field-dressed adiabatic curves then consist of a shallow potential well and a dissociative channel, both of which become open to the molecule:

1. The molecule may undergo a non-adiabatic transition to the upper adiabatic electronic-field state.

2. The molecule may undergo an adiabatic transition from the diabatic dressed bound state to the repulsive states $OD(X^2\Pi) + D(^2S)$. In the adiabatic electronic-field basis, this corresponds to movement on the lower adiabatic surface (lower solid line of fig. 3b). Neutral fragments however, cannot be measured with the present experimental apparatus.

The well formed by the curve crossings is likely near the dissociation limit, where the potential curves in the diabatic representation approach each other. Consequently, the potential well that arises in the laser-induced resonances is at a larger internuclear distance than that of the bare molecule (without field) in its ground state (see fig. 3). With longer bond lengths, the dressed molecule is expanded in size compared to the bare one. We propose that tunnelling occurs from this enlarged, field-dressed molecule. The enlarged molecule is illustrated as a "reduced atom" in the schematic diagram of fig. 4. The term "reduced atom" refers to a picture of the molecule in which the mass is assumed at the centre of mass of the molecule. The potential acting on the peripheral electron is represented by a Coulomb potential, similar to that for atoms. For the same field strength, the potential barrier required by the electron to tunnel through is shorter than that in the bare molecule.

The proposed ionization process depends upon the acquisition of a highly vibrationnally excited molecule. Criteria determining the extent of vibrational excitation consist primarily of: the density of states, the anharmonicity of the excitation ladder,

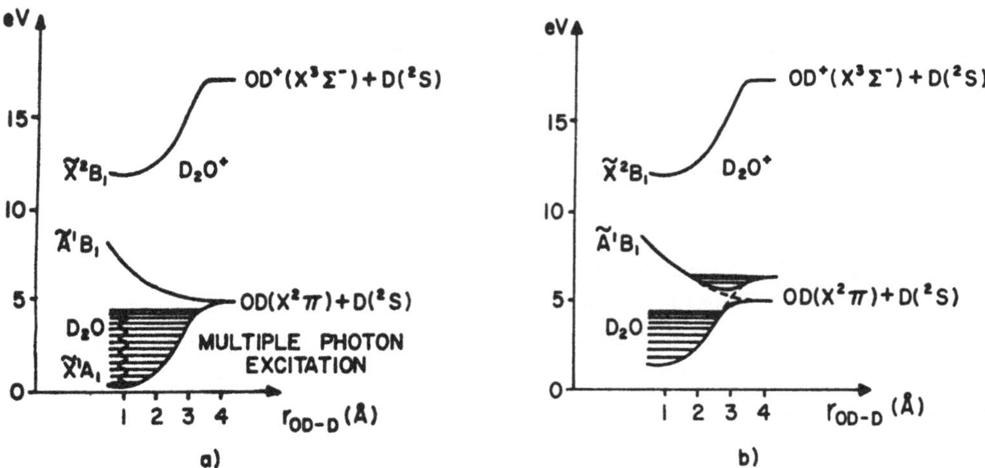

Figure 3. a) Relevant potential energy curves for the D_2O molecule from Tsurubuchi (1975). The internuclear spacing along the horizontal axis is estimated based on values for H_2O from Herzberg (1966a). Potential energy curves of the dressed molecule are illustrated in b). Solid lines represent the adiabatic electronic-field basis.

and the resonance (or lack thereof) of the photon energy with vibrational modes. If the density of states is too low or the excitation ladder is very anharmonic, the chances of attaining a high level of vibrational excitation are small due to bottlenecking[1]. Thus, even if the field is strong enough to deform the electronic surfaces, the molecule is not adequately excited to populate a potential well formed by curve crossings. In this case, much higher laser intensties are required to drive the initial multiphoton vibrational excitation for ensueing tunnel ionization to take place. Conversely, if the vibrational modes are resonant with the photon energy, and the ladder is not too anharmonic or there is a high density of states (larger molecules), it is likely that the molecule will dissociate before the laser field is strong enough to induce electronic state mixings in preparation for tunnel ionization. The present tri-atomic case seems to be an intermediate one that favors the observation of tunnel ionization assisted by photon dressing.

3.3. Ionization of Small Clusters

Experiments investigating clusters frequently rely on the measurement of the cluster ionization products using electron bombardment. In this method, electrons of about 50–100 eV energy are used to ionize clusters from which ionic fragments are detected. Initial cluster sizes are deduced by considering the probable electron-cluster reaction and possible subsequent unimolecular reactions leading to the observed cluster ion. Bombarding molecules with this much energy most likely leads to the excitation of internal modes of vibration. This method likely ionizes the cluster, but leaves it highly excited

[1]Bottlenecking occurs when the excitation manifold is such that the molecule can be excited through multiphoton absorption only to a particular level. Further excitation from this level is inhibited by anharmonicities noncommensurate with the laser photon, in addition to poor coupling with other vibrational or rotational modes. It therefore becomes very difficult to excite the molecule beyond this energy level.

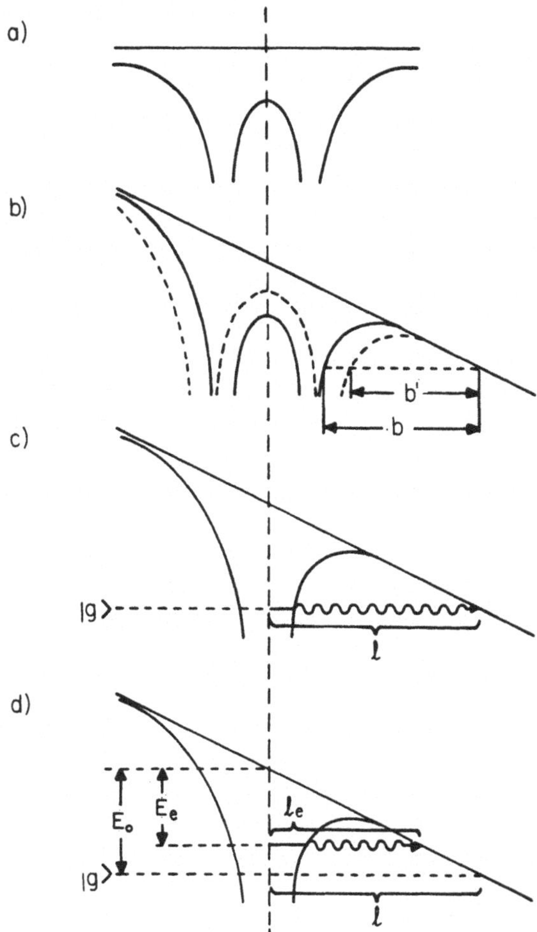

Figure 4. A schematic representation of the Coulomb potential of the D_2O molecule along one of the DO bond directions. (b) The same molecule in the presence of a linearly polarized, quasi-static laser field. The full curve represents the molecule in its ground electronic state. The broken curve represents the molecule in its dressed state in which the internuclear distances become large. b and b' correspond to barrier widths of the bare, and the expanded molecule, respectively. In (c), the Coulomb potential of the reduced atom's peripheral electron is shown. In the reduced atom picture, (d) illustrates the barrier widths and ionization potentials l, E_0 and l_e, E_e corresponding to the bare and dressed molecule, respectively.

so that intramolecular reactions following the ionization are immanent. In the case of water clusters, protonated clusters are often observed (Vernon et al., 1982).

Unprotonated cluster ions were first observed by Shinohara et al., (1986) for $(H_2O)_n^+$ $(2 \leq n \leq 10)$, in addition to the protonated ions, using VUV radiation. Shiromaru et al., (1987) used synchrotron radiation to measure the appearance potentials for water dimer and trimer ions and their protonated counterparts. The ionization process used in these experiments differs from that of electron bombardment in that just enough energy is provided for the ionization; there is very little excess energy involved. Thus, internal modes of the cluster ion are probably not excited and it remains in the same configuration as the neutral cluster. As a result, this "soft" ionization yields primarily parent cluster ions and few fragments. Shiromaru et al., (1987) have noted the difference between water cluster ionization using 40 eV electron bombardment and 14.6 eV synchrotron radiation. In both cases the excess energy is sufficient to produce protonated species[2], however, fewer protonated clusters were observed in the photoionization experiment than in the electron bombardment experiment. They remark that unprotonated species can be formed by even higher energy photons, and suggest that the intracluster proton transfer reaction forming protonated ions may be avoided if the excess energy is removed by the ejected electron. The addition of the energetic electron to the cluster likely induces reactions different from those when only energy is provided, such as in the case of photoionization.

We propose that the ionization by infrared pulses in the present intensity range, namely $\leq 5 \times 10^{12}$ W/cm^2, leaves the cluster ion with very little excess energy as in the soft, photo-ionization case mentioned above. Therefore, the parent dimer ion may be observed with very little evidence of D_3O^+, as indeed it has been in our experiments. Evidence that monomer ions are not produced by the fragmentation of the dimer is provided by correlation calculations performed on the time-of-flight spectra (Paquet et al., 1991). Furthermore, in the plot of ion number vs. laser intensity of fig. 2 the dependence of the ion signal on laser intensity is similar for both the monomer and the dimer. If the monomer was a product of the dimer, the dimer signal would decrease with increasing monomer signal.

4. On the Estimated Dressing Probability

We would like to fit our experimental data with some kind of formula involving tunnel ionization. Our model, however, stipulates that the ionization commences from a neutral, dressed molecule, not a bare one. The initial process will therefore be the dressing of the molecules. In order to estimate the probability of preparing the moleucle in the aforementioned dressed state, we use ionization rate formulae similar to those of Yergeau et al., (1987) for sequential multiple ionization of rare-gas atoms. The solution is briefly reviewed for clarity.

The number of ions created in a photoionization experiment, n_j, is the product of the initial, neutral sample density, N_0, and the total, local probability of ionization as a

[2]The minimum excess energy for proton transfer should be in the neighborhood of 0.52±0.12 eV, which is the internal excitation energy for $(H_2O)_2^+$ to dissociate into H_3O^+, measured in the photoionization experiments of Ng et al., (1977).

function of laser intensity $P_j^L(I)$ in the interaction volume, V, or

$$n_j = N_0 \iiint P_j^L(I)\, dV. \tag{1}$$

The probability is given by:

$$P_j^L(I) = 1 - exp\left[-\int_{-\infty}^{\infty} w_j(I(t))\, dt\right], \tag{2}$$

where $w_j(I(t))$ is the ionization rate (or more generally the rate of process j) as a function of laser intensity. With the intensity distribution separable: $I(r,t) = I_p\, S(r)\, T(t)$, a cylindrical Gaussian spatial distribution is assumed of the form $I_L = I_p exp[-(r/\omega_0)^2]$ where I_p is the peak laser intensity and ω_0 is the spot size at the laser focus. The interaction volume is assumed cylindrical of length δz. The temporal profile is also assumed to be Gaussian such that $I(t) = I_L \exp[-(t/\tau)^2]$, where τ is the laser pulse duration. Hence, the ion number is given by the volume integral of the probability (in cylindrical co-ordinates):

$$n_j = N_0\, \pi\, \omega_0^2\, \delta z \int_0^{F_p} \frac{P_j^L(F)}{F_L}\, 2dF_L. \tag{3}$$

The variable of the volume integration has been changed from the radial distance r, to that of the electric field amplitude, $F = 5.35 \times 10^{-9} \sqrt{I(\mathrm{W/cm^2})}$, in atomic units. Definitions of F_L and F_p follow from those for the intensity. The interaction volume, $V = \pi\, \omega_0^2\, \delta z$. The probability may be expressed in the form of equation (2) and substituted into (3) in order to incorporate the ionization rate into the formula for the resulting ion number:

$$n_j = n_0 \int_0^{F_p} \left(1 - exp\left[-\int_{-\infty}^{\infty} w_j(F(t))\, dt\right]\right) \frac{2dF_L}{F_L}, \tag{4}$$

where the number of neutral molecules in the interaction volume, $n_0 = N_0 V$, and the total probability of ionization (or process j) is represented by the integral in (4).

The total ionization process is separated into two parts: dressing and ionization. Not all of the molecules in the focal volume are dressed by the strong field in the same way due to the spatial and temporal distributions of the laser pulse. Only those molecules that are dressed, and in the potential well formed in the adiabatic electronic-field basis are in a position to be ionized. Therefore, the initial sample density for ionization is that of the "trapped," dressed molecules. The number of these particularly dressed molecules created by the laser pulse is described in a manner similar to equation (1):

$$n_D = N_0 \iiint P_D^L(I_L)\, dV. \tag{5}$$

where P_D^L shall be referred to as the local probability of dressing and P_{TI}^L the local tunnelling probability. The average dressing probability, P_D, is defined in terms of the local probability as:

$$P_D \equiv \frac{\iiint P_D^L(I_L)\, dV}{V}. \tag{6}$$

The average tunnelling probability, P_{TI}, is defined similarly. The dressing rate, w_D, which leads to P_D^L, represents the conglomeration of the rates of vibrational excitation and that of the transition to the potential well of the electronic-field basis. The number of ions produced by tunnelling from the sample of adequately prepared molecules may be written in terms of the initial number of neutral molecules in the sample, n_0. Since the number of dressed molecules is $n_D = N_0 V P_D = n_0 P_D$, the initial sample density for ionization is $N_D = n_D/V$. Thus, the number of ions produced from the sample of dressed molecules may be written as:

$$\begin{aligned} n_{TI} &= N_D \iiint P^L_{TI}(I_L)\, dV \\ &\equiv n_0\, P_D\, P_{TI}. \end{aligned} \tag{7}$$

An approximate tunnel ionization formula for ground state atoms is used to estimate the ionization rate due to tunnelling. Results of Ilkov et al., (1991) are used, which are similar to those of Ammosov, Delone and Krainov (ADK) theory (Ammosov et al., 1986), and have been shown to fit very well with data from the ionization of mercury atoms by similar CO_2 laser pulses. The tunnelling rate formula of Ilkov et al., (1991) is corrected for the dimensionless constant C_{n^*l} which reduces the ionization rate predicted by ADK theory by a factor of three. The following, for which $l = 0$ and $m = 0$, has been employed in our calculations (in a.u.):

$$w_{TI} = \left(\frac{3e}{\pi}\right)^{3/2} \frac{Z^2}{3n^{*9/2}} \left(\frac{4eZ^3}{n^{*4}F}\right)^{2n^*-3/2} \exp\left[\frac{-2Z^3}{3n^{*3}F}\right], \tag{8}$$

where $e = 2.718\ldots$ and Z is the charge of the atomic residue. The principle quantum number $n = Z(2E_0)^{-1/2}$ is replaced by the effective quantum number $n^* = n - \delta$, where δ is the quantum defect $\delta = n - (2E_0)^{-1/2}$.

The dressing probability is estimated by fitting experimental data with the numerical evaluation of the ion number produced in the process using formulae (7) and (4):

$$n_{TI} = \alpha \int_0^{F_p} \left(1 - exp\left[-\int_{-\infty}^{\infty} w_{TI}(F(t))\, dt\right]\right) \frac{2dF_L}{F_L}, \tag{9}$$

where $\alpha = n_0\, P_D$, and w_{TI} is given by (8). The neutral molecule density $N_0 \sim 10^{12}$ cm^{-3} and the interaction volume $V \sim 10^{-5}$ cm^3 are known from the experimental conditions, therefore $n_0 \sim 10^7$. In order to perform the numerical fit of (9), experimental points are plotted on a log-log scale assuming that the minimum ion signal corresponds to the detection of a few ions. Succeeding experimental points are placed relative to this first point. The theoretical curve is fitted by varying the values of α and n^* until a best least squares fit to the perpendicular distance to the slope (Bacon, 1953) is attained. Alterations in n^* correspond to movement of the calculated curve along the horizontal axis, while adjustment of α corresponds to vertical shifts in the calculated curve. The fit shown in fig. 2 for the monomer ions corresponds to $\alpha = 1030$ and $n^* = 2.08$ ($E_0 = 3.14$ eV). The fit of the dimer ion data yields similar results: $\alpha = 1050$, and $n^* = 2.04$ ($E_0 = 3.27$ eV). Therefore, within the experimental uncertainty, the probability of dressing for the monomer is estimated to be $P_D \sim 10^{-4}$, following equation (7). It is emphasized that the dressing rate deduced from this comparison is indeed a very rough approximation since the dressing rate should be a function of laser intensity.

We now address the question of why the measured appearance intensity for D_2O^+ does not conform to that which would be expected for the ionization potential provided by the numerical fit, with respect to known data for tunnelling in atoms. According to atomic data, the appearance intensity for the ionization of an atom with a 3 eV ionization potential should be less than the appearance intensity for potassium ($E_0 = 4.34$ eV), namely 5×10^{11} W/cm^2. The most obvious reason for the discrepancy is that for a given laser pulse, the ionization of molecules is not the same as the ionizaton of atoms. Another reason may be that the comparison of appearance intensities is just not a very precise method of comparing different ionization processes. For a given laser intensity, therefore ionization probability, the minimum ion number collected depends on the initial number of neutral molecules. In the present case, it is the number of dressed, trapped molecules which is relevent. This number in our experiment is certainly much smaller than the number of neutral potassium atoms in the atomic ionization experiment, for a similar focal volume. So the appearance intensity is higher for D_2O^+ than K^+, although the effective ionization potential of D_2O is less than the ionization potential of potassium.

5. Conclusion

Ionization of heavy water vapour by intense CO_2 laser pulses produces D_2O^+ and $(D_2O)_2^+$ ions. The ionization process can be explained by multiphoton absorption to a high level of vibrational excitation followed by strong field dressing of electronic states. Curve crossing produces a potential well which is populated by these high-lying, vibrational states. Tunnel ionization then occurs from the expanded, dressed molecule. Observation of the dimer ion, the lack of fragment ions, and evidence that the dimer does not dissociate into monomers, implies that the ionization process leaves the ion in a low vibrational state with very little energy.

A numerical fit of the experimental data with tunnel ionization formulae allow an estimation of the probability for the D_2O molecule to be prepared in the dressed state prior to ionization of $\sim 10^{-4}$. The ionization potential that the dressed monomer exhibits, from the point of view of tunnel ionization, is estimated to be about 3 eV.

It is our pleasure to acknowledge the technical expertise of S. Lagacé, and helpful discussions with J. Hepburn, T.T. Nguyen-Dang, O. Atabek, A. Bandrauk, P. Corkum, F.A. Ilkov, G. Mainfray and D. Normand. This work is supported in part by grants from the following organizations: NSERC, le Fonds FCAR and NATO.

References

Ammosov M.V., Delone N.B., Krainov V.P., 1986, *Zh. Eksp. Teor. Fiz.*, **91**:2008 (Engl. transl. 1986, *Sov. Phys.-JETP*, **64**:1191).

Bacon R.H., 1953, *Am. J. Phys.*, **21**:428.

Bandrauk A.D., Turcotte G., 1985, *J. Phys. Chem.*, **89**:3039.

Bandrauk A.D., Atabek O., 1989, *Lasers, Molecules and Methods*, eds. J.O. Hirschfelder, R.E. Wyatt, R.D. Coalson, *Advances in Chemical Physics vol. LXXIII* (J. Wiley & Sons, New York) p. 823.

Bucksbaum P.H., Zavriyev A., Muller H.G., Schumacher D.W., 1990, *Phys. Rev. Lett.*, **64**:1883.

Chin S.L., 1977, *Phys. Lett.*, **61A**:311.

Chin S.L., Faubert D., 1978, *Appl. Phys. Lett.*, **32**:303.

Chin S.L., Yergeau F., Lavigne P., 1985, *J. Phys. B: At. Mol. Phys.*, **18**:L213.

Chu S.I., 1981, *J. Chem. Phys.*, **75**:2215.

Codling K., Frasinski L.J., Hatherly P., Barr J.R.M., 1987, *J. Phys. B: At. Mol. Phys.*, **20**:L525.

Decker J.E., Xu G., Chin S.L., 1991a, *J. Phys. B: At. Mol. Opt. Phys.*, **24**:L281.

Decker J.E., Lagace S., Berube J., Beaudoin Y., Chin S.L., 1991b, *Appl. Opt.*, **30**:1888.

Evans D.K., McAlpine R.D., Ivanco M., 1989, *Laser Applications in Physical Chemistry* ed. D.K. Evans (Marcel Dekker Inc., New York) p. 64.

Freeman R.R., Bucksbaum P.H., 1991, *J. Phys. B: At. Mol. Opt. Phys.*, **24**:325.

George T.F., 1982, *J. Phys. Chem.*, **86**:10.

Gibson G., Luk T.S., Rhodes C.K., 1990, *Phys. Rev. A*, **41**:5049.

Guisti-Suzor A., He X., Atabek O., Mies F.H., 1990, *Phys. Rev. Lett.*, **64**:515.

He X., Atabek O., Guisti-Suzor A., 1988, *Phys. Rev. A*, **38**:5586.

Herzberg G., 1966a, *Molecular Spectra and Molecular Structure vol. III. Electronic Spectra and Electronic Structure of Polyatomic Molecules,* (Van Nostrand Reinhold Co., New York) p. 440.

— 1966b, *ibid.,* Ch. 3.

Ilkov F., Decker J.E., Chin S.L., 1991, *J. Phys. B: At. Mol. Opt. Phys., submitted.*

Keldysh L.V., 1964, *Zh. Eksp. Teor. Fiz.*, **47**:1945 (Engl. transl. 1965, *Sov. Phys.-JETP,* **20**:1307).

Lavancier J., Normand D., Cornaggia C., Morellec J., 1990, *J. Phys. B: At. Mol. Opt. Phys.*, **23**:1839.

Luk T.S., Rhodes C.K., 1988, *Phys. Rev. A*, **38**:6180.

Ng C.Y., Trevor D.J., Tiedemann P.W., Ceyer S.T., Kronebusch P.L., Mahan B.H., Lee Y.T., 1977, *J. Chem. Phys.*, **67**:4235.

Paquet E., Decker J.E., Chin S.L., 1991, *to be published.*

Schram B., Boerboom A., Kleine W., Kistemaker J., 1965, *Proceedings of the 7th Int. Conf. on Phenomena in Ionized Gases (Belgrade)* ed. Perovio, Tosic, vol. 1, 170.

Shinohara H., Nishi N., Washida N., 1986, *J. Chem. Phys.,* **84**:5561.

Shiromaru H., Shinohara H., Washida N., Yoo H.S., Kimura K., 1987, *Chem. Phys. Lett.,* **141**:7.

Tsurubuchi S., 1975, *Chem. Phys.,* **10**:335.

Vernon M.F., Krajnovich D.J., Kwok H.S., Lisy J.M., Shen Y.R., Lee Y.T., 1982, *J. Chem. Phys.*, **77**:47.

Xiong W., Yergeau F., Chin S.L., Lavigne P., 1988, *J. Phys. B: At. Mol. Opt. Phys.*, **21**:L159.

Xiong W., Chin S.L., 1991a, *Zh. Eksp. Teor. Fiz.*, **99**:481 (Engl. transl. 1991, *Sov. Phys.-JETP*, **72**:268).

Xiong W., Chin S.L., 1991b, *Laser and Particle Beams*, **10**:1.

Yergeau F., Chin S.L., Lavigne P., 1987, *J. Phys. B: At. Mol. Phys.*, **20**:723.

Zavriyev A., Bucksbaum P.H., Muller H.G., Schumacher D.W., 1990, *Phys. Rev. A*, **42**:5500.

EXCITATION OF MOLECULES AND SOLIDS WITH INTENSE SUBPICOSECOND ULTRAVIOLET RADIATION

T. S. Luk, D. A. Tate, A. McPherson, K. Boyer, and C. K. Rhodes

Department of Physics, University of Illinois at Chicago
P. O. Box 4348, Chicago, Illinois 60680

ABSTRACT

Studies of the interaction of matter with high intensity radiation are leading to the observation of new physical phenomena and the production of new classes of highly excited matter. Comparative studies of molecular coulomb explosions produced in O_2 and CO_2 with subpicosecond 248 nm radiation at intensities above 10^{17} W/cm^2 are revealing features of the multiquantum coupling that are sensitive to the molecular structure. With CO_2, O^{3+} ions with maximum kinetic energies on the order of 100 eV have been observed, a result which is comparable to the outcome of ion–molecule encounters at a collisional energy of ~ 1 MeV/au. Solid matter enables the combination of a high particle density with a high level of excitation. Studies of the excitation of solid surfaces are demonstrating the ability to produce fast, intense and easily controlled sources of kilovolt radiation under circumstances comparable to thermonuclear conditions.

I. INTRODUCTION

The availability of high peak power ultraviolet (UV) lasers[1–7] has revolutionized experimental research on the behavior of matter perturbed by very strong electric fields ($E \geqslant e^2/a_0$). Moreover, the situation is such that presently available theories are not effective in describing many aspects of this new physical regime. This is particularly true for many–electron systems, a class of materials that includes all molecules. Despite our imperfect understanding of the mechanisms involved in these high–field interactions, recent experimental results suggest that these multiphoton interactions can result in a high rate of energy transfer to the irradiated system. For an intensity of ~ 10^{17} W/cm^2, a total energy transfer to an atom or molecule exceeding 1 keV has been observed. Consequently, it is expected that total energy transfers considerably above 10 keV will be characteristic of the higher intensities (> 10^{19} W/cm^2) that will become available in the near future.

Molecules are emerging as a very interesting class of systems for study for at least two reasons. They are (1) the multitude of species readily available, which can involve almost all known atoms, and (2) the experimental finding that the dynamics of the multiphoton coupling is sensitive to the molecular structure.[8–10] Accordingly, data concerning two different molecules, O_2 and CO_2, each representing separate features of the observed interactions, are presented and discussed below.

Solid matter is another important class of materials that has recently come under very active study.[11–15] In particular, it has been shown that the characteristics of the dense plasmas that can be formed with intense subpicosecond irradiation are such that extremely powerful and spatially compact sources[14] of kilovolt radiation can be produced. Certain aspects of this work will be presented below, particularly in relation to issues associated with bright subpicosecond x–ray sources.

II. EXPERIMENTAL APPARATUS

The ultraviolet (248 nm) laser system used in these experiments is shown schematically in Fig. 1. Since this instrument has been described in detail elsewhere, the description given here will be brief. It has three main components; they are (1) an amplified infrared (745 nm) seed laser, (2) wavelength conversion crystals, and (3) ultraviolet (UV) excimer amplifiers. A recent modification to the system incorporates a $Ti:Al_2O_3$ amplifier for amplification of the 745 nm radiation prior to conversion to the UV. This has resulted in a 40–fold increase (to ~40 μJ) in the 248 nm output energy from the crystals, a result which greatly enhances the overall performance of the instrument.

The system has a final output pulse width of ~ 600 fs and a pulse energy of ~ 500 mJ. Stringent measures have been taken throughout the entire system to ensure high spatial quality of the output. This has resulted in a focusability which is within a factor of two of the diffraction limit for a 10 cm diameter beam. Indeed, with the use of a simple F/10 lens, peak intensities approaching a value of $\sim 10^{17}$ W/cm^2 can be produced, even in the presence of significant spherical aberration. For the present studies, the use of an improved focusing system involving an F/6 mirror permits peak intensities above 10^{17} W/cm^2 to be achieved.

III. DISCUSSION OF EXPERIMENTAL RESULTS

A. Atomic Ionization

The measurement of the threshold of ionization of free atoms can be used to give an experimentally based estimate of the peak field strength produced in the focal zone. With such data, the maximum intensity can be obtained with the use of models involving tunneling ionization.[16–22]

Measurements of krypton have been performed in order to appraise the maximum intensity achieved in the present studies. The results, as presented in the ion time–of–flight spectrum, are illustrated in Fig. 2. Clearly seen are signals from Kr^{8+} and Kr^{9+} along with a weak, but discernable, presence of Kr^{11+}. The O^{2+} peak arising from an impurity of H_2O in the chamber, masks the Kr^{10+} signal that occurs in its vicinity. Note also the prominent isotopic splittings that are present in the Kr^{8+} signal.

The representation given in Fig. 3 combines experimental data on the threshold intensities for ion production in Kr with the corresponding predictions of several theoretical models. The measured datum for Kr^{11+} (Z = 11), which is explicitly shown and has an ionization potential[23] $E_p \simeq 318$ eV, gives a peak intensity ~ 2 x 10^{17} W/cm^2. This experimental value is within approximately a factor of two of that predicted by the 1–D picture[16] and gives an intensity produced by the ultraviolet wave with a peak field of ~ 2.4 in atomic units (e/a_0^2). The total energy transfer needed to produce Kr^{11+} from the neutral atom is ~ 1348 eV.

B. Molecular Studies

Ion time–of–flight data, under appropriate conditions, can be used to give direct information on the kinetic energy distributions arising from the molecular

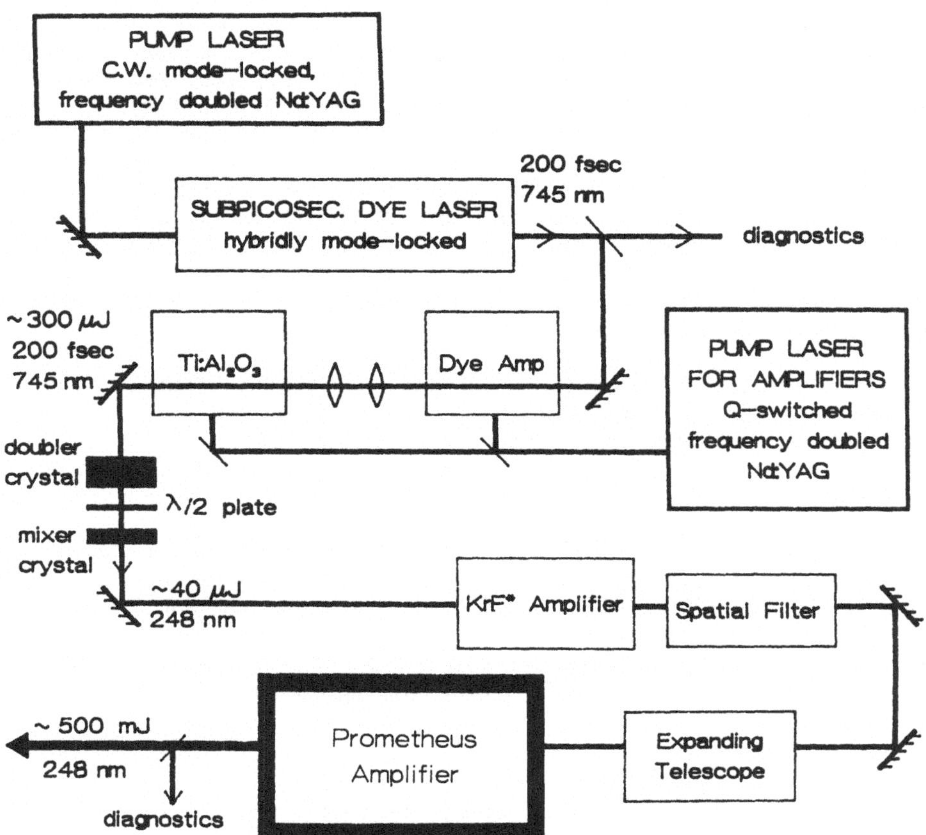

Fig. 1. A schematic of the subpicosecond high intensity KrF (248 nm) laser system.

coulomb explosions.[24,25] The specific temporal shapes of the ion signals in these experiments are determined by both the instrumental response and the kinetic energies of the fragments. The individual ion signals are actually composed of two components which in general, may or may not be fully resolved in the data. One component arises from ions initially formed with velocities parallel to the time-of-flight direction while the other arises from ions whose initial velocity is anti-parallel to that vector. The two resulting peaks generally are of unequal height and width, resolved or unresolved, due to the detailed nature of the trajectories which, in turn, are determined by the extraction and acceleration voltages used in the time-of-flight apparatus.[24,25] The separation of these two components, or an analysis of the shape of either component, provides a measure of the initial kinetic energy of the fragment. Naturally, the total ion yield is given by the integral of the signal over the distribution.

1. CO_2. In previous preliminary work, certain aspects of the fragmentation exhibited by CO_2 have been discussed. These earlier data[26] were taken with a peak intensity of ~ 3 x 10^{16} W/cm^2 and the observed kinetic energy distribution for the O^{3+} ions is shown in Fig. 4(a). In this case, most of the observed ions had kinetic energies considerably below ~ 65 eV.

The intensity dependence[9] of the distributions of ionic kinetic energy can reveal important properties of the molecular coupling. The production of Kr^{11+}, as discussed in paragraph III.A above, demonstrates that the improved focusing system produces a maximum intensity $I_m \sim 2 \times 10^{17}$ W/cm^2, a value considerably higher than available in the prior work.[26] Figure 4(b) illustrates the kinetic energy spectrum for the O^{3+} fragments observed from CO_2 with this higher value for the maximum intensity. Under these conditions, the peak in the kinetic energy distribution occurs at ~ 65 eV and a clearly discernable signal exists even in the 90–100 eV range. A comparison of these data with the results appearing in Fig. 4(a) distinctly illustrates a significant shift of the distribution to higher kinetic energies. The mean energy of the O^{3+} ions is roughly doubled in going from ~ 3 x 10^{16} W/cm^2 to ~ 2 x 10^{17} W/cm^2 in the maximum intensity. This result indicates that considerably more energy has been transferred to the molecule prior to fragmentation by the stronger external field.

An approximate appraisal of the kinetics of molecular dissociation for the O^{3+} data shown in Fig. 4(b) is informative. For this estimate, we will assume that the dissociation occurs as a two-body event of the form

$$(CO_2)^{(3+q)+} \longrightarrow O^{3+} + (CO)^{q+} \tag{1}$$

with the participation of a $(CO)^{q+}$ fragment. In the neutral ground state,[27] CO_2 has equilibrium C–O spacings of 1.177 Å. If we assume that the effective location of the (q+) charge of the CO fragment is located at its midpoint, then the O^{3+} ion is initially at a distance of $r_0 \sim 3.33\, a_0$. This picture assumes, of course, that no relative nuclear displacement develops during the process of ionization. The total Coulomb energy release E_C is then given approximately by

$$E_C \simeq \left(\frac{3q}{3.33}\right)\left(\frac{e^2}{a_0}\right). \tag{2}$$

Consequently, the full kinetic energy E_k associated with the dissociation, given the equal magnitudes of the momenta for the two fragments, is

$$E_k = E_{O^{3+}}\left(1 + \frac{M_O}{M_{CO}}\right) = 1.57\, E_{O^{3+}}, \tag{3}$$

in which $E_{O^{3+}}$ corresponds to the measured O^{3+} kinetic energy with M_O and M_{CO} representing the masses of the O^{3+} and $(CO)^{q+}$ fragments, respectively. For $E_{O^{3+}} \simeq 100$ eV, roughly the maximum ion energy observed, $E_k \simeq 157$ eV. Equating $E_k \simeq E_C$, we obtain $q \simeq 6.5$ which we interpret as $q \simeq 6$ for the CO^{q+} fragment. Since the CO^{q+} fragment involves two atoms, an effective value of $q \simeq 6$ for the diatomic fragment does not appear unreasonable for conditions that would generate O^{3+}. Interestingly, the coulomb energies associated with the motion of the heavy particles, in particular, either $E_{O^{3+}}$ or E_k, significantly exceed the ionization potential[23,28] of the O^{3+} fragment. It is, therefore, possible to imagine the coupling of this kinetic energy, perhaps through appropriate curve crossings,[10,29] to channels involving either excitation or further ionization. An experimental signature of such a mechanism would be the production of electronically excited ions, a question that demands further study.[30,31]

An important basic consideration relates to these molecular data. In the example examined in relation to Fig. 4(b), the nascent system produced is a species of the form $(CO_2)^{q_0^+}$ with a charge q_0 representing several electrons, perhaps, with a value as great as $q_0 \sim 9$. Moreover, this multiply charged system is produced by an interaction causing a negligible transfer of momentum to the molecule.[10] Core state excitation followed by Auger decay can produce multiple ionization, but generally that process generates for this material only doubly ionized material[32] with, perhaps, a small fraction of triply ionized ions. The maximum level of ionization that can be attained by the Auger mechanism for molecule composed of atoms having relatively low atomic numbers is modest[33, 34]

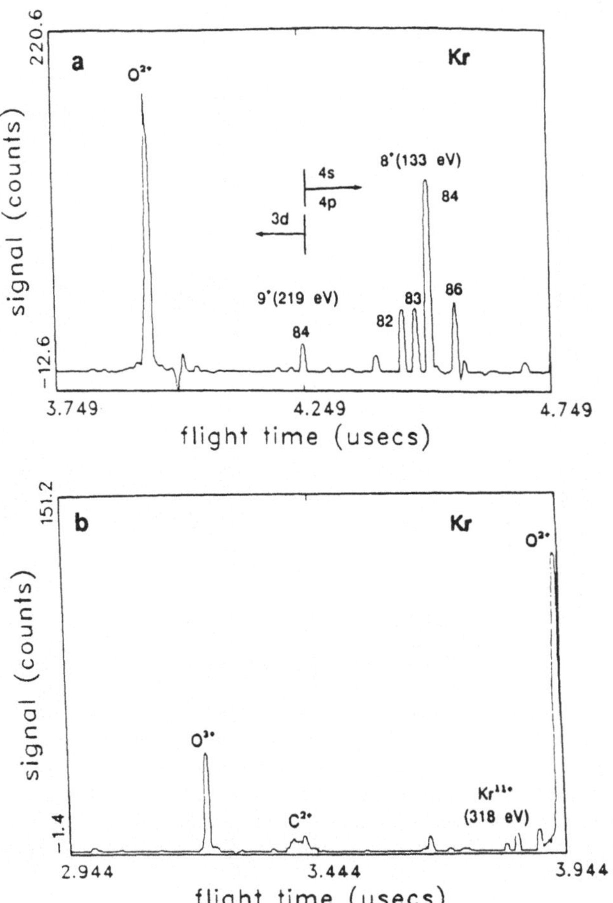

Fig. 2. Time–of–flight ion spectrum of Kr. (a) The region showing Kr^{8+} and Kr^{9+} with appropriate isotopic mass numbers identifying the peaks. Ionization with $q \geqslant 9$ must involve the 3d–shell. The Kr^{10+} signal is obscured by the O^{2+} signal arising from H_2O present as an impurity. (b) The region showing Kr^{11+}. O^{2+}, O^{3+}, and C^{2+} signals stemming from impurities are also identified. The ionization energies associated with the production of Kr^{8+}, Kr^{9+}, and Kr^{11+} species are shown in these panels as ~ 133 eV, ~ 219 eV, and ~ 318 eV, respectively.

Energetic heavy ion collisions[35,36] can generate considerably higher levels of ionization, but this interaction generally imparts a very substantial recoil momentum to the constituents of the molecular targets.[10] However, as the data show, the combination of a recoil-free interaction and a high level of ionization can be quite simply achieved by the strong-field multiquantum mechanism. This method, therefore, enables the unique production of forms of kinetically cold highly excited matter[30] that are essentially impossible to produce by other available methods.[10,26] Basically, these results give evidence that entirely new forms of excited material can be generated by strong-field coupling.

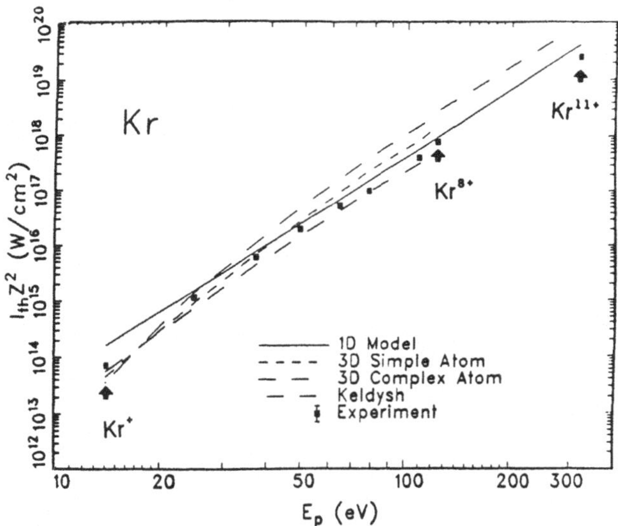

Fig. 3. Scaled threshold ionization intensities I_{th} as a fraction of ionization potential E_p for Kr along with calculated results from various theories. The parameter Z represents the final ionic charge. The theoretical curves correspond to a 1D model, 3D simple atom picture, a 3D complex atom model, and the Keldysh theory, as discussed in Ref. (22). Experimental data are also shown, in particular, for the point corresponding to Kr^{11+} (E_p = 318 eV).

The comparison of the kinetic energies seen for the O^{3+} fragment from CO_2, as shown in Fig. 4(b), with those produced by ion-molecule collisions is also informative. The collisional process that has been studied[37] involved the reaction

$$Ar^{12+} + CO_2 \longrightarrow Ar^{12+} + \text{fragments } (C^{i+}, O^{j+}, e^-). \tag{4}$$

The collisional energy for process (4) corresponded to ~ 1 MeV/au so that the time of interaction is sufficiently small that negligible nuclear motion developed in the molecular frame during the process of ionization. Hence, the resulting coulomb explosion commenced from an intact ionized molecule in a spatial configuration determined by the initially neutral target molecule.

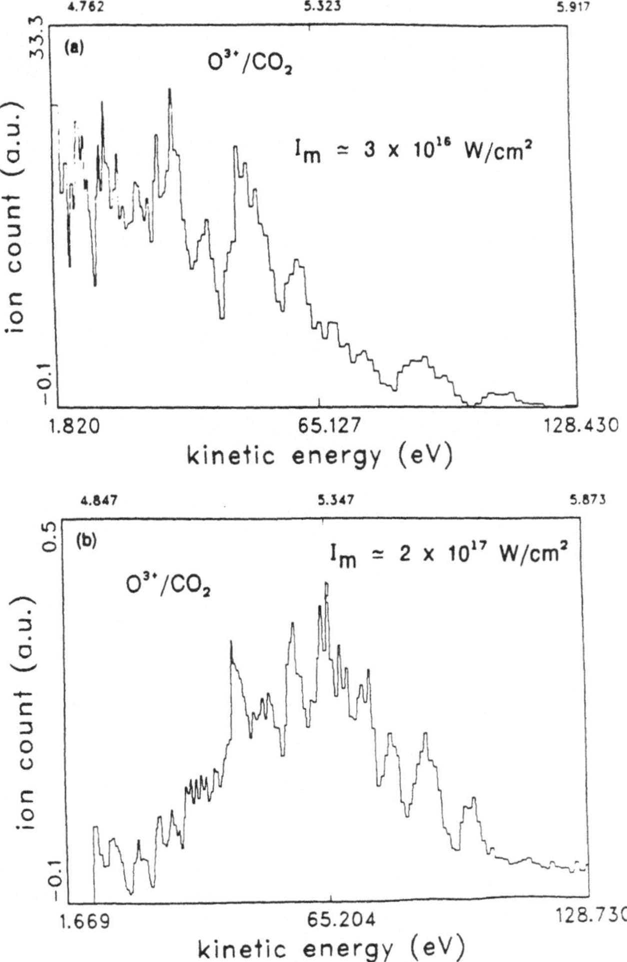

Fig. 4. Kinetic energy spectra for O^{3+} from CO_2. (a) Ion distribution obtained with a maximum intensity $I_m \sim 3 \times 10^{16}$ W/cm². Most ions have an energy below ~ 65 eV. (b) Ion distribution observed with a maximum intensity $I_m \simeq 2 \times 10^{17}$ W/cm². The peak in the kinetic energy distribution occurs at ~ 65 eV.

The collisional data[37] for oxygen ions produced from CO_2 are shown in Fig. 5. Also indicated for comparison with these data is a representation of the kinetic energies characteristic of the O^{3+} ions observed in the multiphoton studies [Fig. 4(b)]. The maximum range of energies seen in the multiphoton case is of the same order as that corresponding to the collisional case. This similarity in the fragment energies implies that the nature of the molecular ion which undergoes the coulomb explosion is essentially the same in both the multiphoton and collisional cases. Specifically, we conclude that the strong–field interaction, in relation to the total energy exchanged between the field and the molecule, has many features in common with ion–atom and ion–molecule collisions, an analogy which has been explored previously in other studies.[38,39]

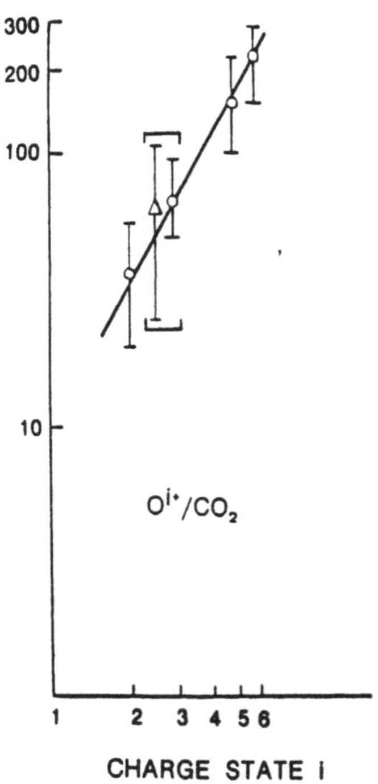

Fig. 5. The data of Ref. (37) on oxygen ions produced in Ar^{12+} ion–molecule collisions in the reaction $Ar^{12+} + CO_2 \rightarrow Ar^{12+}$ + fragments (C^{i+}, O^{i+}, e^+) at a collisional energy of ~ 1 MeV/au. The multiphoton datum (Δ) for O^{3+}, produced with $I_m \simeq 2 \times 10^{17}$ W/cm^2 is shown for comparison. Data of Ref. (37) reproduced with permission.

2. O_2. In comparison with the experimental findings for CO_2, the results observed for O_2 reveal additional features of multiphoton coupling to molecules. We now consider the characteristics of the ionic kinetic energy spectra obtained from O_2 with a maximum intensity $I_m \simeq 2 \times 10^{17}$ W/cm^2.

The spectrum for the O^{3+} is illustrated in Fig. 6(a). It is immediately clear, by comparing these data with that shown in Fig. 4(b), that the kinetic energy associated with the peak of the distribution arising from O_2 is nearly an order of magnitude lower than the corresponded peak for CO_2. Under comparable conditions of irradiation, O_2 produces relatively slow O^{3+} fragments.

The kinetic energy distribution for O^{5+} fragments is shown in Fig. 6(b). Although the shapes of the distributions differ in detail, the O^{5+} ions occur over a range of energies which is essentially the same as that observed for the O^{3+} fragments. This finding shows that the higher charge of the O^{5+} ion does not correlate with an appreciably greater kinetic energy of the particle. This observation, the relatively low values of the kinetic energies produced, and the general similarity of the O^{3+} and O^{5+} distributions, suggests that appreciable separation of atoms in the O_2 system has developed prior to the final steps of ionization yielding the O^{3+} and O^{5+} species. Indeed, the decrease in the O^{3+} signal occurring above ~ 12–15 eV, in light of the presence of a corresponding O^{5+} signal in that same energy range, additionally suggests that an appreciable fraction of the O^{5+} distribution arises from $O^{3+} \rightarrow O^{5+}$ conversion by multiphoton ionization at relatively large internuclear separations.

The picture of the coupling described above is supported by the data on all of the charge states observed with O_2. The overall characteristics of the spectra of kinetic energies for the O^{q+} ($1 \leqslant q \leqslant 6$) fragments are presented in Fig. 7. The vertical bars indicate the ranges of the observed energies while the circles designate the locations of peaks visible in the distributions. It is significant that the ranges shown for q = 3, 4, 5, and 6 are essentially identical. These data indicate that the dissociation of this system mainly occurs while the molecule is in rather low stages of ionization, probably corresponding to total molecular charges not exceeding four.

The electronic structure of O_2, which involves two relatively weakly bound electrons in the anti-bonding $1\pi_g$ orbital,[40] has several relatively low-lying manifolds of states that can couple strongly at 248 nm, in low-order (e.g. two photon) processes,[41] to dissociating pathways. Since the pulse has a finite rise-time, the implication is that anomalously low intensities on the leading edge can be effective in initiating the act of dissociation. This process would tend to cause the ionization arising from the most intense interval of the pulse to occur at internuclear separations appreciably greater than the equilibrium value associated with neutral O_2.

The intensity of 248 nm radiation required to produce a significant rate of excitation of these lower lying levels of O_2 can be estimated[41] from available structural data.[40] If we consider a two quantum process connecting the $X^3\Sigma_g^-$ level of O_2 to a suitable gerade state[40] with an excitation of ~ 10 eV, then a simple estimate of the coupling strength can be made.[41] If we assume for the estimate of this amplitude that the intermediate state is the $B^3\Sigma_u^-$ level, and that the interaction is characterized by an energy denominator of ~ 4 eV, transition dipole matrix elements of ~ 1 Debye, and an effective linewidth of ~ 100 cm^{-1}, the two photon coupling parameter is $\alpha \simeq 2 \times 10^{-32}$ cm^4/W. This value gives an excitation rate of ~ $10^{14} s^{-1}$ at an intensity of ~ 1.4×10^{13} W/cm^2, a value approximately 10^4-fold below the maximum intensity used for irradiation. This estimate suggests that the high density of relatively lowly excited states characteristic of O_2, which can couple in near-resonant two photon processes, can encourage dissociation in this system so that appreciable nuclear displacement develops during the multiphoton ionization. For other molecules, such as N_2, CO, and CO_2, in which the excited states lie characteristically higher, it is expected that the influence of this effect would be significantly reduced. This expectation is consistent with the observations of the relatively fast O^{3+} fragments produced with CO_2 that were discussed above in Section III.B.1.

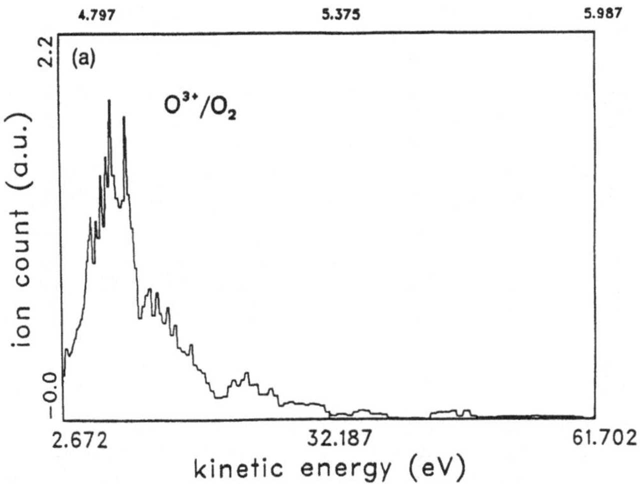

Fig. 6(a). Kinetic energy spectrum of O^{3+} produced from O_2 at a peak intensity $I_m \simeq 2 \times 10^{17}$ W/cm^2.

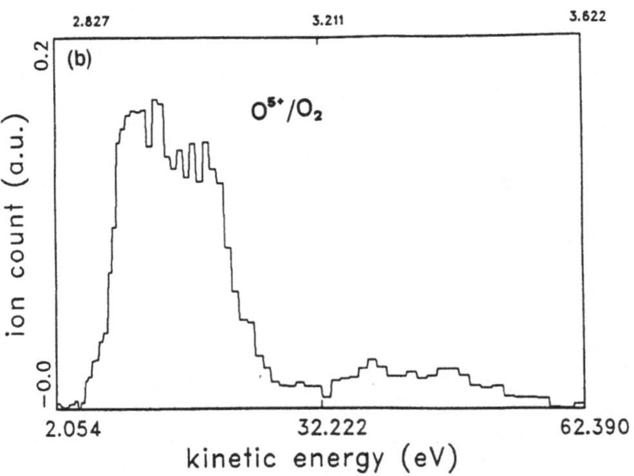

Fig. 6(b). Kinetic energy spectrum of O^{5+} produced from O_2 at a peak intensity $I_m \simeq 2 \times 10^{17}$ W/cm^2.

C. X-Ray Production from Solid Surfaces

Solid materials offer the possibility of combining a high particle density with a very high level of electronic excitation, if the deposition of energy in the solid can be made to occur sufficiently rapidly. High brightness subpicosecond light sources of the type shown in Fig. 1 now enable such rapid excitation to be achieved. Several recent studies[42–46] have probed the physics of these high-density interactions. Given the fact that the studies of ionization described in Section III.A. above indicated total energy investments considerably above 1 keV per atom, at least for a material as heavy as Kr, the immediate suggestion is that a properly irradiated solid could serve as a powerful source of x-rays in the kilovolt region. We now consider certain studies that have examined the kilovolt spectral range.

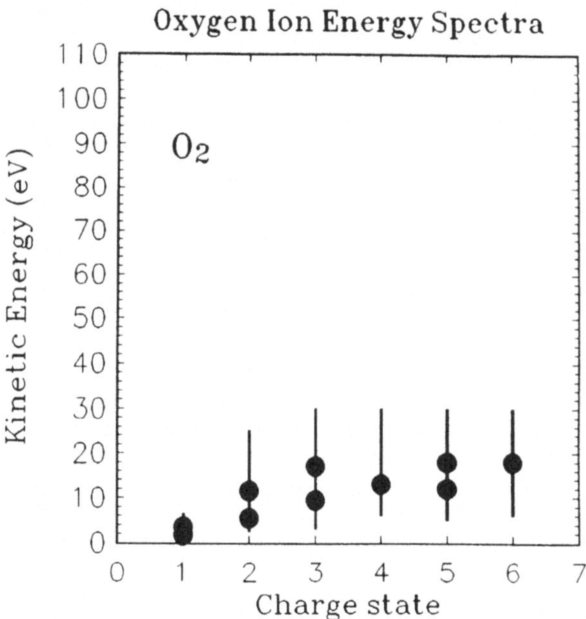

Fig. 7. Ranges of energies (vertical bars) observed for O^{q+} ($1 \leqslant q \leqslant 6$) fragments from O_2. The circles represent the locations of peaks visible in the distribution.

It is experimentally desirable to provide the intense irradiation directly to the high density material presented by the solid surface. Therefore, an experimental concern in studies of the interaction of intense radiation with solid matter is the influence of any low intensity prepulse on the true state of the irradiated system. Such radiation could produce a low density plasma and alter the interaction with the solid material. It has been experimentally found[47] that the effect of the prepulse can be entirely eliminated at 248 nm by the use of a target composed of ultraviolet transmitting materials. For example, MgF_2 and BaF_2 exhibit essentially no absorption up to relatively high intensities and studies have demonstrated that these values are sufficiently high to prevent any significant influence by the prepulse. This enables the intense subpicosecond energy to couple to solid matter and produce dense high-temperature plasmas.[15,47]

For these experiments the KrF^* system illustrated in Fig. 1 delivered a subpicosecond pulse energy of ~ 150 mJ. The accompanying prepulse had a duration of ~ 20 ns, an energy of ~ 10 mJ, and is known to be very poorly focusable. The radiation was focused onto a flat rotating BaF_2 target with a f/10 CaF_2 lens. The peak laser intensity achievable on target was estimated to be ~ 10^{17} W/cm^2, based on direct imaging of the focal zone with a CCD camera. The x-ray emission from the target was collected by flat KAP (2d=26.6Å) and PET (2d=8.7Å) crystal spectrometers and recorded on DEF film. For higher resolution, a convex KAP crystal spectrometer was used. Approximately 25 shots were accumulated for the recording of each spectrum.

The spectrum[14] observed from BaF_2 irradiated at an average intensity of ~ 3×10^{16} W/cm^2 exhibited strong emission in the 9–13 Å range as shown in Fig. 8. In addition to the spectrum, the strength of the radiation is significant and calibration of the film revealed that ~ 3–6 mJ of x-rays were produced on each shot. Subsequent pinhole camera measurements of the x-ray emission from solid targets (MgF_2/SiO_2) indicated a radiating spot size with a diameter in the range of 8–9 μm. This value is quite close to the independently measured focal diameter of the optical system. Finally, experimental results,[11,14] corroborated by calculations, show that the radiation occurs over a time of ~ 1 ps, a value that is consistent with the x-ray pinhole camera measurements. Therefore, assuming isotropic emission from an area of these dimensions, the experimentally radiated intensity I_r was approximately $I_r \sim 2\text{–}3 \times 10^{15}$ W/cm^2.

The radiated intensity in the kilovolt range that was deduced from the spectral measurements is sufficiently high that an estimate of the upper bound that could be produced is of interest. We now estimate this limiting intensity. Consider a slab of unit area with density ρ and thickness δ which, upon excitation with intense subpicosecond radiation, produces radiation in the kilovolt spectral range. Since the time of excitation is short, as a first approximation, we will ignore the hydrodynamical expansion, a process which will occur roughly at a rate of ~ 1–10 Å/fs. An upper limit on the x-ray intensity I_{xm} can then be simply written as

$$I_{xm} \sim \frac{\hbar\omega_x \rho \delta}{2\tau_x}, \tag{5}$$

in which $\hbar\omega_x$ represents a kilovolt quantum and τ_x is the effective radiative rate of the excited material. For the present discussion, we will set ρ as solid density (~ 3×10^{22} cm^{-3}) and take as the lower limit for τ_x a typical radiative rate for kilovolt transitions. For the latter we will use the range $2 \times 10^{-14} \leqslant \tau_x \leqslant 10^{-13}$ s as a rough measure.[48] If we take for δ the recently measured value[15] of the depth of energy penetration of ~ 250 nm for MgF_2/SiO_2 targets irradiated with 248 nm at an intensity of ~ 3×10^{16} W/cm^2, Eq. (5) can be evaluated. With these parameters we find $6 \times 10^{14} \leqslant I_{xm} \leqslant 3 \times 10^{15}$ W/cm^2. This simple upper limit on I_{xm} can now be compared to the experimental result.[14]

We see immediately that the experimental value falls in the range given by the estimate of the upper bound I_{xm}. The measured performance, therefore, appears close to the maximum possible. Although such a source is incoherent, a radiation rate of this magnitude, coupled with the relatively narrow spectrum shown in Fig. 8, constitutes a source of extremely high spectral brightness that can produce unusually high excitation rates in absorbing material.

IV. CONCLUSIONS

The use of recently developed high brightness subpicosecond light sources for the study of the interaction of matter with strong electromagnetic fields ($E > e/a_0^2$) is leading to the observation of new phenomena and new classes of excited states of matter. In particular, it has been found by experiments on both molecules and solids that very high levels of excitation can be easily created by intense subpicosecond irradiation. In this work, the study of molecular coulomb explosions is furnishing direct information on details of the strong field coupling. The variety of molecular systems available, together with the finding that the coupling is sensitive to the molecular structure, leads to the existence of a rich and extensive range of phenomena involving the dynamics of the flow of energy in these complex systems. A topic of particular interest is the potential role of surface crossings in the production of electronically excited fragments, a process that could serve as a selective mechanism for channeling energy into excited levels. The results with solids indicate that a new genre of fast, intense and easily controlled kilovolt x-ray sources may be feasible with power densities comparable to thermonuclear conditions.

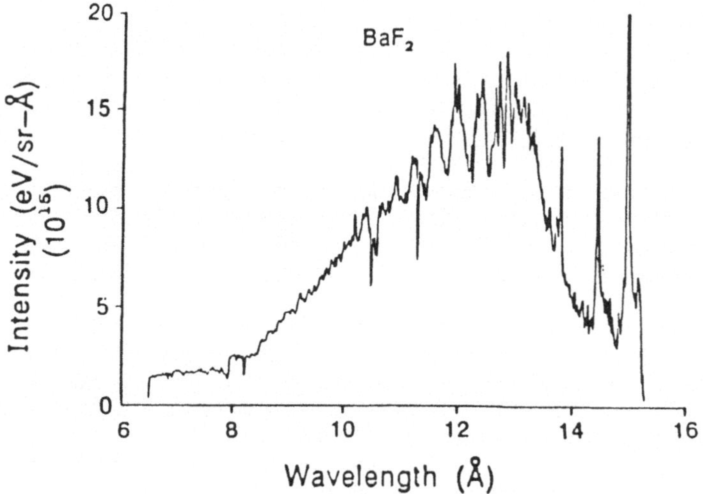

Fig. 8. Spectrum obtained from BaF_2 at an average intensity of ~ 1–2 x 10^{16} W/cm^2.

V. ACKNOWLEDGEMENTS

The technical assistance of P. Noel and J. Wright, and helpful discussions with J. C. Solem and B. E. Bouma, are acknowledged. The fruitful and congenial participation of A. Zigler, P. G. Burkhalter, D. J. Nagel, and M. D. Rosen, in the studies of solid materials is warmly acknowledged. Support for this research was provided by the U.S. Air Force Office of Scientific Research, the U.S. Office of Naval Research, the Department of Energy, the U.S. Army Research Office, and the Strategic Defense Initiative Organization.

VI. REFERENCES

1. S. Szatmári, F. P. Schäfer, E. Müller–Horsche, and W. Mückenheim, "Hybrid Dye–Excimer Laser System for the Generation of 80 fs, 900 GW Pulses at 248 nm," Opt. Comm. 63, 305 (1987).
2. A. Endoh, M. Watanabe, N. Sarukura, and S. Watanabe, "Multiterawatt Subpicosecond KrF Laser," Opt. Lett. 7, 353 (1989).
3. J. R. M. Barr, N. J. Everall, C. J. Hooker, I. N. Ross, M. J. Shaw, W. T. Toner, "High Energy Amplification of Picosecond Pulses at 248 nm," Opt. Comm. 66, 127 (1988).
4. J. G. Glownia, J. Misevich, and P. P. Sorokin, "160–fsec XeCl Excimer Amplifier System," J. Opt. Soc. Am. B4, 1061 (1987).
5. A. P. Schwarzenbach, T. S. Luk, I. A. McIntyre, U. Johann, A. McPherson, K. Boyer, and C. K. Rhodes, "Subpicosecond KrF^* Excimer–Laser Source," Opt. Lett. 11, 499 (1986).
6. J. P. Roberts, A. I. Taylor, P. H. Y. Lee, and R. B. Gibson, "High–Irradiance 248–nm Laser System," Opt. Lett. 13, 734 (1988).
7. T. S. Luk, A. McPherson, G. Gibson, K. Boyer, and C. K. Rhodes, "Ultrahigh–Intensity KrF^* Laser System," Opt. Lett. 14, 1113 (1989).
8. L. J. Frasinski, K. Codling, and P. A. Hatherly, "Covariance Mapping: A Correlation Method Applied to Multiphoton Multiple Ionization," Science 246, 1029 (1989).
9. L. Lavancier, D. Normand, C. Cornaggia, J. Morellec, and H. X. Lin, "Laser–Intensity Dependence of the Multielectron Ionization of CO at 305 and 610 nm, Phys. Rev. A 43, 1461 (1991).
10. T. S. Luk, A. McPherson, G. N. Gibson, K. Boyer, and C. K. Rhodes, "Molecular X–Ray Laser Research," Izvestia Academia Nauk, Seria Fizicheskaya, T. 55, 768 (1991) (in Russian); English Translation to be published in the Proceedings of the International Symposium, "Short Wavelength Lasers and Applications," Samarcand, USSR, 14–18 May 1990.
11. M. M. Murnane, H. C. Kapteyn, and R. W. Falcone, "High–Density Plasmas Produced by Ultrafast Laser Pulses," Phys. Rev. Lett. 62, 155 (1989).
12. H. M. Milchberg, R. R. Freeman, S. C. Davey, and R. M. More, "Resistivity of a simple Metal from Room Temperature to 10^6K," Rev. Lett. 61, 2364 (1989).
13. M. M. Murnane, H. C. Kapteyn, M. D. Rosen, and R. W. Falcone, "Ultrafast X–Ray Pulses from Laser–Produced Plasmas," Science 251, 531 (1991).
14. A. Zigler, P. G. Burkhalter, P. J. Nagel, K. Boyer, T. S. Luk, A. McPherson, J. C. Solem, and C. K. Rhodes, "High Intensity Generation of 9–13 Å X–Ray from BaF_2 Targets," Appl. Phys. Lett. 59, 777 (1991).
15. A. Zigler, P G. Burkhalter, D. J. Nagel, M. D. Rosen, K. Boyer, G. N. Gibson, T. S. Luk, A. McPherson, and C. K. Rhodes, "Measurement of Energy Penetration Depth of Subpicosecond Laser Energy into Solid Density Matter," Appl. Phys. Lett. 59, 534 (1991).
16. S. Augst, D. Strickland, D. D. Meyerhofer, S. L. Chin, and J. H. Eberly, "Tunneling Ionization of Noble Gases in a High–Intensity Laser Field," Phys. Rev. Lett. 63, 2212 (1989).
17. N. H. Burnett and P. B. Corkum, "Cold–Plasma Production for Recombination Extreme–Ultraviolet Lasers by Optical–Field–Induced Ionization," J. Opt. Soc. Am. B 6, 1195 (1989).
18. D. Landau and E. M. Lifshitz, Quantum Mechanics, 3rd ed. (Pergamon, London, 1978).
19. B. M. Smirnov and M. I. Chibisov, "The Breaking Up of Atomic Particles by An Electric Field and by Electron Collisions," Sov. Phys. JETP 22, 585 (1966).
20. A. M. Perelomov, V. S. Popov, and M. V. Terent'ev, "Ionization of Atoms in an Alternating Electric Field," Sov. Phys. JETP 23, 924 (1966).
21. M. V. Ammosov, N. B. Delone, and V. P. Krainov, "Tunnel Ionization of Complex Atoms and of Atomic Ions in an Alternating Electromagnetic Field," Sov. Phys. JETP 64, 1191 (1986).

22. G. N. Gibson, T. S. Luk, and C. K. Rhodes, "Tunneling Ionization in the Multiphoton Regime," Phys. Rev. A 41, 5049 (1990).
23. Thomas A. Carlson, C. W. Nestor, Jr., Niel Wasserman, and J. D. McDowell, "Calculated Ionization Potentials for Multiply Charged Ions," Atomic Data 2, 63 (1970).
24. K. Boyer, T. S. Luk, J. C. Solem, and C. K. Rhodes, "Kinetic Energy Distributions of Ionic Fragments Produced by Subpicosecond Multiphoton Ionization of N_2," Phys. Rev. A 30, 1186 (1989).
25. T. S. Luk and C. K. Rhodes "Multiphoton Dissociative Ionization of Molecular Deuterium," Phys. Rev. A 38, 6180 (1988).
26. T. S. Luk, A. McPherson, B. E. Bouma, K. Boyer, and C. K. Rhodes, "Studies of the Interaction of Molecules and Solids with Intense Subpicosecond 248 nm Radiation," in Multiphoton Processes, edited by G. Mainfray and P. Agostini (CEA, Saclay, 1991) p. 217.
27. G. Herzberg, Molecular Spectra and Molecular Structure III. Electronic Spectra and Electronic Structure of Polyatomic Molecules (D. Van Nostrand Co., Inc., Princeton, NJ, 1966).
28. R. L. Kelly, "Atomic and Ionic Spectrum Lines below 2000 Angstroms: Hydrogen through Krypton," Part I, J. Phys. Chem. Ref. Data 16, Suppl. No. 1 (1987).
29. T. S. Luk, A. McPherson, K. Boyer, and C. K. Rhodes, "Mechanisms of Short Wavelength Generation," to be published in Atoms in Intense Radiation Fields, edited by M. Gavrila (Academic Press, Inc., Harcourt Brace Jovanovich, San Diego, CA, 1991).
30. A. McPherson, T. S. Luk, G. N. Gibson, K. Boyer, and C. K. Rhodes, "Ion Production and Molecular Excitation Occurring in Multiphoton Processes," Nucl. Instr. Methods B 43, 468 (1989).
31. T. S. Luk, R. M. Moriarty, A. Awashti, K. Boyer, and C. K. Rhodes, "Isotopic Studies of Atomic Site–Selectivity in Molecular Multiphoton Ionization of N_2O," Phys. Rev. A, submitted.
32. A. P. Hitchcook, C. E. Brion, and M. J. van der Wiel, "Ionic Fragmentation of Inner Shell Excited States of CO_2 and N_2O," Chem. Phys. Lett. 66, 213 (1979).
33. W. Eberhardt, E. W. Plummer, In Whan Lyo, R. Reininger, R. Carr, W. K. Ford, and D. Sondericker, "Auger Electron–Ion Coincidence Studies to Determine the Pathways in Soft X–Ray Induced Fragmentation of Isolated Molecules," Aust. J. Phys. 39, 633 (1986).
34. C. J. Allen, V. Gelius, D. A. Allison, G. Johansson, H. Siegbahn, and K. Siegbahn, "ESCA Studies of CO_2, CS_2, and COS," J. Electron. Spectrosc. 1, 131 (1972).
35. Schartner, K. –H., "Highly Charged Recoil Ions," in Fundamental Processes in Energetic Atomic Collisions, H. O. Lutz, J. S. Briggs, and H. Kleinpoppen, eds., (Plenum Press, New York, 1983) Series B, Vol. 103, p. 637.
36. A. S. Schlachter, W. Groh, A. Müller, H. F. Beyer, R. Mann, and R. E. Olson, "Production of Highly Charged Rare–Gas Recoil Ions by 1.4 MeV/amu U^{44+}," Phys. Rev. A 26, 1373 (1982).
37. T. Matsuo, T. Tonuma, M. Kase, T. Kambara, H. Kumagai, and H. Tawara, "Production of Multiply Charged Ions from CO and CO_2 Molecules in Energetic Heavy–Ion Impact," Chem. Phys. 121, 93 (1988).
38. K. Boyer, and C. K. Rhodes, "Atomic Inner–Shell Excitation Induced by Coherent Motion of Outer–Shell Electrons," Phys. Rev. Lett. 54, 1490 (1987).
39. K. Boyer, G. Gibson, H. Jara, T. S. Luk, I. A. McIntyre, A. McPherson, R. Rosman, J. C. Solem, and C. K. Rhodes, "Corresponding Aspects of Strong–Field Multiquantum Processes and Ion–Atom Collisions," IEEE Transactions on Plasma Science 16, 541 (1988).
40. Paul H. Krupenie, "The Spectrum of Molecular Oxygen," J. Phys. Chem. Ref. Data 1, 423 (1972).
41. W. K. Bischel, J. Bokor, D. J. Kligler, and C. K. Rhodes, "Nonlinear Optical Processes in Atoms and Molecules Using Rare Gas Halide Lasers," IEEE J. Quantum Electron. QE–15, 380 (1970).
42. M. M. Murname, H. C. Kapteyn, and R. W. Falcone, "High–Density Plasmas Produced by Ultrafast Laser Pulses," Phys. Rev. Lett. 62, 155 (1989).
43 H. M. Milchberg, R. R. Freeman, S. C. Davey, and R. M. More, "Resistivity of a Simple Metal from Room Temperature to 10^6K," Phys. Rev. Lett. 61, 2364 (1988).

44. Falcone, R. W., Kapteyn, H. C., Murnane, M. M., and Rosen, M. D., "Ultrafast X-Ray Pulses from Laser-Produced Plasmas," Science 251, 531 (1991).
45. J. A. Cobble, G. A. Kyrala, A. A. Hauer, A. J. Taylor, C. C. Gomez, N. D. Delamater, and G. T. Schappert, "Kilovolt X-Ray Spectroscopy of a Subpicosecond-Laser-Excited Source," Phys. Rev. A 39, 454 (1989).
46. J. A. Cobble, R. D. Fulton, L. A. Jones, G. A. Kyrala, G. T. Schappert, G. A. Taylor, "The Interaction of a High Irradiance, Subpicosecond Laser Pulse with Aluminum: The Effects of the Prepulse on X-Ray Production," J. Appl. Phys., 69, 3369 (1991).
47. A. Zigler, P. G. Burkhalter, D. J. Nagel, M. D. Rosen, K. Boyer, T. S. Luk, A. McPherson, and C. K. Rhodes, "Plasma Production from Ultraviolet-Transmitting Target Using Subpicosecond Ultraviolet Radiation," Opt. Lett. 16, 1261 (1991).
48. J. H. Scofield, "Radiative Transitions," in Atomic Inner-Shell Processes, edited by Bernd Crasemann (Academic, New York, 1975), Vol. I, p. 265.

COHERENCE IN STRONG FIELD HARMONIC GENERATION

Anne L'Huillier and Philippe Balcou

Service des Photons, Atomes et Molécules
Centre d'Etudes de Saclay 91191 Gif-sur-Yvette FRANCE

Kenneth J. Schafer and Kenneth C. Kulander

Physics Department, Lawrence Livermore National Laboratory
Livermore CA 94550 USA

INTRODUCTION

The high-order harmonics generated when an intense, short-pulsed laser is focused into an atomic gas provide not only a surprisingly strong source of short wavelength, coherent radiation, but also prove to be a sensitive probe of the dynamics of atoms in strong electromagnetic fields. The source of the harmonic radiation is the polarization induced in the atomic medium by the pump laser. This polarization results from the distortion of the electronic charge density in each individual atom, producing a time-dependent dipole moment whose phase is determined by the phase of the driving field at the atom. The Fourier transform of the atomic dipole is directly related to the strengths and phases of the harmonic emission from the individual atoms. Interference between the harmonic fields produced at different points in the medium can be constructive or destructive depending on the phases set by the incident field. The overall harmonic conversion efficiency can depend very strongly on these phase matching effects. Therefore, modelling strong-field harmonic conversion requires both the calculation of the single atom response and the solution of the propagation equations for the harmonic fields.

In this paper we discuss recent studies of harmonic generation in rare gases, including our current understanding of the special circumstances in the intense field regime which result in the observation of very strong high-order harmonics. The combination of a non-perturbative atomic response and enhanced phase matching in the medium is responsible for the high conversion efficiencies. The calculations point to those attributes of the microscopic atomic emission rates which yield the strong macroscopic, coherent harmonic output. We begin with a simple model of a one- dimensional electron in a strong oscillating field to provide some insight into the source of the single-atom emission spectrum. Next, we present a brief description of our calculations for realistic atoms. A review of recent experimental results is followed by a discussion of the method used for calculating the effects of phase matching on the emitted radiation. In the final section we discuss the agreement between the experimental and theoretical results, and describe the important differences

Coherence Phenomena in Atoms and Molecules in Laser Fields
Edited by A.D. Bandrauk and S.C. Wallace, Plenum Press, New York, 1992

between traditional, perturbative harmonic generation and that found in this new, non-perturbative, strong field regime.

A STRONGLY DRIVEN ANHARMONIC OSCILLATOR

We first consider a very simple nonlinear system, the classical one-dimensional quartic oscillator[1] (a Duffing oscillator), whose motion is described by the differential equation:

$$\ddot{x} + \omega_0^2 x + v x^3 = F \cos \omega t. \tag{1}$$

$F \cos \omega t$ is a (strong) external force. The oscillator frequency ω_0 is chosen to be 10 times the frequency ω (=1) of the driving field. The particle (x) moves in the potential $V(x) = \omega_0^2 x^2/2 + v x^4/4$, with an anharmonic term. v is taken to be 500. In Fig. 1, we show the spectrum emitted by this oscillator, i.e. the square of the Fourier transform of $x(t)$ at two values of the driving force F= 50 (Fig. 1(a)) and F= 500 (Fig. 1(b)). The spectra consist of peaks at odd harmonics of the driving frequency ω. In order to get rid of components at frequencies ω_0, $\omega_0 \pm q\omega$, we have introduced a small damping term in Eq. (1). In Fig. 1(a), the amplitudes of the peaks decreases with order. In Fig. 1(b), after a decrease between the fundamental and the third harmonic, the peaks have approximately the same intensity up to the 37th, forming a "plateau", which ends with a sharp cutoff. In order to get more insight into the physics governing these results, we make a Fourier expansion of $x(t) = \sum_{q=1}^{+\infty} x_q \cos q\omega t$ (q odd). Eq. (1) becomes

$$(\omega_0^2 - q^2\omega^2)x_q + \sum_{p,r,t} v x_p x_r x_t/4 = F\delta_{q,1} \tag{2}$$

with $p \pm r \pm t = \pm q$. $\delta_{q,1}$ is the Kronecker symbol equal to zero if $q \neq 1$. Let us assume that the fundamental component of the motion is the dominant one, i.e. $|x_1| \gg \sum_{q=3}^{+\infty} |x_q|$. Then x_1 is the solution of the third degree algebraic equation

$$(\omega_0^2 - \omega^2)x_1 + 3v x_1^3/4 = F. \tag{3}$$

which takes a simple analytical form in two limiting cases:

$$\begin{aligned} x_1 &= F/(\omega_0^2 - \omega^2) \qquad && \text{for } F/\omega_0^2 \ll 1 \\ &= \sqrt[3]{4F/3v} && \text{for } F/\omega_0^2 \gg 1 \end{aligned} \tag{4}$$

For $q \geq 3$, x_q can be deduced from the recursive formula:

$$x_q = -\frac{x_{q-2} + x_{q+2}}{2 + 4(\omega_0^2 - q^2\omega^2)/3v x_1^2}. \tag{5}$$

For a relatively small driving force, F, such that $|v x_1^2|/\omega_0^2 \ll 1$, and $\omega_0 \gg q\omega$, perturbation theory applies. The qth harmonic can be derived from the (q-2)th :

$$x_q = -\frac{3v x_1^2 x_{q-2}}{4(\omega_0^2 - q^2\omega^2)}; \qquad |x_q| \ll |x_{q-2}|. \tag{6}$$

This weak field case is illustrated in Fig. 1(a). The particle moves slightly out of the harmonic part of the potential, thus picking up additional frequencies. However, these nonlinear components decrease rapidly with increasing order. A rather different regime is shown in Fig. 1(b), obtained with a force F = 500. The excited particle interacts strongly with the walls of the anharmonic potential and its motion departs substantially from a purely harmonic one. Here we have chosen a positive anharmonic

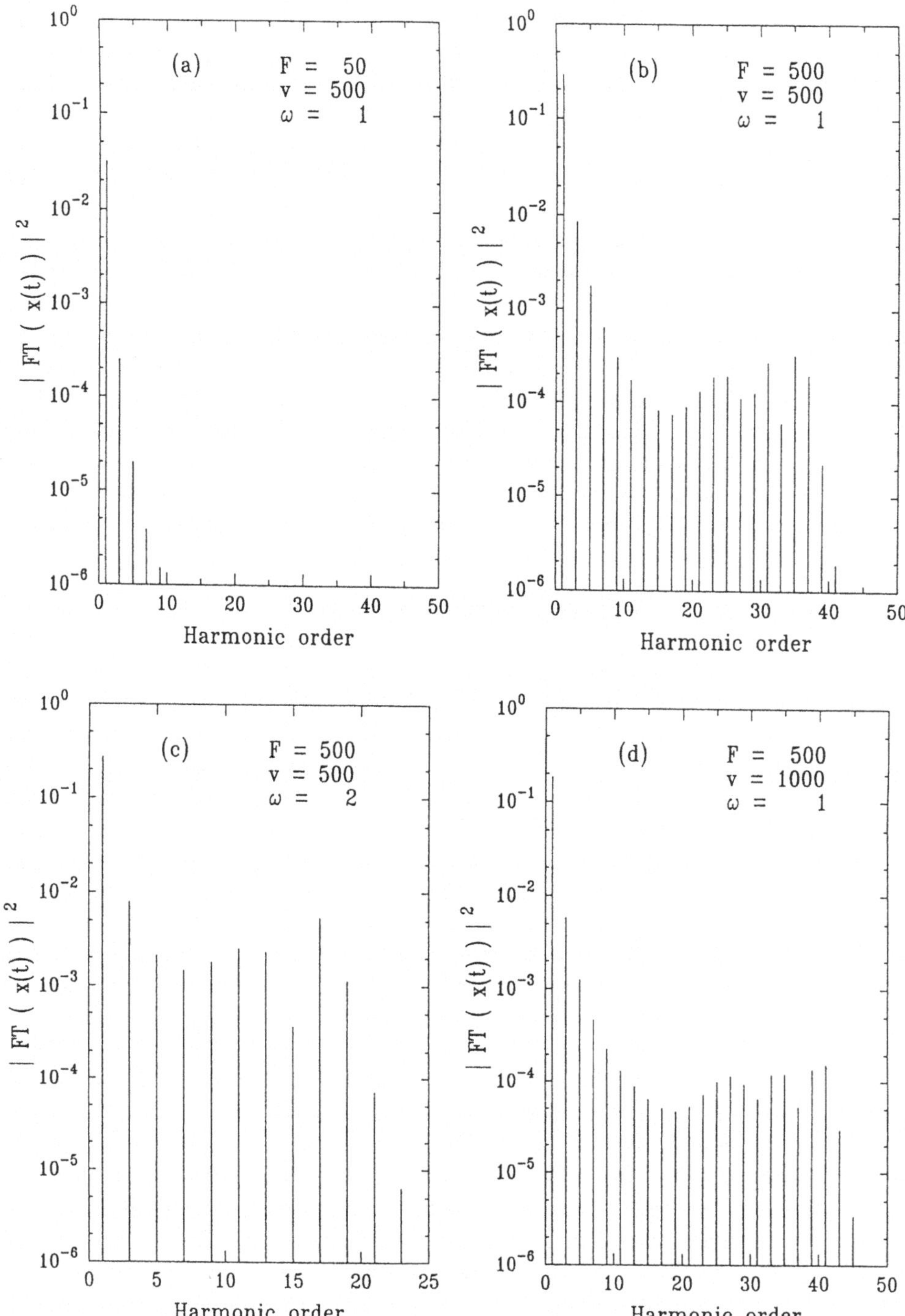

Fig. 1. Spectra emitted by an anharmonic oscillator.

term vx^3, but the results are similar for negative v as long as the force F is not too strong, in which case the trajectory becomes unstable. The nonlinear components of the spectrum first decrease because the value of the denominator in Eq. (5) is large. Then, the denominator becomes small (of the order of unity or less) and the harmonic intensities remain approximately constant in intensity up to about the 37th harmonic for this choice of parameters. One can interpret the plateau as resulting from the superposition of low order harmonics with slightly decreasing amplitudes as in the perturbative regime and a broad resonance for harmonic frequencies such that $\omega_0^2 - q^2\omega^2 + 3vx_1^2/2 \approx 0$. The cut off occurs for high frequencies as one gets away from the resonance. We have studied the variation of the spectrum with the different parameter choices. We show in Fig. 1(c) the spectrum obtained by doubling the force frequency ω (all of the other parameters being held constant). The width of the plateau remains about the same as in Fig. 1(b) provided both figures are compared in photon energy. However, the strength of the harmonic emission increases by about one order of magnitude. The width of the plateau is related to the minimum of the denominator in Eq. (5), which depends only on the photon energy $q\omega$. The intensities of the harmonics in the plateau region depend on how long the "decreasing regime" for the low harmonics is. This regime is half as long for the doubled frequency. Finally, we show in Fig. 1(d) a spectrum obtained by increasing the anharmonic term v from 500 to 1000. The plateau increases but the harmonic intensities decrease.

The interest in this crude model is that most of the trends shown here are reproduced in much more sophisticated and realistic calculations.

A SINGLE ATOM IN A STRONG RADIATION FIELD

We now consider a more realistic system, namely an atom in a strong, linearly polarized radiation field. We are interested in the light spectrum emitted by the atom and, more precisely, in the coherent part of the spectrum. This point (coherent vs. incoherent emission) is discussed by J. H. Eberly in the present volume. The coherent spectrum is simply the square of the Fourier transform of the atomic dipole moment $d(t)$ induced by the laser field. $d(t)$ can be obtained from the wave function of the atomic system, as the expectation value of the dipole operator. This problem has received considerable attention. Various theoretical techniques and approximations have been used: time-dependent methods[2,3], Floquet theory[4], and model calculations[5–7]. Most non-perturbative calculations reproduce qualitatively the experimental results: a decrease in efficiency for the first few harmonics, a plateau and then a cut off. This is apparently a very general property of strongly-driven, nonlinear systems.

In Fig. 2 we present the results of calculations performed by Kulander and coworkers in xenon[2,8]. Fig. 2(a) shows the square of the dipole moment as a function of the harmonic order at 3×10^{13} W.cm^{-2}. The spectrum consists of odd harmonics with the same "strong field" behavior as shown by the anharmonic oscillator in Figs. 1(b-d), i.e. a plateau followed by a cutoff. The harmonics are superimposed on a broad background which has been attributed to Raman effects involving Rydberg states. In Fig. 2(b) the relative harmonic intensities at several field strengths from 10^{13} W.cm^{-2} to 4×10^{13} W.cm^{-2} are displayed. Note how the width of the plateau increases with the pump intensity. The harmonic strengths as a function of the laser intensity exhibit numerous structures and resonances.[9] This is reflected, for example, by the result at 1.3×10^{13} W.cm^{-2} which does not have a simple plateau, instead showing two maxima, one at the 9th and and one at the 15th harmonics.

EXPERIMENTAL RESULTS

We now consider the response of an assembly of atoms to a strong laser field. We first briefly summarize recent experimental results[10–14] particularly those obtained at Saclay.

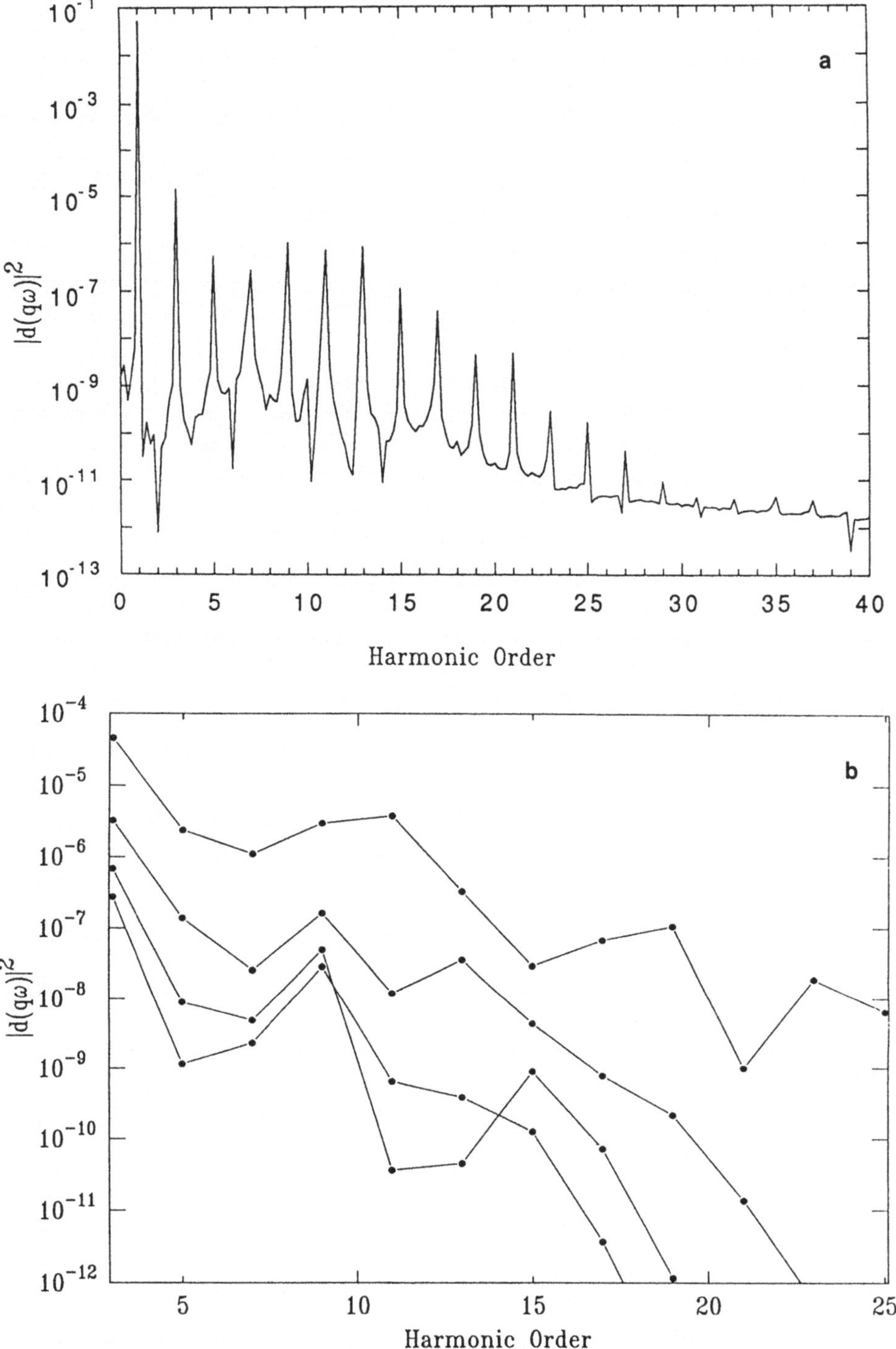

Fig. 2. (a) Photoemission spectrum in xenon. (b) Relative harmonic intensities as a function of the harmonic order q at $10^{13}, 1.3 \times 10^{13}, 2 \times 10^{13}$ and 4×10^{13} W.cm^{-2}.

The first strong-field harmonic generation experiments were performed at the University of Illinois at Chicago by C. K. Rhodes and coworkers[10]. They detected production of up to the 17th harmonic of a KrF pump laser (248 nm) in neon, which represents a wavelength of 18 nm. Recently, the 25th harmonic of a KrF laser was produced also in Ne, by a group at the University of Tokyo[13]. In parallel, experiments were performed using low frequency lasers, of the Nd- YAG or Nd-Glass type ($\lambda \approx 1\mu m$)[11,12]. Very high orders were obtained, up to the 53rd in Neon (20 nm radiation)[14]. The 97th harmonic of a Ti-Al_2O_3 laser system (800 nm) has been recently observed by S.E. Harris and coworkers at Stanford[15].

In this report, we show one typical result obtained at Saclay with a 40 ps 1064 nm Nd-YAG laser. The reader interested in more details about the experimental method or the results is referred to the review by L'Huillier et al.[14] and the original papers mentioned above.

In Fig.3 we show the number of harmonic photons obtained in xenon at a gas pressure of 15 torr for several laser intensities. We observe only odd harmonics which is to be expected for harmonic generation in an isotropic gaseous medium with inversion symmetry. At the lowest laser intensity, the conversion efficiency decreases with increasing order. However, at higher laser intensity, a plateau appears. Its width increases up to the intensity at which the medium becomes ionized during the pulse. The highest harmonic detected is the 21st, corresponding to an energy of 25 eV, a wavelength of 50 nm. The vertical scale is an order-of-magnitude estimate of the number of photons produced at each laser shot. The photon conversion efficiency for the plateau harmonics at the highest intensity is about 10^{-8} – 10^{-9}. This represents a very high brightness, 10^{17} ph/Ås(mrad)^2 .

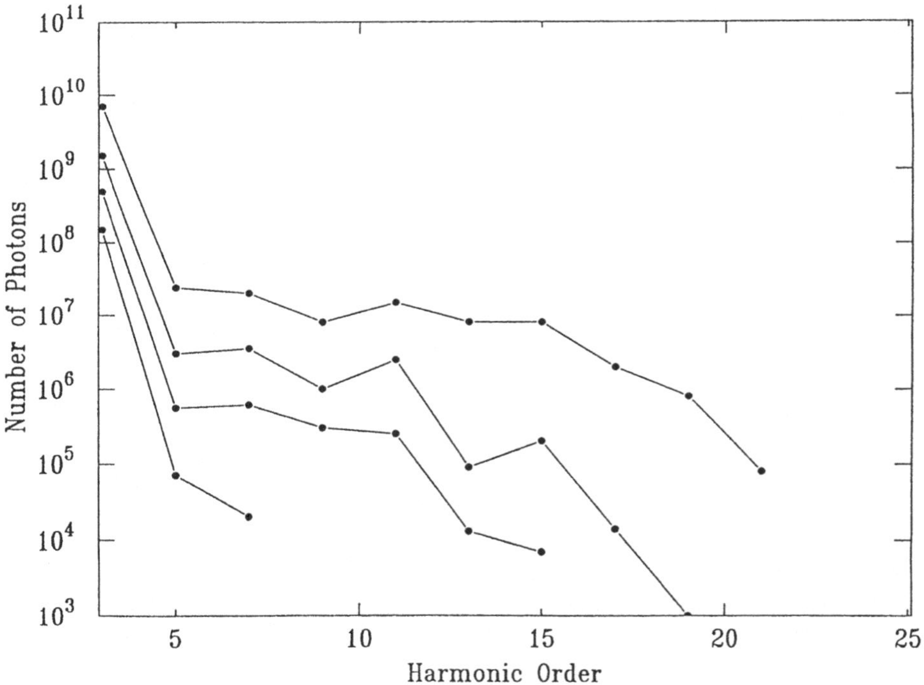

Fig. 3. Number of photons obtained in xenon at a 15 Torr pressure at the same intensities as in Fig.2.

Similar results have been obtained in the other heavy rare gases (krypton and argon). The conversion efficiency decreases from Xe to Ar, which is not surprising, because xenon is more polarizable than the lighter rare gases. However, the maximum order that can be observed increases from 21 in Xe to 29 in Kr to 33 in Ar. The 33rd harmonic (32 nm, 38 eV) is the shortest wavelength radiation that we were able to produce with our 40ps, 20mJ Nd- YAG laser system. Atoms with higher ionization energies are found in general to have a lower conversion efficiency but they produce higher harmonics. This is partly due to the fact that they can experience a higher laser intensity without being ionized. Thus, Ne seems to be a good system to use in order to produce shorter wavelengths. Using a 1 ps 1 μm Nd-Glass laser system we recently observed the 53rd harmonic using neon as the nonlinear medium.

DISCUSSION

As shown above, the harmonic intensity distributions from both the non perturbative single-atom calculations and the phase-matched, multi-atom experimental measurements are remarkably similar. This seems to imply that propagation (coherence) effects alter all harmonics to the same extent, independent of the harmonic order. This is contrary to the weak field limit predictions[16]. To resolve this disagreement we present a complete calculation of harmonic generation in a rare gas medium, involving the solution of the paraxial propagation equation for a non-perturbative polarization field[9,17]. The results show that phase matching is, in fact, the same for all the harmonics in the plateau. Moreover, we give the reason why efficient phase matching of high harmonics is successful for a focused beam geometry as soon as one enters a "strong field regime" whereas it is predicted to fail in a perturbative (weak field limit) picture.

We start from the propagation equation for the harmonic field $\mathcal{E}_q$, obtained by Fourier transforming the time-dependent propagation equation:

$$\nabla^2 \mathcal{E}_q + k_q^2 \mathcal{E}_q = -4\pi (q\omega/c)^2 \mathcal{P}_q. \quad (7)$$

In this equation, ω is the laser frequency, c the speed of light and k_q the wavevector associated to the harmonic field $\mathcal{E}_q$, which we assume to be independent of the laser intensity I. $\mathcal{P}_q$ is the polarization induced by the fundamental field only; we can neglect the influence of the other (lower-order) harmonic fields. We also assume that the laser field is not modified by the interaction with the nonlinear medium, which is true for the low atomic densities used in the experiments. Eq. (7) can be written as

$$\mathcal{E}_q(\mathbf{r}') = \left(\frac{q\omega}{c}\right)^2 \int \frac{e^{ik_q R}}{R} \mathcal{P}_q(\mathbf{r})\, d^3\mathbf{r} \quad (8)$$

with $R = |\mathbf{r}' - \mathbf{r}|$. We introduce the envelope functions $E_q = \mathcal{E}_q \exp(-ik_q z)$ and $P_q = \mathcal{P}_q \exp(-iqk_1 z)$, where z denotes the propagation axis. Making the slowly-varying envelope approximation (i.e. $\partial^2/\partial z^2 \ll k_q \partial/\partial z$), the harmonic spatial profile $E_q(\mathbf{r}')$ in the *far field* (i.e. for $|z' - z| \gg |x' - x|, |y' - y|$, $\mathbf{r} = (x, y, z)$ being a point in the medium) becomes

$$E_q(\mathbf{r}') = \left(\frac{q\omega}{c}\right)^2 \int \frac{P_q(\mathbf{r}) e^{-i\int \Delta k_q(z)\, dz}}{z' - z} \exp\left(\frac{ik_q\left[(x' - x)^2 + (y' - y)^2\right]}{2(z' - z)}\right) d^3\mathbf{r}. \quad (9)$$

$\Delta k_q(z)$ is the (complex) phase mismatch factor which includes the effects of dispersion and absorption throughout the medium. These are found to be small in the present situation where the atomic density is kept rather low (typically 15 Torr, i.e. 5×10^{17} atoms/cm^3). Using the cylindrical symmetry of the problem, Eq. (9) becomes

$$E_q(r', z') = \left(\frac{q\omega}{c}\right)^2 \int \frac{P_q(r,z)e^{-i\int \Delta k_q(z)\,dz}}{z'-z} J_0\left(\frac{krr'}{z'-z}\right) \exp\left(\frac{ik_q(r^2+r'^2)}{2(z'-z)}\right) 2\pi r\,dr\,dz. \tag{10}$$

where J_0 is the zero-order Bessel function. Eqs. (9) and (10) are exact as long as the phase mismatch Δk_q is r–independent. A more general approach consists in making the slowly varying envelope approximation in Eq. (7) before switching to an integral formulation as in Eq. (8). The harmonic field envelope E_q obeys the propagation equation:

$$\nabla_\perp^2 E_q + 2ik_q \partial E_q/\partial z + 2k_q \Delta k_q\, E_q = -4\pi(q\omega/c)^2 P_q. \tag{11}$$

where $\nabla_\perp$ acts on the transverse coordinates. This differential equation can be solved by using finite difference techniques. Both methods are completely equivalent when Δk_q is r–independent. The differential method allows us to account more accurately for the influence of free electrons and also for the depletion of the neutral medium which introduces intensity-dependent densities and refractive indices. Moreover, it gives additional insight into the physics of the problem because it shows the variation of $E_q(r,z)$ as it is generated in the nonlinear medium.

The polarization $P_q(r,z,t)$ is defined by

$$P_q(r,z) = 2\mathcal{N}(r,z)d_q(r,z)\exp\left[-iq\left(\tan^{-1}(2z/b) - \frac{2k_1r^2z}{b^2+4z^2}\right)\right] \tag{12}$$

where $d_q(r,z)$ is the atomic dipole moment calculated for the intensity $I(r,z)$ and $\mathcal{N}(r,z)$ the atomic density. We assume an incident Gaussian beam, with a confocal parameter b. The phase factor in Eq.(12) is q times that of the incident beam. The atomic dipole moment d_q is obtained by Fourier transforming the time-dependent dipole $\langle\Psi(\mathbf{r},t)|z|\Psi(\mathbf{r},t)\rangle$ over a few laser cycles. The wave function $\Psi(\mathbf{r},t)$ is calculated by integrating the time-dependent Schrödinger equation with an ℓ–dependent effective potential describing a single electron in the outer shell of a xenon atom. Eqs.(10) or (11) are solved for a sequence of times t' scanning the pulse length. The number of photons N_q is then obtained by integrating $|E_q(r',z',t')|^2$ spatially and temporally (we assume a Gaussian envelope):

$$N_q = \frac{c}{4\hbar q\omega} \int |E_q(r',z',t')|^2 r'\,dr'dt'. \tag{13}$$

In Fig. 4, we show the numbers of photons obtained in xenon for the same representative intensities as in Figs. 2 and 3. The macroscopic parameters of the calculation have been chosen to mimic the experimental conditions as closely as possible[11]. The atomic density distribution is a (truncated) Lorentzian with a width at half maximum $L = 0.8$ mm. The laser confocal parameter b has been recently measured to be ≈ 1.5 mm (more than a factor of two smaller than the one estimated previously). Finally, the laser pulse width is 36 psec. The agreement between the experiment and our ab initio calculation is quite good. The theoretical "plateau" decreases slightly, whereas the experimental one is approximately flat. Note, however, that the experimental number of photons is determined only to within one order of magnitude.

A more detailed comparison can be made by considering the intensity dependence of the number of photons measured and calculated for each harmonic. We show

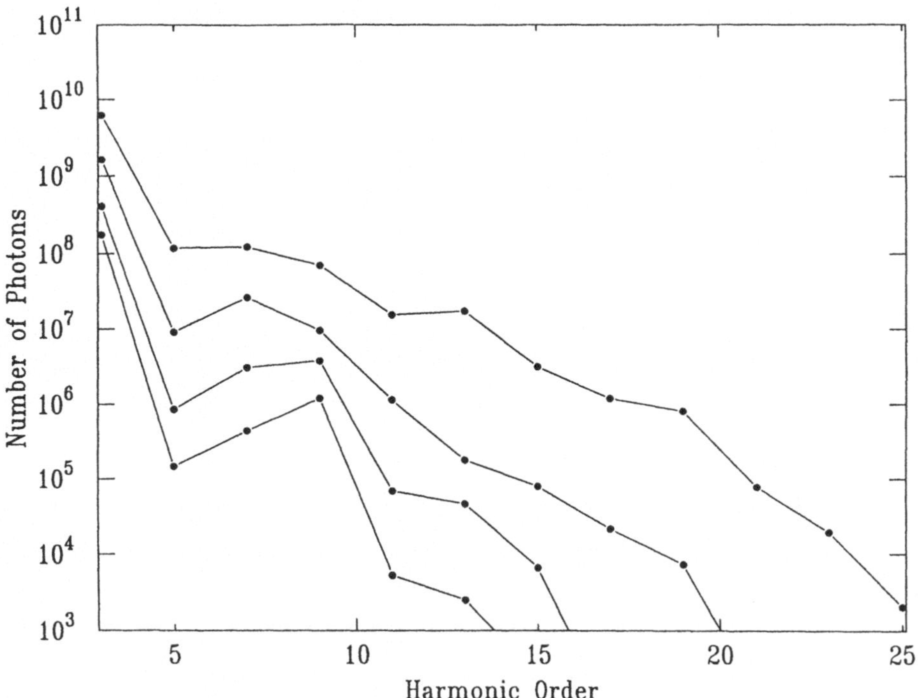

Fig. 4. Calculated number of photons. Same intensities as in Figs. 2 and 3.

in Fig. 5 the results for the 3rd and 15th harmonics. The experimental values are indicated by the open circles, the calculation by the solid line. The 3rd harmonic varies as I^3 up to the intensity 2×10^{13} W.cm^{-2}, when ionization becomes important, thus reducing the harmonic conversion efficiency. The 15th harmonic does not vary as I^{15} but exhibits a rather smooth behavior, much smoother than the single-atom data[9]. The inclusion of propagation effects, and the volume and temporal integrations leads to a significant averaging of all of the structures and resonances. Phase matching improves as the single-atom response decreases, and vice–versa. Consequently, phase matching fills up the resonances and smoothes out the peaks. However, some traces of these features remain: one of them can clearly be seen on the 15th harmonic at low laser intensity.

We illustrate in Fig. 6 how a harmonic field is generated in a nonlinear medium. We simulate a "strong-field" polarization by assuming that d_q varies as the 5th power of the incident field. This is approximately the average power law found in the single-atom data for the plateau harmonics in the intensity range of these experiments. We also neglect dispersion, setting Δk to zero. The first column in Fig. 6 shows how the perturbative, nonlinear polarization P_q, for $q = 5, 13$ and 21 varies over the interaction volume. The second column shows the harmonic field E_q calculated by assuming the polarization to vary as I^5. This is the perturbative limit for the fifth harmonic. The perturbative 13th and 21st harmonic fields are not shown as they are identical in shape to the 13th and 21st perturbative polarizations. In all cases the harmonic field first increases from the left edge of the medium to a maximum and then decreases. The maximum occurs at the position of the focus for the 13th and 21st harmonics. It is shifted to the right of the focus for the 5th harmonic. In Fig. 6, the plots have been normalized to have the same maximum value. However, this normalization closely

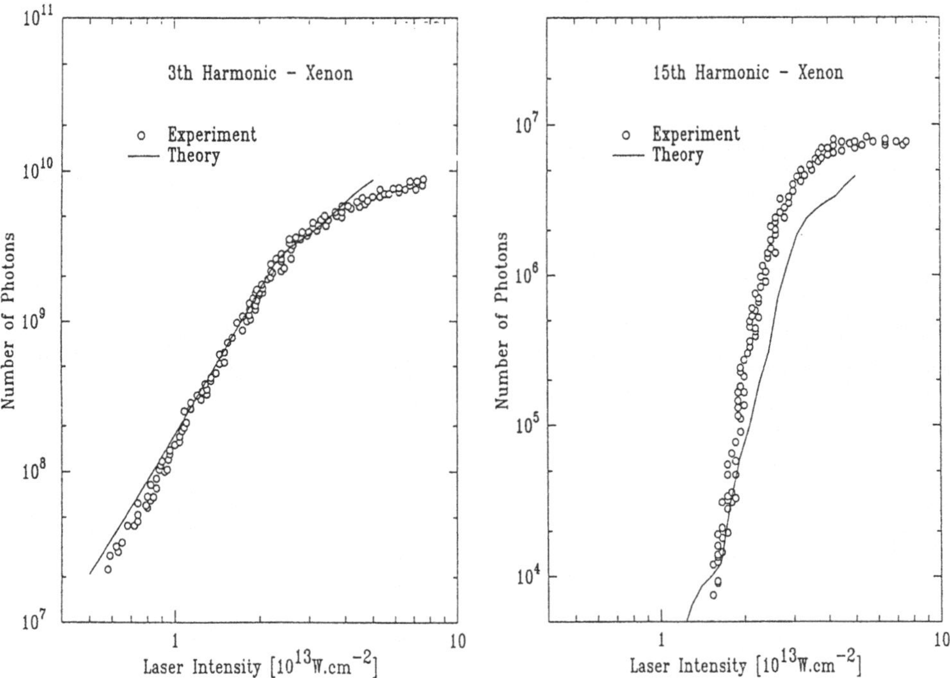

Fig. 5. 3rd and 15th harmonics as a function of the laser intensity.

corresponds to dividing the number of photons by the single atom response at the peak intensity. The integral of the square of the field at the right edge of the medium gives the magnitude of the effect of phase matching (propagation) on the harmonic conversion process. In the perturbative limit, phase matching decreases rapidly as the process order increases. In contrast, the "strong field" 13th and 21st harmonic fields are much more complicated. They oscillate in a rather regular fashion. The harmonic profile is not Gaussian as in the perturbative limit. It is defocused and develops rings. Furthermore, phase matching does not decrease with increasing order. The integrated intensity at the right edge of the medium stays approximately constant (independent of the process order)[16].

In this calculation where dispersion has been ignored, phase matching is determined by focusing, which introduces a geometrical phase lag between the induced and driving fields[9,14]. This difference in phase is approximately equal to $(1-q)\tan^{-1}(2z/b)$. The interference fringes are separated by twice a coherence length, equal to $(b/2)\tan[\pi/(q-1)]$. In the perturbative limit, P_q is concentrated within one coherence length on either side of the focus, so that the interference pattern reduces to one oscillation. The field that is created before the focus is cancelled out by the one generated after. This cancellation becomes more efficient as the process order increases. In contrast, in the model situation where the polarization is assumed to vary as the fifth power of the radiation field, the amplitude of the polarization does not fall off as rapidly away from the focus. The length over which the polarization field has appreciable amplitude spans several coherence lengths. The last two plots on the right side of Fig. 6 clearly show interference effects with a period equal to the coherence length (decreasing with increasing order). The field that is generated in the medium shows strong modulations depending on the position along

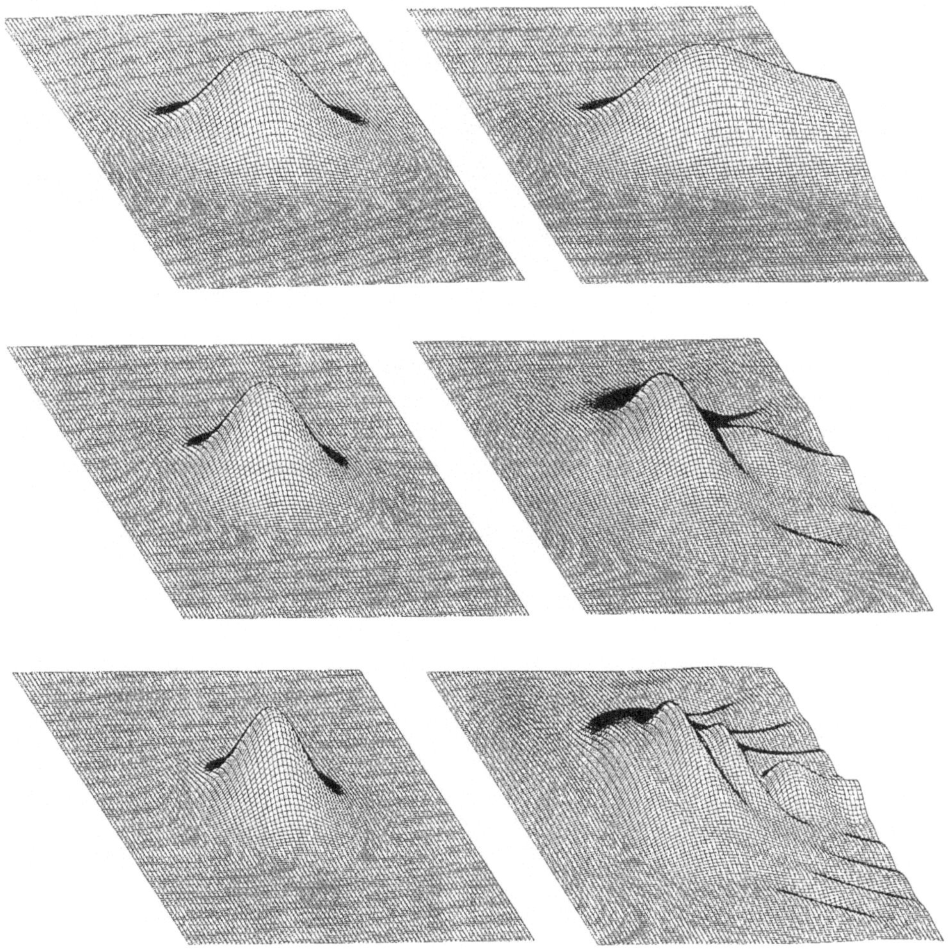

Fig. 6. Graphical representation of $|P_q(r,z)|$ and $|E_q(r,z)|$ in the nonlinear medium (see text). The light propagates along the horizontal axis (z) from the left to the right.

the propagation axis. Therefore, the field that exits the medium can be orders of magnitude more intense in the strong field case than in the weak field limit. Phase matching is found to remain approximately constant for all of the plateau harmonics, because of the slow variation of the polarization field throughout the medium leading to an important defocusing effect and to breaking of the cancellation that occurs in the weak field limit.

This work has been carried out in part under the auspices of the U. S. Department of Energy at the Lawrence Livermore National Laboratory under contract number W-7405-ENG-48.

References

[1] See e.g. P. W. Milonni, M. L. Shih and J. R. Ackerhalt, in "Chaos in Laser-Matter Interactions", Singapore, World Scientific and Refs. therein.

[2] K. C. Kulander and B. W. Shore, Phys. Rev. Lett. **62**, 524 (1989); J. Opt. Soc. Am. B **7**, 502 (1990).

[3] J. H. Eberly, Q. Su and J. Javanainen, Phys. Rev. Lett. **62** 881 (1989); J. Opt. Soc. Am. B **6**, 1289 (1989).
[4] R. M. Potvliege and R. Shakeshaft, Phys. Rev. A, **40**, 3061 (1989).
[5] G. Bandarage, A. Maquet and J. Cooper, Phys. Rev. A **41**, 1744 (1990).
[6] W.Becker, S. Long and J. K . McIver, Phys. Rev. A **41**, 4112 (1990).
[7] B. Sundaram and P.W. Milonni, Phys. Rev. A **41**, 6571 (1990).
[8] J. F. Krause, K. J. Schafer and K. C. Kulander, Phys. Rev. A, in press.
[9] A. L'Huillier, K. J. Schafer and K. C. Kulander, J. Phys. B **24** 3315 (1991).
[10] A. McPherson, G. Gibson, H. Jara, U. Johann, T. S. Luk, I. McIntyre, K. Boyer and C. K. Rhodes, J. Opt. Soc. Am. B **4**, 595 (1987).
[11] M. Ferray, A. L'Huillier, X. F. Li, L. A. Lompré, G. Mainfray and C. Manus, J. Phys. B **21**, L31 (1988); X. F. Li, A. L'Huillier, M. Ferray, L. A. Lompré, G. Mainfray and C. Manus, Phys. Rev. A **39**, 5751 (1989).
[12] L. A. Lompré, A. L'Huillier, M. Ferray, P. Monot, G. Mainfray and C. Manus, J. Opt. Soc. Am. B **7**, 754 (1990).
[13] N. Sarakura, K. Hata, T. Adachi, R. Nodomi, M. Watanabe and S. Watanabe, Phys. Rev. A **43** 1669 (1991).
[14] A. L'Huillier, L. A. Lompré, G. Mainfray and C. Manus, Adv. At. Mol. Opt. Phys., Ed. M. Gavrila, in press.
[15] S. E. Harris, private communication.
[16] A. L'Huillier, X. F. Li and L. A. Lompré J. Opt. Soc. Am. B **7**, 527-536 (1990).
[17] A. L'Huillier, K. J. Schafer and K. C. Kulander, Phys. Rev. Lett. **66**, 2200 (1991).

CHIRP-CONTROLLED INTERFERENCE

IN TWO-PHOTON PROCESSES

B. Broers[1], L.D. Noordam[1], and H.B. van Linden van den Heuvell[1,2]

[1]FOM-Institute for Atomic and Molecular Physics, Kruislaan 407
1098 SJ Amsterdam, The Netherlands
[2]Van der Waals-Zeeman Laboratorium, Valckenierstraat 65
1018 XE Amsterdam, The Netherlands

ABSTRACT

We discuss experimental results for two two-photon processes with chirped pulses: two-photon excitation of Rydberg levels in an atom, and frequency-doubling in a non-linear crystal. The results are explained in terms of interference between different frequency components within the bandwidth of the pulse. We also point out an analogy with Fresnel diffraction from a slit.

INTRODUCTION

Interference, in general, occurs if two or more 'paths' exist along which an initial state of a system can evolve to one particular final state. For example, in an optical excitation process, interference effects in a transition probability can be observed if various (coherent) light fields are present, which can drive that particular transition via two (or more) different 'paths', i.e. different combinations of frequencies, adding up to match the transition frequency. In this case the resulting total transition probability is directly influenced by the relative phases of the driving fields, which determine whether the interference is constructive of destructive.

Recently, a nice observation of this type of interference was reported by Ce Chen et al. [1]. In their experiment the $6s\ ^1S_0 \rightarrow 6p\ ^1P_1$ transition in Hg (frequency difference ω) was driven by both a field of frequency ω, and one of frequency $\omega/3$, so two excitation paths were available: one consisting of absorption of one ω-photon, the other of three $\omega/3$-photons. The total transition probability showed an oscillatory dependence on the phase difference between the two optical fields.

Another demonstration of this two-color interference was given by Muller et al. [2], who studied the photo-electron spectrum of krypton after multi-photon ionization with a two-color laser field. The resulting electron spectra were strongly influenced by the phase difference between the two colors.

In the type of optical interference described above, two (or more) optical fields of different frequency ("different colors") are present. However, for a multi-photon transition it is also possible to observe interference in resulting transition probabilities if only *one, pulsed* field is applied. In this case the interference occurs between the frequencies *within the bandwidth* of the pulse.

Coherence Phenomena in Atoms and Molecules in Laser Fields
Edited by A.D. Bandrauk and S.C. Wallace, Plenum Press, New York, 1992

It is the purpose of this contribution to discuss experimental results of this kind of interference in two two-photon processes: two-photon excitation of Rydberg states in an atom, and second harmonic generation (SHG) in a non-linear crystal. It will be shown that the resulting transition probabilities critically depend on the exact phase relations between the frequency components in the pulse. In other words: the *populations* of excited levels are strongly influenced by its phase profile. In addition, it is pointed out that two-photon excitation with a pulse, of which the phase changes quadratically with the frequency, can be seen as an analogue of Fresnel diffraction from a slit.

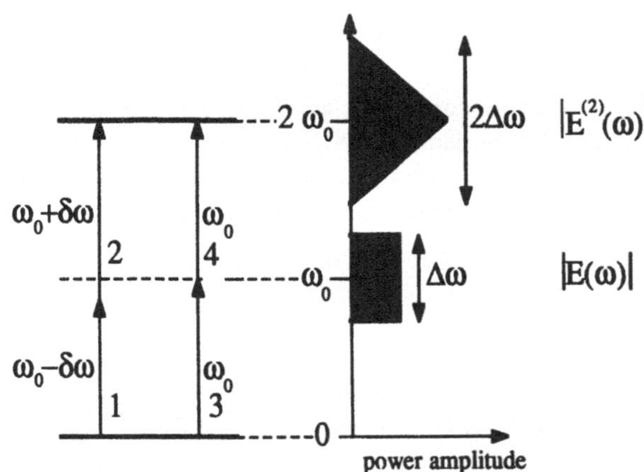

Fig. 1 *For a chirp-free pulse, the power amplitude at the two-photon level, $|E^{(2)}(\omega)|$, can easily be derived from the power amplitude of the excitation pulse, $|E(\omega)|$, by taking combinations of two frequencies ω_i, which lie within the bandwidth of the pulse. As an example, two combinations, both adding up to $2\omega_0$, are drawn.*

THEORY

This interference effect is most easily illustrated for a two-photon process, driven by a pulse with a square-shaped power spectrum (Fig. 1). Let $E(\omega)$ denote the (complex) electric field of the pulse in the frequency domain, $|E(\omega)|^2$ its power spectrum, centered around ω_0 with a bandwidth of $\Delta\omega$. It is evident that frequencies within the bandwidth can be used to add up to frequencies ranging from $(2\omega_0 - \Delta\omega)$ to $(2\omega_0 + \Delta\omega)$ at the two-photon level, and that many combinations of frequencies add up to the same sum, in particular if this sum is close to $2\omega_0$. In Fig. 1, for example, one has $(\omega_1 + \omega_2) = (\omega_3 + \omega_4)$. The effective 'two-photon amplitude spectrum', $|E^{(2)}(\omega)|$ of the pulse can then be found by considering all combinations ω_i, ω_j, within the fundamental bandwidth, that add up to a certain frequency $\omega_f = \omega_i + \omega_j$. This implies that $|E^{(2)}(\omega)|$ is simply given by the self-convolution of the (one-photon) amplitude spectrum of the pulse:

$$|E^{(2)}(\omega)| = \int_{-\infty}^{\infty} d\omega' |E(\omega')||E(\omega - \omega')| \quad (1)$$

This results in a triangular shape of $|E^{(2)}(\omega)|$ for a square-shaped $|E(\omega)|$ (see Fig. 1).

However, the idea sketched above is only correct *if all frequency combinations add up constructively*, so if all are in phase, i.e. $E(\omega) = |E(\omega)|$. Such a pulse is called transform-limited (the Fourier bandwidth-time product is minimal), or chirp-free. If this is not the case, the pulse is often said to be chirped, so $E(\omega) = |E(\omega)|e^{i\phi(\omega)}$, where ϕ is a non-linear function of ω. If ϕ depends linearly on ω, the only consequence of the phase factor is a translation of the pulse in time, i.e. $E(t) \to E(t + \Delta t)$.

For *chirped* pulses, the interference between various paths need not be constructive. In Fig. 1 the total contribution of paths 1 + 2 and 3 + 4 depends on the phases ϕ_i of the frequencies ω_i: if the *total* phase $(\phi_1 + \phi_2)$ (note that $|E(\omega_1)| = |E(\omega_2)|$, giving equal weight to both phases) of the first path differs from the *total* phase of the second path, $(\phi_3 + \phi_4)$, by $2\pi n$ (n integer), their contributions will interfere constructively. On the other hand, a phase difference of $2\pi(n+1)$ will cause their contributions to cancel totally. So Eq. (1) has to be rewritten to include the phases of the frequency components explicitly ($E(\omega) = |E(\omega)|e^{i\phi(\omega)}$):

$$E^{(2)}(\omega) = \int_{-\infty}^{\infty} d\omega' E(\omega')E(\omega - \omega'). \tag{2}$$

The distribution of excited states around $2\omega_0$ is therefore expected to be proportional to $|E^{(2)}(\omega)|^2$. Note that $E^{(2)}(\omega)$ is the Fourier-transform of $E^2(t)$, the 'effective time evolution of the field for a two-photon process'.

EXPERIMENTAL

In this section a brief description will be given of two experiments [3], which were carried out to demonstrate this kind of interference. In one experiment the population of atomic Rydberg states after two-photon excitation with a chirped pulse is studied. In an other experiment the power spectrum of frequency-doubled light, generated in a non-linear crystal, was measured.

In these experiments, use was made of a laser system [4] providing short pulses with tunable wavelength, bandwidth, and chirp, which will be described here only briefly. Pulses from a colliding-pulse modelocked (CPM) dye-laser, with a central wavelength of 620 nm, were amplified in dye cells of the Bethune type [5], pumped by the second harmonic of a Nd:YAG (Yttrium Aluminum Garnet) laser. To make the wavelength tunable, these pulses were focussed in water to generate a wavelength continuum. Selection of the desired wavelength was performed with a pulse shaper (Fig. 2), consisting of a grating, a lens, and a mirror placed in the focal plane of this lens, on which the various frequencies coming off the grating, are separated. By placing a slit in front of the mirror both central frequency and bandwidth of the pulse coming out of the shaper can be chosen arbitrarily. These pulses were amplified again to an energy of $\sim 30\mu J$.

The shaper was also used to select the chirp of the pulses. If the distance between grating and lens in the pulse shaper is varied by Δz with respect to the focal distance of the lens, different frequencies will no longer travel the same distance through the pulse shaper. This implies that the frequency spectrum of a pulse before the shaper, $E(\omega)$, is changed to $E(\omega)e^{i\phi(\omega)}$ after the shaper. A calculation of the frequency-dependent pathlength through the shaper shows that

$$\phi(\omega) = 2(\cos r_0)\frac{\omega}{c}(\cos i + \cos r(\omega))\Delta z. \tag{3}$$

Here i and $r(\omega)$ denote the angle between the incoming and diffracted beam with respect to the grating normal, respectively. Note that i is fixed, while r depends on the wavelength λ according to the grating formula: $\sin i + \sin r = m\lambda/d$, with m the diffraction order and d the groove spacing. The fixed angle between grating normal and the direction perpendicular to the mirror is r_0. Since varying Δz means varying the phase differences between the frequencies, it is possible to perform experiments in which only the chirp of the pulse is changed. If the bandwidth of the pulses is relatively small (< 10 nm), $\phi(\omega)$ is very well approximated by a

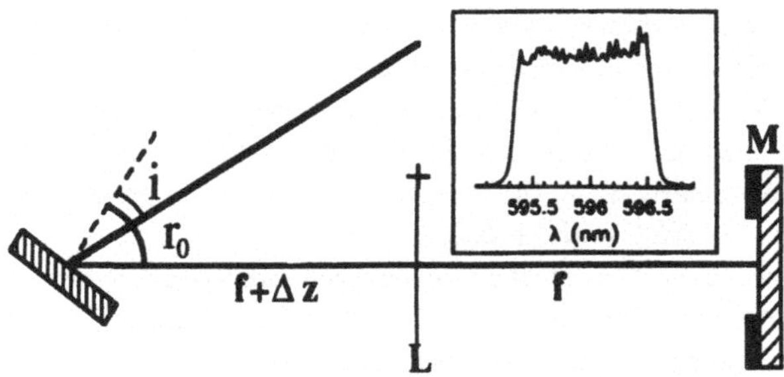

Fig. 2 *The pulse shaper. L=lens (focal distance f), M=mirror with an adjustable slit in front of it; i and r_0 denote the angle between incoming and refracted beams with respect to the grating normal.*
Inset: measured power spectrum of the pulse.

Taylor expansion around ω_0, in which only the second order in $(\omega - \omega_0)$, the first non-trivial order, is kept. This results in

$$\phi(\omega) \approx \frac{1}{2}(\frac{\delta^2\phi}{\delta\omega^2})_{\omega=\omega_0}(\omega - \omega_0)^2 \equiv \alpha(\omega - \omega_0)^2. \tag{4}$$

The effect of shaper on the description of the pulse in the time domain, $E(t)$, can also be considered. Fourier-transforming shows that for a pulse $E(\omega) = |E(\omega)|\, e^{i\alpha(\omega-\omega_0)^2}$, with $|E(\omega)|$ Gaussian: $E(t) = |E(t)|\, e^{i(\omega_0+\phi_0 t)t} \equiv |E(t)|\, e^{i\phi(t)}$ with $|E(t)|$ also Gaussian and ϕ_0 a constant. This can be seen as a time-dependent 'instantaneous frequency'. Since the instantaneous frequency $\omega(t) = \frac{\delta\phi(t)}{\delta t} = \omega_0 + 2\phi_0 t$ changes linearly in time, this pulse is said to have a *linear frequency chirp*. So Gaussian pulses become linearly chirped. For an *arbitrary* pulse envelope $|E(\omega)|$, however, the phase factor $e^{i\alpha(\omega-\omega_0)^2}$ does *not* imply that $\omega(t)$ changes linearly in time. It therefore seems more natural to take a phase factor $e^{i\alpha(\omega-\omega_0)^2}$ as the definition of a linearly-chirped pulse, rather than an instantaneous $\omega(t)$ that changes linear in time, since the first is determined by dispersion only, while the second also depends on the particular pulse shape.

In the first experiment, two-photon excitation of electronic Rydberg states in Rb was investigated, resulting almost exclusively in population of *d*-states [6]. The Rb-atoms in the vacuum system (background pressure $1 \cdot 10^{-4}$ Pa) formed a diverging beam coming out of a tube, which was connected to an oven. By adjusting the slit of the pulse shaper, the bandwidth of the excitation pulse was taken large compared to the spacing between the levels, so several states were excited coherently, around the principal quantum number $n = 30$. The intensity of the pulses was kept low enough to avoid AC-Stark shifts of the Rb-levels [7].

After excitation, the populations of the levels were probed by means of field ionization and the field-ionized electrons were detected on a channel plate. The electric field strength of this ionizing pulse increased in time. Consequently, the electron signal, recorded as a function of time, showed consecutive peaks corresponding to Rydberg levels with decreasing principal quantum number n.

The second experiment was optical second-harmonic generation (SHG) in a non-linear crystal. In this case the power spectrum of the frequency-doubled light was measured with a monochromator (SPEX 1870) and optical multichannel analyzer (EG&G 1453A). The crystal used was KDP and had a length $L = 0.5$ mm, short enough to ensure that all frequencies within the bandwidth of the pulse were properly phase-matched: we calculated that $\Delta k_{max} L \leq 0.4\pi$, where Δk_{max} is the maximal phase mismatch in the experiments.

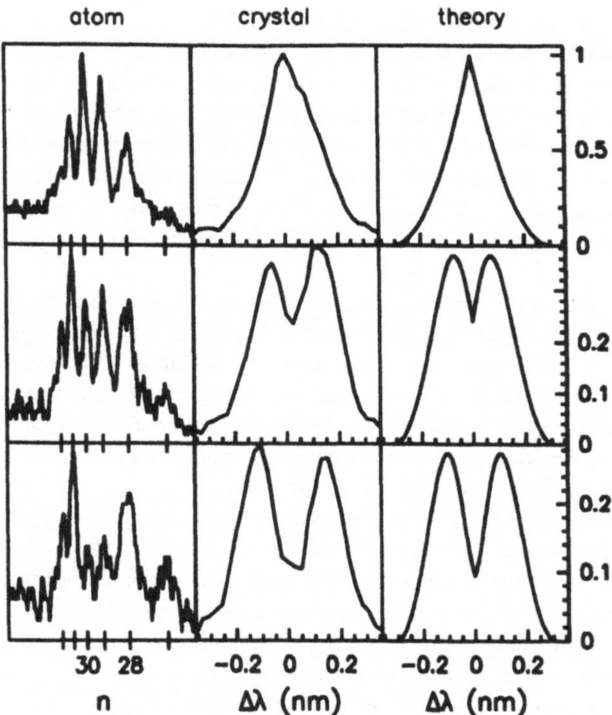

Fig. 3 *Populations of Rydberg levels in Rb (left column), and power spectra of frequency-doubled light (middle column), resulting from pulses with varying chirp. Upper figures result from a chirp-free pulse. Middle and lower figures correspond to pulses with a bandwidth-time product, relative to that of a chirp-free pulse, of* 1.10 *and* 1.27*, respectively. The right column gives the calculated power spectrum* $|E^{(2)}(\omega)|^2$ *of* $E^2(t)$*. The vertical axis is normalized to the top height of the distribution resulting from a chirp-free pulse.*

RESULTS AND DISCUSSION

The pulses used for both experiments had a square-shaped power spectrum (see inset Fig. 2) centered around 596.0 nm with a bandwidth of 1.2 nm, corresponding to a pulse duration of 850 fs. For increasing values of the chirp, experimental results are given in the first two columns of Fig. 3, starting at the top with those of a chirp-free pulse. The middle and bottom row correspond to pulses with a bandwidth-time product, relative to that of a chirp-free pulse, of 1.10 and 1.27, respectively. Note that these pulses are only slightly lengthened! Although the power spectrum of the frequency-doubled light is a continuous energy-probe on the excitation process, whereas the populations of the Rydberg-levels form a discrete probe, it is clearly seen that the results look completely similar.

A few comments should be made on the raw data shown for Rb. First, the horizontal axis is a linear *time* scale and therefore does not represent a perfectly linear *energy* scale due to the characteristics of the field-ionizing pulse. Second, the area under a peak is a better measure for the population of the corresponding level rather than the height of the peak. Third, the transition probability for excitation from the ground state to a state with (high) principal quantum number n varies as n^{-3}. These refinements were not corrected for, since the data already show the important features, resulting from the chirp over the excitation pulse, clearly.

The triangular shape produced by the chirp-free pulse was already discussed in the section THEORY. If the chirp is increased, this triangular shape develops into a symmetric two-peak structure with a large dip in the center of the energy-distribution after excitation. This means that *energy levels at exactly* $2\omega_0$ *are less efficiently populated than levels with a slight detuning with respect to* $2\omega_0$. This is in complete contrast with a one-photon excitation process, where the resulting populations would have been independent of the chirp, since the chirp does not affect the power spectrum of the pulse.

In the third column of Fig. 3 the expected power spectra $|E^{(2)}(\omega)|^2$, calculated from Eqs. (2), (3) and (4), are shown. The vertical scale was normalized to the top height calculated for a chirp-free pulse. Note that the discontinuity at $\omega = 2\omega_0$ is caused by the discontinuities in a square pulse. It is seen that this formulation gives a very good description of the observations. It also proves that Eq.(3) describes the effect of the pulse shaper accurately, since the curves were calculated without any adjustable parameter.

The measurements using SHG in KDP were also extended to a regime of stronger chirping of the pulse. In this case the bandwidth of the pulses was taken 2.8 nm around the central wavelength of 605.6 nm. Results of the experiments are given in the left column of Fig. 4. The asymmetry in the spectra is caused by a small asymmetry in the spectrum of the fundamental pulse. This was not taken into account in the calculations of the expected spectra, shown in the right column. From top to bottom, the (normalized) pulse durations are: 2.9, 3.5, 5.2, 8.0, and 12.0. It is seen that more peaks appear in the excitation profile as the chirp is increased.

This observation can be clarified with the help of Fig. 5, in which the phase of each frequency component in plotted, according to Eq.(4). If one looks at the *range* of frequencies ω_i, ω_j, that can be used to add up to $(\omega_i + \omega_j) = \omega_f = 2(\omega_0 \pm \delta\omega)$, $0 \leq \delta\omega \leq \Delta\omega/2$, one can see that both the available range of frequencies, and the variations of the total phase $(\phi_i + \phi_j)$ over these frequencies, increase with decreasing $\delta\omega$. The maximal phase difference is found between the situation with $(\omega_i = \omega_0, \omega_j = \omega_0)$ and the one with $(\omega_i = \omega_0 - \Delta\omega/2, \omega_j = \omega_0 + \Delta\omega/2)$, both adding up to $(\omega_i + \omega_j) = 2\omega_0$. Using Eq.(4) one finds: $\Delta\phi_{max} = \{\alpha(-\Delta\omega/2)^2 + \alpha(+\Delta\omega/2)^2\} - \{0+0\} = \frac{1}{2}\alpha(\Delta\omega)^2$. If $\Delta\phi_{max} \approx \pi$, one can speak of destructive interference between these two ways and this causes the first dip in the excitation profile (see Figs. 3 and 4). This motivation illustrates that a new peak in the excitation profile shows up each time $\frac{1}{2}\alpha(\Delta\omega)^2$ is increased by approximately π. Consequently, the total number of peaks in this profile can be estimated by $n_{peaks} \approx \frac{\alpha(\Delta\omega)^2}{2\pi} + 1$.

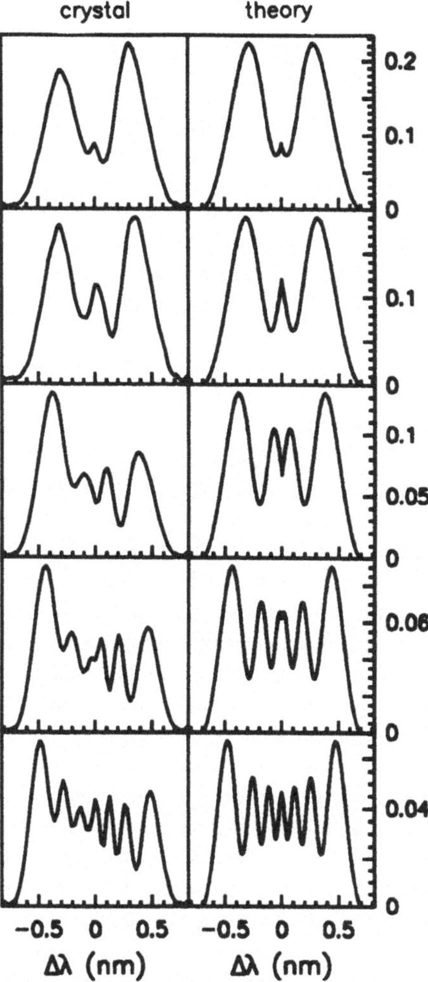

Fig. 4 *Measured (left column) and calculated (right column) power spectra of frequency-doubled light for increasing magnitude of the chirp of the fundamental pulse. Starting at the top, the bandwidth-time product, relative to that of a chirp-free pulse, has the values:* 2.9, 3.5, 5.2, 8.0, *and* 12.0.

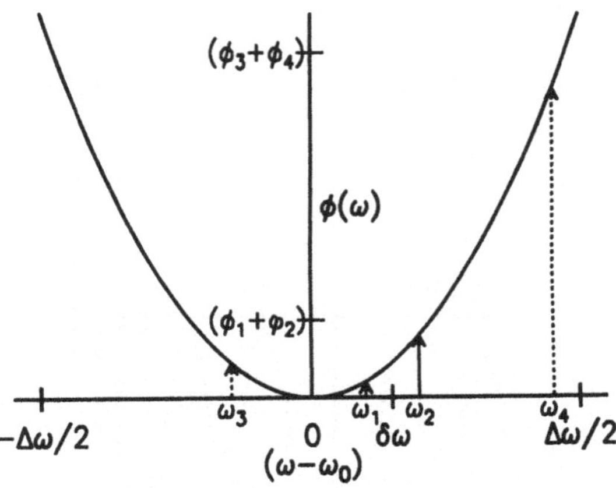

Fig. 5 *Phase of the frequency components in the pulse, according to Eq.(4). The frequency pairs (ω_1, ω_2) and (ω_3, ω_4) both add up to $2\delta\omega$. The sums of their phases, however, are not the same.*

The total power represented by $E^{(2)}(\omega)$ decreases with increasing chirp, as can be seen from the vertical scales in Fig. 4, and eventually becomes zero. This is most conveniently understood in the time domain: increasing the chirp lengthens the pulse, so the intensity $I(t)$ decreases, while the total energy $\int I(t)dt$ is kept constant. Consequently, the 'driving term' for a two-photon process, $\int I^2(t)dt$, vanishes with increasing chirp.

It should be noted that the effects of the chirp discussed above are not at all restricted to two-photon processes in the systems mentioned, nor to a simple linear frequency chirp. Any chirp in any multi-photon process may give rise to similar interference. This is of importance since in many experiments laser pulses are unintentionally slightly chirped. From the examples it is clear that a bandwidth-time product that is only some 10% larger than that of a chirp-free pulse, already is an indication for changed excitation probabilities in a multi-photon process.

It is worth mentioning that the discussion of interference between the frequency components within the bandwidth of the pulse, occuring in a two-photon process, greatly resembles the situation encountered when Fresnel diffraction from a slit is considered. This can be appreciated by looking at Fig. 6, in which a line source S (directed perpendicular to the plane of the paper) illuminates a slit with width Δv, giving a diffraction pattern on the screen P. It is clearly seen that the intensity distribution over the slit is almost uniform, while the phase profile over the slit has the shape of a parabola. So we can make the following identifications:

width of the slit Δv	$\longleftrightarrow$	bandwidth of the pulse $\Delta\omega$
intensity profile over the slit	$\longleftrightarrow$	power spectrum of the pulse
phase profile over the slit	$\longleftrightarrow$	phase profile of the pulse

Calculation of the Fresnel diffracted intensity at $z = 0$, the center of the screen, now is completely equivalent to calculation of the "two-photon power density" at $2\omega_0$, the central frequency of the two-photon process. For the off-center points on the screen ($z \neq 0$), however, the situation is not equivalent to the "two-photon power density" at frequencies with a detuning, $2(\omega_0 \pm \delta)$: ***all points within*** Δv determine the diffraction pattern at $z \neq 0$, while

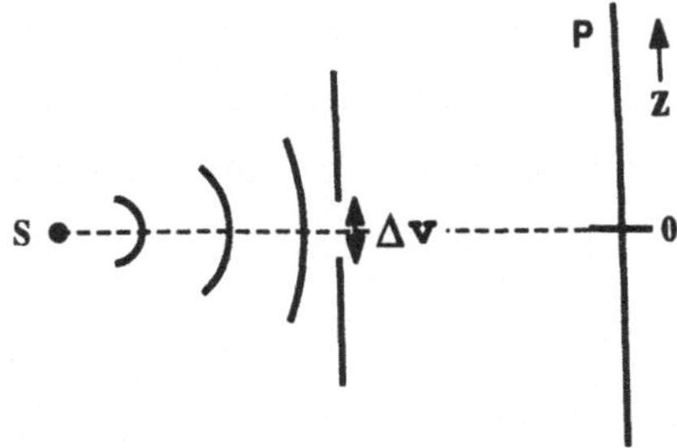

Fig. 6 *Situation for the case of Fresnel diffraction: S=line source (directed perpendicular to the paper), Δv=width of the slit, P=screen where diffraction pattern is observed. Note that the intensity profile over the slit is almost uniform, while the phase changes quadratically.*

conservation of energy ($\omega_1 + \omega_2 = 2(\omega_0 \pm \delta)$) selects only the *part of the bandwidth $\Delta\omega$ for which this condition can be fullfilled.* This would be equivalent to a width of the slit that is decreased linearly with increasing $|z|$. Experiments are under way to show more aspects of the analogy.

CONCLUSIONS

In conclusion, it has been shown, both theoretically and experimentally, that interference can be observed in the transition probabilities for a multi-photon process. The interference is due to the different excitation 'paths', i.e. different combinations of photons adding up to the same energy, which are allowed by the finite bandwidth of the excitation pulse. Since the resulting interference is determined by the phases of the frequency components within the bandwidth, it can directly be related to the chirp of the pulse. For a pulse with a quadratic phase chirp this interference is in close analogy with Fresnel-diffraction from a slit.

ACKNOWLEDGEMENTS

It is a pleasure to acknowledge enthousiastic support from H.Y. Hutter and H.M. Visser during the preparation of the atomic experiments. The work in this paper is part of the research program of the 'Stichting voor Fundamenteel Onderzoek van de Materie' (Foundation for Fundamental Research on Matter) and was made possible by financial support from the 'Nederlandse Organisatie voor Wetenschappelijk Onderzoek' (Netherlands Organisation for the Advancement of Research).

REFERENCES

[1] Ce Chen, Yi-Yian Yin, and D.S. Elliot, Phys. Rev. Lett. **64**, 507 (1990).

[2] H.G. Muller, P.H. Bucksbaum, D.W. Schumacher, and A. Zavriyev, J. Phys. B: At. Mol. Opt. Phys. **23** 2761 (1990).

[3] B. Broers, H.B. van Linden van den Heuvell, and L.D. Noordam, to be published.

[4] L.D. Noordam, W. Joosen, B. Broers, A. ten Wolde, A. Lagendijk, H.B. van Linden van den Heuvell, and H.G. Muller, Opt. Comm. (accepted for publication).

[5] D.S. Bethune, Appl. Opt. **20**, 1897 (1981).

[6] G. Alber and P. Zoller, Phys. Rep. **199**, 231 (1991).

[7] H. Gratl, G. Alber, and P. Zoller, J. Phys. B **22**, L547 (1989).

PHASE EFFECTS IN TWO-COLOR MULTIPHOTON PROCESSES

Kenneth C. Kulander and Kenneth J. Schafer

Physics Department
Lawrence Livermore National Laboratory
P. O. Box 808, Livermore, CA 94550

INTRODUCTION

There has been considerable interest over the past few years in the interactions of atoms and molecules with intense, short pulsed lasers. Measurements and calculations of ionization rates and photoemission rates have been reported for many wavelengths and intensities for single frequency laser fields. More recently, the effects of the presence of a second, intense laser field have been considered. Predictions that ionization dynamics can be controlled using two lasers with a fixed phase relationship between the them have been realized in experiments by Chen and Elliot[1] and by Muller, et al.[2] In both experiments the rate for multiphoton ionization was shown to depend on this phase. Muller, et al. also showed that the ATI (above threshold ionization) photoelectron energy distributions were also affected. Here we report calculations of the two-color ionization of hydrogen that reproduce these observed phase sensitive effects. We also present angular distributions for the photoelectrons and their variation with the relative phase between the lasers. In addition these calculactions address the question of the magnitude of the ponderomotive shift of the ionization potential caused by the combined laser fields.

In the next section, we present a brief description of our theoretical methods. We show in the third section some representative results for the phase dependence of the multiphoton ionization of a hydrogen atom for the two color case 1ω-2ω, where both fields are assumed to have the same intensity. We relate these results to the measurements of Muller, et al.[2] and give our conclusions in the final section.

CALCULATIONS

We solve the time-dependent Schrödinger equation for a hydrogen atom in two laser fields, where one frequency is twice the other. We assume the two fields have the same intensity and vary the phase between the two fields. The ionization rates and photoelectron energy and angular distributions are found to depend strongly on this phase. The intensities are large enough that we can treat the fields classically. At such intensities, perturbation theory is no longer valid, so that the direct solution of the time-dependent equations is necessary.

Our computational methods have been presented in detail elsewhere,[3] so we give only a brief description here. We expand the electronic wave function in spherical harmonics, then discretize the radial coordinate. We consider only linearly polarized fields and the direction of polarization is the same for both lasers. Under these circumstances the azimuthal quantum number is conserved. Thus we can write the wave function for an electron initially in an s-state as

$$\psi(r,\theta,\varphi,t)=\sum_{\ell}\Phi_{\ell}(r,t)\,Y_{\ell}^{o}(\theta,\varphi). \tag{1}$$

The radial function is defined only at a finite set of grid points. Putting this expansion into the time-dependent Schrödinger equation (we use atomic units throughout),

$$i\frac{\partial}{\partial t}\psi(r,\theta,\varphi,t)=\left\{-\frac{1}{2}\nabla^2-\frac{1}{r}+V_I(t)\right\}\psi(r,\theta,\varphi,t) \tag{2}$$

where V_I is the laser-atom interaction, we obtain a set of coupled finite difference equations. The time evolution is accomplished using the Peaceman-Rachford alternating-directions implicit method. This propagator is accurate to second order in the time step and is approximately unitary (exactly so when the laser is off). The laser-atom interaction is given by

$$V_I(t)=-z\varepsilon_o f(t)\left[\sin(\omega t)+\sin(2\omega t+\phi)\right]. \tag{3}$$

ε_o is the amplitude of the individual fields, ϕ is the phase difference between the fields and $f(t)$ gives the overall pulse envelope. (Note that our definition of the phase differs by 90° from the convention used by Muller et al.) For the calculations reported here, we use an envelope function which rises linearly over five optical cycles then is constant for the next 20 cycles. If only ionization rates are of interest, they can be obtained from this constant intensity interval. To calculate photoelectron energy and angular distributions, the pulse is ramped down over an additional five cycles and the final wave function is analyzed for the desired information. The analysis is performed using an energy window function, which is a projection operator that selects the part of the wave function lying within a very narrow energy range about a chosen energy.[5] The distributions are obtained over the energy range of interest by repeating these projections for a sequence of contiguous energy bins.

The factor in the brackets on the right-hand-side of Eq. (3) can be re-expressed as $2\sin(1.5\omega t'-\phi)\cos(0.5\omega t')$ with $t' = t+\phi/\omega$. From this form, we note the following symmetries (ignoring an overall time shift): (i) replacing ϕ with $\phi\pm 2\pi$ does not change the time-dependence of the electric field; (ii) replacing ϕ with $\phi\pm\pi$ and z with $-z$, also leaves the cycle shape unchanged and (iii) replacing ϕ with $-\phi$ and t by $-t$ changes the sign of the field. In the second case above, the angular distribution of the emitted photoelectrons is reflected through the plane perpendicular to the direction of polarization. In the last case, as we will see below, the change of sign of the phase can lead to *completely different* ionization rates, angular distributions and ATI spectra.

In Fig. 1 we show the time dependence of the electric field over three optical cycles of the lower frequency field for four different choices of ϕ from 0° to 90°. The peak electric field varies from approximately $1.76\varepsilon_o$ (at $\phi=0^o$) to $2\varepsilon_o$ (at $\phi=90^o$) depending on this phase. For a

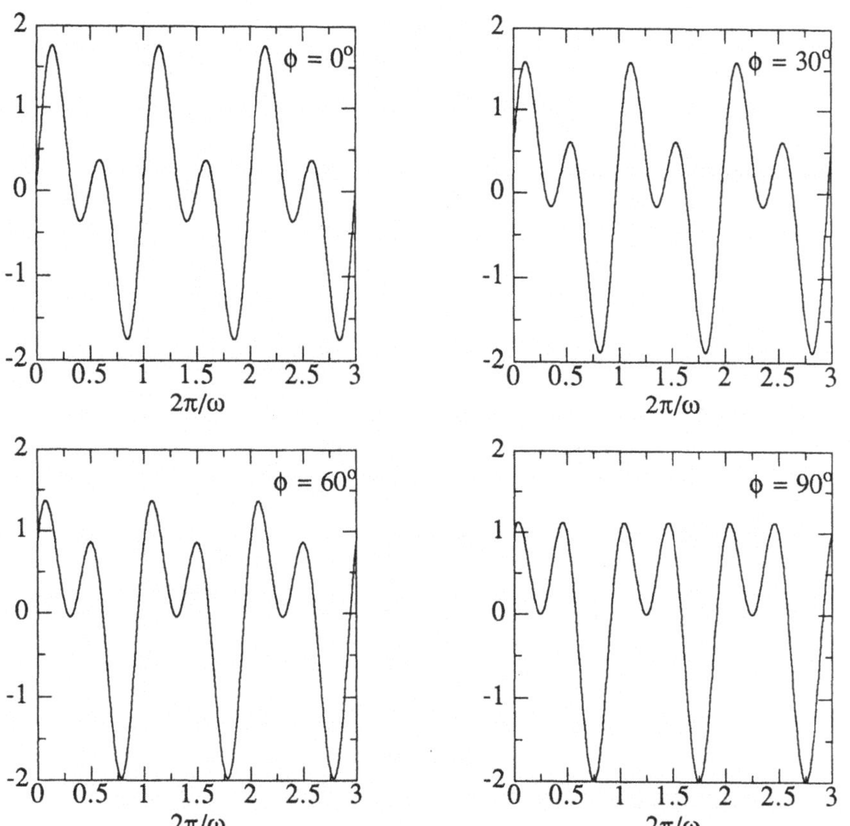

Fig. 1. Time-dependent electric field (in units of ε_o) for four choices of the phase angle ϕ.

strongly nonlinear process this difference can be expected to lead to dramatically different ionization rates. In the tunnelling regime (long wavelengths and high intensities) the ionization rate depends only on the maximum amplitude of the field, so that it depends, for a given ε_o, only on the magnitude of ϕ. The time dependence of the electric field for other choices of phase can be inferred from these plots and the symmetry properties discussed above.

RESULTS

We present multiphoton ionization rates and electron energy and angular distributions for hydrogen atoms subject to two lasers, one at the fundamental frequency of the Nd/YAG laser and the other its second harmonic corresponding to wavelengths of 1064 and 532nm. We consider intensities above 10^{13} W/cm^2, in the nonperturbative regime, and present the emission

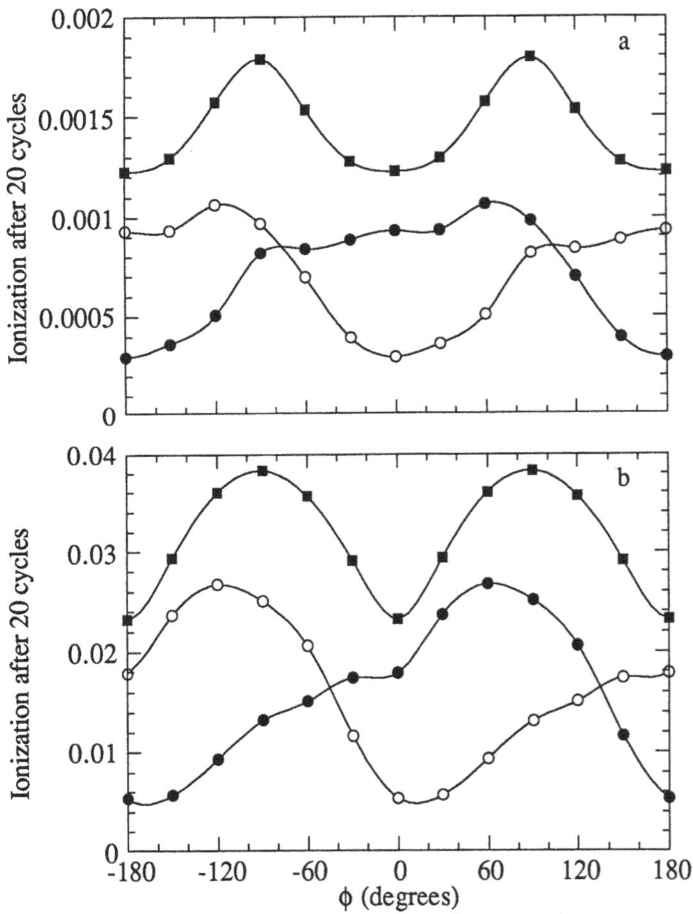

Fig. 2. Total (filled squares) and directional (filled circles, forward; open circles,backward) ionization probability as functions of the phase angle ϕ for (a) $I_o = 1.0\ 10^{13}$ W/cm^2, and (b) $I_o = 2.0\ 10^{13}$ W/cm^2.

rates as functions of ϕ, the phase difference between the two fields. From our previous work on this system,[4] we know that in this intensity range, the ionization dynamics is beginning to enter the tunnelling regime for the longer wavelength.

We show, in Fig. 2, the phase dependence of the total ionization probability for a pulse in which both laser fields rise linearly over five cycles of the lower frequency field then are held constant for the next 20 cycles. Also plotted are the angular distributions for the emitted electrons integrated over either the forward (positive z) or backward (negative z) direction, where z is the axis of polarization. (Note that Muller, et al. measured ionization only along the axis of polarization.) We find that the ionization probability is greatest at the phase difference which produces the maximum total electric field, $\phi = \pm 90^{o}$. (See Fig. 1.) This results in the total rates having a period of π in the relative phase. However, the angle dependent probabilities have a period of 2π, as expected from the symmetry properties discussed above. The emission in the forward direction for ϕ is calculated to be the same as in the backward direction for $\phi \pm \pi$. What is most surprising is that even though the *total* rates are the same for ϕ and $-\phi$, the directional rates are very different. Therefore an experiment set up to measure the phase dependence of the total ionization probability must measure the emission in either the forward or backward direction alone for a range of 2π, or in both directions over π.

As the intensities of the lasers increases the ionization process will be dominated by tunnelling during the interval when the combined fields give the largest amplitude. Therefore, considering the cycle shapes shown in Fig. 1, we would expect the largest asymmetry in the angular distribution for $\phi = \pm 90^{o}$ and no asymmetry for $\phi = 0, \pm 180^{o}$. As can be seen from Fig. 2 this is not the case. We find that the forward and backward distributions are only approximately consistent with the direction of the field at its maximum during the cycle. The emission in the backward direction should be larger when the sign of the field, when it reaches its maximum amplitude, is positive. At this instant the field is pushing the electrons in negative z-direction. This trend is more pronounced in the higher intensity case shown.

We have also determined the variation with ϕ of the photoelectron energy (ATI) distributions. We obtain partial emission rates by dividing the total yield of photoelectrons in a given ATI peak by the period of constant intensity in the pulse. Although the electron energy distributions exhibit approximately the same number of peaks at a given intensity for all phases, they vary in detail with the relative phase . As an example we compare the ATI spectra for ϕ of 30 and -30° in Fig. 3. Although the *total* emission rates for these two cases are identical, the *partial* rates are quite different. Fig. 4 shows angular distributions for the first ATI peak for ϕ of 60 and -120°. We expect these distributions to be mirror images of each other, which is the case within the accuracy of the calculation. For this case the angular distribution is quite strongly peaked in the backward (or forward) direction. In calculations on helium for these same wavelengths but at an intensity of the two fields of 5×10^{14} W/cm^{2} it was found that over

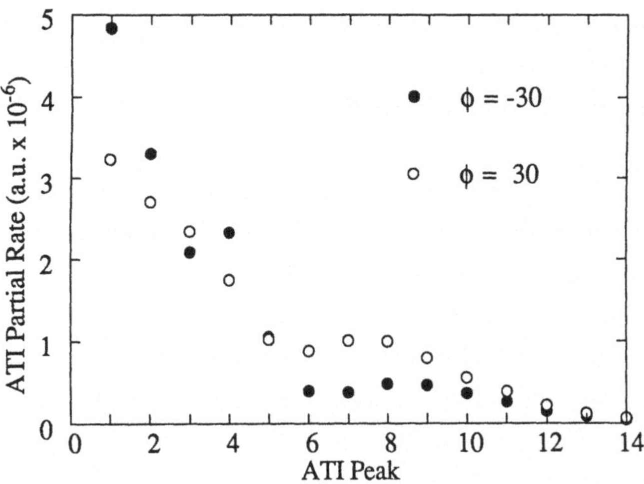

Fig. 3. Partial ATI rates for $I_o = 2.0\ 10^{13}$ W/cm^2 and two choices of the phase angle ϕ.

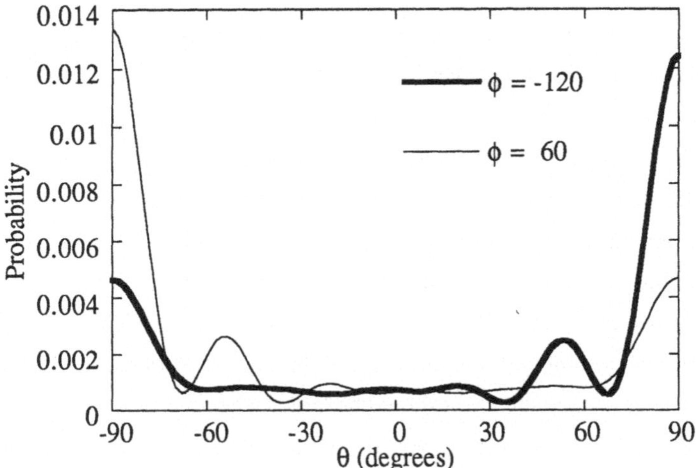

Fig. 4. Angular distribution of the first ATI peak for $I_o = 2.0\ 10^{13}$ W/cm^2 and two choices of the phase angle ϕ.

93% of the electrons were emitted in one direction for $\phi = 90^o$.[6] This was well into the tunnelling regime and is an impressive illustration of coherent control.

Finally in Fig. 5 we show the total ionization rate for hydrogen for single frequen ionization for intensities from 2 10^{13} to 2 10^{14} W/cm^2. The rates for both 1064 and 532 nm plotted. At the higher frequency some strong resonance structures are evident. As the intens rises, the rates for the two frequencies begin to merge, indicating the onset of tunnelling. A shown in this figure are the calculated total rates for 1ω-2ω ionization as functions of an effect "peak" intensity corresponding to the square of the ϕ-dependent peak electric fie $I_{peak}(\phi) \equiv c\varepsilon^2_{max}(\phi)/8\pi$. We have considered three cases where the two lasers have intensitie

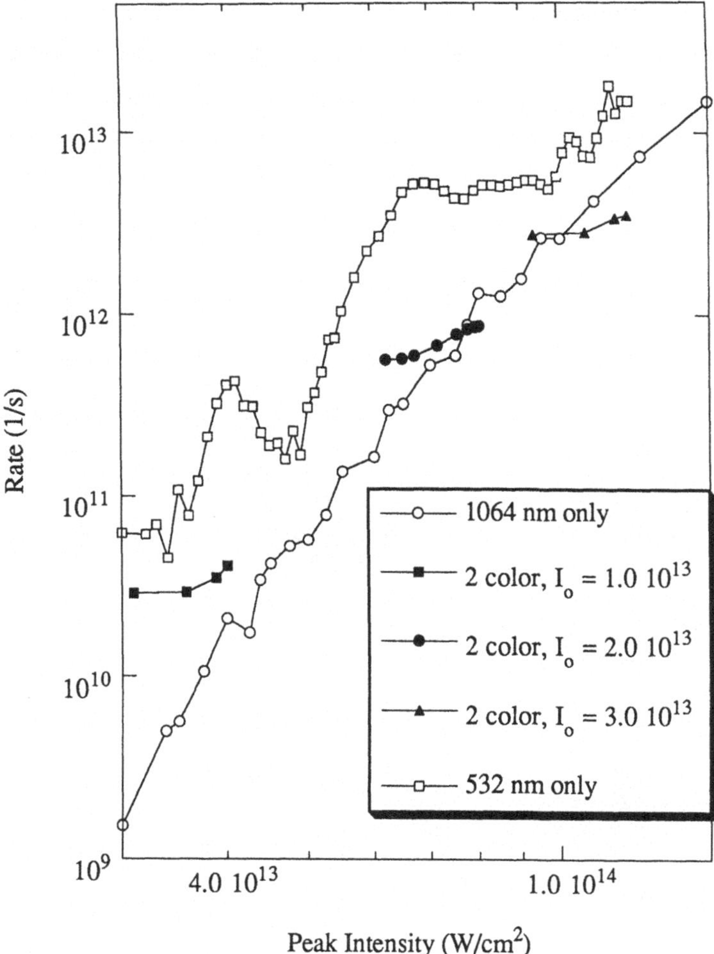

Fig. 5. One- and two-color ionization rates for hydgrogen.

have intensities $I = c\varepsilon_o^2 / 8\pi$ of 1, 2 or 3x10^{13} W/cm^2. The two-color rates fall close to the results for the lower frequency laser.

CONCLUSIONS

In the previous section we have presented some representative results of our calculations of two-color multiphoton ionization of hydrogen. At the intensities and wavelengths chosen here, the ionization is dominated by tunnelling. We found that the total ionization probability depended only on the maximum electric field, which varies as the magnitude of the relative phase of the two lasers, is changed. We showed that the emitted electron angular and energy distributions depended also on the sign of this phase implying that in this regime, while the excitation of the electron into the continuum is only a function of the peak field strength, the distribution of those electrons within the continuum depends also on the details of the shape of the electric field over a cycle. At shorter wavelengths the ionization process occurs by the absorption of a small number of photons so that the peak electric field becomes less important. In Fig. 6 we show the total rates for 1ω-2ω ionization where each laser has an intensity of 2×10^{13} W/cm^2 and the fundamental wavelengths are 1064, 532 or 355 nm. The upper curve repeats the results shown above for the longest wavelength considered.

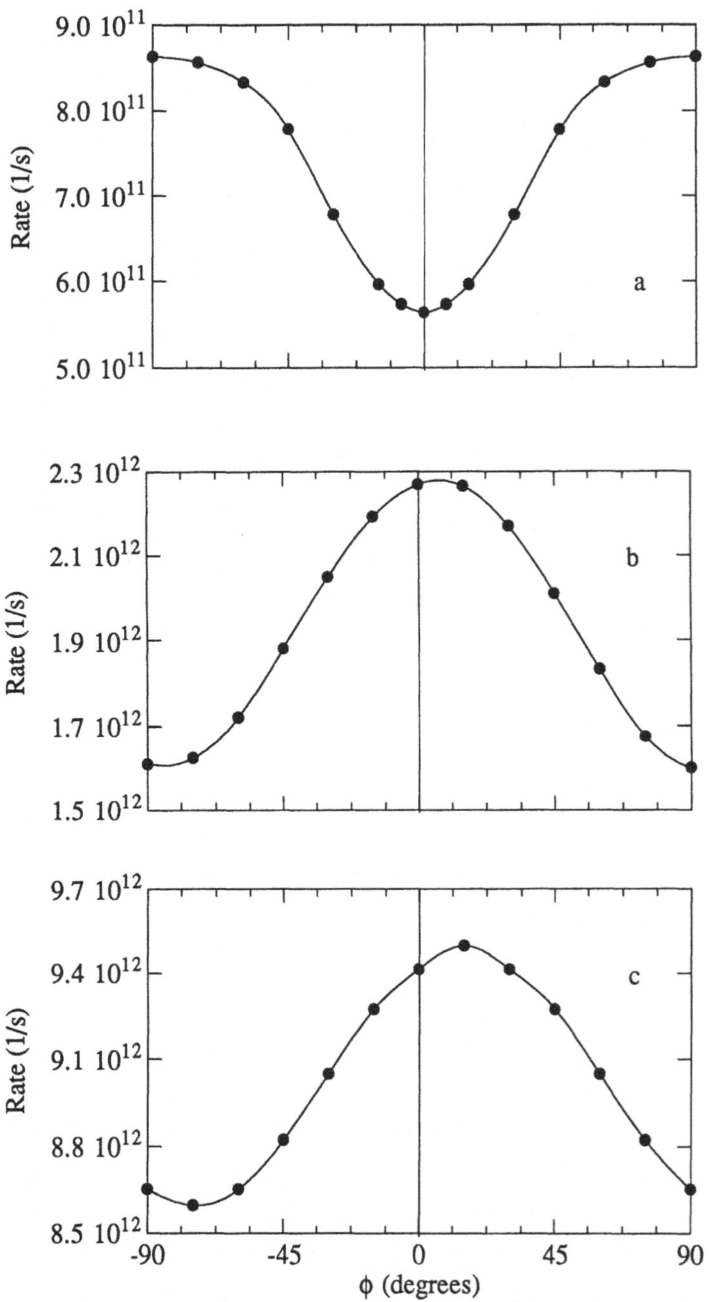

Fig. 6. 1ω-2ω total ionization rates as functions of phase angle ϕ. The fundamental wavelength is (a) 1064 nm, (b) 532 nm and (c) 355 nm.

The two lower panels are dramatically different in that the peak rate does not occur for the phase which produces the highest peak electric field. The symmetry with respect to $\phi = 0^{o}$ is also absent. These results will be discussed further elsewhere[7] but we note that far from the tunnelling regime both the total and partial rates show a period of 2π in the phase as expected from the symmetry of the interaction.

Muller, et al.[2] have conducted experiments in krypton using a Nd/YAG laser and its second harmonic. While they could not assign an absolute phase difference between the lasers, they did demonstrate the periodic dependence of the total and ATI rates on ϕ over several multiples of 2π. They observed that individual ATI peaks varied with the phase as we have found and showed that the total rate depends also on this phase. Therefore there is good qualitative agreement between these experimental results and our calculations.

Finally, although we have not explicitly shown the results here, we find that the ponderomotive shift of the continuum due to the combined fields is simply the sum of the shifts from the two fields as expected classically and as has been observed by Muller, et al.[2] The shift is determined from the calculated energy of the lowest ATI peak. Although there has been some disagreement about this fact in the literature[8,9] our calculations agree with the expected result.

ACKNOWLEDGEMENT

This work has been carried out under the auspices of the U. S. Department of Energy at the Lawrence Livermore National Laboratory under contract number W-7405-ENG-48.

REFERENCES

1. C. Chen and D. S. Elliot, Phys. Rev. Lett. **65**, 1737 (1990).
2. H. G. Muller, P. H. Bucksbaum, D. W. Schumacher and A. Zavriyev, J. Phys. B **23**, 2761 (1990).
3. K. C. Kulander, K. J. Schafer and J. L. Krause, Atoms in Intense Radiation Fields, M. Gavrila, Ed. (Academic Press, New York,1991) in press.
4. K. J. Schafer and K. C. Kulander, Phys. Rev. A **42**, 5798 (1990).
5. K. J. Schafer, Comp. Phys. Comm. **63**, 247 (1991).
6. A. Szöke, K. C. Kulander and J. N. Bardsley, J. Phys. B **24**, 3165 (1991).
7. K. J. Schafer and K. C. Kulander, Phys. Rev. A (to be submitted).
8. R. Trainham, G. H. Fletcher, N. B. Mansour, and D. J. Larson, Phys. Rev. Lett. **59**, 2291 (1987).
9. D. Normond, L. A. Lompre, A. L'Huillier, J. Morrelec, M. Ferray, J. Lavancier, G. Mainfray, and C. Manus, J. Opt. Soc. Am. B **6**, 1513 (1989).

COOPERATIVE ATOMIC EFFECTS IN TWO-PHOTON PROCESSES

Zhidang Chen and Helen Freedhoff

Department of Physics, York University, Toronto, Canada, M3J1P3

1 Introduction

The subject of cooperative radiative transitions by a system of two or more identical atoms (or molecules) has received considerable attention over the years, beginning with an initial study by Dicke [1]. Many authors since have dealt with coherent single-quantum electric-dipole excitations of systems of two [2,3] or more [3,4] atoms. The effects of the atomic interaction include in general the splitting of each atomic energy level into a number of sublevels, the shift of each sublevel from the single-atom energy, and a change in the lifetime of each sublevel from the single-atom radiative lifetime.

With the exception of a single paper dealing with cooperative electric quadrupole transitions in atoms [5], all the work performed to date has dealt with single-photon, electric dipole transitions. It is our purpose in this paper to address for the first time the question of co-operative two-photon (2-γ) spontaneous emission and resonance fluorescence in multi-atom systems. We study the simplest example of such a system: two identical atoms, separated by a distance R (large compared to the atomic diameter so that overlap can be ignored), each atom capable of making a 2-γ transition between two of its levels (of the same parity) via non-resonant intermediate levels (of the opposite parity). In studying the evolution in time of the atoms-plus-field system, we do not make the rotating-wave approximation [6], and we include two types of "essential states": States that are resonant in energy with the initial state, which both alter the decay rates and shift the energies of the atomic single-excitation levels; and non-resonant essential states, which contribute only to the energy shifts through the exchange of virtual photons.

We present the theory of 2-atom, 2-γ spontaneous emission in section 2, and evaluate the decay rates and shifts of the 2-atom energy levels in sections 3 and 4, respectively. In section 5, we present the theory of resonance fluorescence by the 2-atom system irradiated by a weak probe beam. Finally, we conclude the paper in section 6.

2 Spontaneous Emission

The calculations in this paper are based on the Heitler-Ma (HM) treatment of 2-γ transitions [7], developed as part of a program to extend radiation theory to 2-γ processes.

We consider two identical atoms, centered at $\mathbf{R}_s$ ($s = 1, 2$) with $|\mathbf{R}_1 - \mathbf{R}_2| = R$. Atom s can make a 2-γ transition between its levels $|b_s>$ and $|a_s>$ via a non-resonant intermediate level $|m_s>$ [8]. We denote by $|C>$ (with energy E_C) eigenstates of the atomic Hamiltonian,

$H_A = H_{A_1} + H_{A_2}$, of the two-atom system,

$$|C>= \begin{cases} |B>=|b_1>|b_2> & E_B = 2E_b \\ |M>=|m_1>|m_2> & E_M = 2E_m \\ |A>=|a_1>|a_2> & E_A = 2E_a \\ |\pm>=\frac{1}{\sqrt{2}}(|a_1>|b_2>\pm|b_1>|a_2>) & E_\pm = E_a + E_b, \end{cases} \tag{1}$$

and we include as well the non-resonant state $|M>$ $(E_m > E_b > E_a)$, which contributes to the energy shifts. The Hamiltonian of the combined atom-field system is $H = H_0 + V = H_A + H_F + V^1 + V^2$, where H_F is the Hamiltonian of the field and V^s is the interaction of atom s with the field,

$$V^s = -\boldsymbol{\mu}(s) \cdot \mathbf{E}(\mathbf{R}_s), \tag{2}$$

$$\mathbf{E}(\mathbf{R}) = i \sum_{\mathbf{k}\epsilon} \sqrt{\frac{2\pi\hbar ck}{L^3}} \left(\epsilon(\mathbf{k}) a_k e^{i\mathbf{k}\cdot\mathbf{R}} - \epsilon^*(\mathbf{k}) a_k^\dagger e^{-i\mathbf{k}\cdot\mathbf{R}} \right). \tag{3}$$

We will be involved with the states $|C0>$, where $|0>$ is the vacuum state of the field, and the states $|C\,kl>$ (with energy $E_{Ckl} = E_C + \hbar c(k+l)$), in which the atoms are in the state $|C>$ and one photon is present in each of modes k and l. We shall omit $|0>$ from $|C0>$ when no confusion can result.

We take for the initial state $|I>$ of our system the single-excitation state

$$|I>= c_+|+0> + c_-|-0>, \qquad |c_+|^2 + |c_-|^2 = 1. \tag{4}$$

(For example, if one atom is initially in state $|b>$ and the other in state $|a>$, $c_\pm = 1/\sqrt{2}$.) In HM, we study $G(E)|I>$, the Fourier transform of the wavefunction $|\psi(t)>$ of the atom-field system. The basic (exact) equation of the HM method for 2-γ processes is [7]

$$[E - H_0 - K(E)]\ G(E)|I>= \left[1 + V(E - H_0)^{-1}\right]\ |I>, \tag{5}$$

where the effective operator K is given by

$$K(E) \equiv \sum_{s,s'=1}^{2} V^s\ (E - H_A - H_F)^{-1}\ V^{s'}. \tag{6}$$

Our objective is to find the dependence on the interatomic separation R of the decay rates and energy shifts of the system, to the lowest nonvanishing order in V. The diagonal elements of $K(E)$ are found to give second-order R-independent shifts and are absorbed into the corresponding energies. Multiplying Eq.(5) by $<C|$ and by $<Ckl|$ in turn, and denoting $<C|G|I>$ by G_C and so on, we obtain the equations

$$(E - E_C)G_C - \sum_{D\neq C} K_{C,D}G_D - \sum_{Dkl} K_{C,Dkl}G_{Dkl} =< C|I> \tag{7}$$

$$(E - E_{Ckl})G_{Ckl} - \sum_{D\neq C} K_{Ckl,Dkl}G_{Dkl} - \sum_{D} K_{Ckl,D}G_D = 0. \tag{8}$$

These equations are then solved by using the continued-fraction rules of Swain [9], with details presented elsewhere [10]. We obtain the decay rates $\gamma_\pm$ and energy shifts $\Delta_\pm$ of levels $|\pm>$:

$$\gamma_\pm = \frac{2\pi}{\hbar} \sum_{kl} |K_{\pm,Akl}(E_\pm)|^2 \delta(E_\pm - E_{Akl}) \tag{9}$$

$$\Delta_\pm = \frac{|K_{\pm,M}(E_\pm)|^2}{E_\pm - E_M} + \sum_{Ckl} \frac{|K_{\pm,Ckl}(E_\pm)|^2}{(E_\pm - E_{Ckl})}, \tag{10}$$

in which the factors $1/(E_\pm - E_{Ckl})$ are understood as principal values.

Eqs.(9) and (10) describe the effects of the atomic interaction on the decay rates and energy shifts for 2-atom 2-γ spontaneous emission processes. To evaluate these quantities, we need the matrix elements of the effective operator $K(E)$. We define the single-atom operators $K^s(E)$ (s=1,2) by

$$K^s(E) \equiv V^s \frac{1}{E - H_{A_s} - H_F} V^s, \tag{11}$$

and the two-atom operators $K^{ss'}(E)$ ($s \neq s'$) by

$$K^{ss'}(E) \equiv V^s \frac{1}{E - H_A - H_F} V^{s'}. \tag{12}$$

Using Eqs.(1) and (6), we find

$$K_{\pm,Akl}(E_\pm) = \frac{1}{\sqrt{2}}\left[K^2_{b,akl}(E_b) \pm K^1_{b,akl}(E_b)\right] \tag{13}$$

$$K_{\pm,Bkl}(E_\pm) = \frac{1}{\sqrt{2}}\left[K^1_{a,bkl}(E_a) \pm K^2_{a,bkl}(E_a)\right] \tag{14}$$

$$K_{+,\pm kl}(E_+) = \frac{1}{2}\left[K^1_{a,akl}(E_a) + K^2_{b,bkl}(E_b) \pm K^1_{b,bkl}(E_b) \pm K^2_{a,akl}(E_a)\right] \tag{15}$$

$$K_{\pm,Mn'}(E_\pm) = \frac{1}{\sqrt{2}}\sum_{s\neq s'}\left[K^{ss'}_{a_1b_2,\ Mn'}(E_\pm) \pm K^{ss'}_{b_1a_2,\ Mn'}(E_\pm)\right], \tag{16}$$

where $|n'>$ is either the vacuum state $|0>$ or the two photon state $|kl>$. The single-atom matrix elements $K^s_{r,tkl}(E)$ are evaluated in terms of the matrix elements of V in Eq.(2):

$$K^s_{r,tkl}(E) = -\left(\frac{2\pi\hbar c}{L^3}\right)\sum_{ij}\epsilon_j(\mathbf{k})\ \epsilon_i(\mathbf{l})\ \sqrt{kl}\ \alpha^{ji}_{rt}(E,k,l;s)e^{i(\mathbf{k}+\mathbf{l})\cdot\mathbf{R}_s} \tag{17}$$

$$\alpha^{ji}_{rt}(E,k,l;s) \equiv \left(\frac{\mu^j_{rm}(s)\mu^i_{mt}(s)}{E - E_m - \hbar ck} + \frac{\mu^i_{rm}(s)\mu^j_{mt}(s)}{E - E_m - \hbar cl}\right) = \alpha^{ij}_{rt}(E,l,k;s) \tag{18}$$

where $\{r,t\} = \{a,b\}$, i and j are the $\{x,y,z\}$ directions, and $\mu_{rm}(s) =< r_s|\mu^s|m_s >$. The matrix elements of the two-atom operator $K^{ss'}$ are evaluated similarly.

3 Decay Rates

Eq.(9) is now divided into two terms:

$$\gamma_\pm = \gamma_0 \pm \gamma_{12} \tag{19}$$

$$\gamma_0 = \frac{\pi}{\hbar}\sum_{skl}\left|K^s_{b,akl}(E_b)\right|^2 \delta(E_b - E_{akl}) \tag{20}$$

$$\gamma_{12} = \frac{2\pi}{\hbar}Re\sum_{kl}K^1_{b,akl}(E_b)K^2_{akl,b}(E_b)\ \delta(E_b - E_{akl}). \tag{21}$$

The quantity γ_0 in (20) is independent of R; it is recognized as the 2-γ decay rate of the excited state for an isolated atom [7]. The quantity γ_{12}, on the other hand, is R-dependent, and gives the effect of the cooperative atomic interaction on the damping rate of the system:

$$\begin{aligned}\gamma_{12}(R) = {} & \frac{c}{\pi}Re\sum_{ijpq}\int_0^{k_0} dk\ k^3\ (k_0-k)^3\ F_{jq}(kR)F_{ip}\left[(k_0-k)R\right] \\ & \alpha^{ji}_{ba}(E_b,k,k_0-k;2)\times\alpha^{pq}_{ab}(E_b,k,k_0-k;1),\end{aligned} \tag{22}$$

where $k_0 = (E_b - E_a)/\hbar c$. The function $F_{jq}(kR)$ is given by

$$\begin{aligned} F_{jq}(kR) &= (\delta_{jq} - \hat{R}_j\hat{R}_q)\frac{\sin kR}{kR} + (\delta_{jq} - 3\hat{R}_j\hat{R}_q)\left(\frac{\cos kR}{k^2R^2} - \frac{\sin kR}{k^3R^3}\right) \\ &= \delta_{jq}\left[\frac{2}{3}\, j_0(kR) - \frac{1}{3}\, j_2(kR)\right] + \hat{R}_j\hat{R}_q j_2(kR), \end{aligned} \tag{23}$$

where $j_n(x)$ is the spherical Bessel function (of the first kind) of order n, and $\hat{R}_j$ is the jth-component of $\mathbf{R}/R$. The expression $\sum_{jq} k^3\mu^j(2)F_{jq}(kR)\mu^q(1)$ represents the retarded interaction of a transition dipole at $\mathbf{R}_1$, oscillating with frequency kc, with a transition dipole at $\mathbf{R}_2$ [3]. Thus it is seen that γ_{12} is the convolution of two such dipole-dipole interaction expressions, one each at frequencies kc and $(k_0 - k)c$. We evaluate Eq.(22) for γ_{12}, with the result [10]

$$\begin{aligned} \gamma_{12}(R) &= \frac{2\sqrt{2}c}{3\sqrt{\pi}R^7} Re\sum_{ijpq} \alpha_{ba}^{ji}(E_b, \frac{k_0}{2}, \frac{k_0}{2}; 2)\alpha_{ab}^{pq}(E_b, \frac{k_0}{2}, \frac{k_0}{2}; 1) \\ &\times \left\{\delta_{jq}\delta_{ip}\,(k_0R)^{7/2}J_{\frac{7}{2}}(k_0R) + \left(\delta_{jq} - \hat{R}_j\hat{R}_q\right)\left(\delta_{ip} - \hat{R}_i\hat{R}_p\right)\frac{(k_0R)^{11/2}J_{\frac{11}{2}}(k_0R)}{20}\right. \\ &\left. - \left[\delta_{jq}\left(\delta_{ip} - \hat{R}_i\hat{R}_p\right) + \delta_{ip}\left(\delta_{jq} - \hat{R}_j\hat{R}_q\right)\right]\frac{(k_0R)^{9/2}J_{\frac{9}{2}}(k_0R)}{4}\right\} \end{aligned} \tag{24}$$

where $J_n(x)$ is a (cylindrical) Bessel function. In general, γ_{12} involves terms oscillating at frequency k_0R, multiplied by inverse powers of k_0R ranging from $(k_0R)^{-7}$ to $(k_0R)^{-2}$. In the two limiting cases of $k_0R \ll 1$ and $k_0R \gg 1$, we find

$$\gamma_{12} \Rightarrow \begin{cases} \gamma_0\left[1 - O\,(k_0R)^2\right] & (k_0R \ll 1) \\ -D\,(k_0R)^{-2}\cos(k_0R) & (k_0R \gg 1), \end{cases} \tag{25}$$

where the coefficient D is given by

$$D = \frac{ck_0^2}{15\pi} Re\sum_{ijpq} \alpha_{ba}^{ji}(E_b, \frac{k_0}{2}, \frac{k_0}{2}; 2)\alpha_{ab}^{pq}(E_b, \frac{k_0}{2}, \frac{k_0}{2}; 1)(\delta_{jq} - \hat{R}_j\hat{R}_q)(\delta_{ip} - \hat{R}_i\hat{R}_p). \tag{26}$$

Thus the decay rates for the $|\pm>$ states become

$$\gamma_+ \Rightarrow \begin{cases} 2\gamma_0 & (k_0R \ll 1) \\ \gamma_0 & (k_0R \gg 1), \end{cases} \tag{27}$$

$$\gamma_- \Rightarrow \begin{cases} O(k_0R)^2\gamma_0 & (k_0R \ll 1) \\ \gamma_0 & (k_0R \gg 1). \end{cases} \tag{28}$$

We see that the symmetric single-excitation state of the 2-atom system is superradiant and the antisymmetric state subradiant for small atomic separations, exactly as is the case for one-photon processes. For large atomic separations, the effects of the atomic interaction on the decay rates vanish as $(k_0R)^{-2}$, in analogy with the $(k_0R)^{-1}$ behavior of the corresponding quantity for one-photon emission.

4 Level Shifts

We now calculate the contribution of the atomic interaction to the level shifts. All single-atom shifts (which are R-independent) are absorbed into the atomic energies E_a and E_b.

We divide the shifts in Eq.(10) into $\Delta^r_\pm$ and $\Delta^{nr}_\pm$:

$$\Delta_\pm = \Delta^r_\pm + \Delta^{nr}_\pm \tag{29}$$

$$\Delta^r_\pm \equiv \sum_{kl} \frac{|K_{\pm,Akl}(E_\pm)|^2}{(E_\pm - E_{Akl})} \tag{30}$$

$$\Delta^{nr}_\pm \equiv \frac{|K_{\pm,M}(E_\pm)|^2}{E_\pm - E_M} + \sum_{C\neq A}\sum_{kl} \frac{|K_{\pm,Ckl}(E_\pm)|^2}{(E_\pm - E_{Ckl})}. \tag{31}$$

The shifts $\Delta^r_\pm$ arise from 2-γ transitions from $|\pm>$ to the states $|Akl>$, which are resonant for $k+l \simeq k_0$, while $\Delta^{nr}_\pm$ arise from transitions to other non-resonant states. Denoting $k_{b\ (a)}$ by $k_{b\ (a)} \equiv E_{m,\ b\ (a)}/\hbar c \equiv (E_m - E_{b\ (a)})/\hbar c$, we obtain the R-dependent terms in $\Delta^r_\pm$ and $\Delta^{nr}_\pm$:

$$\Delta^r_\pm = \mp\frac{2}{\pi^2\hbar c}\, Re \sum_{ijpq} \int_0^\infty\int_0^\infty dk\, dl\, k^3\, l^3\, F_{jq}(kR)\, F_{ip}(lR)$$
$$\times \left[\frac{\mu^j_{bm}(2)\mu^i_{ma}(2)}{(k+k_b)(k+l-k_0)}\left(\frac{\mu^p_{am}(1)\mu^q_{mb}(1)}{k+k_b} + \frac{\mu^q_{am}(1)\mu^p_{mb}(1)}{l+k_b}\right)\right] \tag{32}$$

$$\Delta^{nr}_\pm = -\left|\sum_{jq}\int_0^\infty \frac{dk\, k^3 F_{jq}(kR)}{\sqrt{2\pi^2\hbar c}\,(k_a+k_b)}\left[\left(\frac{\mu^j_{am}(1)\mu^q_{bm}(2)}{k+k_a} + \frac{\mu^j_{bm}(2)\mu^q_{am}(1)}{k+k_b}\right) \pm (1 \Leftrightarrow 2)\right]\right|^2$$
$$-\frac{2}{\pi^2\hbar c}\, Re \sum_{ijpq}\int_0^\infty\int_0^\infty dk\, dl\, k^3\, l^3\, F_{jq}(kR)F_{ip}(lR)$$

$$\times\left[\frac{2\mu^j_{bm}(2)\mu^i_{mb}(2)}{(k+k_b)(k+l)}\left\{\frac{\mu^p_{am}(1)\mu^q_{ma}(1)}{k+k_a} + \frac{\mu^q_{am}(1)\mu^p_{ma}(1)}{l+k_a}\right\}\right.$$

$$\pm\frac{\mu^j_{am}(2)\mu^i_{mb}(2)}{(k+k_a)(k+l+k_0)}\left\{\frac{\mu^p_{bm}(1)\mu^q_{ma}(1)}{k+k_a} + \frac{\mu^q_{bm}(1)\mu^p_{ma}(1)}{l+k_a}\right\}$$
$$+\frac{\mu^j_{am}(2)\mu^p_{bm}(1)}{(k+k_a)(l+k_b)}\left\{\mu^i_{ma}(2)\mu^q_{mb}(1)\left(\frac{1}{k+k_b}+\frac{1}{l+k_a}\right)\right.$$
$$\left.\left.\pm\mu^i_{mb}(2)\mu^q_{ma}(1)\left(\frac{1}{k+k_a}+\frac{1}{l+k_b}\right)\right\}\right] \tag{33}$$

The integrals in $\Delta^{nr}_\pm$ are similar to those in the calculation of the Casimir-Polder (CP) dispersion interaction energy between two-atoms in the ground state [11]; the only real difference is that one of our atoms is in an excited state, $|b>$. The integral in $\Delta^r_\pm$ is new in that the integrand contains the factor $\mathcal{P}/(k+l-k_0)$, which has poles at $k+l-k_0 \pm i\sigma = 0$. The integrals in $\Delta^{nr}_\pm$ are evaluated routinely by contour integration in the complex k- and l-planes; for the integrals in $\Delta^r_\pm$, additional care must be taken for poles of the integrands. We have succeeded in finding analytical expressions for these integrals for the limiting cases of small and large $k_0 R$ [10].

At small separations, the dominant contributions of the integrals are of order R^{-6}; the resulting shifts $\Delta_\pm$,

$$\Delta_\pm \Rightarrow -\frac{1}{2(E_{m,a}+E_{m,b})R^6}\left|\sum_{jq}\left(\mu^j_{am}(1)\mu^q_{bm}(2) \pm \mu^j_{am}(2)\mu^q_{bm}(1)\right)\left(\delta_{jq} - 3\hat{R}_j\hat{R}_q\right)\right|^2, \tag{34}$$

are the leading results at short distances, at which retardation of the interaction should be negligible. Indeed, if we replace the retarded interaction of Eq.(2) by the instantaneous (Coulomb) dipole-dipole coupling, $V_c = \frac{1}{R^3}\sum_{jq}\mu^j(1)\mu^q(2)\left(\delta_{jq} - 3\hat{R}_j\hat{R}_q\right)$, and use second order perturbation theory, we obtain results identical with Eq.(34), as expected. For large atomic separations, it is found (as with CP) that $\Delta_{\mp}^{nr}$ yields shifts whose leading terms are proportional to R^{-7}, which are found to be negligible compared to $\Delta_{\pm}^{r}$, which is of order R^{-2} at large R. The shifts $\Delta_{\pm}^{r}$ are absent in the CP theory of the dispersion interaction; they are analogous to the first-order (resonance) dispersion energies arising between an atom in an excited state and an identical atom in the ground state in one-photon transitions (see e.g., Stephen, in reference 2). The contribution of the shifts $\Delta_{\pm}^{r}$ at large atomic separations is much larger than that of the $\Delta_{\mp}^{nr}$ for reasons which can be viewed in the following way: Photons which fail to conserve energy by an amount ΔE can survive only for a time $\hbar/\Delta E$, which is of order of magnitude $(k_0 c)^{-1}$ in the present case. In that time, they can propagate a distance of order k_0^{-1}. Thus, for distances $R \gg k_0^{-1}$, the contribution of these virtual photons is negligible, and the shifts are dominated by $\Delta_{\pm}^{r}$,

$$\Delta_{\pm} \quad \Rightarrow \quad \mp \frac{2\hbar c k_0^7}{15\pi} \frac{\sin(k_0 R)}{(k_0 R)^2} Re \sum_{ijpq} \frac{\mu_{bm}^{j}(2)\mu_{ma}^{i}(2)\left[\mu_{am}^{p}(1)\mu_{mb}^{q}(1) + \mu_{am}^{q}(1)\mu_{mb}^{p}(1)\right]}{(E_{m,a} + E_{m,b})^2}$$
$$\times \left(\delta_{jq} - \hat{R}_j \hat{R}_q\right)\left(\delta_{ip} - \hat{R}_i \hat{R}_p\right). \tag{35}$$

These shifts vanish as $(k_0 R)^{-2}$, and are analogous to those occurring in the corresponding one-photon case, which involve functions oscillating at frequency $k_0 R$ and vanishing as $(k_0 R)^{-1}$ for large atomic separations.

We have also calculated [10] the probability of excitation as a function of time of the atomic system, and the spectrum of the emitted photons. As in the one-photon case, we see that the lifetime of the excitation in the 2-atom system is apparently lengthened. For small atomic separations, γ_- is very small, and the antisymmetric state $|- >$ is metastable. For times t such that $\gamma_+^{-1} \ll t \ll \gamma_-^{-1}$, (approximately) half of the energy still remains in the atomic system, and half is emitted into the field. For $t \gg \gamma_{\pm}^{-1}$, all the energy is emitted into the field. In studying the frequency distribution of the emitted radiation of the 2-atom system, the spectrum is found to be proportional to the factor $\omega^3(\Omega_0 - \omega + \Delta_{\pm}/\hbar)^3$ for the initial state $|\pm >$. For general initial states, the spectrum is a superposition of terms proportional to $\omega^3(\Omega_0 - \omega + \Delta_+/\hbar)^3$ and $\omega^3(\Omega_0 - \omega + \Delta_-/\hbar)^3$.

5 Resonance Fluorescence

We have further studied [10] the atomic system in the ground state $|A >$, irradiated by a nearly-2-γ-resonant weak probe beam. The atoms are excited to $|\pm >$ by the absorption of 2 photons from the beam, and subsequently emit 2 photons into modes k and l, returning to $|A >$. At the same time, the probability also exists for ground-state atoms to emit virtual photons and become "excited". To calculate the shift of the ground state properly, we must take both these processes into account. (We continue, of course, to include emission of virtual photons from $|\pm >$ as well, so that their shifts too are included properly.)

We separate the interaction Hamiltonian into two parts: $V = \sum_s V^s$, as defined in Eqs.(2) and (3) (with all modes but that of the probe), and the interaction of the atoms with the laser beam, $W = \sum_s W^s$, defined as in Eqs.(2) and (3), with mode k replaced by the laser mode. The effective 2-γ interaction operator of the atoms with the whole field is given by

$$K(E) = (V + W)\,(E - H_0)^{-1}\,(V + W), \tag{36}$$

where $H_0 = H_A + H_L + H_F$, with H_L the Hamiltonian of the laser beam, and H_F the Hamiltonian of the vacuum field, including all modes but that of the (laser) probe beam. The basic equation of the HM method for 2-γ resonance fluorescence becomes

$$[E - H_0 - K(E)]\; G(E)|I >= \left[1 + (V + W)(E - H_0)^{-1}\right]\; |I > . \tag{37}$$

where $|I\rangle = |An\rangle$, in which the atomic system is the state $|A\rangle$ while n photons are present in the probe field.

The application of Eq.(37) shows that the width of level A, γ_A, consists of two terms, corresponding to the probabilities of absorption from level $|A\rangle$ to the levels $|\pm\rangle$. For small R, the absorption to level $|+\rangle$ is predominant; for large R, γ_A becomes the absorption probability for two noninteracting atoms. The shift of level A becomes the sum of the laser-induced shift and the shift due to (virtual) 2-γ exchange between the two (ground-state) atoms; the latter produces exactly the London-van der Waals dispersion energy for small R and the Casimir-Polder energy for large R [10].

6 Conclusions

We have used the Heitler-Ma method to study a system of 2 atoms undergoing cooperative 2-γ transitions for two cases: 1. the atomic system initially in a single- excitation state, interacting with the vacuum field (2-γ spontaneous emission), and 2. the atomic system initially in the ground state, interacting with the vacuum field and with a weak probe field (2-γ resonance fluorescence). In both cases, cooperative atomic effects are found analogous to those which are well-known for single-photon transitions:

1. The decay rates of the $|\pm\rangle$ states are altered by the atomic interaction. The cooperative decay rate γ_{12} consists of oscillating functions of k_0R, multiplied by inverse powers of k_0R ranging from $(k_0R)^{-7}$ to $(k_0R)^{-2}$. For small atomic separations, the $|+\rangle$ state is superradiant, with $\gamma_+ \rightarrow 2\gamma_0$, while the $|-\rangle$ state is subradiant, with $\gamma_- \rightarrow O(k_0R)^2\gamma_0$; we thus predict for the first time the existence of superradiance and subradiance in 2-γ transitions in multiatom systems. For large separations, γ_{12} vanishes as $(k_0R)^{-2}\gamma_0$.

2. The energies of the $|\pm\rangle$ sublevels are shifted from the single-atom energy by an amount $\Delta_\pm$, which in general also consists of oscillating functions of k_0R multiplied by inverse powers of k_0R. The leading terms in $\Delta_\pm$ are proportional to $(k_0R)^{-6}$ for small separations, and to $(k_0R)^{-2}$ for large separations. The shift of the ground state $|A\rangle$ in 2-γ resonance fluorescence is given by the sum of the laser-induced shift and the Casimir-London energy.

Acknowledgments

This research was supported in part by the Natural Sciences and Engineering Research Council of Canada, to whom the authors extend their thanks.

References

1. R.H. Dicke, Phys. Rev., **93**, 99 (1954).
2. M.J. Stephen, J. Chem. Phys., **40**, 669 (1964); D.A. Hutchinson and H.F. Hameka, J. Chem. Phys., **41**, 2006 (1964); P.R. Fontana and D.D. Hearn, Phys. Rev. Lett., **19**, 481 (1967); R.H. Lehmberg, Phys. Rev. A, **2**, 889 (1970); T. Richter, Ann. Phys. Lpz., **36**, 266 (1979) and references therein.
3. H.S. Freedhoff and J. Van Kranendonk, Can. J. Phys. **45**, 1833 (1967).
4. M.R. Philpott, J. Chem. Phys. **63**, 485 (1975); M.R. Philpott, and P.G. Sherman, Phys. Rev. B**12** 5381 (1975); H.S. Freedhoff, J. Chem. Phys. **55**, 5140 (1971); ibid., J. Phys. C**13**, 5329 (1980); ibid., J. Chem. Phys. **85**, 6110 (1986); H.S. Freedhoff and W. Markiewicz, J. Phys. C**13**, 5315 (1980).
5. H.S. Freedhoff, Phys. Rev. A**5**, 126 (1972).
6. P.L. Knight and L. Allen, Phys. Rev. A**7**, 368 (1973).
7. Z. Chen and H.S. Freedhoff, J. Phys. B**24**, 1935, (1991).

8. For simplicity, only a single intermediate level is included explicitly in our calculations. The theory can be extended to a set $\{|m>\}$ of intermediate levels simply by summing.
9. S. Swain, in *Advances in Atomic and Molecular Physics* **22**, 387, (Academic Press, 1986).
10. Z. Chen and H.S. Freedhoff, Phys. Rev. A**44**, 546, (1991).
11. H.B.C. Casimir and D. Polder, Phys. Rev. **73** 360 (1948); F. London, Z. Phys. **63**, 245 (1930).

COHERENT DEFLECTION OF ATOMS BY ADIABATIC PASSAGE IN MULTILEVEL SYSTEMS

P. Marte, P. Zoller, and J. L. Hall*

Joint Institute for Laboratory Astrophysics, National Institute of Standards and Technology and Univ. of Colorado, Boulder, Colorado 80309-0440 USA

Introduction

The study of mechanical light effects on atoms has been the subject of considerable theoretical and experimental interest during the past few years.[1] In particular, progress in building an atomic interferometer (AI) has stimulated recent interest in *coherent* deflection of atoms by laser light.[2-6]

In an AI the center-of-mass wave packet corresponding to a single atom is transformed by a *beam splitter* into a macroscopic superposition state corresponding to two center-of-mass wave packets propagating in different spatial directions. An *atomic mirror* deflects these wave packets so that the matter waves traveling along two paths of the interferometer can be brought to interference and analyzed. This interference will, of course, be observed only if these scattering processes are coherent. In particular this implies that spontaneous emission must be suppressed to preserve the coherence of the atomic wave function.

In comparison to a neutron interferometer,[7] an AI promises enhanced sensitivity for high precision experiments in both gravitational and quantum physics, and a new generation of frequency standards and rotation sensors. In addition, one of the new perspectives of atomic interferometry is the wealth of possible experiments to manipulate and probe atoms with laser radiation: this includes scattering of atomic wave packets from light waves (e.g., the Kapitza–Dirac effect) to build atomic beam splitters and mirrors,[2,3,8,9] laser cooling techniques to prepare slow atomic beams,[1] and applications to laser spectroscopy. Recently, several groups have reported first experimental observations of atomic interference fringes by scattering atoms from mechanical gratings,[4,5] or by applying a sequence of short $\pi/2$, π and $\pi/2$ laser pulses to atoms in an atomic fountain[6] similar in concept to optical Ramsey experiments.

Here we discuss and analyze a new scheme for coherent atomic beam deflection from laser light waves which combines several attractive features, close to the require-

* Staff member, Quantum Physics Division, National Institute of Standards and Technology

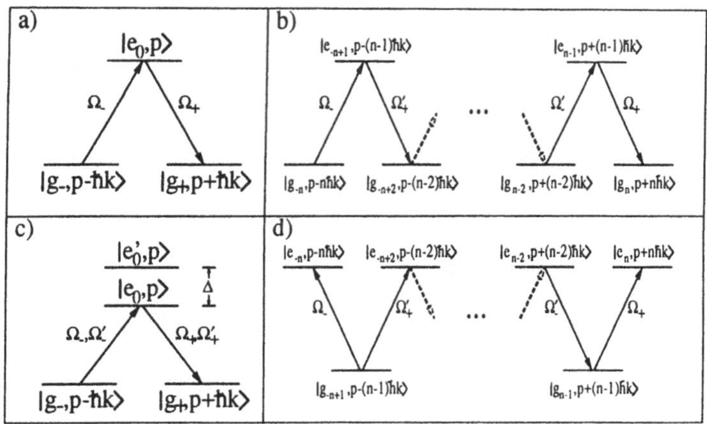

Fig. 1. a) Three–level system with Zeeman ground states $|g_{\pm 1,0}\rangle$ which are coupled by $\sigma^{\pm}$–polarized laser light to the upper state $|e_0\rangle$, b) Multi level system with chain of Λ transitions, c) Λ transition with two excited states, d) W system.

ments of an "ideal" atomic beam splitter and mirror for atomic interferometry. The proposed scheme is based on the concept of coherent adiabatic population transfer in Raman processes with *time-delayed laser pulses.* This has been demonstrated in the context of molecular spectroscopy by Bergmann and collaborators[10,11] and discussed theoretically by Oreg *et al.*,[12] Hioe and Carroll,[13] and Kuklinski *et al.*[11]

The basic idea of this mechanism is the following. Consider the three–level Λ–system in Fig. 1a, where the left and right ground state are coupled to the excited state by a first and a second laser, respectively. If the atoms are prepared initially in the left ground state, a sequence of two overlapping laser pulses where the second laser precedes the first laser in time leads to an adiabatic population transfer to the right ground state. This corresponds to an adiabatic passage along a dressed atom + field state which connects the left and right ground state in the Λ–system. It is an essential element of this process that the transfer occurs *without populating the intermediate excited state*, so that spontaneous emission is suppressed. Note that for two counter–propagating lasers this Raman process is associated with a momentum transfer of the sum of the two photon momenta of the lasers.

The attraction of this scheme as a coherent atomic beam splitter and mirror for atomic interferometry is based on the following features:

1. We have (close to) complete scattering from the intial state to a *single* final state corresponding to a fixed momentum transfer.

2. Since the process is adiabatic it is, within the validity of the adiabaticity of the process condition (see below), quite robust against changes in laser parameters (Rabi frequencies and detunings), interaction time with the laser (atomic velocity), etc.

3. Although the interaction can be resonant, the scattering process is immune to spontaneous decay as the excited state is never populated during the transition. This avoids the common approach of eliminating spontaneous decay by detuning the laser far off-resonance to reduce the excited state population, at the expense of introducing a weak atom–laser coupling.

This scheme can be generalized to a chain of Raman multiphoton transitions as indicated in Fig. 1b. Again one can show in this case that complete population transfer from the leftmost to the *single* rightmost groundstate is possible without populating the excited states. Due to the high order of multiphoton absorption large coherent momentum transfer from the laser to the atoms becomes possible.

Coherent momentum transfer by adiabatic passage in multilevel systems

Three-level system

We consider a stationary collimated beam of atoms propagating along the x-axis of our coordinate system. The atoms are scattered from two counterpropagating σ^+ and σ^- light waves which both have frequency ω and wave vectors k directed along the $+z$ and $-z$ axis, respectively. The atomic configuration is a three–level system with two Zeeman–ground states $|g_{m=\pm 1}\rangle$ coupled to an excited state $|e_{m=0}\rangle$ corresponding to a $J_g = 1$ to $J_e = 1$ transition as illustrated in Fig. 1a. An example of this configuration is the $2s\,^3S_1$ to $2p\,^3P_1$ transition in metastable helium.[14] The incident atoms are deflected from the light waves by absorption and subsequent reemission of laser photons. For small deflection angles (when the longitudinal atomic momentum $p_z = Mv_z$ is larger than the photon recoil $\hbar k$) our problem can be reduced to solving the one-dimensional Schrödinger equation for the transverse atomic motion[2,3]; the corresponding atomic Hamiltonian is[14]

$$
\begin{aligned}
H_{0A}(t) &= \frac{\hat{p_z}^2}{2M} + \hbar\omega_{eg}\,|e_0\rangle\langle e_0| \\
&\quad -\frac{1}{2}\hbar(\Omega_+(t)e^{-ik\hat{z}-i\omega t}\,|e_0\rangle\langle g_1| + \Omega_-(t)e^{ik\hat{z}-i\omega t}\,|e_0\rangle\langle g_{-1}| + h.c.).
\end{aligned}
\tag{1}
$$

The time coordinate is related to the atomic motion $x = v_x t$ along the x–axis. In particular, the time dependence of the Rabi frequencies $\Omega_\pm(t)$ (see Fig. 1) corresponds to the atom moving through the laser interaction zones. We choose $\Omega_\pm$ real. ω_{eg} is the atomic transition frequency. The Hamiltonian $H_{0A}(t)$ has the property that it only couples states within the family $\{|g_-, p_z - \hbar k\rangle, |e_0, p_z\rangle, |g_+, p_z + \hbar k\rangle\}$ with p_z the transverse momentum. In addition to the laser–induced scattering we allow for spontaneous decay during the interaction. Thus the total Hamiltonian is

$$
H = H_{0A}(t) + H_{0F} + V_{AF} \tag{2}
$$

with $H_{0A}(t)$ the atomic Hamiltonian Eq. (1), H_{0F} the Hamilton operator of the quantized free radiation field, and V_{AF} the dipole coupling of the vacuum modes to the atom.[3] The state vector $|\Psi(t)\rangle$ of the combined atom–radiation field consists of contributions where the atom has emitted no spontaneous photon, one photon, two photons, etc.[3] Thus the state vector can be expanded according to

$$
\begin{aligned}
|\Psi(t)\rangle &= |vac\rangle \otimes \int dp_z(|g_-, p_z - \hbar k\rangle a_-(p_z,t) + |e_0, p_z\rangle a_0(p_z,t)e^{-i\omega t} + \\
&\quad |g_+, p_z + \hbar k\rangle a_+(p_z,t)) \\
&\quad + \text{(1-photon contribution)} + \ldots
\end{aligned}
$$

where $a_{\pm,0}(p_z,t)$ are the atomic vacuum amplitudes. Since the momentum transfer to the atom by spontaneous emission corresponds to random momentum kicks, the vacuum contribution in the above is the (coherent) part of the state vector responsible for interference fringes in the AI. It can be shown that these amplitudes obey the Schrödinger equation[3]

$$
i\frac{\partial}{\partial t}a_\pm = (\frac{p_z{}^2}{2\hbar M} + \omega_R \pm kv_z)a_\pm - \frac{1}{2}\Omega_\pm a_0(p_z,t) \tag{3}
$$

$$
i\frac{\partial}{\partial t}a_0(p_z,t) = \left(\frac{p_z{}^2}{2\hbar M} - \Delta - i\frac{1}{2}\kappa\right)a_0 - \frac{1}{2}\Omega_- a_+(p_z,t) - \frac{1}{2}\Omega_+ a_-(p_z,t) \tag{4}
$$

with $\Delta = \omega - \omega_{eg}$ the detuning from resonance, and κ the spontaneous emission rate of the upper state. The ground states are shifted from the two–photon Raman resonance condition due to the Doppler detunings $\pm kv_z$ (with $v_z = p_z/M$); $\omega_R = \hbar k^2/2M$ is the recoil shift. In the following we assume that the atom is initially prepared in the $|g_-\rangle$ state with center–of–mass distribution corresponding to a well–collimated atomic beam. Note that the population transfer $\{|g_-, p_z - \hbar k\rangle \rightarrow |g_+, p_z + \hbar k\rangle\}$ corresponds to a momentum transfer $2\hbar k$, i.e., to a deflection of the atom.

Equation (3) is analogous to the equations studied in Refs. 11-13 in the context of optimizing population transfer in three–level (molecular) systems. Adopting the arguments presented in these papers in our present case, we see that an initial state $|g_-, p_z - \hbar k\rangle$ can be *adiabatically* transferred to $|g_+, p_z + \hbar k\rangle$ provided the two pulses $\Omega_\pm(t)$ are time–delayed with respect to each other (but still overlapping) such that the σ^- wave, i.e., the light wave acting on the second transition in Fig. 1, *precedes* the σ^+ pulse. This population transfer (and hence momentum transfer) is illustrated in Fig. 2 where we have plotted the population of the final state $|g_+\rangle$ (i.e., $|a_+|^2$) as a function of the time delay τ between two pulses. The curves in Fig. 2 were obtained by numerically integrating the Schrödinger equation (3) for two Gaussian pulses with equal pulse durations $T_1 = T_2 = T$ (FWHM) and intensities ($\Omega_+ = \Omega_-$), and the initial condition that the atoms are prepared in the $|g_-\rangle$–state. The dashed and solid curves correspond to $\Omega T = 50$, with $\kappa = 0$ and $\kappa T = 5$ ($\Omega/\kappa = 10$), respectively. In both cases we find a broad maximum for $\tau \approx -1.57T$ which is fairly insensitive to the presence of spontaneous decay even if the interaction time is long compared with the lifetime of the upper state ($\kappa T = 5$, solid curve). For $\tau \geq 0$ on the other hand, the upper state population shows strong (Rabi) oscillations for the undamped case (dashed line), and is close to zero in the presence of spontaneous emission (solid curve).

These results can be explained in an adiabatic dressed state picture.[11–13] For $v_z \approx 0$ the Hamiltonian matrix in Eq. (3) has an adiabatic dressed–state eigenvalue $E = 0$. The associated eigenvector is

$$
|E=0\rangle = \frac{\Omega_-}{\sqrt{\Omega_+^2 + \Omega_-^2}}|g_-, p_z - \hbar k\rangle - \frac{\Omega_+}{\sqrt{\Omega_+^2 + \Omega_-^2}}|g_+, p_z + \hbar k\rangle \tag{5}
$$

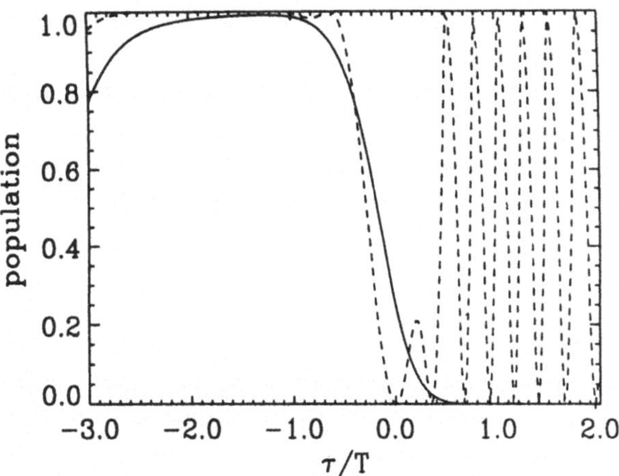

Fig. 2. Population of the final $|g_{+1}\rangle$ state as a function of the time delay τ between the Gaussian laser pulses. The Rabi frequencies are $\Omega_{\pm}T = 50$. The solid line corresponds to a spontaneous decay rate $\kappa T = 5$, while the dashed curve is for $\kappa T = 0$. T is the pulse duration of both laser pulses.

which is not contaminated by admixtures from the (decaying) excited state, and is independent of both Δ and κ. Furthermore it follows from Eq. (5) that

$$|E=0\rangle \rightarrow |g_-, p_z - \hbar k\rangle \quad \text{for} \quad \Omega_+(t)/\Omega_-(t) \rightarrow 0 \tag{6}$$

$$|E=0\rangle \rightarrow -|g_+, p_z + \hbar k\rangle \quad \text{for} \quad \Omega_-(t)/\Omega_+(t) \rightarrow 0. \tag{7}$$

Thus a pulse sequence where the $\Omega_+(t)$-pulse is time-delayed with respect to $\Omega_-(t)$ will satisfy the condition (6) at the beginning, and (7) at the end of the interaction with the pulses and thus transform $|g_-, p_z - \hbar k\rangle$ adiabatically into $|g_+, p_z + \hbar k\rangle$. For $\Delta = 0$ the other two dressed state eigenvalues are $E_{\pm} = \pm\frac{1}{2}\sqrt{\Omega_+^2 + \Omega_-^2}$. The condition for the initial state to follow $|E=0\rangle$ adiabatically is

$$\Omega_{\pm}T >> 1, \quad \Omega_{\pm} >> \kappa\,, \tag{8}$$

and the pulses must have significant overlap in time, so that the dressed energy level $|E=0\rangle$ is well separated from the states $|E_{\pm}\rangle$ for all times during the interaction. For transverse Doppler shifts the above argument is still valid if $\Omega_{\pm} >> kv_z$: typically we expect v_z to be of the order of a few recoil velocities $v_r = \hbar k/M$ so that this assumption is extremely well fulfilled (Raman Nath approximation).

The attraction of the above scheme for atomic interferometry is based on the following features, as summarized in the introduction: We have close to 100% scattering to a *single* final state corresponding to a fixed momentum transfer $2\hbar k$. Thus there is no splitting of the incident wave packet into a superposition corresponding to many momentum peaks $\pm\hbar k$, $\pm 2\hbar k$, etc., as for scattering of a two-level system from a standing light wave. Since the process is adiabatic it is, within the validity of the adiabaticity condition Eq. (8), quite robust against changes in laser parameters (Rabi frequencies and detunings), interaction time (atomic velocity), etc.; this is in contrast to transfer by π-pulses with Rabi oscillations which is very sensitive to the exact value of the pulse area. Furthermore, although the interaction is resonant, the scattering process is immune to spontaneous decay as the excited state is never populated during the transition. This avoids the common approach of eliminating spontaneous decay by detuning the laser far off-resonance to reduce the excited state population, at the expense of introducing a weak atom-laser coupling. Finally, the Λ-configuration of Fig. 1 has the advantage that it requires only a single laser since the σ^- wave can be derived by reflecting the σ^+ laser light.

Multilevel systems

The scheme outlined above for three-level systems is readily generalized to a chain of Raman transitions. This leads to an increased deflection angle due to the momentum transfer associated with the multiphoton transition (Fig. 1b). As an example consider the five-level system corresponding to an $J_g = 2$ to $J_e = 2$ transition with transfer of $4\hbar k$. We assume again that the atoms are initially prepared in the $|g_{-2}\rangle$ state. Two counterpropagating circularly polarized waves of frequency ω couple a chain of two Λ-transitions from $|g_{-2}\rangle$ to $|g_{+2}\rangle$. We can show again that there is an adiabatic dressed state eigenvalue $E = 0$ with eigenvector

$$|E=0\rangle = N\left(\Omega_-^2\Omega_-^{2'}\,|g_{-2},p_z-2\hbar k\rangle - \Omega_+^2\Omega_-^2\,|g_0,p_z\rangle + \Omega_+^2\Omega_+^{2'}\,|g_{+2},p_z+2\hbar k\rangle\right) \quad (9)$$

with N a normalization constant, and $\Omega_\pm$, $\Omega_\pm'$ Rabi frequencies related by appropriate Clebsch-Gordan coefficients (compare Fig. 3). From (9) it follows again that a delay of the σ^+ pulse with respect to the σ^- wave gives complete adiabatic population transfer from $|g_{-2}\rangle$ to the final state $|g_{+2}\rangle$ with the atom absorbing a momentum $4\hbar k$. In particular, we emphasize that there is no population left in the middle groundstate $|g_0\rangle$ after the interaction. We have again no admixture from the excited states during the process and find the transfer to be insensitive to variations of the Rabi frequencies and detunings within the validity of the adiabaticity condition. This adiabatic 4-photon process is illustrated in Fig. 4, which shows the time evolution of the atomic populations (the modulus squared of the vacuum amplitudes) during the pulse, obtained by numerical integration of the Schrödinger equation for two time-delayed Gaussian pulses. The parameters are $\Omega T = 50$ and $\kappa T = 5$ with time-delay $\tau = -1.2T$ and $T = T_1 = T_2$ the pulse duration. In both analytical and numerical work we have found that the adiabatic population transfer works extremely well even for very high-order Raman transitions achieving high momentum transfer in a single laser interaction zone.

Modularity

Another attractive feature of adiabatic passage is the possibility to achieve large momentum transfer by deflection of atoms in several successive interaction zones. We discuss the idea for the case of a Λ-transition interacting with $\sigma^\pm$ light waves (discussed in the context of Figs. 1a and 2). Again we consider an incident atomic wave packet prepared in a $|g_-,p_z\rangle$ state. The wave packet is scattered in a first interaction zone into the $|g_+,p_z+2\hbar k\rangle$ state with the $\sigma^\pm$-waves propagating in the $\pm z$-directions, respectively. In a second interaction zone with the lasers propagating in opposite directions the electron is transferred back to the $m = -1$ state, corresponding to $|g_-,p_z+4\hbar k\rangle$. Thus we have the sequence illustrated in Fig. 4. Note that after two laser zones the atom is in the same internal atomic state $|g_-\rangle$ that

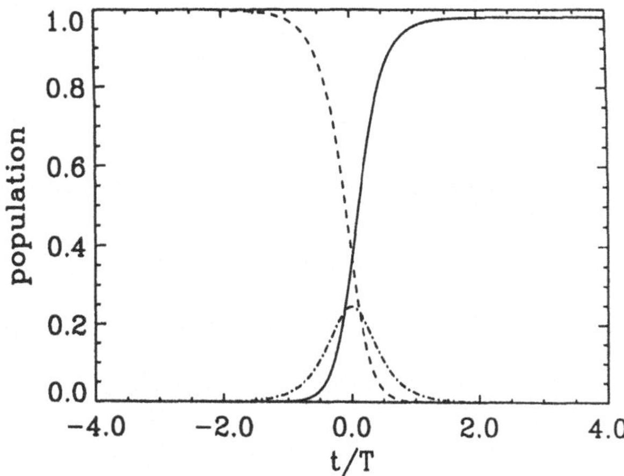

Fig. 3. Time evolution in the five–level system (Fig. 3) corresponding to a $J_g = 2$ to $J_e = 2$ transition. According to the Clebsch–Gordan coefficients the Rabi frequencies are related by $\Omega\prime_{\pm}(t) = \sqrt{\frac{3}{2}}\,\Omega_{\pm}(t)$. The population of the initial state $|g_{-2}\rangle$ (dashed line), the middle state $|g_0\rangle$ and final state $|g_{+2}\rangle$ is plotted as a function of the interaction time for a Gaussian laser pulses with duration $T = T_1 = T_2$. The parameters are $\Omega_{\pm}T = 50$, $\tau = -1.2\,T$, $\Delta T = 0$, $\kappa T = 5$.

it occupied initially, but it has received a momentum kick of $4\hbar k$. An advantage of adiabatic passage is that its efficiency factor for the transfer is close to unity, which allows us to combine a large number of these interaction zones. Finally, replacing the Λ–transitions (Fig. 1) by a chain of Raman transitions in a single zone (Fig. 3) promises very high order accumulated momentum transfer into a single final momentum state of the atom.

Beam splitter

So far our discussion has concentrated on coherent beam deflection corresponding to an atomic mirror. There are several possibilities to realize a beam splitter. Ideally, one expects an atomic beam splitter to produce a superposition state of two wave packets which differ by a large center–of–mass momentum. As a first possibility, Eq. (5) suggests that in a three-level Λ–system (Fig. 1) a coherent superposition can be formed by pulse shapes $\Omega_+(t)/\Omega_-(t) \to 1$,

$$|E=0\rangle \;\;\rightarrow\;\; \frac{1}{\sqrt{2}}(|g_-, p_z - \hbar k\rangle + |g_+, p_z + \hbar k\rangle)\,. \tag{10}$$

A second possibility is to create a coherent superposition of two atomic ground states prior to the interaction with the laser light using a radio frequency field (RF). Then *one* component of this superposition state is deflected *selectively* by adiabatic passage through a sequence of laser interaction zones. Consider again the Λ–system $\{|g_-\rangle, |e_0\rangle, |g_+\rangle\}$ described in Fig. 1. Typically, $|g_m\rangle$ will be hyperfine structure components for a particular F–value. Consider now a situation where in addition to

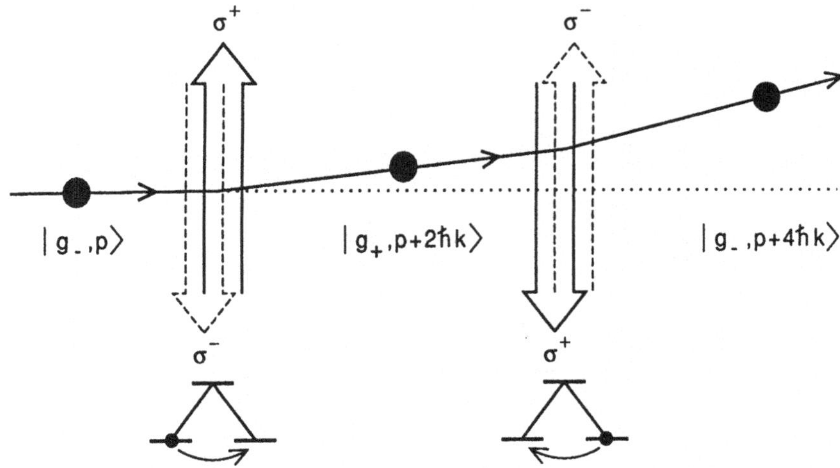

Fig. 4. Deflection of a Λ–system in several consecutive field zones illustrating the modularity of adiabatic passage to achieve large deflection angles.

F we have a second hyperfine structure state F' with states $|g'_m\rangle$ separated from the F–states by a RF transition. Then a possible sequence of transitions to generate a superposition state is

$$\begin{aligned} |g_-, p_z\rangle \quad &\xrightarrow{\text{RF}} \quad \cos\Theta\,|g_-, p_z\rangle + \sin\Theta\,\left|g'_-, p_z\right\rangle \\ &\xrightarrow{\text{zone 1}} \quad -\cos\Theta\,|g_+, p_z + 2\hbar k\rangle + \sin\Theta\,\left|g'_-, p_z\right\rangle \\ &\xrightarrow{\text{zone 2}} \quad \cos\Theta\,|g_-, p_z + 4\hbar k\rangle + \sin\Theta\,\left|g'_-, p_z\right\rangle \quad \text{etc.} \end{aligned} \tag{11}$$

Here Θ is the mixing angle due to the RF field. $\left|g'_-\right\rangle$ denotes a ground state which is assumed non–resonant with the laser light and thus remains undeflected. The last line in Eq. (11) corresponds to a macroscopic superposition state of two wave packets differing by a center–of–mass momentum $4\hbar k$.

Limitations

The discussion of the previous sections is based on the assumption of a three or multilevel system with Λ–transitions (Figs. 1a and 1b). Deviations from these "ideal" configurations lead to incomplete transfers in the adiabatic passage. A first example illustrating a nonideal system is given in Fig. 1c where the ground states are coupled to two excited states: in this case there is, in general, no eigenvalue $E = 0$ – except the Rabi frequencies satisfy the special conditions $\Omega_+/\Omega_- = \Omega'_+/\Omega'_-$. Note, however, that in general this condition is not fulfilled for a pair of hyperfine Zeeman levels. Furthermore an $E = 0$ eigenvalue does not exist for a W–system illustrated in Fig. 1d. An example is a transition $F \rightarrow F'$ with $F' > F$.

Conclusions

We have shown that adiabatic passage in multilevel systems leads to atomic beam deflection with the possibility of achieving high momentum transfer from the laser to the atomic wave packet to prepare a single momentum final state with high efficiency, while at the same time avoiding the momentum diffusion associated with spontaneous decay. An analysis of an atomic interferometer in a triple Laue configuration with interaction zones using counterpropagating $\sigma^{\pm}$–light to couple Zeeman levels by Raman transitions has been given in Ref. 2a.

Acknowledgment: The work at JILA is supported in part by the NSF, ONR and NIST.

References

1. See articles in: "Laser Cooling and Trapping," J. Opt. Soc. Am. B **6** (1989).
2. V. P. Chebotayev, B. Y. Dubetskii, A. P. Kazantsev, and V. P. Yakovlev, J. Opt. Soc. Am. B **2**, 1791 (1985); J. F. Clauser, Physica B **151**, 262 (1988) Ch. J. Borde, Phys. Lett. A **140**, 10 (1989); E. M. Wright and P. Meystre, Opt. Commun. **75**, 388 (1990); S. Glasgow, P. Meystre, M. Wilkens, and E. M. Wright, Phys. Rev. A **43**, 2455 (1991).
3. a) M. A. M. Marte, J. I. Cirac, and P. Zoller, J. Mod. Opt., in press; b) R. Graham, D. F. Walls, and P. Zoller, Phys. Rev. A, in press.
4. O. Carnal and J. Mlynek, Phys. Rev. Lett. **21**, 2689 (1991).
5. W. Keith, C. R. Ekstrom, Q. A. Turchette, and D. E. Pritchard, Phys. Rev. Lett. **21**, 2693 (1991).
6. M. Kasevich and S. Chu, preprint.
7. See contributions in "Proceedings of the International Workshop on Matter Wave Interferometry," Physica B+C **151**, (1988); *Neutron Interferometry*, edited by U. Bonse, and H. Rauch, Clarendon, Oxford (1979).
8. P. J. Martin, P. L. Gould, B. Oldaker, A. H. Miklich, and D. E. Pritchard, Phys. Rev. A **36**, 2495 (1987); P. J. Martin, B. Oldaker, A. H. Miklich, and D. E. Pritchard, Phys. Rev. Lett. **60**, 515 (1988).
9. V. I. Balykin, V. S. Letokhov, Yu. B. Ovchinnikov, and A. I. Sidorov, Phys. Rev. Lett. **23**, 2137 (1988).
10. U. Gaubatz, P. Rudecki, S. Schiemann, and K. Bergmann, J. Chem. Phys. **92**, 5363 (1990).
11. J. R. Kuklinski, U. Gaubatz, F. T. Hioe, and K. Bergmann, Phys. Rev. A **40**, 6741 (1989).
12. J. Oreg, F. T. Hioe, and J. H. Eberly, Phys. Rev. A **29**, 690 (1984).
13. C. E. Carroll and F. T. Hioe, Phys. Rev. A **42**, 1522 (1990); C. E. Carroll and F. T. Hioe, J. Opt. Soc. Am. B **5**, 1335 (1988); F. T. Hioe and C. E. Carroll, Phys. Rev. A **37**, 3000 (1988).
14. A. Aspect, E. Arimondo, R. Kaiser, N. Vansteenkiste, and C. Cohen-Tannoudji, Phys. Rev. Lett. **61**, 826 (1988).
15. E. Arimondo and G. Orriols, Lett. Nuovo Cimento **17**, 333 (1976); H. R. Gray, R. M. Whitley, and C. R. Stroud, Opt. Lett. **3**, 218 (1978).

INTERFERING OPTICAL INTERACTIONS: PHASE SENSITIVE ABSORPTION

Ce Chen,[+] Yi-Yian Yin,[+] A. V. Smith[*] and D. S. Elliott[+]

[+] School of Electrical Engineering, Purdue University, West Lafayette, IN 47902 USA
[*] Sandia National Laboratories, Div. 1124, Albuquerque, NM 87185 USA

Interference effects have been especially important in the development of new frontiers in physics. In this paper we will discuss a relatively new type of interference in laser-atom interactions, and our experimental studies of its properties, limitations, and applications[1,2]. The interference which we have been studying is between different optical processes which can individually lead to the excitation of the same atomic or molecular state. This interference has been identified[3-5] as the cause of the suppression of multiphoton excitation of various atomic species when the density of the nonlinear medium is relatively high. The experiments we will discuss here can be extended in many new directions, including studies of propagation effects in nonlinear optical media, control of reaction products in laser induced ionization or dissociation, or control of photoelectron angular distributions.

The history of interference between optical processes starts as early as the mid-1960's when Manykin and Afanasév[6], and Gurevich and Khronopulo[7], considered theoretically the effect of resonantly enhanced multi-photon absorption in a nonlinear atomic vapor. These authors predicted a type of "bleaching" of the nonlinear medium, in that the medium would become transparent to the input laser radiation. This effect is distinctly different from a saturation effect, however, in which the medium becomes transparent due to a population transfer induced by the laser. Rather, this self-induced transparency is a result of the nonlinear generation of new optical fields by the medium, which can then interact with the medium in such a way as to suppress the transition amplitude for excitation of the atom. Suppression of multiphoton ionization[8] and multiphoton absorption[9] at high vapor densities were first observed experimentally in 1977. The connection between these observations and the earlier theoretical work of references 6 and 7 was not recognized immediately. In later works, the correlation between the suppression of multiphoton excitation or ionization and generation of harmonic waves in the medium[10-13] was studied, as well as the competition between four-wave mixing and amplified spontaneous emission[14], the suppression of multiphoton ionization by amplified spontaneous emission[15], and the suppression of two-photon excitation in sum frequency mixing processes[16].

In all of these works prior to ours, the common feature is that the nonlinear generation of the new wave and the multiphoton absorption or ionization take place in the same region. Our goal in the experiments we will describe next was to separate the harmonic generation process from the multi-photon absorption process such that we could control the phase of the interference. This would allow for the continuous variation from conditions leading to constructive interference (enhancement) to those leading to destructive interference (inhibition).

We chose to observe the interference in atomic mercury (see figure 1), using two ionization processes each leading to the same continuum state and each resonantly enhanced by the 6p $^1P_1$ intermediate state. The first process is a five-photon process using light at λ = 554 nm. This process is resonantly enhanced due to the proximity of three times the laser frequency to the atomic transition frequency. The second process is a three-photon process using one photon at $\lambda/3$ and two photons at λ. Thus the interference which we observe is due to the parallel pathways between the 6s $^1S_0$ and the 6p $^1P_1$. This interference is clearly revealed by measuring the dependence of the ionization probability of the atomic mercury on the relative phase between the two fields. For purposes of clarity, we discuss the excitation process to the 6p state rather than the ionization process directly. The arguments are easily extended to the latter process. Under conditions of concurrent excitation by the two fields, E^{uv} and E^{vis}, the net transition rate is expected to be given by

$$W_{1,3} = \frac{2\pi}{\hbar^2}\left|\mu E^{uv} e^{i\phi_1} + \mu^{(3)}\left(E^{vis} e^{i\phi_2}\right)^3\right|^2 g\left(\Omega_{fg} - 3\omega\right). \quad (1)$$

Here μ and $\mu^{(3)}$ are the transition moments for linear and three-photon processes, respectively, and g $(\Omega_{fg} - 3\omega)$ is a lineshape function. The phase, ϕ_i (i = 1,2), of each field must be retained in this expression, leading to a term in the transition probability which varies as cos $(3\phi_2 - \phi_1)$.

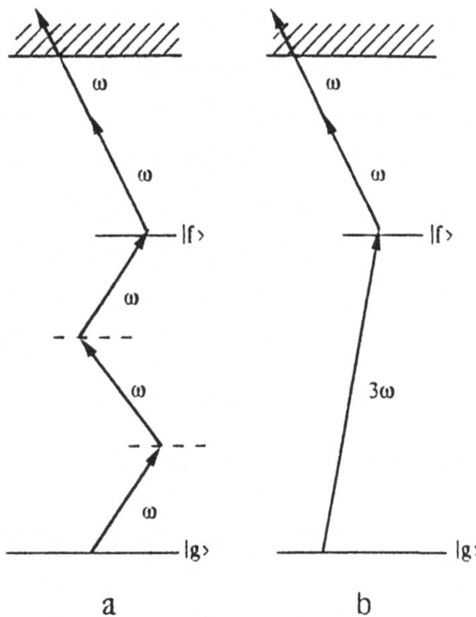

Figure 1. The two processes which interfere in this observation. The transition |g> → |f> is three- and one-photon allowed, as shown in (a) and (b). In our experiment in mercury, |g> is the ground 6s $^1S_0$ state, and |f> is 6p $^1P_1$ state.

In the case of excitation by plane waves, the single-photon and three-photon transition amplitudes can be matched everywhere by the adjustment of the field amplitudes E^{uv} and E^{vis}, leading to a maximum depth of modulation of one. When using focussed laser beams to photoionize the atoms, we must consider that the relative amplitudes of the two beams, as well as their relative phase, will vary through the focal region. This will limit the depth of modulation of the interference fringes to a value of less than one. If we assume lowest-order Gaussian modes, the transition rate at a distance z from the focus is proportional to:

$$W \propto \left| \frac{E^{uv} \cdot \mu e^{i\phi_1}}{\sqrt{1+(z/z_o)^2}} e^{-3r^2/w^2} e^{-i(k^{uv}z-\tan^{-1}z/z_o)} + \frac{\left(E^{vis} e^{i\phi_2}\right)^3 \mu^{(3)}}{\left[1+(z/z_o)^2\right]^{3/2}} \left(e^{-r^2/w^2} e^{-i(k^{vis}z-\tan^{-1}z/z_o)}\right)^3 \right|^2 \quad (2)$$

where w (z) is the $1/e^2$ intensity beam radius, k is the propagation constant and $z_o = \pi w^2(o)/\lambda$ is one half the confocal parameter. We set $\Delta k = 3k^{vis} - k^{uv} = 0$, valid at low mercury densities. Under our experimental conditions z_o is the same for the two beams, while the beam radius for the uv beam is $1/\sqrt{3}$ that of the visible beam. Since the three-photon process depends on the cube of the visible field, the z-dependence of the magnitude of this term of the transition amplitude varies as $[1 + (z/z_o)^2]^{-3/2}$ while the phase varies as $\exp(3i \tan^{-1} z/z_o)$. Therefore the magnitude of the two processes can be matched at at most two locations symmetrically placed about the focus, and the phase of the two transition amplitudes varies through the focal region over a range of 2π. The interference can still be quite visible however because of the high-order intensity dependence of the process, confining ionization to the center of the focal region. This effect can be calculated by integrating the expression for the ionization rate (eqn. 2) over the entire focal region. Since two additional photons must be absorbed to ionize the mercury, we must weight equation (2) by an additional factor to reflect the intensity dependence of the entire photoionization process. In the absence of saturation effects, we should multiply equation two by the square of the visible beam intensity (retaining its z-dependence) before integrating, yielding an average ionization rate of

$$W \propto \left(1 + \frac{5}{8}M^2\right) + M\cos(3\phi_2 - \phi_1) \tag{3}$$

where $M = (E^{vis})^3 \mu^{(3)} / E^{uv}\mu$ represents the relative contribution of the two processes at the beam waist on axis. When $M = 1$, the magnitude of the transition amplitudes are matched at the center of the focus, but the linear process will be stronger at all other locations. For $M = 2$, the magnitudes are matched at $z = \pm z_0$. Inside the focus the three-photon process is stronger, while for $|z| > z_0$, the linear process is stronger. The depth of modulation has a maximum value of $\sqrt{2/5} \sim 0.63$ at $M = \sqrt{8/5}$, and is within 10% of this maximum value over the range from $M = 3/4$ to $M = 2$. Thus the depth of modulation depends only weakly on matching the intensities in this range. We add that this is an upper bound for the depth of modulation, subject to our assumption of the lack of saturation. From measurements we have made using a multi-electrode configuration, to be discussed later, it appears that the ionization step is strongly saturated, suggesting that the I^2 factor used to obtain equation (3) is too strong. This leads to a much smaller depth of modulation since the effective length of the interaction region is increased. The interference pattern would thus be less visible, since a maximum to one side of the focus would be partially negated by a minimum to the other side. For example, if we weight equation (2) by a factor of I, rather than I^2, before integrating we calculate that the maximum depth of modulation is reduced to $\sqrt{1/6} \sim 0.41$. For even stronger saturation of the ionization step, the fringe visibility decreases further. In any case, however, the depth of modulation of the fringes depends only weakly on M.

The experiment is carried out in a cell consisting of three chambers, shown in figure 2. Coherent ultraviolet radiation at 185 nm is generated in the first chamber, containing mercury at a relatively high pressure of ~100 mtorr (cold finger temperature = 80° C) by focusing in laser radiation at 554 nm using a 20 cm focal length lens. The radiation produced by third harmonic generation, resonantly enhanced by the nearby 6p state, has a well-determined phase with respect to the laser fundamental. The visible and ultraviolet beams are collimated and refocussed into the third chamber using a pair of spherical

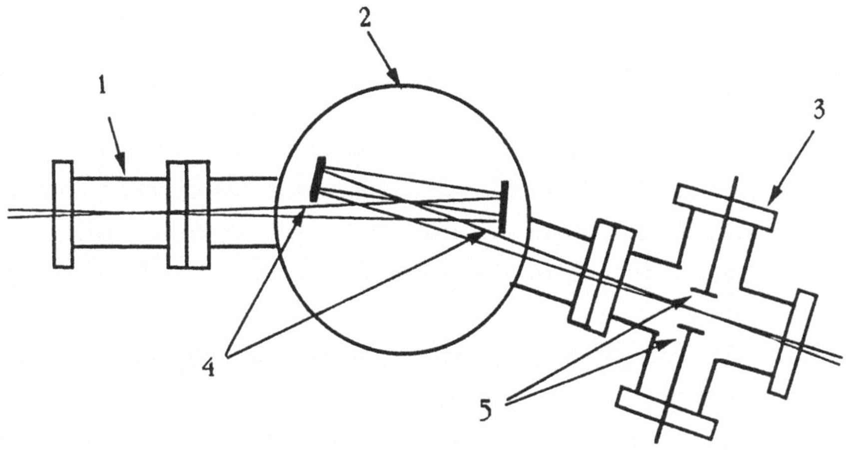

Figure 2. The three chamber cell. The first chamber (1) contains mercury at a high density (~100 mtorr), the second (2) contains argon gas at a variable pressure (0-38 torr), and the third (3) contains mercury at a low density (~2-3 mtorr). Chamber 2 contains (4) a pair of uv enhanced aluminum coated spherical mirrors, and the ionization signal is measured using (5) a pair of biased platinum collection plates.

mirrors with a uv enhanced aluminum coating (focal length = 25 cm). The third chamber also contains mercury vapor, but at a much lower pressure (~ 2-3 mtorr) than that of the first cell by virtue of the cold finger temperature of ~ 30° C. The relatively low vapor density in this cell is important to minimize the amount of third harmonic generated there. The second cell is used to introduce a phase shift between the two laser beams. By varying the density of the argon gas in this cell, the phases of the visible and uv beams undergo a shift of magnitude $\phi_2 = 2\pi l\Delta\rho n^{554}/\lambda$ and $\phi_1 = 6\pi l\Delta\rho n^{185}/\lambda$, respectively, where l is the path length in the argon chamber, $\Delta\rho$ is the change in density in amagats and n^{554} and n^{185} are the refractive indices of argon gas at STP conditions at 554 nm and 185 nm respectively. The relative phase between the two transition amplitudes thus varies as

$$3\phi_2-\phi_1 = \frac{6\pi l\Delta\rho}{\lambda}(n^{554} - n^{185}) . \tag{4}$$

Observation of the interference fringes depends critically on several important features of the experiment, such as spatial coherence of the laser and beam overlap in the focal region. The beam produced by the homemade Littman style short cavity laser is a nearly TEM_{00} Gaussian beam operating on one to three longitudinal modes separated by the free spectral range of the laser of ~3.3 GHz. The Nd:YAG laser used to pump the dye laser operated on several modes producing intensity beats on the laser pulse. After two stages of amplification in longitudinally pumped Brewster angle dye cells, the laser beam has a pulse energy of ~4.5mJ and a pulse duration of less than 15 nsec, and is somewhat elliptical in shape with beam diameters of 2.2 mm and 1.3 mm along the major and minor axes, respectively.

A lens cannot be used to focus the two laser beams into the third chamber because of the chromatic aberration it would introduce. Astigmatism due to the off-axis reflection from the mirrors, if present, would not change the depth of modulation of the interference since the overlap of the wavefronts would not be affected but would serve to decrease the total ionization signal. The degree of astigmatism is made negligible by keeping the angle of incidence of the laser less than 3°. The fused silica windows between the two focal regions have wedge angles less than 3 arc seconds, limiting the separation of the uv and visible beams at the second focus due to dispersion to less than 1μm. The beam radius at the focus is calculated to be ~20μm. (Measurements of the transmission of the focussed beam by a 50μm diameter pinhole provide an upper bound of 40μm. There appear to be shot-to-shot fluctuations of the beam position which limit our measurements of the size of the focussed beam.) The two fields can also be displaced by the windows if the incidence angle is not sufficiently small. We calculate that a 1° angle of incidence will result in a 10μm beam separation at the focus, so beam alignment must be maintained to less than this value for good fringe visibility.

The cell chambers containing the mercury were constructed of stainless steel, with o-ring seals for the windows. The vapor pressure of the mercury was maintained by controlling the temperature of a cold finger with the cell body held at a slightly higher temperature. Temperature fluctuations during a data run were typically <±0.3° C. Temperature stability in the first cell is important to maintain the magnitude and phase of the third harmonic radiation produced there, while in the last cell stability is important since the signal strength is proportional to the mercury density.

The ionization signal is measured by collecting the electrons produced in the focal region by a pair of platinum parallel plates with a cross sectional area of 1 cm x 1 cm. One plate is grounded to the cell body, and electrons are accelerated toward the other plate by a +24 volt bias. The plate separation is 1 cm. The electron pulse is ac coupled into a fast (3 nsec rise time) transimpedance preamplifier with a gain of 25mV/μA, and integrated by a gated integrating ADC. Data is accumulated by a PC AT laboratory computer. The laser pulse energy is monitored, and only data for laser energies within a ± 5% range are accepted. The data are shown in figure 3. Each data point represents the average of 60 - 80 laser shots. The error bars, shown for a few of the data points, represent one standard deviation of the mean. The pressure of the argon gas in the second cell

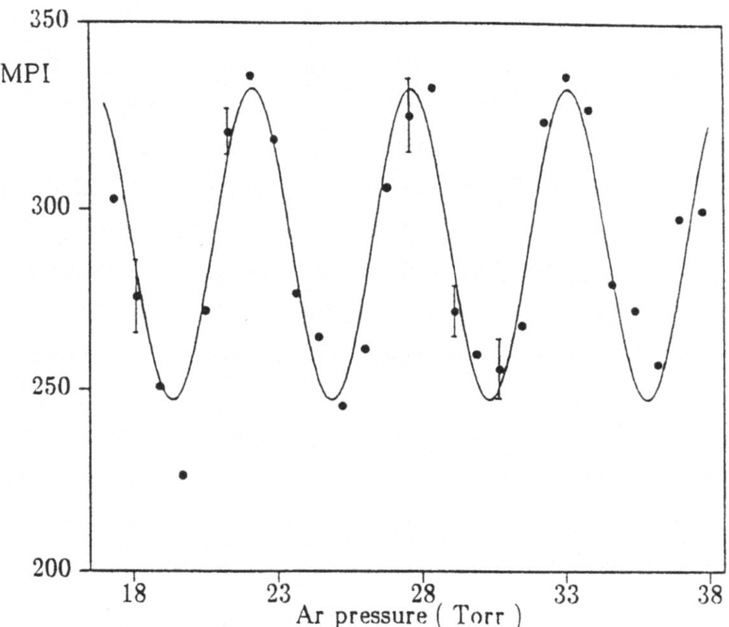

Figure 3. Ionization signal measured as a function of argon pressure in second chamber. Solid line indicates a best fit to the data. Error bars showing one standard deviation of the mean are shown for a few data points.

is measured using a barometer filled with diffusion pump oil whose specific gravity is 1.07. The modulation of the ionization signal is clearly seen in the figure. The solid line is fitted to the data by adjusting the period, amplitude, average value and phase. The depth of modulation (the ratio of the amplitude of the sine curve to the average value) is 15%. Deviation from the maximum possible value of 63% could be due to a number of factors such as imperfect beam overlap (transverse and/or longitudinal), non-optimal ratio of the transition amplitudes at the focus (M), saturation of the optical transition or a.c. Stark shifts of the 6p level leading to an intensity dependence other than I^5, or locally generated third harmonic radiation. The pressure difference of the argon required to change the phase by 2π is approximately 6.1 ± .6 torr, with the uncertainty limited by slow drifts in the experiment. Using equation (4), this yields a refractive index difference of 4.6 ± 0.4 times 10^{-5}, if we do not correct for variations in the temperature across the argon chamber. This is in reasonable agreement with extrapolations of refractive index data reported previously[17] for argon, yielding $\Delta n = 4.5 \times 10^{-5}$.

Our next set of measurements were designed to demonstrate the potential of using the interference process to measure phase variations of the optical fields. To this end we exploited the phase variation of a focussed beam due to diffraction effects, represented by $\exp(i\tan^{-1}z/z_o)$ in equation (2). The experimental system was very similar to that just discussed. The only significant variation was the collection electrodes used, which in the present system consisted of a ground plane and a set of eight biased collection electrodes. A schematic diagram of the experimental cell is shown in figure 4. Each electrode was constructed from a 1.27 mm diameter stainless steel rod. They were aligned transverse to the direction of propagation of the laser beam, side-by-side, in a plane parallel to the ground plane with a center-to-center spacing of 1.65 mm. Each electrode collected the photo-electrons generated in the region directly between it and the ground plane. (The spatial resolution of the detector was limited by the spacing between electrodes, with a smaller contribution from the trajectory of the free electrons in the collection field.) The total charges detected by each of the electrodes were determined concurrently by an eight channel gated integrator, and recorded using the laboratory PC. In this way we were able to measure the number of photo-electrons generated in the laser beam at varying distances from the laser focus.

The z-dependence of the ionization rate is easily calculated from equation (2), subject to assumptions of the overall intensity dependence of the interaction. Since each electrode was sensitive to all electrons generated at a certain distance, z, from the focus, we need to integrate equation (2) over the transverse

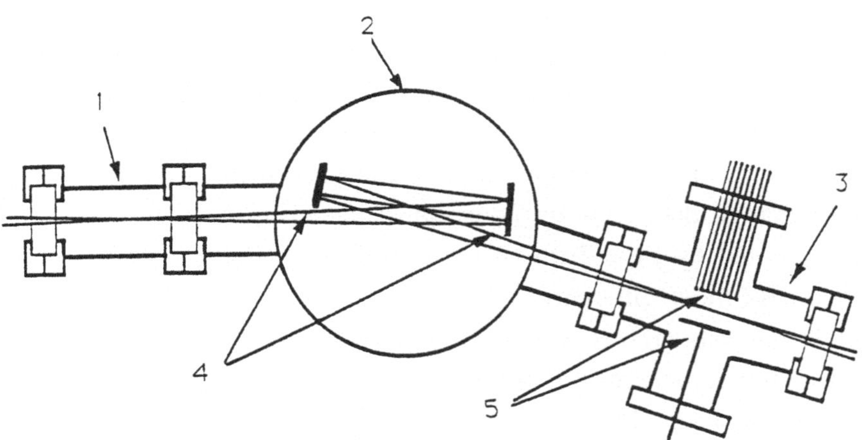

Figure 4. Schematic diagram of the experimental cell for measuring phase variations of the optical fields. Labels 1-4 are as in figure 1. The ionization signals are measured using (5) a set of biased collection electrodes. The electrodes are parallel to each other, normal to the plane of the page. The bottom electrode is a grounded plane.

dimensions. We generalize our result to that for an elliptical Gaussian beam, yielding an ionization rate proportional to

$$W(z) \propto 1 + \frac{M^2}{\left[1+\left(\frac{z}{z_{0x}}\right)^2\right]\left[1+\left(\frac{z}{z_{0y}}\right)^2\right]}$$

$$+ \frac{2M}{\left[1+\left(\frac{z}{z_{0x}}\right)^2\right]^{1/2}\left[1+\left(\frac{z}{z_{0y}}\right)^2\right]^{1/2}} \cos\left\{\phi_1 - 3\phi_2 - \tan^{-1}\left(\frac{z}{z_{0x}}\right) - \tan^{-1}\left(\frac{z}{z_{0y}}\right)\right\} . \quad (5)$$

Since the intensity of the laser varies through the focal region of the laser beam, and each electrode is sensitive to electrons generated in different regions, single shot intensity dependence measurements in a non-interfering case (with the first mercury cell evacuated) are attainable from the z-dependence of the ionization signal. For example, if the 5-photon ionization process is unsaturated, the ionization rate is proportional to I^5, leading to a z-dependence of $\left[\left(1+(z/z_{ox})^2\right)\left(1+(z/z_{oy})^2\right)\right]^{-2}$. On the other hand if the final step (6p→continuum) is saturated, so that the transition rate is proportional to I^3, then a z-dependence of $\left[\left(1+(z/z_{ox})^2\right)\left(1+(z/z_{oy})^2\right)\right]^{-1}$ would be observed. The data we have obtained are in relatively good agreement with the latter. We note, however, that this single shot intensity dependence measurement is hampered somewhat by variation of the efficiency of the different electrodes, possibly due to electrode geometry.

In spite of the need for refinement of the intensity dependence measurements, the phase variation with z, as shown in equation (5), is very evident with the current electrode configuration. Figure 5 shows the ionization signal as a function of the argon pressure in the delay cell for a typical measurement set for six of the eight collection electrodes. Each data point represents the average of the ionization signal over 60-80 laser shots. Electrodes 1 and 8 were positioned sufficiently far from the focus of the laser that only a weak ionization signal was detected, resulting in an insufficient signal to noise ratio. For each electrode data set, the average ionization rate has been subtracted leaving only the part which varies with the argon pressure. Each data set is seen to vary sinusoidally with the argon pressure with a period of around 6-7 torr. This is in accord with the results discussed previously. In the present data, a phase shift of the signal from one electrode to the next can be observed. This shift is due to the $\tan^{-1}\frac{z}{z_0}$ terms in equation (5). By fitting a sinusoidal curve to

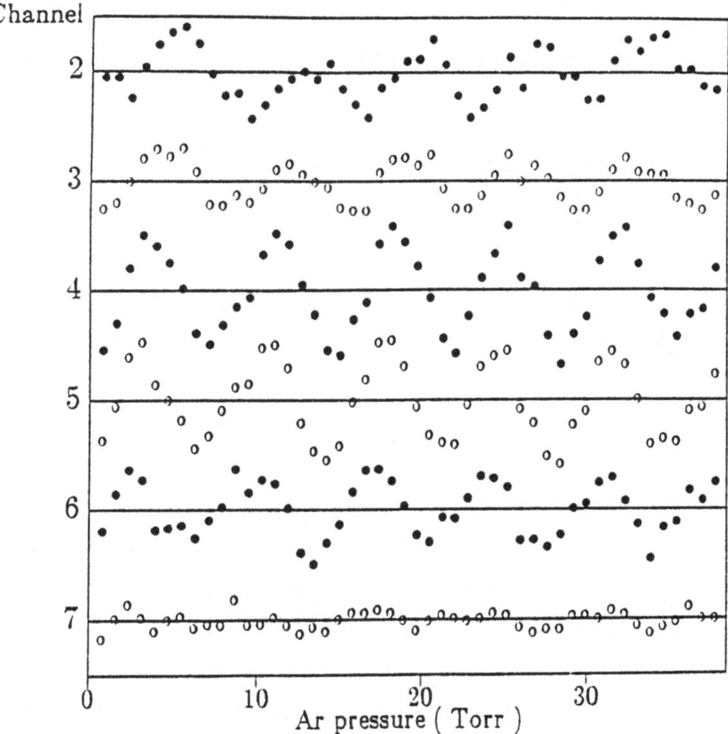

Figure 5. Ionization signal as a function of argon pressure in the delay cell for individual electrodes. The average ionization signal has been subtracted from each data set, leaving only the part which varies with argon pressure.

each data set in figure 5, the relative phase shift of each electrode signal can be determined. These relative phases are shown in figure 6. Only about one half of the total 2π phase shift is visible since the ionization probability falls off rapidly for $z > z_0$. The error bars represent the estimated error in the phase, as determined from the scatter in the data shown in figure 5. The solid line represents the inverse tangent phase terms of equation (2). The values of z_{0x} and z_{0y} were derived from measurements of the beam radius of the nearly-Gaussian beam before being focused into the first mercury vapor cell. The location of the focus ($z = 0$ point) was determined to within 1 mm from the z dependence of the ionization signal. The only adjustable parameter in figure 6, therefore, is a vertical offset of the data, since only a relative phase is determined through these measurements. Data points and calculated results (solid curves) for two different focussing conditions are shown in figure 6. In figure 6a, a 20 cm focal length lens was used, resulting in $z_{0x} = 3.65$ mm and $z_{0y} = 7.05$ mm. Figure 6b shows data for a 17 cm focal length lens, yielding $z_{0x} = 2.64$ mm and $z_{0y} = 5.10$ mm. In each case,the data and calculations are in excellent agreement.

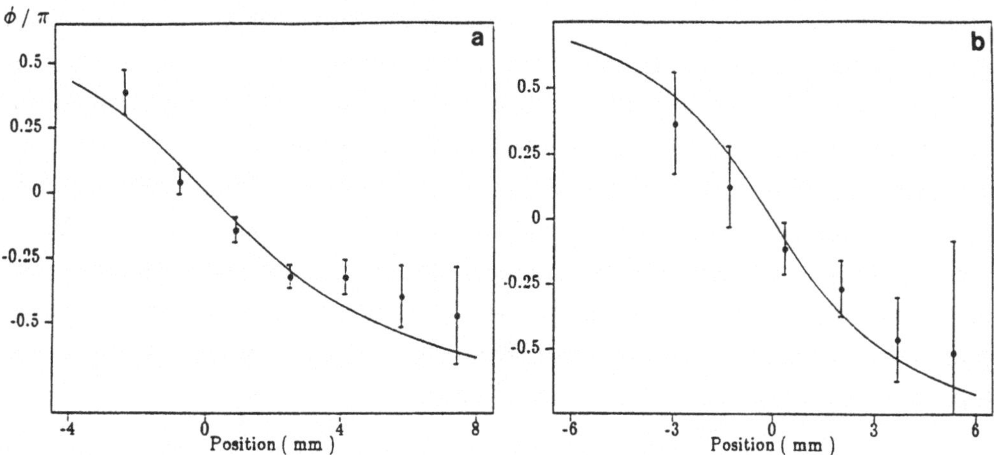

Figure 6. The relative phase of the ionization signal as a function of z, the distance from the laser beam focus. The data in figures (a) and (b) correspond to two different confocal parameters of the focussed elliptical Gaussian beam. In figure (a), z_{ox} = 3.65 mm and z_{oy} = 7.05 mm, while in figure (b), z_{ox} = 2.64 mm and z_{oy} = 5.10 mm. The solid line represents a calculation of the phase variation based on equation (5).

The depth of modulation of the interference signal, defined as the ratio of the amplitude of the sine wave to the average value had a maximum value of 45% in this work. We have observed that the depth of modulation decreases along the z axis. Possible limitations of the depth of modulation include those listed previously, as well as an additional factor related to finite resolution of the electrode array. This decrease in the depth of modulation may also be a signature of phase locking of the different waves in the nonlinear medium. Measurements of this effect are currently underway.

We will conclude this discussion with a brief description of current and future work. We are now working on measuring the effect of interference on photo-electron angular distributions. We are using atomic rubidium, which we can ionize using two photons at 560 nm or one photon at 280 nm. The two-photon process produces a continuum d-state (with a little s-state), while the linear process excites a p-continuum electron. The actual final state is therefore a linear combination of these two wavefunctions, whose precise form depends on the relative magnitude and phase of the waves exciting them. Since we are using an even-ordered process and an odd-ordered process, and our laser intensities are relatively low, the total cross section for photoionization does not vary with phase, but the angular distribution does. To date we have succeeded in modulating the number of photoelectrons ejected along the direction of laser polarization. These data are shown in figure 7. Measurement of a complete angular distribution as a function of phase has remained elusive, however, due to difficulties with rotating the electron detector and the laser polarization. Modulation of the photoelectron current in the direction of laser polarization has also been observed in two other systems. Muller et. al.[18] have studied photoionization of Krypton when irradiated with 1.06μm light from a Nd:YAG laser and its second harmonic. The interference in this case was between a six-photon process and seven-photon process. In addition, Baranova, et. al.[19] have observed modulation of the photo current emitted by a photocathode when illuminated by a laser field and its second harmonic.

A topic for future research is the investigation of the potential of this interference for accomplishing laser controlled molecular dissociation. Paul Brumer and Robert Gordon have discussed this earlier in this workshop. Brumer and Shapiro[20] first suggested that interference can be used to this end. The product set produced by the optical process can be selected by setting the relative phase and amplitude for one product set to interfere destructively. Calculations by Brumer and Shapiro[20] indicate that remarkable selectivity by this process is possible.

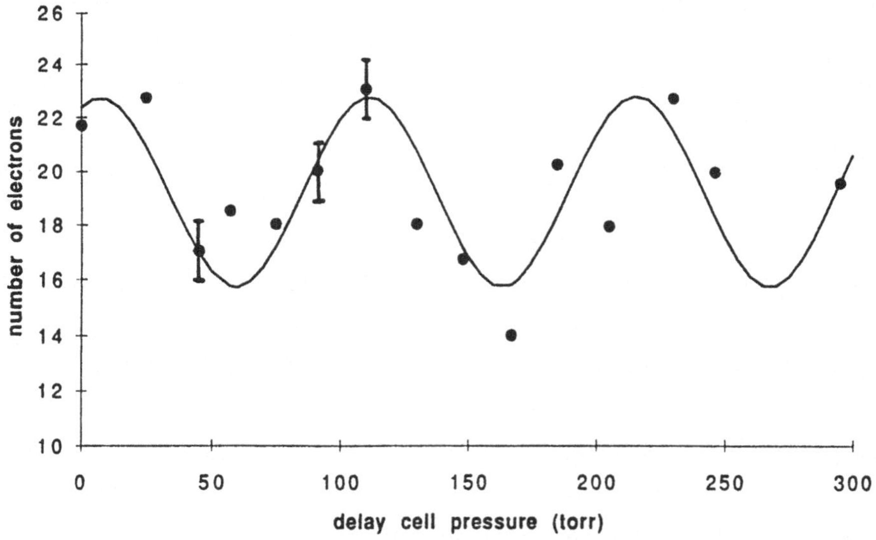

Figure 7. Interference in the photoionization partial cross-section of rubidium. The number of photoelectrons ejected in the direction of the laser polarization is modulated with changing gas pressure in the delay cell.

In summary, we have reviewed our recent experimental results demonstrating phase sensitive absorption by an atomic medium. We have applied this to measurement of phase variations of optical beams, and are currently investigating the effect of interference on photoelectron angular distributions and on wave propagation and phase locking of waves.

Helpful discussions with P. Zoller have been especially valuable. This work was supported by a Presidential Young Investigator Award from the National Science Foundation and a David Ross Grant from Purdue University.

References

1. Ce Chen, Yi-Yian Yin, and D. S. Elliott, Phys. Rev. Lett. **64**, 507 (1990).

2. Ce Chen and D. S. Elliott, Phys. Rev. Lett., 65, 1727 (1990).

3. M. G. Payne, W. R. Garrett, and H. C. Baker, Chem. Phys. Lett. **75**, 468 (1980); M. G. Payne and W. R. Garrett, Phys. Rev. A **26**, 356 (1982); **28**, 3409 (1983).

4. D. J. Jackson and J. J. Wynne, Phys. Rev. Lett. **49**, 543 (1982); D. J. Jackson, J. J. Wynne, and P. H. Kes, Phys. Rev. A **28**, 781 (1983); J. J. Wynne, Phys. Rev. Lett. **52**, 751 (1984); in *Multiphoton Processes*, Proceedings of the Fourth International Conference on Multiphoton Processes, Boulder, CO, 1987, edited by S. J. Smith and P. L. Knight (Cambridge Univ. Press, Cambridge, 1987), p. 318.

5. M. Poirier, Phys. Rev. A **27**, 934 (1983).

6. E. A. Manykin and A. M. Afanas'ev, Zh. Eksp. Teor. Phys. **48**, 931 (19650; **52**, 1246 (1967) [Sov. Phys. JETP **21**, 619 (1965); **25**, 828 (1967)].

7. G. L. Gurevich and Yu. G. Khronopulo, Zh. Eksp. Teor. Phys. **51**, 1499 (1966) [Sov. Phys. JETP **24**, 1012 (1967)].

8. K. Aron and P. M. Johnson, J. Chem. Phys. **67**, 5099 (1977).

9. F. H. M. Faisal, R. Wallenstein, H. Zacharias, Phys. Rev. Lett. **39**, 1138 (1977).

10. R. N. Compton, J. C. Miller, and A. E. Carter, Chem. Phys. Lett. **71**, 87 (1980); J. C. Miller, R. N. Compton, M. G. Payne, and W. R. Garrett, Phys. Rev. Lett. **45**, 114 (1980); J. C. Miller and R. N. Compton, Phys. Rev. A **25**, 2056 (1982); M. G. Payne, W. R. Garrett, and W. R. Ferrell, Phys. Rev. A **34**, 1143 (1986); W. R. Garrett, W. R. Ferrell, M. G. Payne, and J. C. Miller, Phys. Rev. A **34**, 1165 (1986).

11. J. H. Glownia and R. K. Sander, Appl. Phys. Lett. **40**, 648 (1982); Y. I. Geller and A. V. Shvabouskas, Opt. Spektrosk. **53**, 385 (1982) [Opt. Spectrosc. (U.S.S.R.) **53**, 227 (1982)]; S. A. Bakhramov, I. Kirin, P.K. Khabibullaev, and N. S. Shaabdurakhmanova, Kvantovaya Elektron. (Moscow) **9**, 2386 (1982) [Sov. J. Quantum Electron. **12**, 1557 (1982)].

12. D. Normand, J. Morellec, and J. Reif, J. Phys. B **16**, L277 (1983).

13. J. H. Glownia and R. K. Sandner, Phys. Rev. Lett. **49**, 21 (1986).

14. M. S. Malcuit, D. J. Gauthier, and R. W. Boyd, Phys. Rev. Lett. **55**, 1086 (1985); R. W. Boyd, M. S. Malcuit, D. J. Gauthier, and K. Rzazewski, Phys. Rev. A **35**, 1648 (1987).

15. S. J. Bajic, R. N. Compton, J. A. D. Stockdale, and D. D. Konowalow, J. Chem. Phys. **89**, 7056 (1988).

16. A. V. Smith, W. J. Alford, G. R. Hadley, J. Opt. Soc. Am. B 5, 1503 (1988).

17. C. Cuthbertson and M. Cuthbertson, Proc. Roy. Soc **84**, 13 (1910); M. Rusch, Annalen der Physik **70**, 373 (1923); and B. Quarder, Annalen der Physik **70**, 255 (1924).

18. H. G. Muller, P. H. Bucksbaum, D. W. Schumacher and A. Zavriyev, J. Phys. B 23, 2761 (1990).

19. N. B. Baranova, A. N. Chudihov, and B. Ya. Zeldovich, Opt. Commun. 79, 116 (1990).

20. M. Shapiro, J. W. Hepburn and P. Brumer, Chem. Phys. Lett. **149**, 451 (1988); and C. K. Chan, P. Brumer and M. Shapiro, Phys. Rev. Lett., **64**, 3199 (1990), Comments Section.

COHERENT INTERACTIONS WITHIN THE ATOMIC CONTINUUM

P. Lambropoulos*, Jian Zhang and X. Tang

Department of Physics, University of Southern California
Los Angeles, CA 90089-0484
*Also with Foundation for Research and Technology - Hellas
Institute of Electronic Structure & Laser
P.O. Box 1527, Heraklion 71110, Crete, Greece

I. INTRODUCTION

In the traditional view of photoionization of a one-valence-electron atom, the absorption of the photon raises the electron from a bound state into the continuum instantly, so to speak, and irreversibly. There is no characteristic time (analogous to the spontaneous lifetime of a discrete excited state) which we can associate with "how long" the electron "stays" in the continuum energy state to which it was raised. Another side of the same picture is that the dependence of the bound-free matrix element that determines the cross-section on the photon energy is smooth, exhibiting a slow variation originating from the oscillatory behavior of the wavefunctions. It shows no resonance-like structure. The situation changes significantly when the photoabsorption raises two electrons into the continuum. Then we encounter doubly excited discrete states embedded in (degenerate with) the single-electron continuum[1, 2]. The process (at least as long as the field is not too strong) is still irreversible, but there is now a characteristic time which can be viewed as the lifetime of the discrete state that has been formed in the continuum. The dependence of the photoionization cross section on the photon energy is no longer smooth but exhibits maxima and minima reflecting the interference between the amplitudes of the transition to the continuum and discrete parts of the wavefunction. By exciting the appropriate superposition of discrete and continuum wavefunction, we achieve a temporary localization and stabilization of an electron whose energy is above the ionization threshold. These so-called autoionizing states (or resonances) can have lifetimes ranging from less than a pico-second to microseconds, or in rare cases even be metastable against autoionization[3], depending on the atom and the configuration.

The above description tacitly assumed photoexcitation with traditional, weak and incoherent radiation sources. The availability of coherent, more or less monochromatic and strong sources (lasers) led to the idea of creating autoionizing-like features in the photoionization of even one-valence-electron atoms by using a second laser to couple an additional discrete (bound) state to the continuum. The essence of the idea in its simplest form, is depicted by the solid-line arrows in Fig.1. The atom initially is in its ground state $|1>$ with two lasers of frequencies ω_a and ω_b present. If we have only ω_a, we will simply have photoionization with a smooth dependence on ω_a. If on the other hand we also introduce ω_b, so that $E_2 + \hbar\omega_b$ lies above the ionization threshold and we scan ω_a around values that satisfy $E_1 + \hbar\omega_a \simeq E_2 + \hbar\omega_b$, with Ω_b fixed, we may expect to observe some resonance-like structure in the amount of ionization. One of the many ways of explaining this expectation qualitatively, is to view state $|2>$ as embedded in the continuum because of the presence of ω_b which causes the discrete "atom+ photon" state of energy $E_2 + \hbar\omega_b$ to be degenerate with the continuum at that energy. Then the absorption of ω_a encounters a superposition

Coherence Phenomena in Atoms and Molecules in Laser Fields
Edited by A.D. Bandrauk and S.C. Wallace, Plenum Press, New York, 1992

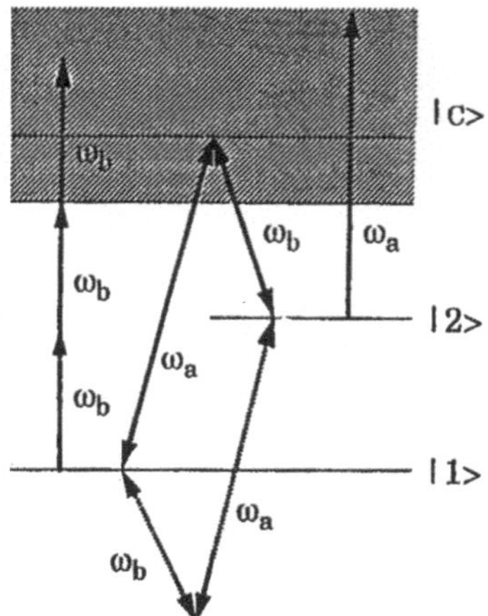

Figure 1. The simplest illustration for laser-induce continuum structure (LICS). $|1>$ and $|2>$ are two bound states of a one-valence-electron atoms, $|c>$ represents the structureless continuum. ω_a, ω_b represent two different lasers.

of discrete and continuum states (much like in autoionization), and depending on the details of the strength of the couplings, a more or less pronounced peak and/or dip ought to be observed in ionization as a function of ω_a. The first publication of this idea appeared in the papers of Heller and Popov[4]–[7] fifteen years ago. Due to the conceptual similarity with autoionization, and since in appearance it would suggest some structure in an otherwise smooth continuum, it goes by a number of names, such as "autoionizing-like resonance" or state, "pseudo-autoionizing resonance", "laser-induced continuum structure" (LICS), etc. In this paper, we shall use the term LICS for all phenomena involving the coupling of a discrete (bound) state to the respective continuum.

The solid-line arrows of Fig.1 correspond to only part of the important interactions that come into play when both lasers are present. The rest are indicated, by dashed-line arrows and represent an additional (Raman-type) coupling of the states $|1>$ and $|2>$ (without a pole in the continuum), as well as ionization of state $|2>$ by photon ω_a and (multiphoton) ionization of $|1>$ by ω_b. The latter two represent non-interfering decays of the atom into the continuum, a consequence of which can be the obliteration of the structure, if ionization is monitored through the collection of ions or electrons but without kinetic energy analysis. As has been shown by Dai and Lambropoulos [8], these additional couplings are as important as the basic solid-line paths and can not be ignored in any realistic assessment of the observability of the effect.

An obvious but significant generalization of the basic idea is to replace ω_a by a considerably smaller frequency, so that near resonance is achieved through the absorption of 3 photons, satisfying $E_1 + 3\hbar\omega_a \simeq E_2 + \hbar\omega_b$. The LICS would now be expected either in the observation of 3-photon ionization as a function of ω_a (with ω_b fixed), or in the spectrum of third harmonic generation $3\omega_a$. The latter is particularly appealing, if it can lead to the enhancement of the harmonic due to the oscillator strength that the embedding of state $|2>$ lends to the continuum. As we shall see later on, the number of additional (Raman) paths is now increased, but the fundamental idea remains the same.

In order to best pursue the analogy with autoionization, one would want to keep the intensity of ω_a weak compared to that of ω_b, so that the process, (whether it be ionization or harmonic generation) can be viewed as a "probe" of the structure created by ω_b. Recall that in autoionization of a two-valence-electron atom, the structure is created by the intraatomic (electron-electron) interaction which comes with the atom and the state. Thus photoabsorption around an autoionizing resonance represents a probe of the intraatomic structure. In LICS, we have more flexibility in that the value of ω_b and its intensity can be controlled externally, making the distinction between probe and coupling interaction only a matter of definition. There is in fact nothing that prevents us from keeping ω_a fixed and scanning

ω_b. In so far as harmonic generation is concerned, it should be noted that a third order process requires some intensity to be observable at all, with the consequence that, at least for observational purposes, the intensity of ω_a can not be arbitrarily weak.

There is one more variety of the manifestation of LICS that has appeared in the literature. It involves the absorption of a weak (probe) beam from the ground state $6s(^2S_{1/2})$ of caesium into the $\epsilon P_{1/2,3/2}$ continuum. The probe beam is linearly polarized. A right-circularly polarized coupling beam couples the excited state $8s(^2S_{1/2})$ to the same energy in the continuum, but due to the selection rules the $8s$ is coupled only to the $\epsilon P_{3/2}$ continuum. The probe beam, which due to its linear polarization is coupled to both continua, sees one continuum with and the other ($\epsilon P_{1/2}$) without structure. Since the linearly polarized radiation can be decomposed into a superposition of left and right circular polarizations, its left component sees a smooth continuum while its right a structured one. The net result is that the two components are refracted unequally, leading thus to a rotation of the polarization of the probe beam as monitored after it has emerged from the interaction region. This effect has been predicted by Heller and Popov[7] and its observation reported by Heller *et al*[9]. It does in fact represent the first, and to this day one of the few, documented and correctly interpreted observations of LICS in any context.

A rather voluminous theoretical literature[10] has accumulated since 1976, the time of appearance of the first papers by Heller and Popov. Most of such theoretical papers have dealt with idealized versions of the system involving two discrete levels and one continuum and usually excluding the additional but inevitable couplings discussed above. In spite of the extensive body of theoretical work, only very few experimental attempts have proven fruitful. One was mentioned above in connection with the rotation of polarization of the probe beam. The second, in chronological order has been published by Pavlov *et al*[11, 12]. who reported structure and enhancement of third harmonic generation (THG) in sodium vapor. In that experiment, the second harmonic (532 *nm*) of a *Nd : glass* laser was used to couple the excited state $5s$ to the continuum, and at the same time to pump a tunable dye laser whose output was used to couple the ground state $3s$ to the continuum via a 3-photon process which also produced third harmonic. The spectrum of the third harmonic as a function of the frequency of the fundamental showed a resonance and clear enhancement around that energy in the continuum at which the $5s$ was coupled by the 532 *nm* photon.

After a gap of a few years, Feldmann *et al*[13] reported observations showing some effects of LICS in 2-photon resonant 3-photon ionization of atomic sodium in the presence of a second laser. While some of the resonances observed can be attributed to LICS, as Feldmann *et al*[13] recognized and Dai and Lambropoulos[8] later confirmed quantitatively, the Raman-type coupling dominated, leading to symmetric peaks which is always the case when the couplings through the continuum are not dominant. Still that was the first observation of the effect of LICS on the ionization signal. It is useful to note here that Feldmann *et al* reported observable structure only when the excited state $4d$ was coupled to the continuum but not when $4d$ was replaced by either the $8s$ or the $6d$. In all cases the ground state was coupled to the continuum via 3-photon absorption, which as noted earlier and discussed by Dai and Lambropoulos[8] in detail introduces a number of additional couplings. It is thus not too surprising that some excited states did not produce LICS. The lesson there is that the simple idea of Fig.1, does not guarantee observability without a quantitative assessment that includes accurate atomic parameters.

The only other report of LICS in ionization was published by Hutchinson and Ness[14] who observed a peak in 2-photon resonant 3-photon ionization of *Xe* while a second laser coupled the excited state $10p$ to the appropriate energy in the continuum. Unfortunately, the reported structure is in serious disagreement with a theoretical calculation by Tang *et al*[15]. Until another experiment is performed, the result of Hutchinson and Ness can not be counted among the documented and properly interpreted observations of LICS.

The initial motivation for the work in this dissertation came from the results of a relatively recent experiment by Baldwin, Chapple and Bachor[16] at the Australian National University who attempted to reproduce and extend the results of Pavlov *et al* [11, 12] on the effect of LICS on THG. The disturbing surprise was that not only were the results of Pavlov *et al* not reproduced, but no LICS was observed in THG. The attempt to understand this contradiction led us to an elaborate theory of THG in the presence of LICS whose details are presented in ref.[21]. Our results demonstrate that there is really no contradiction and

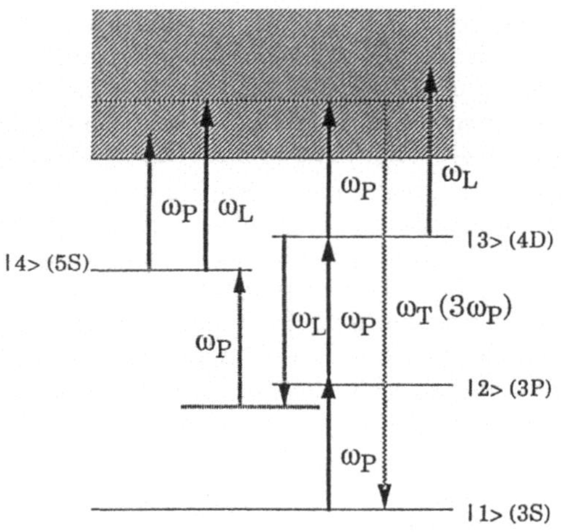

Figure 2. Energy levels of Na involved and the relevant couplings.

that the difference in the conditions of the two experiments fully explains the difference in the experimental outcome.

We then proceeded to solve a more general problem, namely the generation of third harmonic radiation in the vicinity of 3-photon resonance with an autoionizing state of a two-valence-electron atom, in the presence of LICS when an additional bound state is coupled to the vicinity of the autoionizing state by a second laser. We are in other words considering the effect of LICS on a continuum that already has structure due to intraatomic interactions. Even without LICS, the theory of THG in the vicinity of an autoionizing state has not been given rigorous attention, although some experimental data had been reported a number of years ago. By rigorous attention, we mean a theory going beyond formal developments[17] and providing quantitative predictions for a specific atom. Of course the additional presence of LICS constitutes a new problem. As we will see later, the combination of autoionization with LICS provides added flexibility as to where the enhancement occurs. In a sense, we have double enhancement because both the autoionizing resonance and LICS enhance the oscillator strength in the continuum which otherwise is diluted.

Our choice of this topic was also motivated by recent developments[18, 19, 20] on the possibility of radiation amplification and hopefully lasing without population inversion. Autoionizing states or similar structures in the continuum (including LICS) are important ingredients of such schemes, especially in contemplating amplification at short wavelengths, shorter than UV.

II. LASER-INDUCED CONTINUUM STRUCTURE AND THIRD HARMONIC GENERATION IN SODIUM

The energy levels involved in our specific problem are shown in Fig.2.

There are two lasers involved, the pump laser ω_P and the coupling laser ω_L. The pump laser is tuned near the 3S→4D two-photon resonance, while the third ω_P photon couples the 4D state to the continuum. At the same time, the coupling laser couples the 5S state to the same continuum. Laser-induced continuum structure is expected to be seen in the ionization spectrum when $\hbar\omega_L + E_{5s} \simeq 3\hbar\omega_P + E_{3s}$. Now the question is whether LICS can be also observed in the third harmonic spectrum as well, and whether one can take advantage of LICS to substantially enhance the third harmonic generation. Theoretical work on this topic published in the past consists of qualitative studies of the effect of LICS on the third order nonlinear susceptibility expressed in terms of photoionization widths and q parameters of the relevant states[6]. The q parameter is defined as the ratio of the real and the imaginary part of the Raman type two-photon dipole transition moment. However, in real experiments, the pulsed laser beams make the process more complicated. As mentioned in the introduction, two apparently similar experiments did not agree with each other. In order to be able to

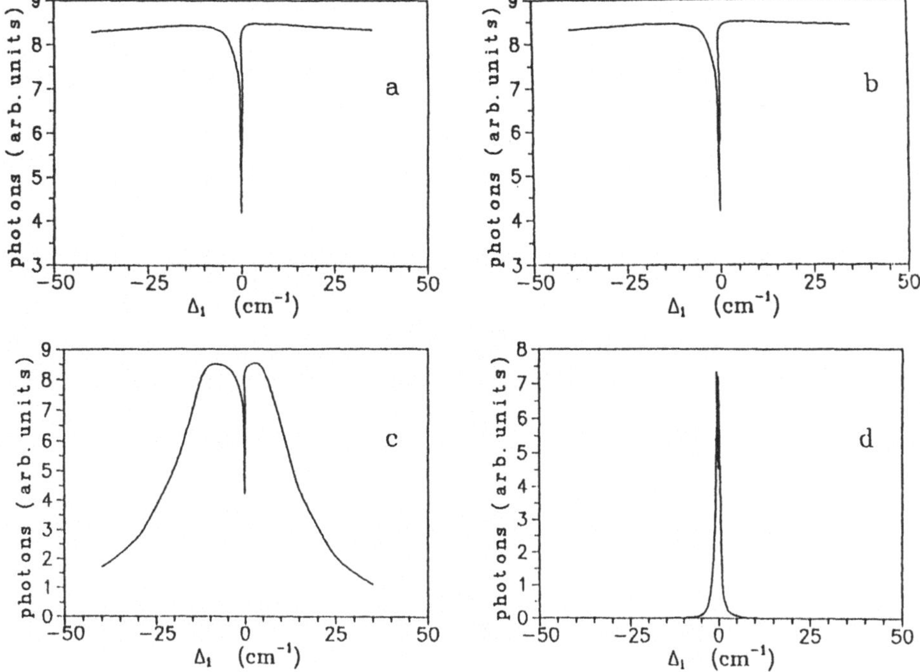

Figure 3. Calculated third harmonic spectrum for single-laser experiment with different laser intensities. $\Delta_1 = 0$ corresponds to $2\omega_P = E_{4d}$.(a)$I_P = 2.2 \times 10^{10} W/cm^2$.(b)$I_P = 5.0 \times 10^9 W/cm^2$. (c)$I_P = 1.0 \times 10^9 W/cm^2$. (d)$I_P = 1.0 \times 10^8 W/cm^2$. Arbitrary units are used for the number of the third harmonic photons.

interpret the experimental results and understand better the behavior of LICS in relation to third harmonic generation, we have developed a theoretical approach[21] to handle the problem through the solution of the time-dependent density matrix equations[22]. A similar problem in lithium has been studied by Ritchie[23] through the solution of time dependent probability amplitude equations and the polarizability.

II-1 Single-laser experiment

All atomic parameters used in our calculations are obtained through single-channel quantum defect method.[24]

Typical values of the experimental parameters used in the work of ref.[16] are: $I_P = I_L = 2.2 \times 10^{10} W/cm^2, \tau = 25ns$, the cell length $L = 10cm$, and the density of atoms $N = 10^{22}/m^3$. In the single-laser experiment, we have $I_L = 0$, and the third harmonic signal is measured as a function of the detuning from the two-photon resonance with $4D$. A peak has been observed at resonance, as well as a sharp dip at the center of the peak. Fig.3 shows a series of calculated third harmonic spectra with different laser intensities.

It can be seen that the widths of the spectra depend very sensitively on the laser intensity. Moreover, the ionization is saturated at intensities above $10^8 W/cm^2$. As a result, the spatial average plays a very important role. Now, we discuss the role of the possible reabsorption of the generated third harmonic signal by the system.

For a very short time duration after the laser pulses are turned on, before the ground state population is depleted, the reabsorption of the third harmonic photons can be simply described by an transition rate. As the ground state population gets depleted, the reabsorption of the third harmonic photons is also reduced. The reabsorption profile when I_L=0, is shown in Fig.4.

It can be seen that this process is only important at the resonant frequency, therefore will not affect the larger scale THG profile shown in Fig.3.

Figure 5 shows one example of the resulting third harmonic spectrum obtained after the spatial average.

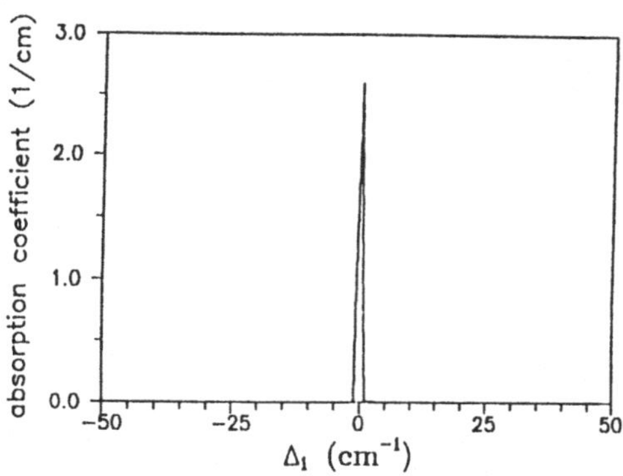

Figure 4. The reabsorption profile of the third harminic photons in the single-laser experiment.

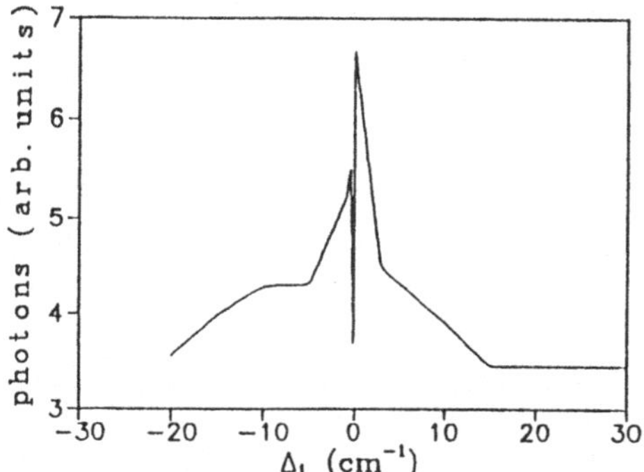

Figure 5. One example of the resulting third harmonic spectrum obtained after the spatial average in r-direction. Arbitrary units are used for the number of the third harmonic photons.

The width and the dip agree with the experimental result. Since, for computational simplicity, we have performed the spatial average part only in the r direction, a small discrepancy in detail between theory and experiment (see ref.[16]), such as the relative heights of the two peaks, has no significant implications. We would need to have much more detailed information on the spatial distribution of intensity, which was not obtained in the experiment, if we were to attempt a finer tuning between theory and experiment.

The dip near the center of the spectrum results from the competition between the ionization and the THG. As a matter of fact, this dip corresponds to a peak in the ionization spectrum. It appears only when the ionization peak is saturated, and represents loss of THG due to the loss of atoms into the continuum.

II-2 Two-laser experiments

For the case in which the first laser is fixed at the two-photon resonance with the $4D$ level, results of the third harmonic spectra versus the detuning of the second laser at different combinations of intensities are shown in Fig.6.

First we note the small peak near the resonance, which is the result of LICS. There is, however, a broad peak even at intensities more than two orders of magnitude lower than the peak intensity. At even lower intensities, this peak disappears. The small LICS peak is so small and narrow that can be easily buried by the experimental background. But at this point, it is surprising that the experiment[16] does not show the broad feature, since we find that its peak does not vanish even after spatial averaging identical to that of the single-laser case A above. Fig.6(f) shows the reabsorption of the third harmonic signal, as discussed before. Again, it is obvious that this reabsorption will not affect the overall lineshape of Figs.6(a)-(e).

Considering that we have calculated the spectra assuming that both lasers have a complete spatio-temporal overlap, and that one possibility for our calculation to differ from the real experimental situation is that the two lasers used in the experiment are not exactly

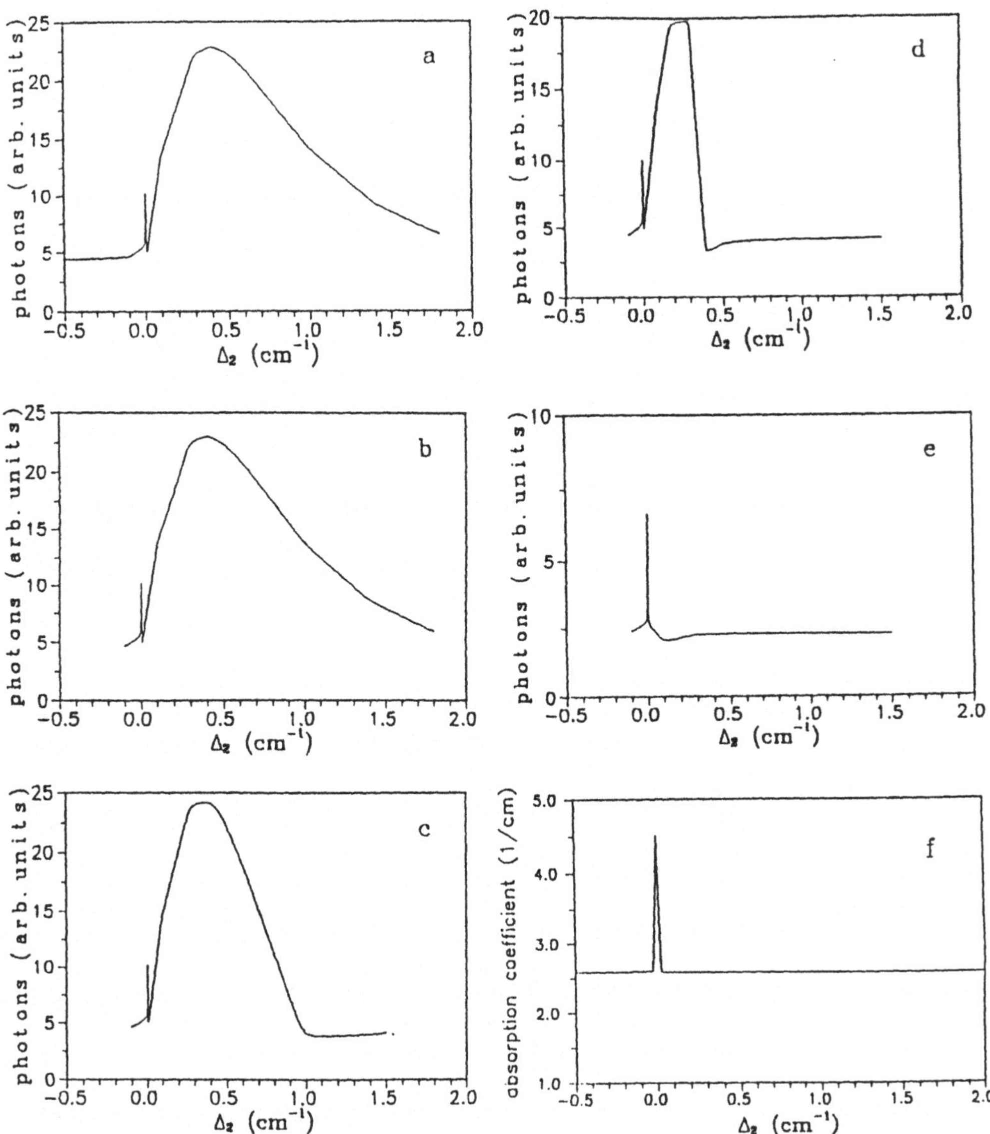

Figure 6. Calculated third harmonic spectrum versus the detuning of the second laser for two-laser experiment at different intensities. (a)$I_P = I_L = 2.2 \times 10^{10} W/cm^2$. (b)$I_P = I_L = 1.0 \times 10^9 W/cm^2$. (c)$I_P = I_L = 1.0 \times 10^8 W/cm^2$. (d)$I_P = I_L = 5.0 \times 10^7 W/cm^2$. (e)$I_P = I_L = 1.0 \times 10^7 W/cm^2$. Arbitrary units are used for the number of the third harmonic photons. (f) The corresponding reabsorption profile of the third harmonic profile.

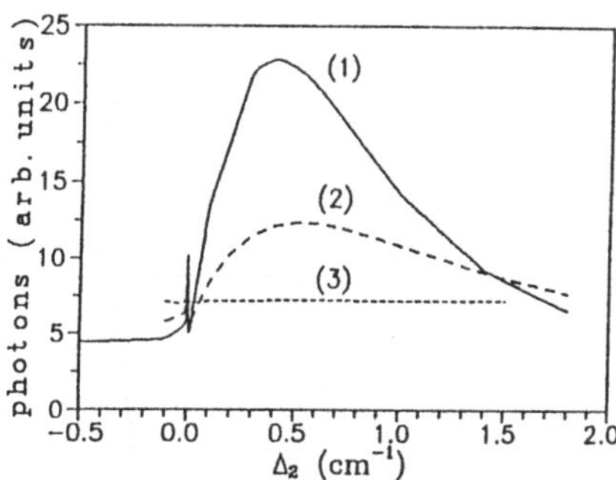

Figure 7. Results of third harmonic spectrum with different combinations of intensities. Curve (1) taken from FIG.5(a). Curve (2) is calculated at $I_P = 2.2 \times 10^{10} W/cm^2$, $I_L = 1.0 \times 10^{10} W/cm^2$. Curve (3) is calculated at $I_P = 2.2 \times 10^{10} W/cm^2$, $I_L = 1.0 \times 10^9 W/cm^2$.

the same and may have not had such a complete overlap, the intensities of the two lasers in a portion of the interaction region may not have been equal. To test whether this can offer a clue about the absence of the broad feature, we have calculated the third harmonic spectrum for different combinations of intensities of the two lasers, as shown in Fig.7.

Clearly, the broad peak disappears even at rather high intensities, as the second laser intensity is reduced. Through further investigation, we have found that this broad peak is most pronounced when the second laser intensity is so strong that the two-photon Rabi frequency between $4D$ and $5S$ is equal to the one photon Rabi frequency between $3P$ and $4D$. Intuitively, one can see that when the two Rabi frequencies are equal, the coherence among the related three energy levels is the strongest. The fact that the intensity of the second laser in the experiment was known to have been somewhat lower than that of the first one lends additional support to our argument.

As for the physical origin of this double-peak structure, we believe that it is due to the AC Stark splitting of the $4D$ state due to the two-photon coupling to the $3S$. To test this interpretation, we have calculated a series of line shapes for different intensities of the first laser and have convinced ourselves that the splitting is proportional to the intensity of the first laser, as it should be for a 2-photon Rabi frequency.

Finally, as a further test of our theory, we turn to the experimental result published by Pavlov *et al*[11] and Dimov *et al* [12]. The difference from the previous experiment[16] is that the first laser is never scanned through the two-photon resonance with the $4D$ level, because the second laser frequency in that work is about $240cm^{-1}$ larger than the one used in the previous experiment. The pulse duration of both lasers was about $2ns$ instead of $25ns$. The intensities of both lasers were $10^9 W/cm^2$. Also, in that experiment, the laser bandwidths of $\sim 0.1cm^{-1}$ (being considerably larger than those of ref.[16]) are comparable with the width of the resulting LICS. Fig.8 shows the result calculated under the conditions described above. The width of the spectrum is comparable with that of the experiment ($\sim 0.5cm^{-1}$), while the asymmetry is somewhat more visible than that in the experiment.

Considering that the experimental data must have had some error bars and the possible background noise, our calculation compares fairly well with the experimental line shape (Fig.3 of ref.[11] or Fig.4 of ref.[12]).

Our theory and calculations[17] have shown that, due to the saturation of ionization, high laser intensities do not always enhance the third harmonic generation, and that the laser-induced structure in the continuum is very sensitive to the relative intensities of the two lasers. We have also shown that the theoretical model we have developed is capable of explaining the apparent puzzles and contradictions presented by the existing experimental results. Further, more detailed experimental investigations would be very useful in clarifying the role of laser-induced continuum structure in nonlinear interactions of atoms with strong radiation fields.

It should be evident from our analysis of the various effects entering these phenomena that large intensity is not necessarily useful. Large must of course be evaluated in the context of the pulse duration because the saturation (due to the inevitable ionization) increases with time as well as intensity. Owing to the nonlinear nature of the processes, one does not have

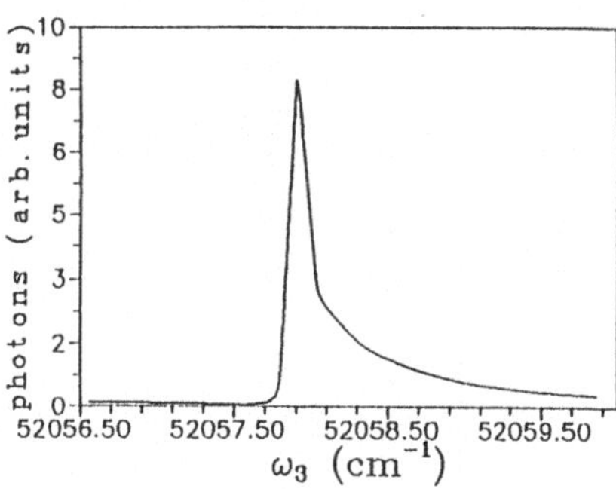

Figure 8. Calculated third harmonic spectrum for two-laser experiment. $I_P = I_L = 1.0 \times 10^9 W/cm^2$, $\tau_1 = \tau_2 = 2.0ns$, $2b_1 = 2b_2 = 0.1cm^{-1}$ and $\beta = 1.0cm^{-1}$. $\lambda_2 = 530nm$. Arbitrary units are used for the number of the third harmonic photons.

simple rules that determine the behavior in terms of, say, the product of duration and intensity. We know, on the other hand, that a minimum of intensity is necessary, if observable LICS is to develop; which means that delicate balances between conflicting requirements determine the observability of these effects. A quantitative analysis, such as the one undertaken here, will thus be necessary for the reliable prediction and/or interpretation of experimental results.

III. LASER-INDUCED COTINUUM STRUCTURE AND THIRD HARMONIC GENERATION IN THE PRESENCE OF AUTOIONIZATION: APPLICATION TO CALCIUM

In the previous section, we have discussed the LICS and THG in atomic sodium. The enhancement of the third harmonic generation has been shown to be possible under certain experimental conditions when LICS is present. In this section, we discuss the same problem in a more complicated system – atomic calcium. One of the complications encountered involves the calculation of the dipole transition moment in an atom with two valence electrons. The method[25] we use to calculate the atomic dipole transition moments is the multiconfiguration Hartree-Fock with a set of finite B spline basis. We concentrate here on the new effects of the combination of autoionizing resonance with LICS and related problems, which are absent in the case of atoms with a single valence electron.

In the past, both experimental and theoretical work concerned with the effect of an autoionizing resonance on third order sum-frequency generation in a two-valence-electron atom has been published [26, 27, 28]. A detailed formal theoretical study of the third order sum frequency generation and the multiphoton ionization near an autoionizing resonance has been published by Alber and Zoller[18]. In that work, much discussion was devoted to the effect of the autoionizing resonance, the two-photon resonance, the intensity-saturation, as well as the relationship between the ionization and the harmonic generation. Our present interest is, however, concentrated on the question of whether any new enhancement in the third harmonic spectrum will appear when the LICS is combined with the autoionizing resonance.

The third harmonic generation in a medium is determined by the third order nonlinear polarizability of the atom. Therefore, our first objective is to calculate the time-dependent polarizability of our atomic system through the solutions of the time-dependent density matrix equations, then calculate the number of third harmonic photons generated through a very short medium. The reason we consider here only a very short medium is to avoid the complication of the problem of propagation of the generated third harmonic field. Since in our system, the third harmonic photon is generated near the resonant frequency between the ground state and the autoionizing state, the linear susceptibility depends on the frequency of this photon. As a result, we can not neglect this effect when we consider the propagation of the third harmonic photon in the medium. This problem is equivalent to the problem of a lasing medium which is pumped by an external source; in our case, it is the 3-photon process which couples the ground state to the autoionizing resonance. In other words, the third harmonic field can be either attenuated or amplified through this medium, which is pumped by the fundamental field. This is an interesting and important problem in itself. It is related to

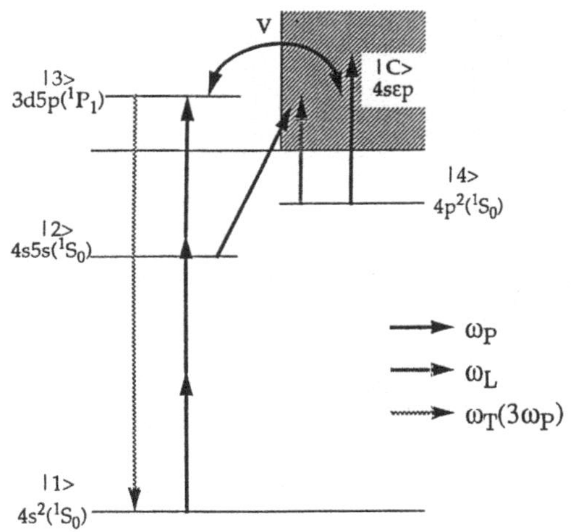

Figure 9. Energy levels of Ca and the relevant couplings involved in our problem.

the possibility of radiation amplification without population inversion [18, 19, 20] which has become one of the interesting topics in this research area. The use of autoionizing resonances is one of the first candidates under consideration. We have in fact chosen this system because it looks like a good choice for this idea to be realized. Hence, it is an interesting problem to study the absorption/emission of a weak probe with frequency $\omega_T \sim 3\omega_P$ in our system under the strong coherent 3-photon pumping with a laser photon of frequency ω_P, and optionally under the influence of another laser of frequency ω_L, which couples another bound state into the continuum. We have in fact work on this problem in progress but it is not part of this paper. Here, we will concentrate only on the third harmonic generation process, therefore we always have $\omega_T = 3\omega_P$.

The theoretical approach for the dynamic properties of our atomic system is similar to the one we employed in the previous section. But if we want to calculate the gain/loss of the ω_T photon propagating in the medium pumped by ω_P, we need to consider the wave equations in the medium in much greater detail, which can not be accommodated in the space available here. [For detail, see J. Zhang, Ph. D dissertation.]

The energy levels involved in our problem are shown in Fig.9.

Two lasers, ω_P and ω_L, are introduced into the problem. We are interested in investigating the third harmonic signal generated by the pump laser near the autoionizing state $3d5p^1P_1$ when the second laser couples the $4s6s^1S_1$ near-resonantly with this autoionizing resonance.

In all the calculations of this section, we have assumed that both lasers have the same temporal shape (namely $I = I_0 sech(t/\tau)$), and that the two lasers have perfect temporal and spatial overlap. To avoid lengthy computations and in the absence of experimental data on this problem we do not include here integration over the spatial distribution of the laser intensity. We have shown how this is accomplished in ref.[21] and it will be only a matter of further computation when experimental data become available. The typical pulse duration employed here is $10ps$, and the intensity of the pump laser is $10^9W/cm^2$.

Fig.10 shows the third harmonic spectrum of ω_P in the absence of the coupling laser. It is evident that the autoionizing resonance enhances the THG process, by almost one order of magnitude. Note that the stronger and stronger enhancement in third harmonic generation as the wavelength increases is due to the fact that it gets closer and closer to the two-photon resonance between ground state and $4s5s(^1S_0)$ state.

Fig.11(a) shows the possibility of using the LICS to enhance the third harmonic generation at different wavelengths. The coupling laser intensity is $10^9W/cm^2$ in this example. Each LICS peak corresponds to a different coupling laser wavelength. The broad structure on the background is the spectrum in the absence of the coupling laser (part of Fig.10(a)). As we can see, LICS plays the same role as the autoionizing resonance itself, but with the

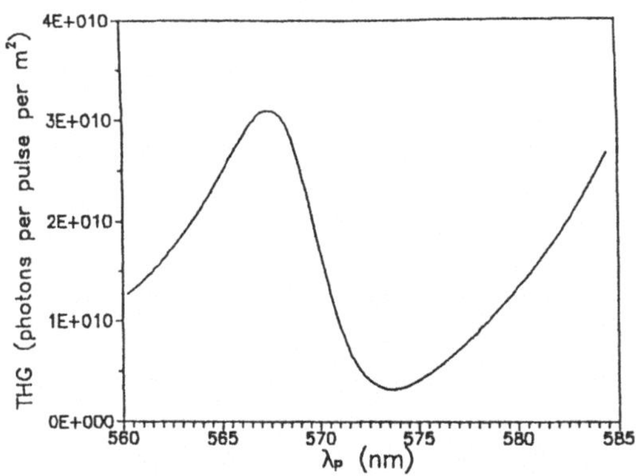

Figure 10. Third harmonic spectrum of ω_P in the absence of the coupling laser. $I_P = 10^9 W/cm^2$, $\tau = 10ps$.

advantage that its position can be changed by changing the wavelenth of the coupling laser, and can be arranged to cause enhancement within a much narrower frequency band. For example, we can see that even near the minimum of the background spectrum, LICS produces a strong enhancement in the third harmonic generation.

Fig. 11(b) shows the corresponding ionization spectra of Fig.11(a). It is interesting to notice that the asymetry in the LICS lineshape reverses as we scan the wavelength of both lasers through the background minimum. This is because of the interference between the autoionizing resonance and LICS.

It is anticipated that the width of the LICS peak will depend on the intensities of both the pumping and coupling lasers. The examples in Fig.12 show the intensity effect. Fig.12(a) shows the THG spectra calculated at three different coupling laser intensities. As we can see, the higher intensity makes the peak broader and the enhancement to the third harmonic generation stronger. A similar effect is shown in the corresponding ionization spectra of Fig.12(b) as well.

It can be seen that the intensity changes the shape of the peak as well. The effects of strong field and of the interaction time on autoionization lineshapes have been discussed in detail by Lambropoulos and Zoller[29]. It is interesting to notice that similar effects can be seen in the laser induced continuum structure as well. In Fig.13, we show the THG spectra and the ionization spectra with different laser pulse durations. It is clear that a sufficiently long pulse duration ($\tau > \Gamma_4^{-1}$) is necessary for the LICS lineshape to fully develop, as has also been the case in the results of ref.[29] for autoionization.

Our results in this section have demonstrated that the LICS in the vicinity of an autoionizing resonance can be a useful method for the enhancement of the THG at tunable wavelengths. The unique property of minimum absorption near an autoionizing resonance makes it possible to reduce the reabsorption or even amplify the generated third harmonic signal. Therefore, the work discussed here can serve as a springboard for further developments in relation with other interesting phenomena in multiphoton excitation and autoionization[19, 20, 21].

IV. PROSPECTS FOR FUTURE WORK

The topics and results of sections II and III have set the stage for a number of further developments.

The question of LICS in ionization and harmonic generation is now quite clear in one-valence-electron atoms. We can obtain very reliable atomic parameters through single-channel quantum defect theory. A great deal of work must be invested in incorporating experimental parameters such as spatio-temporal pulse shapes. For this to be possible, however, careful control of these parameters is necessary. In fact most recently two independent experiments[30, 31] in Na have produced asymmetric structures in ionization employing the scheme of Fig.1, with the states $4S$ and $5S$ as excited states providing for the first time the experimental result that has been sought for 15 years. One of the experiments was performed

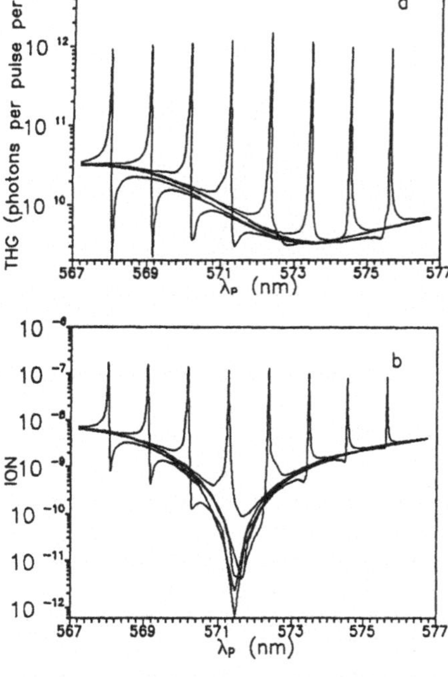

Figure 11. (a) LICS in THG spectra. From the left to the right the corresponding λ_L for each LICS peak is: $906.58nm$, $914.87nm$, $923.32nm$, $931.92nm$, $940.69nm$ and $949.62nm$. (b) Corresponding ionization spectra of Fig.11(a).

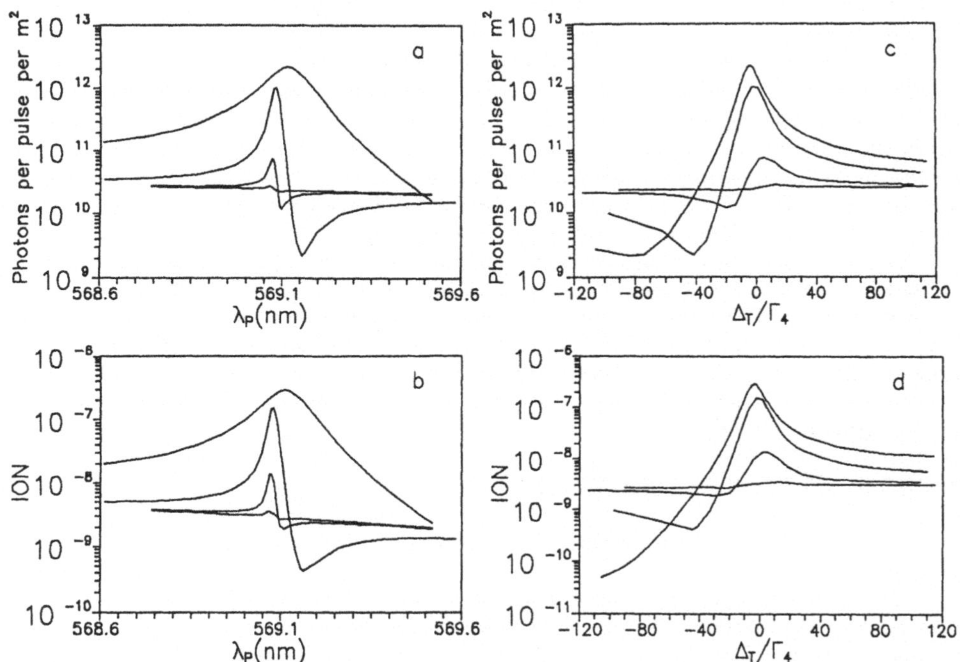

Figure 12. (a)The third harmonic spectra with the coupling laser intensity of $10^{10}W/cm^2$, $10^9W/cm^2$, $10^8W/cm^2$ and $10^7W/cm^2$ (from top to bottom) near the resonant peak of LICS. (b) The corresponding ion spectra of Fig.12(a) at the end of the pulse. (c) The spectra in Fig.12(a) plotted as a function of the normalized detuning. (d) The corresponding ionization spectra of Fig.12(c) at the end of the pulse.

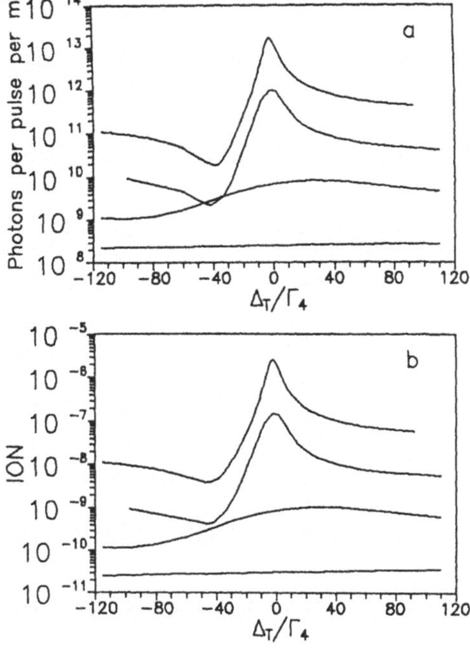

Figure 13. (a) Third harmonic spectra with different pulse durations. $\tau = 100ps$, $\tau = 10ps$, $\tau = 1ps$ and $\tau = 0.1ps$ (from top to bottom) as the functions of the normalized detuning near the resonant peak of LICS. (b) Corresponding ion spectra of Fig.13(a) at the end of the pulse.

in a beam[30] while the other in a heat pipe[31]. We expect to see many more data in the near future, in ionization as well as harmonic generation that will provide a fruitful context for the application of our theory.

The problem of THG in the vicinity of an autoionizing state provides the starting point for a systematic approach to the problem of amplification without population inversion[1-3]. We have constructed a model that enables us to tackle the complete problem including pumping that has either been ignored or left unspecified in all papers until now. We have in fact the capability of going even further by investigating the amplification of the generated third harmonic. This provides a scheme that includes a completely specified pumping mechanism as well as the radiation that is to be amplified. This process is interwoven with four-wave mixing and its separation from amplification through a population difference is a very interesting problem. We need to consider here these processes in a medium taking into account propagation aspects. All of the above is to be explored first without LICS in the context of an autoionizing state, as envisioned in the scheme that are being discussed in the literature for the last two years. As we have seen in section III, the addition of LICS introduces new structure which will also have influence on the amplification process. We already have a number of results on these questions and further work is in progress.

---

* Research sponsored by the National Science Foundation under Grant No. PHY-9013434 and by the Department of Energy under Grant No. DE-FG03-87ER60504.

References

[1] U. Fano, Phys. Rev. 124, 1866(1961).

[2] W. E. Cooke, T. F. Gallagher, S. A. Edelstein, and R. M. Hill, Phys. Rev. Lett. 40, 178(1978); W. E. Cooke and C. L. Cromer, Phys. Rev. A 32, 2725(1985).

[3] C. A. Nicolaides and D. R. Beck, Phys. Rev. A 17, 2116(1978).

[4] Yu.I. Heller and A.K. Popov, Sov. J. Quant. Electron. 6, 606(1976).

[5] Yu.I. Heller and A.K. Popov, Op. Comm. 18, 8 (1976).

[6] Yu.I. Heller and A.K. Popov, Op. Comm. 18, 449 (1976).

[7] Yu.I. Heller and A.K. Popov, Zh. Eksp. Teor. Fiz. 78, 506 (1980), [Sov. Phys. JEPT 51, 255 (1980).

[8] Bo-nian Dai and P. Lambropoulos, Phys. Rev. A36, 5205 (1987).

[9] Yu.I. Heller, V.F. Lukinykh, A.K. Popov and V.V. Slabko, Phys. Lett. A 82, 4 (1981).

[10] For a review see, for example, P.L. Knight, M.A. Lauder, and B.J. Dalton, Phys. Rep. **190**, 1 (1990).

[11] L.I. Pavlov, S.S. Dimov, D.I. Metchkov, G.M. Mileva, and K. V. Stamenov, Phys. Lett. 89A, 441 (1982).

[12] S.S. Dimov, L.I.Pavlov, and K.V. Stamenov, Appl. Phys. B 30, 35 (1983).

[13] D. Feldmann, G. Otto, D. Petring and K. H. Welge, J. Phys. B: At. Mol. Phys. 19, 269 (1986).

[14] M.H.R.Hutchinson and K.M.M. Ness, Phys. Rev. Lett. 60, 105 (1988).

[15] X. Tang, Anne L'Huillier, and P. Lambropoulos,Phys. Rev. Lett. 62, 111 (1989).

[16] K.G.H. Baldwin, P.B. Chapple, H.-A. Bachor, J. Zhang and P. Lambropoulos, Nonlinear Optics, Kauai, HI, 16-20 July, 1990; also P. B. Chapple, Ph.D. thesis, Australian National University, (1988), (unpublished).

[17] J. Zhang, P. Lambropoulos, Phys. Rev. A (to be published).

[18] G. Alber and P. Zoller, Phys. Rev. A 26, 1373 (1983).

[19] V. G. Arkhipkin and Yu. I. Heller, Phys. Lett. 98A,12(1983).

[20] S. E. Harris, Phys. Rev. Lett. 62,1033(1989).

[21] A. Lyras, X. Tang, P. Lambropoulos, and J. Zhang, Phys. Rev. A 40, 4131(1989).

[22] A.T. Georges, P. Lambropoulos and J.H. Marburger, Phys. Rev. A, 15, 300 (1977), and references therein.

[23] B. Ritchie, Phys. Rev. A31, 823 (1985).

[24] M. Edwards, X. Tang, P. Lambropoulos, and R. Shakeshaft, Phys. Rev. A33, 4444(1986).

[25] T. N. Chang, Phys. Rev. A39, 4946(1989).

[26] R. T. Hodgson, P. P. Sorokin, and J. J. Wynne, Phys. Rev. Lett. 32, 343(1974).

[27] J. A. Armstrong, and J. J. Wynne, Phys. Rev. Lett. 33, 1183(1974).

[28] Lloyd Armstrong, Jr. , and Brian Lee Beers, Phys. Rev. Lett. 34, 1290(1975).

[29] P. Lambropoulos and P. Zoller, Phys. Rev. A 24,379(1981).

[30] Y. L. Shao, D. Charalambidis, C. Fotakis, J. Zhang and P. Lambropoulos, Phys. Rev. Lett. (submitted).

[31] S. Cavalieri, Manlio Matera, and Francesco Pavone, Phys. Rev. Lett. (submitted).

THE ROLES OF DIPOLE CORRELATION FUNCTION AND DIPOLE EXPECTATION VALUE IN LIGHT-SCATTERING SPECTRA*

J. H. Eberly

Department of Physics and Astronomy
University of Rochester
Rochester, N.Y.

INTRODUCTION

Recently the observation of high-order harmonics in light scattered from atoms undergoing multiphoton ionization has been reported [1-2]. Theoretical attempts to explain various novel features of the observed spectra have been partially successful [2-5].

Nearly universal reliance in the theoretical work to date has been placed on a simplistic definition of spectrum based on one-time dipole expectation values. However, a note by Sundaram and Milonni [5] has served as a reminder that spectra are fundamentally related to two-time correlation functions. In this note we examine the relative roles of one-time and two-time functions in the theory of spectra. We show that the simplistic definition is likely to be satisfactory in most but not all cases of recent interest [6].

To begin, we recall results from the theory of light detection, and familiar formulas for dipole radiation. According to the standard theory of photo-detection [7], the observed signal at a photo-detector is proportional to the normally ordered local light intensity at the detector (normally ordered because the standard detector is triggered by photon absorption via atomic ionization). Thus we can write:

$$\begin{aligned} \text{observed signal} &\sim \langle :I(\mathbf{R},t): \rangle \\ &\sim \langle \mathbf{E}^{(-)}(\mathbf{R},t) \bullet \mathbf{E}^{(+)}(\mathbf{R},t) \rangle \end{aligned} \quad (1.1)$$

where $E^{(\pm)}$ are the positive- and negative-frequency components of the electric field at the detector.

Let us consider the simplest case: dipole radiation by a single atom exposed to an exciting pump laser, as in Figure 1.

Coherence Phenomena in Atoms and Molecules in Laser Fields
Edited by A.D. Bandrauk and S.C. Wallace, Plenum Press, New York, 1992

In this case the electric field at the detector can be written in terms of the retarded dipole acceleration [8]:

$$\mathbf{E}^{(\pm)}(\mathbf{R},t) \sim (1/R)\ \hat{\mathbf{R}} \times \hat{\mathbf{R}} \times [\ddot{\mathbf{d}}^{(\pm)}(t)] \qquad \text{(far zone)} \tag{1.2}$$

where square brackets [...] will be used to indicate a retarded time argument. We have assumed R is much greater than the emitted wavelength.

If we suppose that the dipole is a two-level atom, for example, then the positive-frequency part of the atom's dipole acceleration, due to weak excitation at frequency ω, is approximately given by

$$\ddot{\mathbf{d}}^{(+)}(t) \approx -\omega^2\ \mathbf{d}\ \sigma_{12}(t)\ e^{i(\mathbf{k}_0 \cdot \mathbf{r}_n - \omega t)}, \tag{1.3}$$

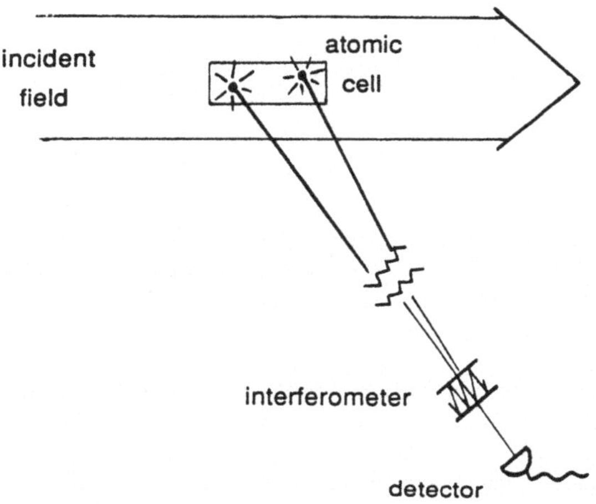

Fig. 1. A sketch showing light being scattered by atoms and then spectrally resolved in an interferometer before being registered at a detector.

where $\sigma_{12}(t)$ is the slowly-varying RWA atomic lowering (i.e., $E_2>E_1$) operator [9]. To be specific, we take x-polarization for the incident wave travelling along the axis: $\mathbf{d} = \hat{\mathbf{x}}d$ and $\mathbf{k}_0 = \hat{\mathbf{z}}k$. Then we have

$$\mathbf{E}^{(+)}(\mathbf{R},t) \sim (d/R)\ \omega^2\ (\hat{\mathbf{R}} \times \hat{\mathbf{R}} \times \hat{\mathbf{x}})\ [\sigma_{12}(t)\ e^{i(kz-\omega t)}] \tag{1.4}$$

Here the retardation bracket gives

$$[\sigma_{12}(t)]\ [e^{i(kz-\omega t)}] = \sigma_{12}\{(t-|\mathbf{R}-\mathbf{r}|/c)\}$$

$$\times \exp(-i\omega t)\ \exp\{i(kz-\omega|\mathbf{R}-\mathbf{r}|/c)\} \tag{1.5}$$

and since $R >> r$ we can simplify matters by using

$$|\mathbf{R}-\mathbf{r}| \approx R - \hat{\mathbf{R}}\cdot\mathbf{r} \tag{1.6}$$

In this far-zone approximation the electric field radiated by the dipole has the form

$$\mathbf{E}^{(+)}(\mathbf{R},t) \approx (d/R)\ \omega^2\ (\hat{\mathbf{R}} \times \hat{\mathbf{R}} \times \hat{\mathbf{x}})\ e^{i(kR-\omega t)} \times [\sigma_{12}(t)]\ e^{ik(z-\hat{\mathbf{R}}\cdot\mathbf{r})} \tag{1.7}$$

SIGNAL STRENGTH OF COHERENT AND INCOHERENT SCATTERING

According to (1.1) and (1.7) the signal strength at the detector is given by:

$$\text{signal} \sim (d^2/R^2)\ \omega^4\ \sin^2\theta\ <[\sigma_{21}(t)\sigma_{12}(t)]> \tag{2.1}$$

In the case of scattering by N different atoms $j = 1, \ldots, N$, we could factor the expectation value of the operator product for different atoms $j \neq k$:

$$<\sigma_{21}^{j}\ \sigma_{12}^{k}> = <\sigma_{21}^{j}><\sigma_{12}^{k}>, \tag{2.2}$$

This is always justifiable when the wave functions of the jth and kth atoms are not correlated.

The factored dipole product (2.2), valid for $j \neq k$, gives the dominant contribution to light scattering in many cases, since there are only N terms of the "diagonal" type with $j=k$ but $N(N-1) \approx N^2$ terms with $j \neq k$, when N atoms participate. Directional considerations also enter the picture when many atoms are radiating and the $j \neq k$ contributions tend to concentrate their emission in the forward direction and to contribute relatively little to side-scattering because of phase cancellations. The factored-dipole form (2.2) is thus responsible for what is called the "coherent" part of the scattered light intensity [6].

The incoherent part of the scattered signal can be quite different. It arises from the $j=k$ terms, in which case one must use the correct operator identity $\sigma_{21}^{j}\ \sigma_{12}^{j} \equiv \sigma_{22}^{j}$. This is valid for any atom or even for any transition:

$$\sigma_{ba}^{j}\ \sigma_{ab}^{j} \equiv \sigma_{bb}^{j}. \tag{2.3}$$

There are only N such terms so they make little contribution to forward scattering. However, since the dipole phase does not survive the operator contraction in (2.3), there are no direction-dependent phases to cancel each other, and the incoherent part often dominates the side-scattering. Since $<\sigma_{22}^{j}>$ is the probability of the jth atom being in the upper level the incoherent signal is just the same as fluorescence.

SPECTRUM OF SCATTERED LIGHT

The spectrum of the light coming from an atom is measured by detecting the various frequency components separately. We assume this is done by passing the light through a frequency-tunable dispersive device, a spectrometer. The light is still detected in the same way, by receiving a signal from a photo-detector [recall (1.1)]:

$$\text{signal} \sim \langle \mathbf{E}^{(-)}(\mathbf{R},t) \bullet \mathbf{E}^{(+)}(\mathbf{R},t) \rangle \tag{3.1}$$

In the case of the scattered spectrum, however, the detected fields $\underline{E}^{(\pm)}(\mathbf{R},t)$ have not come directly from the atom but have undergone whatever generalized filtering is provided by the spectrometer, which we assume to be a linear and causal instrument. Thus we must write

$$\mathbf{E}^{(+)}(\mathbf{R},t) = \int_{-\infty}^{t} H(t-t_1;\omega)\, \mathbf{E}_A{}^{(+)}(\mathbf{R},t_1)\, dt_1 \tag{3.2}$$

where $H(t-t_1;\omega)$ is the temporal transmittance function of the filter when set at the frequency ω, and where $E_A{}^{(+)}(\mathbf{R},t_1)$ now denotes the field entering the filter directly from the atom.

The filter input and output fields are obviously linearly related and causality is guaranteed by the time order explicitly incorporated into the filter integral: $t \geq t_1$. The detected signal, which now carries spectral information, is given by a double integral [11]:

$$\text{signal} \sim \int dt_1 \int dt_2\, H^*(t-t_1;\omega)\, H(t-t_2;\omega)$$
$$\times \langle \mathbf{E}_A{}^{(-)}(\mathbf{R},t_1) \bullet \mathbf{E}_A{}^{(+)}(\mathbf{R},t_2) \rangle \tag{3.3}$$

The integrand in (3.3) contains a two-time field correlation function rather than the field intensity. This is because the light emitted by the atom at a given time t' is "held" in the filter for a time roughly equal to the filter's inverse bandwidth and while in the filter it interferes with fields emitted by the atom both earlier and later than time t'. The integral in (3.2) has a finite upper limit so the signal given in (3.3) is also time dependent as well as ω-dependent. We refer to this as the physical spectrum [11] of the detected light.

Calculation of the physical spectrum (3.3) can usually be done only approximately. The full ω and t dependences have been worked out in only a few cases [12]. In any case (3.3) can be written (assuming x-polarization):

$$\text{spectrum} \sim (1/R^2)\, \omega^4 \sin^2\theta \int dt_1 \int dt_2\, H^*(t-t_1;\omega) H(t-t_2;\omega)$$
$$\times \langle [ex^{(-)}(t_1)]\ [ex^{(+)}(t_2)] \rangle \tag{3.4}$$

It is clear that the spectrum depends on the properties of the normally ordered dipole correlation function $\langle x^{(-)}(t_1) x^{(+)}(t_2) \rangle$ in a very basic way.

Just as for the radiated intensity, we can consider the spectrum to have two distinct parts, if there are actually N atoms emitting scattered light. The "coherent" part comes from $N(N-1)$ terms $(j \neq k)$ that radiate collectively almost entirely in the forward direction and the "incoherent" part comes from N terms $(j=k)$ that radiate individually and almost isotropically and are observable, for all practical purposes, only in side-scattering. In the next section we will focus on a simple model exhibiting incoherent fluorescence as well as coherent scattering, and determine their contributions to the spectrum.

Here we present an explicit example in which we can show some regard for the quantum mechanical aspects of radiative dynamics. We consider a collection of two-level atoms being weakly driven by a far off-resonance laser field. In this case we can use an approximate first-principles solution for the dipole operators that are required by (3.4). It is assumed that the initial density matrix is given by

$$\rho = p_1|1\rangle\langle 1| + p_2|2\rangle\langle 2|, \tag{4.1}$$

that is, a mixture of upper and lower levels with non-zero probabilities p_1 and p_2. One consequence is that all dipole expectation values are initially zero.

In the Heisenberg picture the slowly varying atomic lowering operator obeys the equation:

$$i d/dt\, \sigma_{12}(t) = \Delta\sigma_{12}(t) - (\Omega/2)w(t) \tag{4.2}$$

where $w(t) = \sigma_{22}(t) - \sigma_{11}(t)$ is the atomic inversion operator, $\Omega = 2d_{21}\xi/\hbar$ is the Rabi frequency associated with the external laser field whose amplitude is ξ, and d_{21} is the transition dipole moment between levels 2 and 1. The detuning Δ is an important parameter:

$$\Delta \equiv E_2 - E_1 - \omega \tag{4.3}$$

because only when $\Delta\neq 0$ can we distinguish between scattered light and fluorescence. We will assume Δ to be large enough in relation to the Rabi frequency Ω to justify an adiabatic approximation for the inhomogeneous part of the solution to (4.2). The approximate slowly-varying operator solution can then be written:

$$\sigma_{12}(t) = \sigma_{12}(0)e^{-i\Delta t} + (\Omega/2\Delta)w(0) \tag{4.4}$$

Here we have used $w(0)$ instead of $w(t)$, on the prior assumption that $\Delta >> \Omega$, which means that the time development will be mostly dispersive, with very little induced change in level populations. Note that the action of the external field induces a finite dipole expectation even if there is none at $t=0$. From the specified initial density matrix (4.1) we easily find

$$\langle\sigma_{12}(t)\rangle = (\Omega/2\Delta)(p_2-p_1) \tag{4.5}$$

From this result we can expect a coherent component in the scattered light.

The spectrum of light associated with this atom is given by (3.4), so we will now evaluate the required dipole correlation function, using the solution (4.4) above:

$$\langle\sigma_{21}(t_1)\sigma_{12}(t_2)\rangle = \langle\sigma_{21}(0)\sigma_{12}(0)\rangle\, e^{i\Delta(t_1-t_2)} + (\Omega/2\Delta)^2 \langle w(0)w(0)\rangle \tag{4.6}$$

Here we have dropped the cross terms because they are zero, given the relations $\sigma_{21}(0)w(0) = -\sigma_{21}(0)$ and $w(0)\sigma_{12}(0) = -\sigma_{12}(0)$ and the absence of coherence in the initial density matrix: $\langle\sigma_{21}(0)\rangle = \langle\sigma_{12}(0)\rangle = 0$.

To obtain the actual dipole correlation function we must multiply (4.6) by the time-dependent factor $e^{i\omega(t_1-t_2)}$ which was removed at the outset to work with slowly-varying operators. We find a particularly simple result for the dipole correlation:

$$\langle\sigma_{21}(t_1)\sigma_{12}(t_2)\rangle\, e^{i\omega(t_1-t_2)} = p_2\, e^{i(E_2-E_1)(t_1-t_2)}$$

$$+ [\Omega/2\Delta]^2\, e^{i\omega(t_1-t_2)} \tag{4.7}$$

The spectrum can now be obtained from the double integrals in (3.4) with the aid of a specific spectrometer function. However the final result can be seen by inspection of the correlation function, which is purely sinusoidal. The spectrum obviously consists of two peaks, located at the expected positions -- the natural transition frequency E_2-E_1, and the external laser frequency ω. The first peak represents fluorescence and other types of line radiation (which are not present in lowest order in such a simple atom), and the second peak represents coherent scattering (Rayleigh scattering). We can associate the two peaks directly with spontaneous and stimulated emission if we wish.

SUMMARY

We have sketched the elements of the theory of scattered light spectra in order to exhibit the role of dipole correlation functions. It is important to remember that a really correct spectrum follows only from a correct treatment of the dipole dynamics. We have avoided the main difficulties of a full solution in the previous section by adopting "frozen" initial populations σ_{11} and σ_{22}. If the spontaneous emission rate of level 2 is comparable to Δ and Ω this is not a good approximation. Very few complete treatments of scattered light spectra have been worked out, and none for real atoms in strong external fields.

Acknowledgements: I am pleased to thank M. V. Fedorov, A. L'Huillier and P. W. Milonni for conversations that led to this paper. Discussions with R. Grobe, D. Lappas and K. Rzazewski are also appreciated. This work was supported by the National Science Foundation under grant PHY 8822730.

References

[1]. C. K. Rhodes, Physica Scripta T17 (1987) 193; M. Ferray, A. L'Huillier, X. F. Li, L. A. Lompré, G. Mainfray and C. Manus, J. Phys. B 21 (1988) L31; and X. F. Li, A. L'Huillier, M. Ferray, L. A. Lompré and G. Mainfray, Phys. Rev. A 39 (1989) 5751.
[2]. A review has been presented by A. L'Huillier, and G. Mainfray, in Atoms in Strong Radiation Fields, edited by M. Gavrila (Academic Press, Orlando, in press).
[3]. K. C. Kulander and B. W. Shore, Phys. Rev. Lett. 62 (1989) 524; and A. L'Huillier, K. J. Schafer and K. C. Kulander, Phys. Rev. Lett. 66 (1991) 2200.
[4]. J. H. Eberly, Q. Su and J. Javanainen, Phys. Rev. Lett. 62 (1989) 881.
[5]. B. Sundaram and P. W. Milonni, Phys. Rev. A 41 (1990) 6571.
[6]. A more general examination of these questions has been given by J.H. Eberly, Phys. Rev. A (submitted).
[7]. See R. J. Glauber, in Quantum Optics and Electronics, edited by C. DeWitt, A. Blandin and C. Cohen-Tannoudji, (Gordon and Breach, 1965), pp. 78-88.

[8]. See, for example, Lasers, P. W. Milonni and J. H. Eberly, (Wiley, New York, 1988), Sec. 2.5.
[9]. Let us use the notation $\tau_{12} = |1\rangle\langle 2|$ for the elementary "transition operator" between levels 1 and 2, and $\tau_{12}(t)$ for its Heisenberg-picture counterpart. Here $|m\rangle$ is the bare atomic eigenstate in the Schrödinger picture. The corresponding slowly-varying rotating-frame operator σ_{12} is defined by $\sigma_{12}(t)\ \exp(-i\omega t) = \tau_{12}(t)$. The sign in front of ω is the same as the sign of $E_1 - E_2$, taken to be negative here since E_2 has been assumed to be the upper level.
[10.] See ref. 8, Sec. 17.6, for example.
[11.] J. H. Eberly and K. Wodkiewicz, J. Opt. Soc. Am. 67, 1252 (1977).
[12.] See, for example, J. H. Eberly, C. V. Kunasz and K. Wodkiewicz, J. Phys. B: Atom. Molec. Phys. 13, 217-239 (1980); X. Y. Huang, J. D. Cresser and J. H. Eberly, J. Opt. Soc. Am. B 2, 1361 (1985); M. Florjanczyk, K. Rzazewski and J. Zakrzewski, Phys. Rev. A 31, 1558 (1985); N. Lu, P. R. Berman, Y. S. Bai, J. E. Golub and T. W. Mossberg, Phys. Rev. A 34, 319 (1986).

FEMTOSECOND PULSE SHAPING AND EXCITATION OF MOLECULAR COHERENCES

A.M. Weiner*, D.E. Leaird*, G.P. Wiederrecht**, and K.A. Nelson**

*Bellcore
Red Bank, NJ

**M.I.T.
Dept. of Chemistry
Cambridge, MA

INTRODUCTION

For many years chemical physicists have sought to achieve mode-selective chemistry through laser irradiation. To date, however, attempts at using high-power laser pulses to control molecular reaction pathways have been unsuccessful. The principal limitation to laser mode-selective chemistry seems to be the rapid intramolecular vibrational redistribution (IVR) associated with anharmonic vibrational couplings at high excitation levels[1].

At present an interest in the possibility of laser-control of molecular motions and reactions is reemerging. This recurring interest arises for three reasons. First, as a result of advances in ultrafast laser technology during the last decade, femtosecond lasers with pulse durations shorter than the time required for many nuclear motions are now widely available. Ultrashort pulses are used in many laboratories for direct time-resolved observations of molecular motions as well as physical and chemical transformations involving these motions[2,3,4,5,6,7,8]; ultrashort pulses may also permit laser irradiation to occur on a time scale fast compared to IVR. Second, due to increased appreciation of the importance of intramolecular vibrational redistribution, a number of theoretical ideas[9,10,11,12,13] for laser control of reactions now include IVR in the molecular model. Many of these ideas involve the use of specially shaped femtosecond laser waveforms. Finally, the recent development of techniques for controlling the shapes of ultrashort laser pulses[14,15,16] now makes it feasible to generate specially designed femtosecond pulse sequences and waveforms for exercising molecular control.

In this paper we first describe the femtosecond pulse shaping technique[14] we have developed at Bellcore and then describe an initial, simple application of this technique for controlling molecular motion. In particular, we discuss the use of timed sequences of femtosecond pulses to drive selected vibrations in an organic crystal[17,18]. The build-up of coherent vibrational motion in these experiments is closely analogous to the way a child on a swing is pushed repetitively to achieve an increased oscillatory motion.

FEMTOSECOND PULSE SHAPING

The pulse shaping apparatus is depicted in Fig. 1. As described previously[14], pulse shaping is achieved by passing a single femtosecond pulse through a simple optical system consisting of two gratings, two lenses, and spatially varying phase and amplitude masks.

Coherence Phenomena in Atoms and Molecules in Laser Fields
Edited by A.D. Bandrauk and S.C. Wallace, Plenum Press, New York, 1992

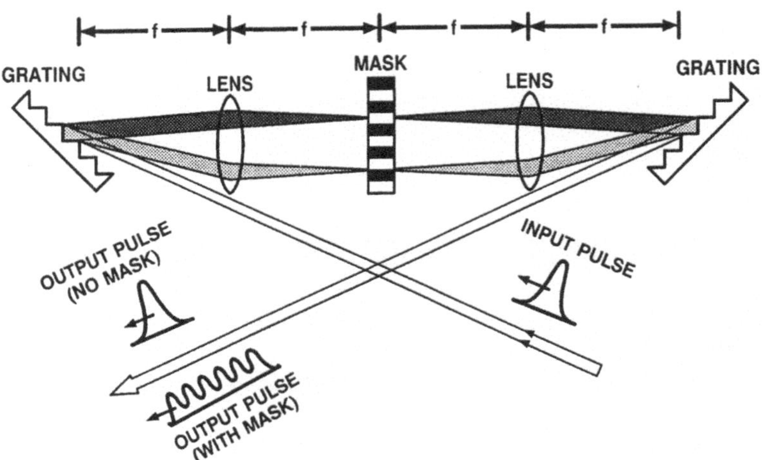

Fig. 1. Femtosecond pulse shaping apparatus.

The 1800-line/mm gratings are placed at the (outside) focal planes of the 15-cm-focal-length lenses, which are set up as a unit magnification telescope, and the masks are placed midway between the lenses. After the first grating and lens spatially separate the optical frequencies contained within the incident ultrashort pulse, the masks alter the amplitudes and phases of the spatially dispersed frequency components. The second lens and grating recombine the optical frequency components into a single collimated beam. The pulse shape at the output of the apparatus is Fourier transform of the patterns transferred by the mask onto the spectrum.

The input pulses for our experiments are obtained from a dispersion-compensated CPM ring dye laser[19]. The pulse duration is typically 75-fsec at a wavelength of 0.62 μm. For the impulsive stimulated Raman scattering experiments these pulses are amplified to μJoule energy levels at an 8.6 KHz repetition rate by using a copper vapor laser pumped dye amplifier system[20]. Phase and amplitude masks are fabricated on fused silica substrates by using standard microlithographic techniques, as described previously[14]. The intensity profiles of shaped pulses are measured by cross-correlation, using 75-fsec pulses split off from before the pulse shaper as the probe.

As a very simple example of pulse shaping, we describe the generation of terahertz-repetition-rate pulse trains by using amplitude masking[14]. The mask is opaque except for two narrow slits which pass two isolated spectral components. The two frequencies interfere, producing a high-frequency tone burst. In essence, this experiment is a time domain analog of Young's two slit diffraction experiment. A cross-correlation measurement of a fsec tone burst is plotted as the solid line in Fig. 2 for a frequency separation of 3.37 nm (2.6 THz). The trace shows some twenty distinct temporal peaks, thereby demonstrating that even a simple mask operating in the frequency domain can yield a rich and detailed temporal structure. As a simple example of phase control, a phase mask is introduced in order to phase shift one of the spectral components by π. As a result, the positions of the peaks and the nulls are interchanged, as shown by the dotted line in Fig. 2. Thus the phase of the temporal beat note is determined by the relative phase of the two selected frequency components.

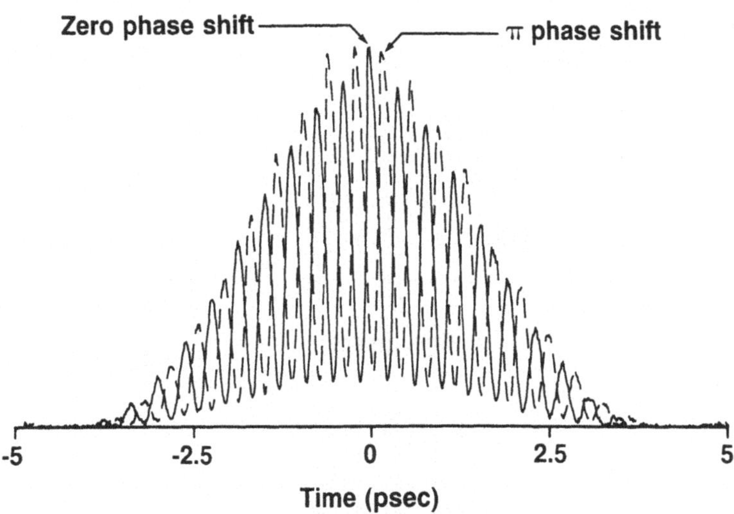

Fig. 2. Cross-correlation measurement of 2.6-THz bursts of femtosecond pulses. Solid line: The two frequencies are in phase. Dotted line: The two frequencies are out of phase by π.

For our multiple-pulse ISRS experiments, we are interested not only to generate high frequency tone bursts, as shown in Fig. 2, but also to maintain high peak intensities. Therefore, the amplitude filters discussed above are not ideal, since they reject most of the incident energy. High frequency tone bursts can be generated with good efficiency by using spatially periodic phase-only filters which do not attenuate the incident energy[21]. The spatial periodicity of the phase mask imparts a periodic modulation to the optical spectrum, and this gives rise to a THz-rate pulse train, with the pulse repetition rate equal to the inverse of the spectral modulation frequency. The shape and duration of the pulse train envelope, as well as the optical phases of the individual pulses within the train, depend on design of the phase-only filter within a single spatial period. For the ISRS experiments, the phase was patterned according to periodic repetitions of a pseudorandom binary phase code called an M-sequence. Details of the mask design are given elsewhere[18,21]. These masks produce sequences consisting of about 15 evenly spaced pulses under a smooth envelope, as shown in Fig. 3(a) for a 4.0-THz repetition rate. A large number of phase-only filters, corresponding to pulse repetition periods ranging from 170 fsec to 700 fsec, were designed and fabricated, so that essentially any pulse repetition rate needed to match phonon frequencies in the range 1.4-6.0 THz (50-200 cm^{-1}) could be selected. We note that more complex pulse sequences, such as pulse trains with "square" envelopes, have also been produced[21], as shown in Fig. 3(b), but these were not used in the ISRS experiments.

In addition to periodic trains of femtosecond pulses, a variety of other types of shaped pulses have been synthesized for experiments other than multiple-pulse ISRS. These include "dark" pulses used in studies of dark soliton propagation in optical fibers[22], square pulses used to achieve improved switching characteristics in femtosecond all-optical switching[23], and "encoded" pseudonoise waveforms with potential applications to high-rate, optical code-division multiple-access communication networks[24]. The success of all these experiments underscores the power and precision of the pulse shaping technique, which allows dependable generation of shaped femtosecond pulses and waveforms under user control.

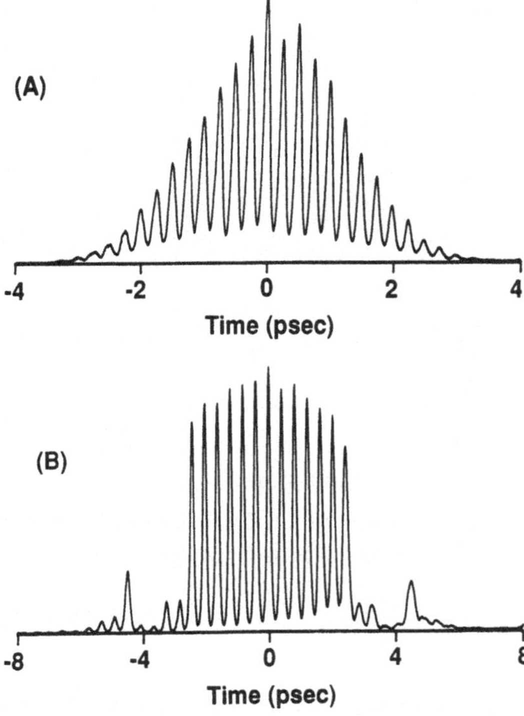

Fig. 3. Femtosecond pulse sequences produced by using phase-only filtering. (A) 4.0-THz pulse train with a smooth envelope. (B) 2.5-THz pulse train with a "square" envelope.

Despite the power and flexibility which current pulse shaping technology already provides, the technology is still evolving. For example, we have recently demonstrated an electronically programmable pulse shaping technique based on the use of a multielement liquid crystal phase modulator[25]. This apparatus makes possible modification of the pulse shape on a millisecond time scale under computer control. An example of real time pulse shaping is shown in Fig. 4, in which the output of the shaper is switched from a single pulse to a pulse doublet at a 20 Hz rate[25]. The use of the multielement modulator also makes possible continuously variable phase control which is not readily available with fixed masks and which is required in order to generate arbitrarily shaped pulses. Other programmable pulse shaping techniques have also been extended into the femtosecond regime[15]. Furthermore, a preliminary demonstration of pulse shaping with input pulses shorter than 30 fsec in duration has recently been reported[26]. As pulse shaping technology continues to improve, it will play an increasingly important in manipulating molecular coherences and molecular motion.

Finally, we discuss two broad constraints that govern the range of waveforms which can be generated[14]. First, the shortest temporal feature that can be realized is inversely related to the total available bandwidth. Since pulse shaping itself does nothing to increase the input bandwidth, temporal features within the shaped pulse can be no shorter than the duration of a bandwidth-limited input pulse. Second, the width of the temporal envelope

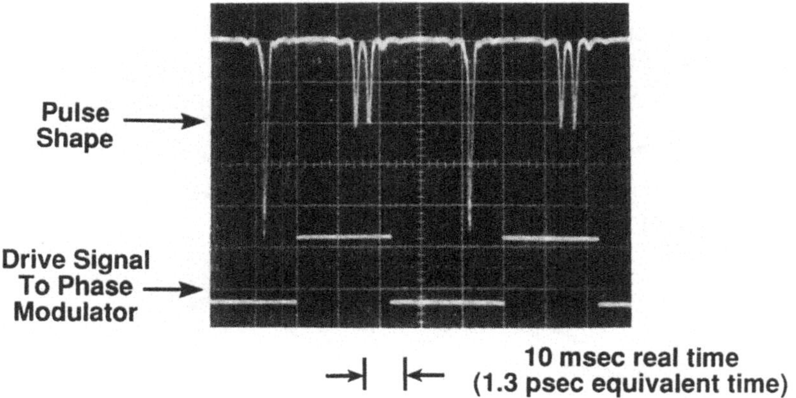

Fig. 4. Real-time pulse shaping operation. Top trace: Cross-correlation measurement of the shaped pulses, demonstrating switching from a single pulse to a pulse doublet at a 20 Hz rate. Bottom trace: An electrical drive signal to the multielement modulator.

which shaped waveforms can occupy is related inversely to the spectral resolution of the shaper. As in any spectrometer, the spectral resolution depends on the angular dispersion of the grating and the divergence of the input beam. Increased spectral resolution can be obtained with finely ruled gratings and large input beams. With typical experimental parameters, pulse shaping can be achieved over a 14-psec temporal window. With 75-fsec input pulses, this means that shaped pulses can consist of nearly two hundred, separately controllable 75-fsec features. By optimizing the parameters, a one order of magnitude improvement in spectral resolution should be possible. In that case highly complex output pulses consisting of several thousand independent features could be generated!

MULTIPLE-PULSE IMPULSIVE STIMULATED RAMAN SCATTERING

The layout of our impulsive stimulated Raman scattering (ISRS) experiments[17,18] is depicted in Figure 5. The sample is a 1-mm thick α-perylene molecular crystal, mounted on a cold finger and cooled to temperatures as low as 5 K. Note that the absorption edge of α-perylene occurs near 0.48 μm; therefore, the perylene is transparent to the 0.62 μm laser light. Amplified laser pulses are converted into THz-rate pulse sequences in the pulse shaping apparatus and then split to yield two identical excitation beams. The excitation beams cross in the sample, with a full angle of 5° and with zero relative time delay, so that they interfere to produce a "standing-wave" pattern of coherent vibrational excitation through ISRS[5,11]. The repetition rate of the excitation pulses is chosen to match the phonon frequency, so that each pulse within the sequence amplifies the phonon motion. A small portion of the output from the amplifier is split off before the pulse shaper to serve as a probe beam. The time-dependent vibrational response is monitored by measuring the coherent scattering (or diffraction) intensity of the temporally delayed probe pulse from the standing wave grating.

In ISRS, ultrashort laser pulses initiate coherent vibrational motion by exerting a sudden ("impulse") driving force on Raman-active vibrational modes[5,11]. For this to occur, the pulse duration must be short compared to the vibrational oscillation cycle. ISRS is a completely time-domain analogue to the more usual frequency-domain coherent Raman excitation techniques, in which the frequency difference between two tunable lasers is adjusted to match the vibrational frequency. One advantage of the time-domain approach

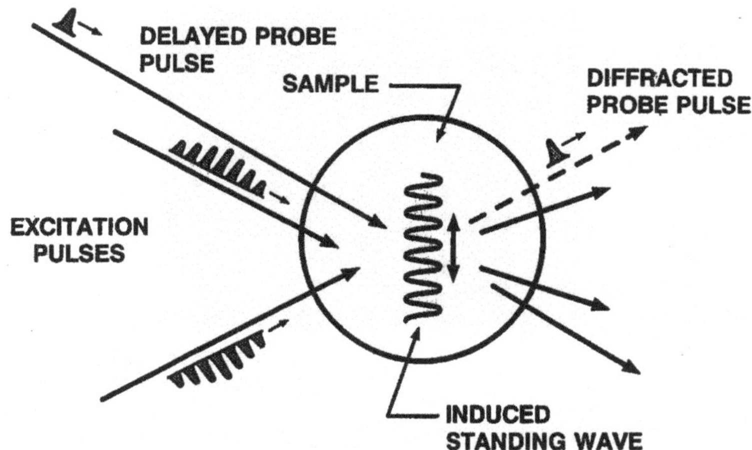

Fig. 5. Schematic diagram of the multiple-pulse ISRS experiment in the transient grating geometry.

is the ability to directly observe time-dependent phonon oscillations on a femtosecond time scale. Furthermore, ISRS can be utilized in principle to drive large vibrational amplitudes, which would allow spectroscopic characterization of nonequilibrium lattice structures and might permit optical control of chemical reactions[11]. α-perylene molecular crystal is of special interest in this regard due to an excimer formation reaction that may be phonon assisted[27]. We note that in practice vibrational amplitudes excited through ISRS have not been large, typically on the order of 10^{-4} Å for translational optic phonon modes.

As a control we recorded ISRS data from α-perylene excited by a single pair of (unshaped) excitation pulses. The results, shown in Fig. 6, are similar to those published previously[5]. The data consist of a sharp peak at time $t=0$, due to the essentially instantaneous nonlinearity associated with distortion of the electronic wave functions, followed by a quasi-periodic signal due to optic phonons excited through ISRS. Two points are of note. First, the intensity of the ISRS signal is only about 4% of that due to the electronic response (which serves as an internal calibration of the ISRS signal). Second, the ISRS diffraction intensity presents an irregular pattern caused by beating between several simultaneously excited phonon modes. A Fourier analysis of the scattering data (Fig. 7, solid line) shows a complicated series of lines corresponding to sum and difference frequencies of known α-perylene modes at 33, 56, 80, and 104 cm^{-1}. These results illustrate two main limitations of ISRS using single-pulse excitation -- namely, small vibrational amplitudes and lack of mode selectivity.

The dashed line in Fig. 7 shows the Fourier transform of the main, electronic scattering peak in the data. This curve represents the instrumental response in our experiments, showing that with 75-fsec pulses the ability to excite and monitor coherent vibrational motion rolls off for frequencies appreciably above 3 THz (100 cm^{-1}). Despite this roll-off, well-resolved features can be observed out to 5.5 THz.

Multiple-pulse excitation (multiple-pulse ISRS) can help improve the control over molecular motion available with ISRS. We have demonstrated this point by using multiple-pulse ISRS for selective amplification of individual phonon modes in α-perylene[17,18].

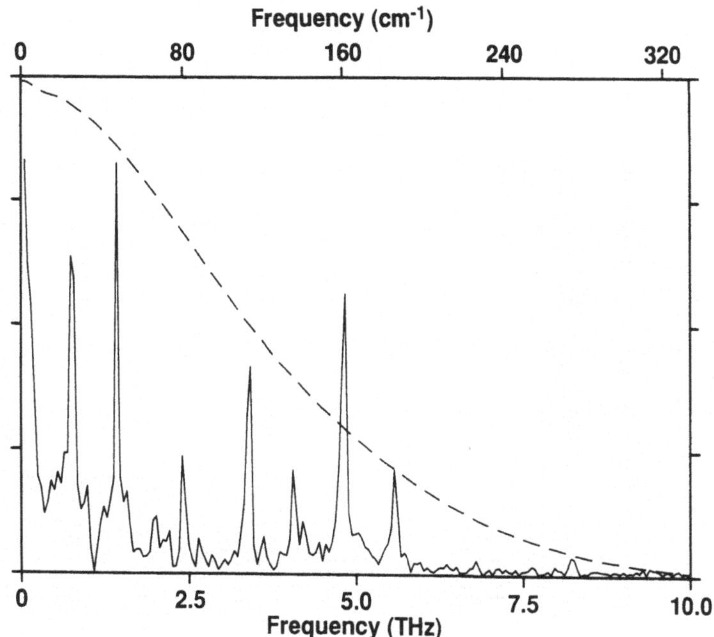

Fig. 6. Single-pulse ISRS data recorded from α-perylene with T < 10 K. The fsec excitation pulses drive several phonon modes which interfere to produce the observed beating pattern. The peak at t=0 is due to the electronic response of the crystal.

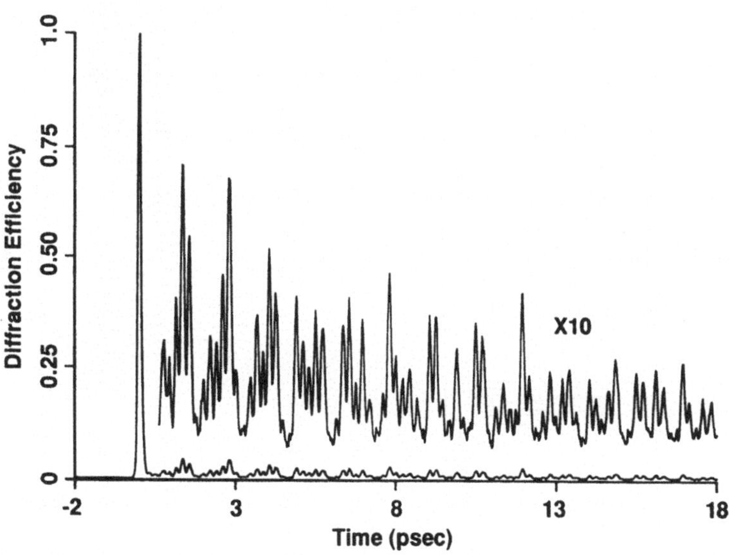

Fig. 7. Solid line: Fourier transform of the scattering data in Fig. 6. Dashed line: Fourier transform of the electronic scattering peak, showing the instrumental response.

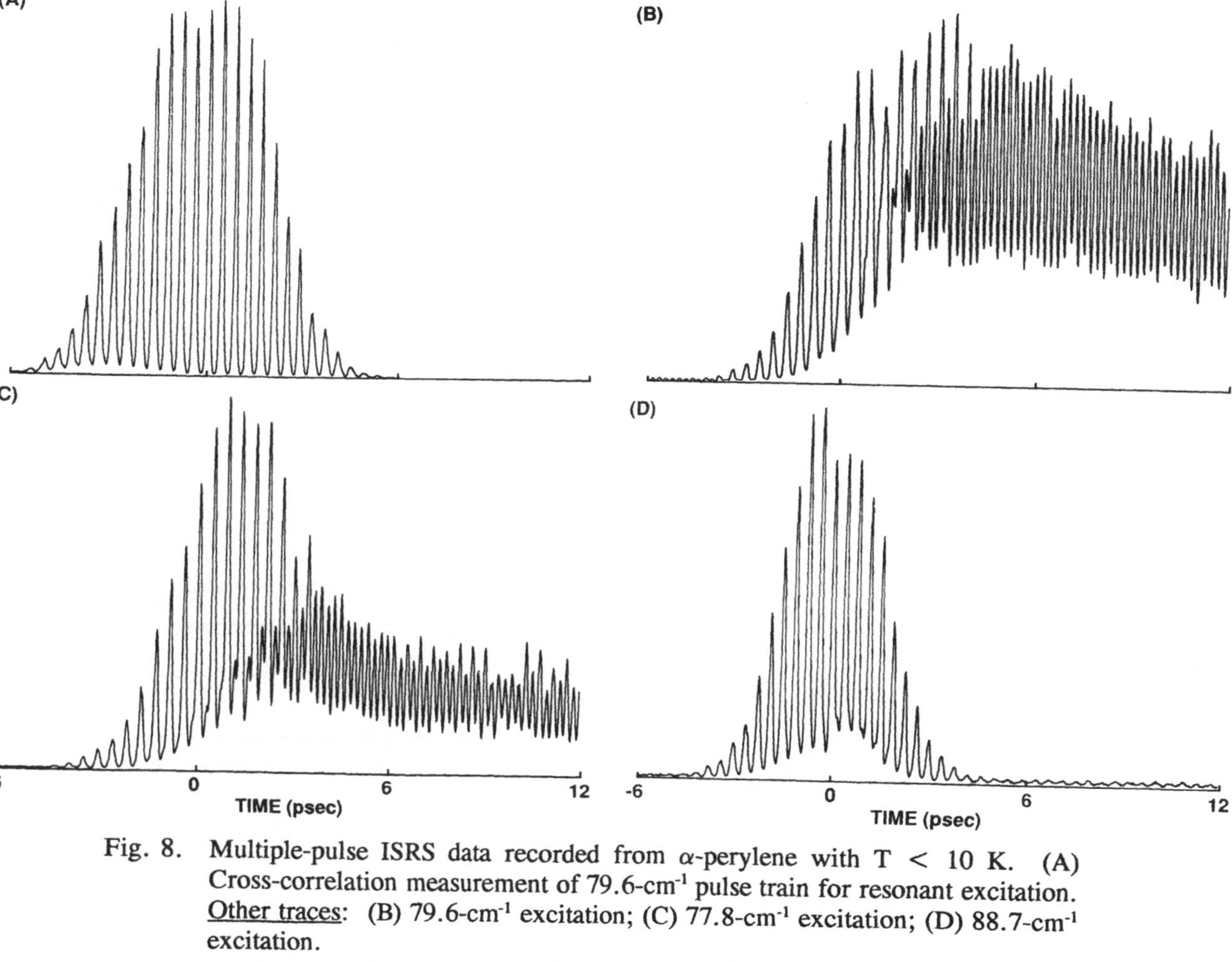

Fig. 8. Multiple-pulse ISRS data recorded from α-perylene with $T < 10$ K. (A) Cross-correlation measurement of 79.6-cm^{-1} pulse train for resonant excitation. Other traces: (B) 79.6-cm^{-1} excitation; (C) 77.8-cm^{-1} excitation; (D) 88.7-cm^{-1} excitation.

Figure 8(a) shows the intensity profile of a 2.39-THz (79.6 cm^{-1}) excitation pulse train, timed to match the vibrational period of the 80-cm^{-1} librational mode, and Fig. 8(b) shows the resulting ISRS signal. Initially, the signal is dominated by the crystal's electronic responses to each of the pulses in the sequence. However, as the pulse train progresses, the vibrational signal builds up. By the end of the input pulse train, the ISRS signal is as intense as the strongest electronic scattering peak. Furthermore, the ISRS signal remaining subsequent to the excitation pulse train oscillates at a single frequency (equal to twice the vibrational frequency, since the ISRS scattering signal is proportional to the square of the coherent vibrational amplitude[5,11]). This observation is confirmed by the Fourier transform of the scattering data (Fig. 9), which shows a single sharp at 160 cm^{-1} (4.8 THz). These observations clearly demonstrate considerable vibrational amplification and enhanced mode selectivity.

We have also studied the detuning behavior of multiple-pulse ISRS. As the detuning is increased, the strength of the phonon response gradually decreases, but the oscillations remain monochromatic with the same 160-cm^{-1} frequency. Examples of our data are shown in Figs. 8(c) and (d). In Fig. 8(c) the excitation repetition rate is changed to 2.33 THz (77.8 cm^{-1}), a relatively small detuning. Vibrational amplification is still achieved, although the strength of the vibrational signal is reduced by about a factor of two compared to resonant excitation. For substantially larger detuning (Fig. 8(d), 2.66 THz excitation), the vibrational signal is strongly suppressed, although instantaneous electronic scattering peaks remain. These observations are consistent with a simple picture of a forced harmonic oscillator. Indeed, the data in Fig. 8, including the build-up of the phonon amplitude during the excitation pulse train as well as the detuning characteristics, can be fit quantitatively using a forced harmonic oscillator model to represent the phonons[18].

We have also conducted other related measurements[18]. For example, since multiple-pulse ISRS results in selective amplification of a single phonon mode, the phonon decay dynamics can be determined easily by measuring the data over a sufficiently long time

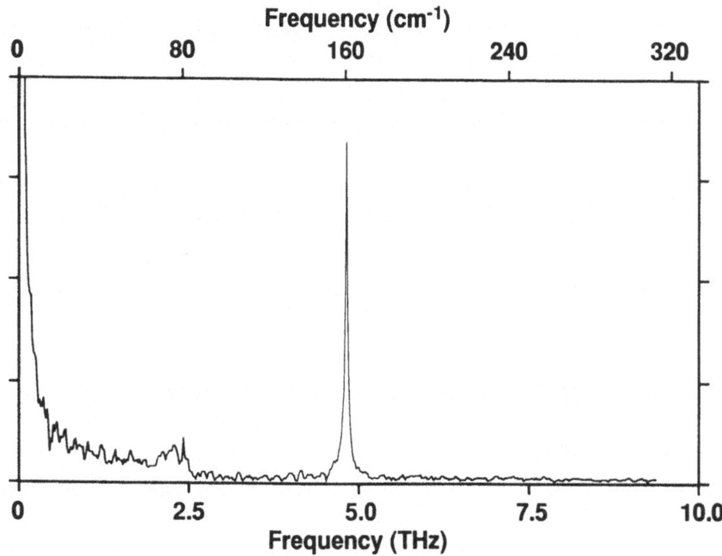

Fig. 9. Fourier transform of ISRS data, corresponding to resonant excitation of the 80-cm^{-1} librational mode, as in Fig. 8(b).

interval. For the 80-cm^{-1} mode in α-perylene, the phonon dephasing time decreases from ~29 psec at 10 K to ~1 psec at room temperature. At higher temperatures where the phonon dephasing time becomes comparable to or shorter than the duration of the driving pulse train, the vibrational amplification becomes less effective, and the phonon response becomes harder to distinguish from the electronic responses. In addition, we have also performed multiple-pulse ISRS experiments in the pump-probe geometry. By using a bandpass filter placed after the sample, we detect periodically varying spectral shifts of the probe pulse induced by the coherent vibrational motion. When the probe arrives at the sample in phase with the excitation pulse train and amplifies the phonon motion, it gives up energy to the vibrational mode and emerges from the sample "Doppler-shifted" to the red. Conversely, when the probe is timed to oppose the coherent vibrational motion, it removes energy from the sample and emerges "Doppler-shifted" to the blue. These measurements allow precise determination of the timing relationship between the excitation pulse train and the induced vibrational motion.

DISCUSSION

These multiple-pulse ISRS experiments constitute a simple, first demonstration of the use of specially designed femtosecond pulse waveforms to improve control over molecular motion. The success of these experiments arises due to two factors. First, because the vibrations can be modeled as classical harmonic oscillators, the concept is simple and intuitive. Second, the ISRS experiment is well matched to the capabilities of current pulse shaping technology. Visible light (as opposed to infrared light for infrared absorption) can be utilized. At least for heavy molecules with relatively low frequency vibrations, shaping of pulses on a "slow" time scale (longer than 100 fsec) is adequate. The required shaped pulses are characterized by slowly varying envelopes, so direct modulaton of the optical carrier itself is not required. Finally, since ISRS excitation depends only on the intensity profile of the excitation pulses, phase-only filtering can be used to maintain high optical intensities.

Some words are now in order concerning the vibrational amplitudes which can be achieved using multiple-pulse ISRS. In the case of single-pulse ISRS, the vibrational amplitude is proportional to the pulse energy, provided that the pulse width is much shorter than the phonon period. In the case of resonant multiple-pulse ISRS, the final vibrational amplitude is proportional to the energy in the entire pulse train, provided in addition that the pulse train lasts less than the phonon dephasing time. To the extent that large vibrational amplitudes are desired, the main limitations are unwanted optical responses, such as multiphoton absorption, continuum generation, and sample damage, which occur at the very high peak intensities required. Splitting up a single pulse into a sequence of less intense pulses is one way of channeling the light energy into ISRS excitation and not into unwanted processes. Therefore, the ability to drive larger vibrational amplitudes by multiple pulse ISRS will depend on the extent to which multiple pulse excitation makes possible increased excitation energies.

In the current experiments with α-perylene, the achieved coherent vibrational oscillations are still in the small amplitude limit. The vibrational amplitude can be estimated from the measured diffraction efficiency and the known experimental parameters[5,11]. For the 104-cm^{-1} phonon mode, which corresponds to vibration of neighboring, planar perylene molecules against each other, the peak diffraction efficiency for multiple pulse excitation is roughly 0.02%, and the estimated peak to peak change in molecular separation is on the order of 10^{-4} Å, compared to an equilibrium molecular separation of 0.2 Å. For the 80 cm^{-1} librational mode, in which the molecule pairs are librating in phase, the diffraction efficiency is approximately 0.12%, and the estimated peak to peak angular deviation is 0.001°. The magnitude of the induced vibrational motion was limited by sample damage, which occurred at total fluences of 3 mJ/cm^2 and 5 mJ/cm^2 for single pulse and multiple pulse excitation, respectively. Based on the measured damage thresholds, in these experiments multiple pulse excitation makes possible only a modest (65%) increase in the coherent vibrational amplitude.

A quantum mechanical analysis of multiple-pulse ISRS excitation has recently been published[28]. The analysis shows that in addition to small vibrational amplitudes, the populations of vibrational eigenstates other than the ground vibrational level are also small. Significant populations in excited vibrational levels will occur only when vibrational amplitudes comparable to the size of the potential well (e.g., 0.2 Å for the 104-cm^{-1} mode) are achieved.

There are several possible approaches for trying to drive larger vibrational motions. First, one can select the best suited materials. Many materials have damage thresholds substantially higher than that of the molecular crystal α-perylene, and in such materials it might be possible to induce correspondingly stronger vibrational amplitudes through ISRS. Materials which combine a high damage threshold with a large Raman cross-section would be especially favorable. Multiple pulse ISRS will be particularly helpful in driving large amplitude vibrational motion in cases where unwanted nonlinear processes limit the optical intensity rather than the energy. This might occur in liquids and high damage threshold solids where continuum generation or multiphoton absorption might constitute the limiting process. Second, one can tune the laser wavelength near (but below) the absorption edge in order to resonantly enhance the Raman cross-section[11]. Finally, one can perform true resonant experiments in which the laser photons are absorbed. Due to the change in equilibrium nuclear coordinates which often accompanies an electronic transition, an absorption event can automatically induce large amplitude vibrational motions in suitable systems[6,10].

CONCLUSION

In summary, we have described femtosecond pulse shaping technology and an initial, simple application to molecular control. In particular, we have discussed multiple-pulse impulsive stimulated Raman scattering, a new laser excitation technique in which timed sequences of femtosecond pulses repetitively drive and selectively amplify optical phonon modes in a manner closely analogous to the way a child on a swing may be pushed repetitively to reach an increased amplitude of oscillatory motion.

Progress in pulse shaping technology, as well as new theoretical schemes for molecular control, will both contribute to achievement of more sophisticated manipulation over molecular behavior. On the theoretical side, a number of ideas for laser control of molecular and chemical processes have been put forth. An increased awareness of technological capabilities and constraints as well as an increased consideration of robustness will facilitate practical realization of proposed control schemes in the laboratory. On the technological side, electronically programmable pulse shaping techniques will provide greater flexibility in producing femtosecond pulse waveforms, which could, for example, lead to compensation of anharmonicities at large vibrational amplitudes (by allowing generation of femtosecond pulse sequences with appropriately varying repetition rates). Extension of pulse shaping technology to ultraviolet and infrared wavelengths will facilitate the ability to control molecular processes through electronic and infrared absorptions, respectively. Generalization to shorter pulse widths will allow interaction with higher frequency vibrational modes. Although much work remains to be done, we can expect that pulse shaping technology will continue to evolve and proliferate; as this occurs, pulse shaping will become an increasingly important tool for laser manipulation of molecular motion.

ACKNOWLEDGEMENTS

G.P. Wiederrecht and K.A. Nelson received support from National Science Foundation grant CHEM-8901722 and contributions from DuPont, Perkin Elmer, and the Alfred P. Sloan Foundation.

REFERENCES

1. N. Bloembergen and A.H. Zewail, J. Phys. Chem. **88**, 5459 (1984).
2. H.L. Fragnito, J.-Y. Bigot, P.C. Becker, and C.V. Shank, Chem. Phys. Lett. **160**, 101 (1989).
3. R.A. Mathies, C.H. Brito Cruz, W.T. Pollard, and C.V. Shank, Science **240**, 777 (1988).
4. Y.-X. Yan, L.-T. Cheng, and K.A. Nelson, in Advances in Nonlinear Spectroscopy, (Adv. in Spectrosc. series, v. 16), R.J.H. Clark and R.E. Hester, Eds. (Wiley, Chichester, 1988), pp. 299-355.
5. S. De Silvestri, J.G. Fujimoto, E.P. Ippen, E.B. Gamble, L.R. Williams, and K.A. Nelson, Chem. Phys. Lett. **116**, 146 (1985).
6. J. Chesnoy and A. Mokhtari, Phys. Rev. **A38**, 3566 (1988).
7. F.W. Wise, M.J. Rosker, and C.L. Tang, J. Chem. Phys. **86**, 2827 (1986).
8. N.F. Scherer, C. Sipes, R.B. Bernstein, and A.H. Zewail, J. Chem. Phys. **92**, 5239 (1990).
9. S. Shi, A. Woody, and H. Rabitz, J. Chem. Phys. **88**, 6870 (1988).
10. D.J. Tannor, R. Kosloff, and S.A. Rice, J. Chem. Phys. **85**, 5805 (1986); R. Kosloff, S.A. Rice, P. Gaspard, S. Tersigni, and D.J. Tannor, Chem. Phys. **139**, 201 (1989).
11. Y.-X. Yan, E.B. Gamble, and K.A. Nelson, J. Chem. Phys. **83**, 5391 (1985).
12. P. Brumer and M. Shapiro, Chem. Phys. Lett. **126**, 541 (1986).
13. S. Chelkowski, A. Bandrauk, and P.B. Corkum, Phys. Rev. Lett. **65**, 2355 (1990).
14. A.M. Weiner, J.P. Heritage, and E.M. Kirschner, J. Opt. Soc. Am. **B5**, 1563 (1988).
15. M. Haner and W.S. Warren, Appl. Phys. Lett. **52**, 1458 (1988); W.S. Warren, Science **242**, 878 (1988).
16. A.J. Ruggiero, N.F. Scherer, M. Du, and G.R. Fleming, in Ultrafast Phenomena VII, C.B. Harris, E.P. Ippen, G.A. Mourou, and A.H. Zewail, eds., (Springer, Berlin, 1990), p. 334.
17. A.M. Weiner, D.E. Leaird, G.P. Wiederrecht, and K.A. Nelson, Science **247**, 1317 (1990).
18. A.M. Weiner, D.E. Leaird, G.P. Wiederrecht, and K.A. Nelson, J. Opt. Soc. Am. **B8**, 1264 (1991).
19. R.L. Fork, B.I. Greene, and C.V. Shank, Appl. Phys. Lett. **38**, 671 (1981); J.A. Valdmanis, R.L. Fork, and J.P. Gordon, Opt. Lett. **10**, 131 (1985).
20. W.H. Knox, M.C. Downer, R.L. Fork, and C.V. Shank, Opt. Lett. **9**, 552 (1984).
21. A.M. Weiner and D.E. Leaird, Opt. Lett. **15**, 51 (1990).
22. A.M. Weiner, J.P. Heritage, R.J. Hawkins, R.N. Thurston, E.M. Kirschner, D.E. Leaird, and W.J. Tomlinson, Phys. Rev. Lett. **61**, 2445 (1988).
23. A.M. Weiner, Y. Silberberg, H. Fouckhardt, D.E. Leaird, M.A. Saifi, M.J. Andrejco, and P.W. Smith, IEEE J. Quantum Electron. **25**, 2648 (1989).
24. A.M. Weiner, J.P. Heritage, and J.A. Salehi, Opt. Lett. **13**, 300 (1988); J.A. Salehi, A.M. Weiner, and J.P. Heritage, J. Lightwave Technology **8**, 478 (1990).
25. A.M. Weiner, D.E. Leaird, J.S. Patel, and J.R. Wullert, Opt. Lett. **15**,, 326 (1990).

26. D.H. Reitze, A.M. Weiner, and D.E. Leaird, "High-power femtosecond pulse compression using spatial solitons", Optical Society of America Annual Meeting, San Jose, CA, Nov. 1991.
27. J. Tanaka, T. Kishi, and M. Tanaka, Bull. Chem. Soc. Jpn. **47**, 2376 (1974).
28. Y.J. Yan and S. Mukamel, J. Chem. Phys. **94**, 997 (1991).

COHERENCE IN THE CONTROL OF MOLECULAR PROCESSES

Paul Brumer

Chemical Physics Theory Group
Department of Chemistry
University of Toronto
Toronto, Ontario M5S 1A1, Canada

Moshe Shapiro

Chemical Physics Department
Weizmann Institute of Science
Rehovot, Israel

ABSTRACT. We describe the current status of coherent radiative control, a quantum-interference based approach to controlling molecular processes by the use of coherent radiation. In addition to providing an overview of proposed laboratory scenarios, ongoing experimental studies and recent theoretical developments, we call attention to recent theoretical results on symmetry breaking in achiral systems.

1. Introduction

This NATO workshop deals with the rapidly developing area of coherence phenomena in Chemistry and Physics. Of particular interest are the phase characteristics of molecular and optical states, and new phenomena which emerge when one can manipulate the quantum phase of material systems. In this paper we discuss a theory, based upon quantum interference effects, which predicts virtually total control over branching ratios in isolated molecular processes.

Control over the yield of chemical reactions is the *raison d'etre* of practical chemistry and the ability to use lasers to achieve this goal is one primary thrust of modern Chemical Physics. The theory of coherent radiative control of chemical reactions, which we have developed over the last five years [1-16], affords a direct method for controlling reaction dynamics using coherence properties of weak lasers, with a large range of yield control expected in laboratory scenarios. In addition, the theory of coherent control provides deep insights into the essential features of reaction dynamics, and of quantum interference, which are necessary to achieve control over elementary chemical processes. Below we provide a schematic overview of coherent radiative control including an example which

Coherence Phenomena in Atoms and Molecules in Laser Fields
Edited by A.D. Bandrauk and S.C. Wallace, Plenum Press, New York, 1992

emphasizes the general principles (Section 2), a survey of several experimental scenarios which implement the principles of coherent control (Section 3), an interesting application to symmetry breaking in achiral systems (Section 4) and a summary of the current status of theory and experiment. A qualitative introduction to coherent control is provided in reference [9]. An alternative approach, based upon the use of shaped light pulses has been advocated by Tannor and Rice [17] and Rabitz[18].

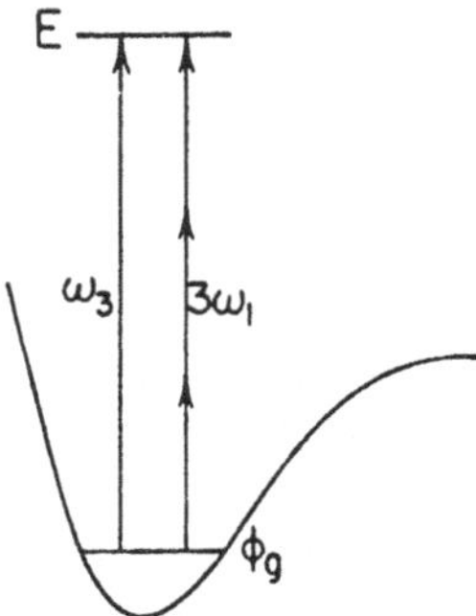

Figure 1. One-photon plus three-photon coherent control scenario.

2. Interference and Control: An Example

Consider a chemical reaction which, at total energy E produces a number of distinct products. The total Hamiltonian is denoted $H = H_q^0 + V_q$, where H_q^0 is the Hamiltonian of the separated products in the arrangement channel labeled by $q, (q = 1, 2, ...)$ and V_q is the interaction between products in arrangement q. We denote eigenvalues of H_0^q by $|E, n, q^0\rangle$, where n denotes all quantum numbers other than E; eigenfunctions of H, which correlate with $|E, n, q^0\rangle$ at large product separation, are labeled $|E, n, q^-\rangle$. By definition of the minus states, a state prepared experimentally as as superposition $|\Psi(t = 0)\rangle = \sum_{n,q} c_{n,q}|E, n, q^-\rangle$ has probability $|c_{n,q}|^2$ of forming product in channel q, with quantum numbers n. As a consequence, control over the probability of forming a product in any asymptotic state is equal to the probability of initially forming the appropriate minus state which correlates with the desired product. The essence of control lies, therefore, in forming the desired linear combination at the time of preparation. The essential *modus operandi* of coherent radiative control is to utilize phase and intensity properties of laser excitation to alter the character of the prepared state so as to enhance production of the desired product.

Many of the proposed coherent control scenarios rely upon a simple way of achieving active control over the prepared state and product. Specifically, active control over products is achieved by driving an initially pure molecular state through two or more independent coherent optical excitation routes. The resultant product probability displays interference terms between these two routes, whose magnitude and sign depend upon laboratory parameters. As a consequence, product probabilities can be manipulated directly in the laboratory.

The scheme outlined above has, as a well-known analogy, the interference between paths as a beam of either particles or of light passes through a double slit. In that instance source coherence leads to either constructive or destructive interference, manifest as patterns of enhanced or reduced probabilitites on an observation screen. In the case of coherent control the overall coherence of a pure state plus laser source allows for the constructive or destructive manipulation of probabilities in product channels.

We note, as an aside, that practitioners of "mode-selective chemistry" often advocate the use of lasers to excite specific bonds or modes in the reactant molecule in order to drive a reaction in a desired direction. The brief remarks above, elaborated upon elsewhere[10], make clear that the proper modes to excite in order to produce product in arrangement q are the eigenfunctions $|E, n, q^-\rangle$, the system's "natural modes", if you will. Thus the extent to which excitation of some zeroth order state $|\chi\rangle$ is successful in promoting reaction to the desired product q depends entirely on the extent to which $|\chi\rangle$ overlaps $|E, n, q^-\rangle$.

Consider, as an example of coherent control, a specific scenario for unimolecular photoexcitation[5,13] (Figure 1) where a system, initially in pure state ϕ_g (or $|E_i\rangle$ below), is excited to energy E, by simultaneous application of two CW frequencies ω_3 and $3\omega_1$ ($\omega_3 = 3\omega_1$), providing two independent optically driven routes from ϕ_g to $|E, n, q^-\rangle$.

Straightforward perturbation theory, valid for the weak fields under consideration, gives the probability $P(E, q; E_i)$ of forming product at energy E in arrangement q as:

$$P(E,q;E_i) = P_3(E,q;E_i) + P_{13}(E,q;E_i) + P_1(E,q;E_i)\,. \tag{1}$$

with terms defined as follows:

$$P_3(E,q;E_i) = (\frac{\pi}{\hbar})^2 \epsilon_3^2 \sum_n |\langle E,n,q^- \,|\, (\hat{\underset{\sim}{\epsilon}}_3 \cdot \underset{\sim}{\mu})_{e,g} \,|\, E_i \rangle|^2\,. \tag{2}$$

Here $\underset{\sim}{\mu}$ is the electric dipole operator, and

$$(\hat{\underset{\sim}{\epsilon}}_3 \cdot \underset{\sim}{\mu})_{e,g} = \langle e \,|\, \hat{\underset{\sim}{\epsilon}}_3 \cdot \underset{\sim}{\mu} \,|\, g \rangle\,, \tag{3}$$

where $|\,g\,\rangle$ and $|\,e\,\rangle$ are the ground and excited electronic state wavefunctions, respectively. The second and third terms in Eq. (1) are

$$P_1(E,q;E_i) = (\frac{\pi}{\hbar})^2 \epsilon_1^6 \sum_n |\langle E,n,q^- \,|\, T \,|\, E_i \rangle|^2\,, \tag{4}$$

with

$$T = (\hat{\underset{\sim}{\epsilon}}_1 \cdot \underset{\sim}{\mu})_{e,g}(E_i - H_g + 2\hbar\omega_1)^{-1}(\hat{\underset{\sim}{\epsilon}}_1 \cdot \underset{\sim}{\mu})_{g,e}(E_i - H_e + \hbar\omega_1)^{-1}(\hat{\underset{\sim}{\epsilon}}_1 \cdot \underset{\sim}{\mu})_{e,g}\,. \tag{5}$$

and

$$P_{13}(E,q;E_i) = -2(\frac{\pi}{\hbar})^2 \epsilon_3 \epsilon_1^3 \cos(\theta_3 - 3\theta_1 + \delta_{13}^{(q)})|F_{13}^{(q)}| \tag{6}$$

with the amplitude $|F_{13}^{(q)}|$ and phase $\delta_{13}^{(q)}$ defined by

$$|F_{13}^{(q)}|\exp(i\delta_{13}^{(q)}) = \sum_n \langle E_i \,|\, T \,|\, E,n,q^- \rangle \langle E,n,q^- \,|\, (\hat{\underset{\sim}{\epsilon}}_3 \cdot \underset{\sim}{\mu})_{e,g} \,|\, E_i \rangle\,. \tag{7}$$

The branching ratio $R_{qq'}$ for channels q and q', can then be written as

$$R_{qq'} = \frac{P(E,q;E_i)}{P(E,q';E_i)} = \frac{\epsilon_3^2 F_3^{(q)} - 2\epsilon_3\epsilon_1^3 \cos(\theta_3 - 3\theta_1 + \delta_{13}^{(q)})|F_{13}^{(q)}| + \epsilon_1^6 F_1^{(q)}}{\epsilon_3^2 F_3^{(q')} - 2\epsilon_3\epsilon_1^3 \cos(\theta_3 - 3\theta_1 + \delta_{13}^{(q')})|F_{13}^{(q')}| + \epsilon_1^6 F_1^{(q')}}, \tag{8}$$

where

$$\begin{aligned} F_3^{(q)} &= (\tfrac{\hbar}{\pi})^2 \frac{P_3(E,q;E_i)}{\epsilon_3^2}, \\ F_1^{(q)} &= (\tfrac{\hbar}{\pi})^2 \frac{P_1(E,q;E_i)}{\epsilon_1^6}, \end{aligned} \tag{9}$$

with $F_3^{(q')}$ and $F_1^{(q')}$ defined similarly.

The numerator and denominator of Eq. (8) each display what we regard as the canonical form for coherent control. That is, independent contributions from two routes, modulated by an interference term. Since the interference term is controllable through variation of laboratory parameters (here the relative intensity and relative phase of the two lasers), so too is the product ratio R.

This 3-photon + 1-photon scenario has now been experimentally implemented[19] in studies of HCl ionization through a resonant bound Rydberg state. Specifically, HCl is excited to a selected rotational state in the $^3\Sigma^-(\Omega^+)$ manifold using $\omega_1 = 336$ nm; ω_3 is obtained by third harmonic generation in a Krypton gas cell. The relative phase of the light fields was then varied by passing the beams through a second Krypton cell and varying the cell gas pressure. The population of the resultant Rydberg state was interrogated by ionizing to HCl^+ with an additional photon. This REMPI type experiment showed that the HCl^+ ion probability depended upon both the relative phase and intensity of the two exciting lasers, in accord with the theory described above. Note that in this case, control is achieved by simultaneous excitation to a nondegenerate bound state. As a consequence there is no sum over n in Eqs. (2), (4) and (7) and hence $|F_3^{(q)} F_1^{(q)}| = |F_{13}^{(q)}|^2$. Satisfaction of this Schwartz equality suffices to ensure that the probability of forming the HCl Rydberg state can be varied over the full range of zero to unity[1]. Specifically, setting $\theta_3 - 3\theta_1 = -\delta_{13}^{(q)}$ and $\epsilon_3^2/\epsilon_1^6 = F_1^{(q)}/F_3^{(q)}$ gives zero probability in channel q.

A similar phase control experiment has been performed on atoms[20], in the simultaneous 3-photon + 5-photon ionization of Hg. Although the effect of the relative laser intensity was not studied, the Hg^+ ionization probability was shown to be a function of relative phase of the two lasers. Since in this case excitation is nonresonant, this experiment shows the feasibility of control over continuum channels via interfering optical excitation routes.

These experiments clearly show that coherent control of simple molecular processes through quantum interference of multiple optical excitation routes is both feasible and experimentally observable. Further experimental studies designed to show control over processes with more than one product channel are in progress by a number of experimental groups. Further, our recent computational results [13] on photodissociation of IBr, to produce I + Br and I + Br* suggest that such experiments should show huge control over the product ratio. For example, 1-photon plus 3-photon photodissociation of the ground vibrotational state in the ground electronic state of IBr showed variations in the Br*/Br ratio of 45% to 95% with variations of the relative laser intensity and phase. Similar results were obtained for higher J, including $J = 42$ where extensive M_J averaging was required.

3. The Essential Principle and Assorted Coherent Control Scenarios

The 3-photon plus 1-photon case is but one example of a scenario which embodies the essential principle of coherent control, i.e. that ***coherently driving a pure state through***

multiple optical excitation routes to the same final state allows for the possibility of control. Given this general principle, numerous scenarios may be proposed to obtain control in the laboratory. Such proposals must, however, properly account for a number of factors which reduce or eliminate control. Amongst these are the need to (a) adhere to selection rule requirements, (b) minimize extraneous and parasitic uncontrolled satellites, (c) insure properly treated laser spatial dependence and phase jitter and (d) allow maintainance of coherence over time scale associated with any true relaxation processes, e.g. collisions. Briefly:

(a) control requires that the interference term (e.g. $F_{13}^{(q)}$ above) arising between optical routes is nonzero. In general, this is the case only if the various optical excitation routes satisfy specific rules regarding conserved integrals of motion. For example[21], consider the 1-photon + 3-photon case discussed above. Here excitation from, e.g. $J = 0$, where J is the rotational angular momentum, yields $J = 1$ for the 1-photon route and $J = 1$ and $J = 3$ for the three photon route. In this case selection rules are such that only final states with the same J, here $J = 1$, interfere;

(b) the $J = 3$ state created by three photon absorption in this scenario is an example of a satellite state, i.e. an uncontrolled (and hence undesired) product state arising in the course of the photoexcitation. Effective scenarios must insure that such contributions are small compared to the controlled component;

(c) since the ability to accurately manipulate the laser relative phase is crucial to coherent control, one must account for all laboratory features which affect the phase. For example, scenarios must be designed so as to eliminate effects due to the $\mathbf{k} \cdot \mathbf{R}$ spatial dependence of the laser phase. Not doing so would results in the dimunition of control resulting from the variation of this term over the molecule beam-laser beam intersection volume.

(d) proposed scenarios must maintain phase coherence in the face of possibly dephasing effects (e.g. collisions or partial laser coherence). Studies indicate[7,22] that control can survive moderate levels of such phase destructive processes.

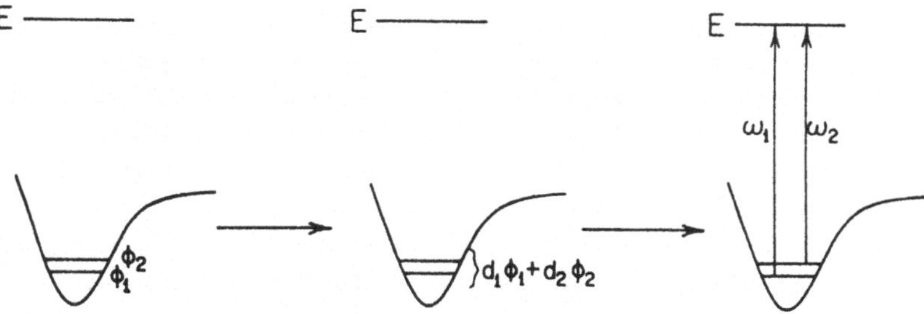

Figure 2. Basic control scenario.

We have, thus far, proposed a number of different scenarios for coherent control. Below we comment on some of them, with the intent of providing only the briefest of "roadmaps" for the proposed schemes. Detailed discussions are provided in the literature, along with extensive computations. The original coherent control scenario is shown in Figure 2 wherein a superposition of two bound states is subjected to two cw frequencies which lift the system to energy E. An analysis of this very basic scenario shows that the control parameters are the relative intensity and relative phase of the two indicated electromagnetic fields. This basic mechanism has been examined in the gas phase, both in the presence and in the absence of collisions and this scenario has been adapted to control currents in semiconductors. In all cases control was extensive.

Developments in pulsed laser technology may be used to good advantage in a straightforward modification of the above scenario. Specifically, the superposition state preparation and subsequent excitation with frequencies ω_1, ω_2 may be both carred out using pulsed lasers in accord with Figure 3:

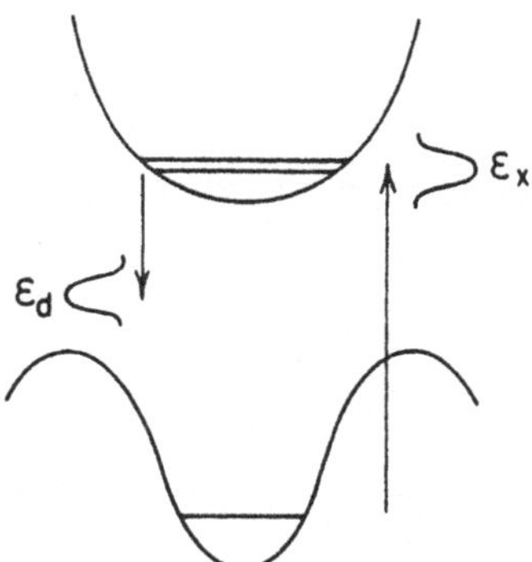

Figure 3. Pump-Pump scenario

In this instance the frequencies required to excite to energy E are contained within the second excitation pulse, which also creates product, and associated interference contributions over a wider energy range. Multiple excitation paths are contained within this overall pump-pump excitation scheme and serve to introduce the necessary interfering coherent paths. The straightforward analysis[8,11] of this scenario consists of a perturbation theory treatment of the molecule in the presence of two temporally separated, sequential pulses. The first pump step, in which the system is excited from ground state $|E_g\rangle$, yields a superposition state $\sum_k c_k |E_k\rangle$ with c_k proportional to $\langle E_k|\mu|E_g\rangle$; the subsequent pulse causes dissociation. Perturbation theory gives the probability, $P(E, m_j, q)$, required in the section below, of forming product in arrangement channel q at energy E with total fragment angular momentum projection m_j along a space fixed axis as,

$$P(E, m_j, q) = \sum_n{}' |B(E, n, q|t = \infty)|^2$$

$$= (2\pi/\hbar^2) \sum_n{}' |\sum_{k=1,2} c_k \langle E, n, q^-|\mu|E_k\rangle \epsilon_d(\omega_{EE_k})|^2 \tag{10}$$

where $\omega_{EE_k} = (E - E_k)/\hbar$. Here $\epsilon_d(\omega_{EE_k})$ is the amplitude of the electric field of the second pulse at the frequency ω_{EE_k} and the prime denotes summation over all quantum numbers n (including scattering angles) other than m_j.

Expanding the square, for the case where c_k is nonzero for two states, and using a gaussian pulse shape gives:

$$P(E,m_j,q) = (2\pi/\hbar^2)[|c_1|^2\mu_{1,1}^{(q)}\varepsilon_1^2+|c_2|^2\mu_{2,2}^{(q)}\varepsilon_2^2+2|c_1c_2^*\varepsilon_1\varepsilon_2\mu_{1,2}^{(q)}|\cos(\omega_{2,1}(t_d-t_x)+\alpha_{1,2}^{(q)}(E)+\phi)] \tag{11}$$

where $(t_d - t_x)$ is the temporal delay between pulses, $\varepsilon_i = |\epsilon_d(\omega_{EE_i})|$, $\omega_{2,1} = (E_2 - E_1)/\hbar$ and the phases ϕ, $\alpha_{1,2}^{(q)}(E)$ are defined by

$$\langle E_1 |\mu| E_g \rangle\langle E_g |\mu| E_2 \rangle \equiv |\langle E_1 |\mu| E_g \rangle\langle E_g |\mu| E_2 \rangle| e^{i\phi}$$

$$\mu_{i,k}^{(q)}(E) \equiv |\mu_{i,k}^{(q)}(E)| e^{i\alpha_{i,k}^{(q)}(E)} = \sum_n{}' \langle E,n,q^- |\mu| E_i \rangle\langle E_k |\mu| E,n,q^- \rangle \tag{12}$$

Note the generic form of Eq. (11) in which there are two direct routes plus a term representing interference between them. Integrating over E to encompass the width of the second pulse, and forming the ratio $Y(m_j) = P(q=1,m_j)/[P(q=1,m_j)+P(q=2,m_j)]$, gives the controllable ratio of products in fixed m_j states. This particular quantity is of interest in the case of chirality control discussed in Section 4. Of greater interest in general is the ratio of products in each arrangement channel, i.e.
$Y = P(q=1)/[P(q=1)+P(q=2)]$ where $P(q) = \sum_{m_j} P(q,m_j)$.

These results show that in this case the laboratory control parameters which affect the relative yield are the time delay between the pulses and the detuning of the initial excitation pulse which, in turn, determines the relative population and phase of the intermediate superposition state. We have applied this scenario to both model DH_2[8] and to IBr photodissociation [1]. In both cases relatively small changes in the detuning or in the time delay, which is in the convenient picosecond domain, resulted in extremely large variations of the yield (e.g. variations in the Br yield in IBr dissociation case over a range of 4% to 96%).

We have also applied this pump-dump type scheme to control bimolecular reactions [14]. In this case the initial excitation is from the continuum ground electronic state to a bound excited electronic state, followed by a dump back down to the ground electronic state. Significant control results only in the energy domain which is essentially below reaction threshold. Above threshold the interference term does not effectively compete with the energetically allowed reaction.

It is important to note that the scenarios involving two bound levels described above are not "restricted-two-state problems". Rather, we recognize that two levels are sufficient to carry the necessary phase information to allow control over the relative yield. That is, lasers with narrow frequency profiles (i.e. long pulses in time) are sufficient. Indeed as the molecules increase in size, and the density of states increases, longer and longer pulses could be used, and two intermediate levels retained. There appears little advantage to the inclusion of more than two levels, a situation which would result if the pulses were considerably shorter in time.

4. Symmetry Breaking and Chirality Control

The results above make clear that two arrangement channels, whose photodissociation amplitudes differ, can be controlled through manipulation of external laser fields. A particularly interesting example results from consideration of the photodissociation of a molecule schematically written as :

$$(q=2), B' + AB \overset{h\nu}{\leftarrow} BAB' \overset{h\nu}{\rightarrow} B + AB', (q=1) \tag{13}$$

where q defines the product arrangement channel and BAB' denotes a molecule, or a molecular fragment, which is symmetric with respect to reflection σ across the plane which interchanges the enantiomers B and B' (e.g., Ref. 23). The existence of this plane implies that the point group of the reactant is at least C_s.

Consider then the molecule $B'AB$ subjected to the two pulse scenario sketched above. The probability of forming product in arrangement channel q with products in the m_j state is given by Eq. (11). Active control over the products $B + AB'$ vs. $B' + AB$, attained by manipulating the time delay or the pulse detuning will result as long as $P(q=1, m_j)$ and $P(q=2, m_j)$ have different functional dependences on laboratory parameters.

It is surprising that $P(q=1, m_j)$ and $P(q=2, m_j)$ differ in the $B'AB$ case[24]. To see this consider the behavior of c_k and $\mu_{i,k}^{(q)}$, for $|E_1\rangle$ and $|E_2\rangle$ of different symmetry with respect to the reflection σ. For simplicity we restrict attention to BAB' belonging to point group C_s, the smallest group possessing the required symmetry plane. Further, we focus upon transitions between electronic potential energy surfaces of similar species, e.g. A' to A' or A'' to A'' and assume the ground vibronic state to be of species A'. Similar arguments apply for larger groups containing σ, to ground vibronic states of odd parity, and to transition between electronic states of different species.

Excitation Coefficients c_k: Components of μ lying in the symmetry plane, denoted μ_s, transform as A' and are symmetric with respect to reflection whereas the component perpendicular to the symmetry plane, denoted μ_a, is antisymmetric (A''). Hence $\langle E_k|\mu|E_g\rangle = \langle E_k|\mu_a + \mu_s|E_g\rangle$, and hence c_k, is nonzero for transitions between vibronic states of the same symmetry due to the μ_s component, and nonzero for transitions between vibronic states of different symmetry due to the μ_a component. The latter is common in IR spectroscopy. In electronic spectroscopy such transitions result from the non Franck-Condon Herzberg-Teller intensity borrowing[25] mechanism. Thus, the excitation pulse can create an $|E_1\rangle, |E_2\rangle$ superposition consisting of two states of different reflection symmetry and hence a state which no longer displays the reflection symmetry of the Hamiltonian.

Cumulative matrix elements $\mu_{i,k}^{(q)}$: In contrast to the bound states, the continuum states of interest $|E, n, q^-\rangle$ are neither symmetric nor antisymmetric. Rather, $\sigma|E, n, q=1^-\rangle = |E, n, q=2^-\rangle$ and vice-versa. Such a choice is possible because of the exact degeneracies which exist in the continuum. To examine $\mu_{i,k}^{(q)}$ we introduce symmetric and antisymmetric continuum eigenfunctions of σ, $|\psi^s\rangle = (|E, n, q=1^-\rangle + |E, n, q=2^-\rangle)/2$ and $\psi^a = (|E, n, q=1^-\rangle - |E, n, q=2^-\rangle)/2$. Assuming $|E_1\rangle$ is symmetric and $|E_2\rangle$ antisymmetric, rewriting $|E, n, q^-\rangle$ as a linear combination of $|\psi^a\rangle$ and $|\psi^s\rangle$, and adopting the notation $A_{s2} = \langle\psi^s|\mu_a|E_2\rangle$, $S_{a1} = \langle\psi^a|\mu_s|E_1\rangle$, etc. we have, after elimination of null matrix elements,

$$\begin{aligned}
\mu_{11}^{(q)} &= \sum{}' \left[|S_{s1}|^2 + |A_{a1}|^2 \pm A_{a1}S_{s1}^* \pm A_{a1}^*S_{s1}\right] \\
\mu_{22}^{(q)} &= \sum{}' \left[|A_{s2}|^2 + |S_{a2}|^2 \pm A_{s2}S_{a2}^* \pm A_{s2}^*S_{a2}\right] \\
\mu_{12}^{(q)} &= \sum{}' \left[S_{s1}A_{s2}^* + A_{a1}S_{a2}^* \pm S_{s1}S_{a2}^* \pm A_{a1}A_{s2}^*\right]
\end{aligned} \tag{14}$$

where the plus sign applies for $q=1$ and the minus sign for $q=2$. Equation (14) displays two noteworthy features:

(1) $\mu_{kk}^{(1)} \neq \mu_{kk}^{(2)}$. That is, the system displays ***natural symmetry breaking*** in photodissociation from state $|E_k\rangle$, with right and left handed product probabilities differing by

$2\sum' \text{Re}(S_{s1}^* A_{a1})$ for excitation from $| E_1 \rangle$ and $2\sum' \text{Re}(A_{s2} S_{a2}^*)$ for excitation from $| E_2 \rangle$. Note that these symmetry breaking terms may be relatively small since they rely upon non Franck-Condon contributions. However:

(2) $\mu_{12}^{(1)} \neq \mu_{12}^{(2)}$. Thus laser controlled symmetry breaking, which depends upon $\mu_{12}^{(q)}$ in accordance with Eq. (11), is therefore possible, allowing enhancement of the enantiomer ratio for the m_j polarized product.

To demonstrate the extent of control we considered a model of the enantiomer selectivity, i.e. HOH photodissociation in three dimensions, where the two hydrogens are assumed distinguishable. The computation is in accord with state-of-the art formalism and computational machinery and details are provided elsewhere[24]. Typical results showed that it was possible to vary the product ratio, in this completely symmetric case, over a range of 61% to 39%. As expected the control disappears if the products are not m_j selected. Specifically, the natural symmetry breaking between right and left handed products seen with fixed m_j is lost upon m_j summation, both channels $q = 1$ and $q = 2$ having equal photodissociation probabilities. In addition, control over the enantiomer ratio is lost since the interference terms no longer distinguish the $q = 1$ and $q = 2$ channels.

5. The Status of Coherent Control

We have provided a capsule summary of the current status of coherent control of chemical reactions and molecular processes. Three essential points should be emphasized:

a) theoretical studies show that control over molecular processes, via simple quantum interference, is possible;

b) computational results show that such control can be extensive, with achievable variations in yield close to the full possible range of 0% to 100%;

c) experimental studies have shown that the proposed coherent control scenarios are feasible in the laboratory.

In summary, the age of mode selective experiments, based upon the principle of coherent control, is upon us.

Acknowledgements

We acknowledge support from the U.S. Office of Naval Research under contract numbers N00014-87-J-1204 and N00014-90-J-1014. Work carried out in conjunction with R.J. Gordon was supported by NSF PHY-8908161.

References

1. P. Brumer and M. Shapiro, "Control of Unimolecular Reactions Using Coherent Light", Chem. Phys. Lett. **126**:541 (1986).

2. P. Brumer and M. Shapiro, "Coherent Radiative Control of Unimolecular Reactions: Three Dimensional Results", Faraday Disc. Chem. Soc. **82**:177 (1987).

3. M. Shapiro and P. Brumer, "Laser Control of Product Quantum State Populations in Unimolecular Reactions", J. Chem. Phys. **84**:4103 (1986).

4. C. Asaro, P. Brumer and M. Shapiro, "Polarization Control of Branching Ratios in Photodissociation", Phys. Rev. Lett. **60**:1634 (1988).

5. M. Shapiro, J.W. Hepburn and P. Brumer, "Simplified Laser Control of Unimolecular Reactions: Simultaneous (ω_1, ω_3) Excitation", Chem. Phys. Lett. **149**:451 (1988).

6. G. Kurizki, M. Shapiro and P. Brumer, "Phase Coherent Control of Photocurrent Directionality in Semiconductors", Phys. Rev. B**39**:3435 (1989).

7. M. Shapiro and P. Brumer, "Laser Control of Unimolecular Decay Yields in the Presence of Collisions", J. Chem. Phys. **90**:6179 (1989).

8. T. Seideman, M. Shapiro and P.Brumer, "Coherent Radiative Control of Unimolecular Reactions: Selective Bond Breaking with Picosecond Pulses", J. Chem. Phys. **90**:7132 (1989).

9. For an introductory discussion see P. Brumer and M. Shapiro, "Coherence Chemistry: Controlling Chemical Reactions With Lasers", Accounts Chem. Res. **22**:407 (1989).

10. P. Brumer and M. Shapiro, "One Photon Mode Selective Control of Reactions by Rapid or Shaped Laser Pulses: An Emperor Without Clothes?", Chem. Phys. **139**:221 (1989).

11. I. Levy, M. Shapiro and P. Brumer, "Two-Pulse Coherent Control of Electronic States in the Photodissociation of IBr: Theory and Proposed Experiment", J. Chem. Phys. **93**:2493 (1990).

12. M. Shapiro and P. Brumer (to be published).

13. C.K. Chan, P. Brumer and M. Shapiro, "Coherent Radiative Control of IBr Photodissociation via Simultaneous (ω_1, ω_3) Excitation", J. Chem. Phys. **94**: 2688 (1991).

14. J. Krause, M. Shapiro and P. Brumer, "Coherent Control of Bimolecular Reactions", J. Chem. Phys. **92**:1126 (1990).

15. M. Shapiro and P. Brumer (to be published).

16. P. Brumer, X-P. Jiang and M. Shapiro (to be published).

17. See, e.g., S.H. Tersegni, P. Gaspard and S.A. Rice, " On Using Shaped Laser Pulses to Control the Selectivity of Product Formation in a Chemical Reaction", J. Chem. Phys. **93**:1670 (1990) and references therein.

18. See, e.g., S. Shi and H. Rabitz, "Optimal Control of Selective Vibrational Excitation of Harmonic Molecules: Analytic Solution and Restricted Forms for the Optimal Field", J. Chem. Phys. **92**:2927 (1990) and references therein.

19. S.M. Park, S-P. Lu and R.J. Gordon, "Coherent Laser Control of the Resonant Enhanced Multiphoton Ionization of HCl", J. Chem. Phys. (in press).

20. C. Chen, Y-Y. Yin and D.S. Elliott, "Interference Betweeen Optical Transitions", Phys. Rev. Lett. **64**:507 (1990).

21. As another interesting case consider the use of two components of laser polarization to excite a bound state $|E_i\rangle$ to the continuum[4]. These two components will interfere constructively and destructively and allow control over product ratios in the differential cross section. However, the interference contribution to the *total* cross section vanishes since these two routes reach different values of M_J.

22. X-P. Jiang, P. Brumer, and M. Shapiro (to be published).

23. An example, although requiring a slight extension of the BAB' notation, is the Norrish type II reaction: $D(CH_2)_3CO(CH_2)_3D'$ dissociating to $DCHCH_2 + D'(CH_2)_3COCH_3$ and $D'CHCH_2 + D(CH_2)_3COCH_3$ where D and D' are enantiomers.

24. M. Shapiro and P. Brumer, "Controlled Photon Induced Symmetry Breaking: Chiral Molecular Products from Achiral Precursors", J. Chem. Phys. (in press).

25. J.M. Hollas, "High Resolution Spectroscopy", (Butterworths, London, 1982).

COHERENT PHASE CONTROL OF THE RESONANCE ENHANCED MULTIPHOTON IONIZATION OF HCl AND CO

Shao-Ping Lu, Seung Min Park, Yongjin Xie, and Robert J. Gordon

Department of Chemistry (m/c 111), University of Illinois at Chicago, Chicago IL 60076

I. INTRODUCTION

A long-standing goal of photochemistry is the use of radiation to control the distribution of products in chemical reactions. For collisionless processes, the products are determined entirely by the wave function which results from absorption of a photon. The transition dipole matrix elements connecting the ground and excited states determine the subsequent evolution of the excited parent molecule. In conventional photochemistry the product distribution can be controlled only "passively" by varying the wavelength of the exciting light. When more than one bond can be broken, it is usually the weakest one that ruptures, and it is only by serendipity that one can find a wavelength that causes a stronger bond to break. For each chemical pathway the internal states of the products are similarly determined by the transition dipole matrix elements, and one has only limited freedom in altering the product populations.

The use of lasers has introduced the possibility of more active control over the reaction dynamics. The properties of laser radiation, including monochromaticity, high intensity, short pulse duration, and coherence, offer many possible strategies for achieving this goal. One such method, first proposed by Brumer and Shapiro[1] (BS), is based on the principle of quantum mechanical interference between competing paths. Suppose that two different excitation paths lead to an excited intermediate state. Let P_1 and P_2 be the reaction probabilities for the two paths. Then the total reaction probability is given by

$$P = P_1 + P_2 + P_{12}, \qquad (1)$$

where P_{12} is a cross term. The key point in the BS method is that P_{12} is experimentally controllable. Pathways 1 and 2 are assumed to be induced by light sources which have a coherent and adjustable phase difference, $\Delta\theta$. The interference term P_{12} is proportional to $\cos\Delta\theta$. By experimentally varying $\Delta\theta$ it is therefore possible to

Coherence Phenomena in Atoms and Molecules in Laser Fields
Edited by A.D. Bandrauk and S.C. Wallace, Plenum Press, New York, 1992

modulate the total transition probability, P. We refer to this variation of the total reaction probability as "population control."

Suppose now that there are two (or more) possible sets of products for each pathway. The product branching ratio can also be controlled if the two excitation paths have different product distributions. If we label the product channels by L, then the probability of exiting in each channel is given by

$$P^L = P_1^L + P_2^L + P_{12}^L, \quad (2)$$

where P_i^L is the probability of obtaining product distribution L by optical path i. Control over the distribution is possible so long as the optimal phase difference $\Delta\theta$ for maximizing the product yield is different for the two reaction paths.

There are many ways of selecting the excitation paths. The method we have used in this study, first proposed by Shapiro, Hepburn, and Brumer[2] (SHB), consists of excitation by three photons of frequency ω_1 along the first path, and excitation by one photon of frequency $\omega_3=3\omega_1$ along the second path. The two optical paths are shown schematically in Figure 1a, where the two reaction channels are labelled A and B. Elliott and coworkers[3] previously used this scheme to control the population of Hg^+ ions in a bulb.

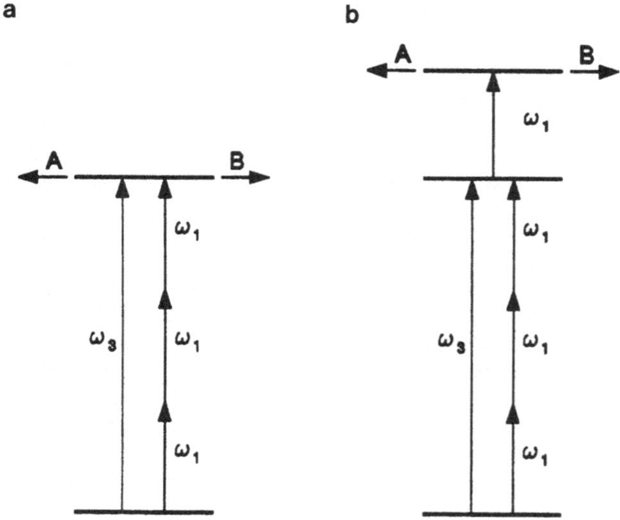

Figure 1. Schematic diagram of the SHB control mechanism. In case (a) one photon of frequency ω_3 or three photons of frequency ω_1 excite an upper level, which reacts to produce products A or B. The product branching ratio can be controlled by varying the phase difference between ω_1 and ω_3. In case (b) an intermediate level is excited, and the branching ratio is not controlled.

In the present study we have used the SHB method to control the transition rate between bound states of diatomic molecules in a molecular beam.[4] An additional photon was then used to populate a higher level which can either autoionize or dissociate. This excitation scheme is depicted in Figure 1b. Since only the intermediate bound state is actively controlled in this scheme, we expect to achieve only population and not channel control. In this paper we have examined a variety of transitions to demonstrate that population control in bound-to-bound transitions of a diatomic molecule is a general phenomenon.

2. EXPERIMENTAL

The experiment was performed in a pulsed molecular beam machine equipped with a time-of-flight mass spectrometer. Two coaxial laser beams intersected the molecular beam between the repeller and extractor electrodes of the spectrometer. The wavelengths of the laser beams were typically 336 nm (UV) and 112 nm (VUV). The VUV beam was generated by third harmonic generation (THG) in a Kr gas cell.[5] Ionic products were collected with a set of Wiley-McLaren electrodes,[6] accelerated down a one meter long time-of-flight tube, and detected with a multichannel plate. Details of the laser and molecular beams are presented elsewhere.[4,7]

A key part of the experiment is control of the phase difference between the two beams. This was accomplished by passing the UV and VUV beams through a gas which has a different index of refraction at ω_1 and ω_3. The phase-tuning chamber also contains two concave mirrors which were used to focus the laser beams into the reaction chamber. The phase difference for three photons of frequency ω_1 and one photon of ω_3 is given by

$$\Delta\theta = 3\theta_1 - \theta_3 = \ell\omega_3(n_3 - n_1)/c, \qquad (3)$$

where ℓ is the path length, n_i is the index of refraction at frequency ω_i, and c is the speed of light. The difference n_3-n_1 is proportional to the pressure of the tuning gas. The phase difference was continuously varied during an experiment by slowly filling the phase-tuning chamber to several Torr.

3. RESULTS

Although in principle phase control is possible for any bound-to-bound transition, there are practical considerations which limit the choice of molecules and transitions which can be used. These stem from the requirement that the rates of the two excitation paths be comparable. For one- vs. three-photon excitation, the tradeoff is between a one-photon excitation which typically has a large absorption cross section but a weak laser intensity (VUV THG typically has a quantum efficiency of only 10^{-6}), and a three photon-excitation which has a much smaller excitation cross section but a much higher laser intensity. In favorable cases it is possible to balance the two rates by varying the pressure of the gas in the THG cell. In practice this may not always be possible. Frequently it happens that the three-photon transition is weak or even absent, either because the matrix elements are very small or because a competing process such as predissociation intervenes at the two-photon level. In other cases the one-photon VUV absorption is too weak. Improving the THG conversion efficiency by increasing the UV intensity is not helpful since this would in turn increase the three-photon rate.

So far we have controlled the resonance-enhanced multiphoton ionization (REMPI) of three Rydberg states of HCl and one of CO. In each case the VUV transition lies between 110 and 115 nm. For HCl we studied the $j^3\Sigma^-(0^+)$, $H^1\Sigma^+$, and $m^3\Pi$ states. All three have an HCl^+ $X(^2\Pi)$ core, with a $3d\pi$ Rydberg electron in the first two cases and a $5s\sigma$ electron in the third case. For CO we studied the $B^1\Sigma^+$ state, which has a CO^+ $X^2\Sigma^+$ core and a $3s\sigma$ Rydberg electron.

Our results are illustrated for the $j^3\Sigma^-(0^+)$ state of HCl. The REMPI spectrum for the P and R branches of the 0-0 band with both the ω_1 and ω_3 beams on is shown in Figure 2.

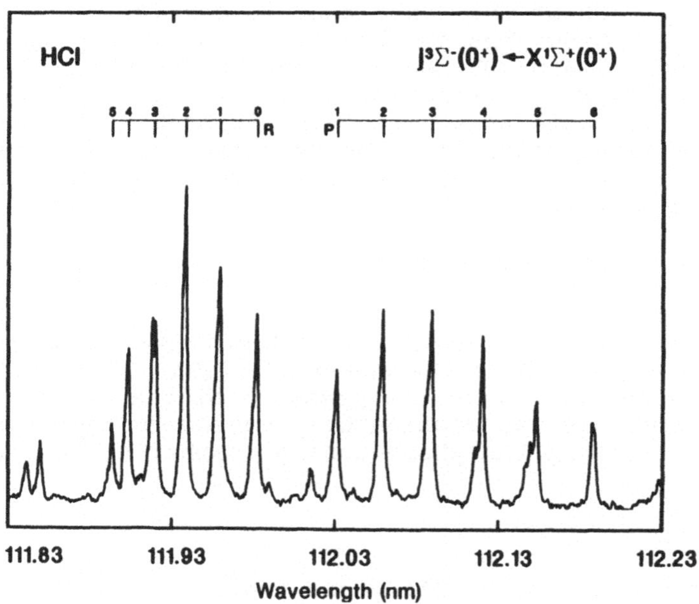

Figure 2. REMPI spectrum of the $j^3\Sigma^-(0^+)$ state of HCl. (Reproduced from ref. 4 by permission of the American Physical Society).

Control of the ionization rate was obtained for all of the assigned lines. For each transition, the Kr pressure in the tripling cell was adjusted so that the 1+1 and 3+1 MPI rates were nearly equal. The tuning gas pressure was then continuously scanned. The results for the R(2) transition are shown in Figure 3 using Ar and H_2 as tuning gases.

In Figure 4 the modulation curves are shown for a number of rotational transitions of the j state. In every case the

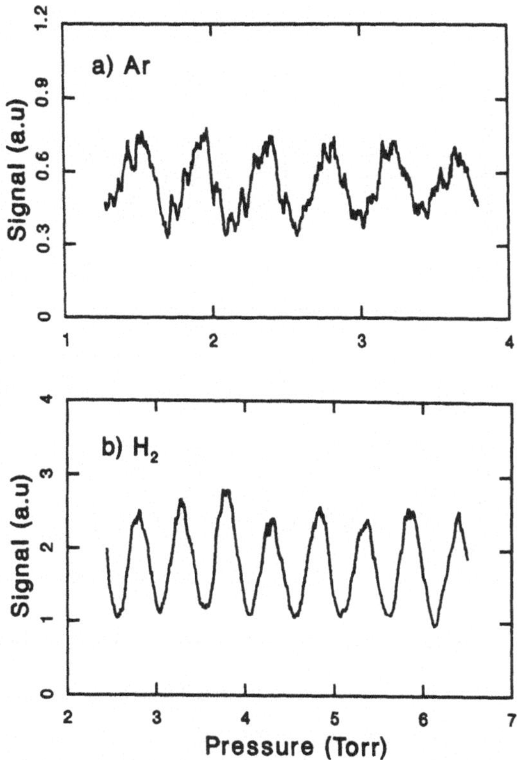

Figure 3. Ionization signal for the R(2) transition as a function of pressure in the tuning cell, using either (a) Ar or (b) H_2 to control the relative phases of ω_1 and ω_3. (Reproduced from ref. 4 by permission of the American Physical Society.)

modulation depth was close to 40%. Similar behavior was observed for the H and m states of HCl. For the CO(B) state, however, the modulation depth was only 20%.

For the $j^3\Sigma^-(0^+)$ state of HCl we also studied the competition between ionization and dissociation. A plausible

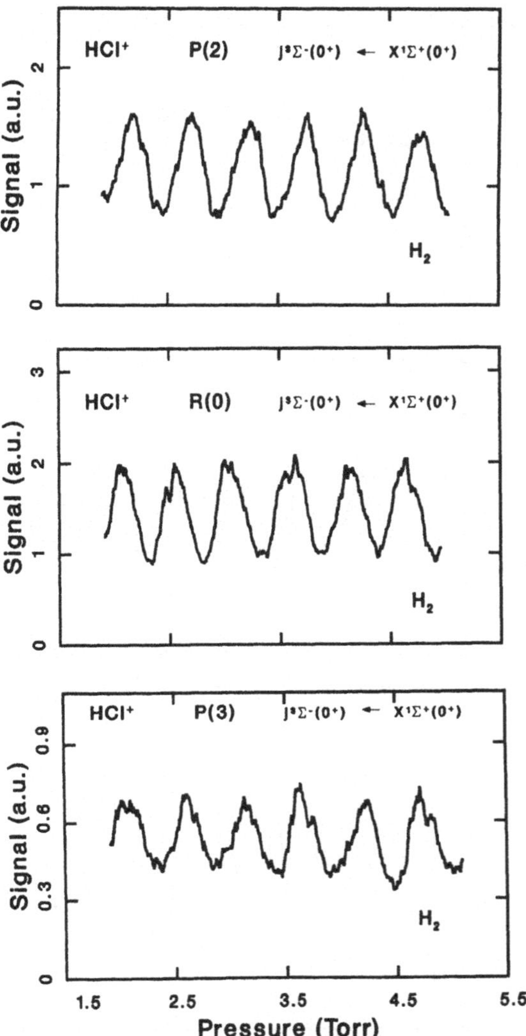

Figure 4. Coherent phase control of the ionization of the $j^3\Sigma^-(0^+)$ state of HCl for different rotational transitions. The phase tuning gas in this case was H_2.

mechanism which explains our data is

$$HCl(X^1\Sigma^+) + \omega_3 \rightarrow HCl(j^3\Sigma^-(0^+)) \quad (4a)$$

and

$$HCl(X^1\Sigma^+) + 3\omega_1 \rightarrow HCl(j^3\Sigma^-(0^+)), \quad (4b)$$

followed by

$$HCl(j^3\Sigma^-(0^+)) + \omega_1 \rightarrow HCl^{**}. \quad (5)$$

The doubly excited molecule HCl^{**} can then either autoionize,[8]

$$HCl^{**} \rightarrow HCl^+ + e^-, \quad (6)$$

or dissociate,

$$HCl^{**} \rightarrow H^*(n=2) + Cl. \quad (7)$$

The excited hydrogen atom produced in the last step can absorb an additional UV photon to produce a proton,

$$H^* + \omega_1 \rightarrow H^+. \quad (8)$$

An alternative mechanism for producing H^+ by photodissociation of HCl^+ by ω_1 can be ruled out on energetic grounds.[7]

Experimentally we observed an H^+ to HCl^+ ion ratio of 10%, with the signals for the two masses showing the same modulation depth and phase dependence. This result was anticipated since coherent control occurs only in the intermediate excitation step (4). Population rather than channel control is all that can be expected in this situation.

4. DISCUSSION

Modulation of the ion intensities is caused by population control of the resonant intermediate state. The rate of excitation of this state is proportional to the square of the matrix element of the transition operator, summed over initial and final magnetic quantum numbers; i.e.,

$$\Gamma(J'\Omega',J''\Omega'') = \Sigma_{M',M''}|\langle J'M'\Omega'u'v'|\mu\mathscr{E}_3+T\mathscr{E}_1^3|J''M''\Omega''u''v''\rangle|^2. \quad (9)$$

In Eq. (9), μ is the component of the transition dipole operator along the polarization direction of the laser beam, and is derived by a rotational transformation from the molecular frame,[9]

$$\mu = \Sigma_q D^1_{0q}{}^* \mu_q. \quad (10)$$

Also in Eq. (9), T is the three-photon operator, and $\mathscr{E}_3$ and $\mathscr{E}_1$ are the VUV and UV electric field amplitudes, which are factored into spatial and temporal parts

$$\mathscr{E}_s(r,z,t) = E_s(r,z)\ \exp\{i(\omega_s t+\theta_s)\}. \tag{11}$$

The initial and final state vectors are designated by double and single primes, with the quantum numbers J, M, Ω, u, and v representing the total angular momentum, the projection of J along the laboratory polarization axis and the molecular axis, and the electronic and vibrational states, respectively.

Eq. (9) is expanded to give

$$\begin{aligned}\Gamma(J'\Omega',J''\Omega'') = {} & E_3^2\ \Sigma_{M',M''}|\langle J'M'\Omega'u'v'|\mu|J''M''\Omega''u''v''\rangle|^2 \\ & +E_1^6\ \Sigma_{M',M''}|\langle J'M'\Omega'u'v'|T|J''M''\Omega''u''v''\rangle|^2 \\ & +E_3E_1^3\ \Sigma_{M',M''}[\langle J'M'\Omega'u'v'|\mu|J''M''\Omega''u''v''\rangle \\ & \cdot\langle J''M''\Omega''u''v''|T|J'M'\Omega'u'v'\rangle\ \exp\{i(\theta_3-3\theta_1)\} \\ & +\langle J'M'\Omega'u'v'|T|J''M''\Omega''u''v''\rangle \\ & \cdot\langle J''M''\Omega''u''v''|\mu|J'M'\Omega'u'v'\rangle\ \exp\{-i(\theta_3-3\theta_1)\}].\end{aligned} \tag{12}$$

Evaluating the sum in the first term of Eq. (12) gives the familiar result for the single-photon line strength,

$$\Gamma_3 = E_3^2 S_3(J'\Omega',J''\Omega''), \tag{13}$$

where

$$S_3 = \mu_3^2 F_3^2\{(2J'+1)[\langle J'\Omega'1\ -\Delta\Omega|J''\Omega''\rangle]^2\}. \tag{14}$$

In Eq. (14) μ_3^2 is the square of the transition dipole moment, F_3 is the Franck-Condon factor, which may in general be complex,

$$F_3 = |\langle v'|v''\rangle|\ \exp(i\delta_3), \tag{15}$$

and the expression in curly brackets is the usual Hönl-London factor.

The second term in Eq. (12) is the three-photon line strength. Writing the T operator explicitly gives[10]

$$\Gamma_1 = E_1^6 S_1(J'\Omega',J''\Omega'') \tag{16}$$

and

$$\begin{aligned}S_1 = {} & \Sigma_{M',M''}|\Sigma_{i,j}\langle J'M'\Omega'u'v'|\mu|J_iM_i\Omega_iu_iv_i\rangle \\ & \cdot\langle J_iM_i\Omega_iu_iv_i|\mu|J_jM_j\Omega_ju_jv_j\rangle \\ & \cdot\langle J_jM_j\Omega_ju_jv_j|\mu|J''M''\Omega''u''v''\rangle \\ & \cdot[(E_i-2\hbar\omega)(E_j-\hbar\omega)]^{-1}|^2.\end{aligned} \tag{17}$$

In Eq. (17) subscripts i and j refer to the intermediate electronic states having energies E_i and E_j. There are in general 19 types of terms in this sum. One of these corresponds to the case where all four electronic states are distinct, producing a product of three transition dipole matrix elements. Six more correspond to the case of three distinct electronic states, with the possibilities of $u_j = u'$, $u_j = u_i$, and $u_j = u''$. For the transitions between the identical electronic states either v and J both change or only J changes. The remaining 12 types of terms have only two electronic states, with the options of $u_i = u_j = u'$, $u_i = u_j = u''$, and $\{u_i = u', u_j = u''\}$. Following the example of Halpern et al.,[10] we treat only the case of four distinct electronic states, without loss of generality.

Using Eq. (10) to evaluate the transition dipole matrix elements, we obtain

$$S_1 = (2J'+1)^{1/2}(2J'+1)^{1/2}|\Sigma_{p,q,r}\mu_p\mu_q'\mu_r'' \ \Sigma_{i,j}F_{ij}[(E_i-2\hbar\omega)(E_j-\hbar\omega)]^{-1} \cdot\Sigma_{M',M''}[A_{pqr}'\langle J'\Omega'1\ -\Delta\Omega|J''\Omega''\rangle\langle J'M'1\ 0|J''M''\rangle + B_{pqr}'\langle J'\Omega'3\ -\Delta\Omega|J''\Omega''\rangle\langle J'M'3\ 0|J''M''\rangle]|^2, \quad (18)$$

where A_{pqr}' and B_{pqr}' are products of Clebsch-Gordan coefficients and

$$F_{ij} = |\langle v'|v_i\rangle\langle v_i|v_j\rangle\langle v_j|v''\rangle| \ \exp(i\delta_{ij}) \quad (19)$$

is a Franck-Condon factor. A crucial assumption in writing this expression is that F_{ij} is independent of rotational quantum numbers. Taking advantage of the orthonormality properties of the Clebsch-Gordan coefficients and lumping the remaining factors into the A and B coefficients, we obtain the very simple result

$$S_1 = (2J'+1)|\Sigma_{p,q,r}A_{pqr}\mu_p\mu_q'\mu_r''|^2\ [\langle J'\Omega'1\ -\Delta\Omega|J''\Omega''\rangle]^2 + (2J'+1)|\Sigma_{p,q,r}B_{pqr}\mu_p\mu_q'\mu_r''|^2\ [\langle J'\Omega'3\ -\Delta\Omega|J''\Omega''\rangle]^2. \quad (20)$$

The first term in Eq. (20) corresponds to a transition with $\Delta J=1$ and the second to $\Delta J=3$. Since the latter cannot interfere with single-photon $\Delta J=1$ transitions, the occurrence of $\Delta J=3$ transitions will reduce the control over the final products.

In the present study, where only discrete levels of the resonant state are excited, the second term in Eq. (20) vanishes. Since S_1 and S_3 are then proportional to the square of the same coupling coefficient, we find that the cross term in Eq. (12) is just the geometric mean of the first two terms, apart from the crucial phase factor. That is, we find that the total transition rate is proportional to the quantity

$$\Gamma(J'\Omega',J''\Omega'') = E_3^2S_3 + E_1^6S_1 + 2E_3E_1^3(S_1S_3)^{1/2}\cos(\Delta\theta+\delta_{13}), \quad (21)$$

where

$$\Delta\theta = \theta_3 - 3\theta_1 \quad (22)$$

and

$$\delta_{13} = \delta_3 - \delta_1. \quad (23)$$

In Eq. (23) δ_1 is the total phase of the three-photon Franck-Condon factor. Since S_1 and S_3 have the same J dependence, the modulation depth is independent of rotational number. The interference period is also independent of J since the difference in index of refraction in Eq. (3) changes negligibly in the frequency interval of the rotational bands.

Defining the experimentally observable ratio

$$R = \Gamma_3/\Gamma_1, \quad (24)$$

we find that the fractional modulation f_M is given by

$$f_M = 2R^{1/2}/(1+R). \quad (25)$$

From Eq. (25) it follows that 100% modulation occurs for R=1. This condition is satisfied, however, only at isolated points where the intensities of the UV and VUV beams have the proper ratio. Since the field- of-view of our detector is a 0.75 mm radius circle,[7] it is necessary to spatially average the numerator and denominator of Eq. (25) before taking the ratio.

Assuming that the laser beams have Gaussian cross sections, the amplitude of the electric field is given by

$$E_s(r,z) = E_{s0}[1+(z/z_0)^2]^{-1/2} \exp\{i \tan^{-1}(z/z_0)\} \exp(-r^2/w_s^2), \quad (26)$$

where z_0 is one half the confocal parameter and w_s is the Gaussian radius, with $w_1^2=3w_3^2$. Although we have not measured these parameters, we can be confident that w_s is much smaller than 0.75 mm, while z_0 is comparable to or larger than this length.

In averaging over r and z we make the reasonable assumption that the ionization and dissociation rates of the controlled level are saturated, so that the spatial variation of the additional UV photons should not be included. Performing first the integral over r we find that the difference in spot sizes of the two beams has no effect on the modulation. That is, the condition that $w_1^2=3w_3^2$ is exactly offset by the fact that $\omega_3=3\omega_1$.

The axial average is another matter. In the limit that z_0>>0.75 mm averaging over z has no effect on f_M. In the other limit (achieved in the experiment of Chen et al.[3a]), averaging over all z gives

$$f_M = 4R^{1/2}/(3 + 8R). \quad (27)$$

Eq. (27) gives a maximum modulation of $6^{-1/2}$=0.41 at R=3/8. Experimentally we observe a maximum modulation much closer to R=1, which is a further reason why axial averaging is not very important in our experiment. At most spatial averaging could reduce f_M to about 60%.

Our observation of f_M=40% for HCl is not yet completely understood. Chromatic aberration of the entrance and exit windows of the tuning cell axially separates the focal points of the UV and VUV beams, but we estimate that this effect is much smaller than z_0. A more likely cause of the loss of modulation is a radial separation of the focal points. This could be due to the imperfect flatness of the windows, or to slight deviation from normal incidence of the beams. Future experiments are planned to test these possibilities. In the case of CO the still lower value of f_M is due to the much smaller three-photon line strength. In order to get an acceptable signal-to-noise ratio it was necessary to operate with an intensity ratio of $R>1$, which lowers f_M from its maximum value.

5. CONCLUSIONS

We have shown that population control in bound-to-bound transitions is a general phenomenon. This effect has been demonstrated for the following cases:

- Different molecules (HCl and CO)
- Different electronic configurations (j, H and m states of HCl)
- Different rotational transitions (P and R branches)
- Competing processes (ionization and dissociation)

In future experiments we hope to apply the phase-control technique to bound-to-continuum transitions. This is a crucial step for eventually being able to control product distributions.

ACKNOWLEDGEMENTS

We wish to thank Profs. Paul Brumer and Moshe Shapiro for many fruitful conversations and for originally suggesting the one- vs. three-photon excitation scheme. Support by the National Science Foundation under Grant no. PHY-8908161 is gratefully acknowledged.

REFERENCES

1. P. Brumer and M. Shapiro, Accts. Chem. Res. 22: 407 (1989).

2. M. Shapiro, J.W. Hepburn, and P. Brumer, Chem. Phys. Lett. 149: 451 (1988).

3. (a) C. Chen, Y.-Y. Yin, and D.S. Elliott, Phys. Rev. Lett. 64: 507 (1990); (b) C. Chen and D.S. Elliott, Phys. Rev. Lett. 65: 1737 (1990).

4. S.M. Park, S.-P. Lu, and R.J. Gordon, J. Chem. Phys. 94: 8622 (1991).

5. G.C. Bjorklund, IEEE J. Quant. Elec. 6: 287 (1975).

6. W.C. Wiley and I.H. McLaren, Rev. Sci. Instrum. 26: 1150 (1955).

7. S.-P. Lu, S. Min Park, Y. Xie, and R.J. Gordon, (submitted).

8. H. Lefebvre-Brion and F. Keller, J. Chem. Phys. 90: 7176 (1989).

9. R.N. Zare, "Angular Momentum," Wiley, New York, 1988.

10. J.B. Halpern, H. Zacharias, and R. Wallenstein, J. Molec. Spectroscopy 79: 1 (1980).

OPTIMAL CONTROL OF MOLECULAR MOTION

Herschel Rabitz

Chemistry Department
Princeton University
Princeton, NJ 08544

ABSTRACT

This paper presents an overview of recent developments concerning the introduction of optimal control techniques into the molecular domain. The goal of this research is to identify the degree to which molecular dynamics may be controlled by external optical fields and provide a systematic means for designing optical fields for this purpose. Preliminary illustrations of molecular optimal control theory have been considered involving rotational, vibrational, and electronic degrees of freedom. With the ultimate objective being laboratory implementation of the designed fields, the robustness of the designs is a critically important issue. These various topics are reviewed in this paper.

INTRODUCTION

The objective of breaking chemical bonds or otherwise manipulating chemical structure is at the heart of chemistry. In the earliest days of laser development, three decades ago, it was suggested that precisely controlled optical fields might be amenable for producing a pair of "molecular scissors" capable of selectively cutting or manipulating molecular structure. This dream of using photons as a source of energy for precise molecular manipulation has remained a challenging and frustrating goal in the chemical physics community. Many experimental and theoretical studies have

Coherence Phenomena in Atoms and Molecules in Laser Fields
Edited by A.D. Bandrauk and S.C. Wallace, Plenum Press, New York, 1992

been pursued over the years with little promise that success might be achieved. A recognized chief difficulty is the typically strong coupling amongst the molecular degrees of freedom and the ultra-short time scales involved. These difficulties still remain, but in the last few years the true nature of the problem has become clearer and a rigorous foundation for approaching it has been put forth[1,2] as summarized in this paper.

Generally the manipulation of details on the molecular scale, regardless of whether optical means are employed, must recognize that motion on this scale is governed by the laws of quantum mechanics. Classical mechanics may still have a useful role in this regime, but its approximate nature must be accepted. A critical issue is that control over molecular motion translates to control over coherence phenomena whether viewed as quantum mechanical waves or the motion associated with classical coupled oscillators. The problem of precise optical manipulation of molecules reduces to a task of design and control at the molecular scale. Putting aside the molecular scales involved, problems of this type have been studied for many years in the engineering disciplines, particularly drawing on the tools of optimal control theory.[3] Indeed, it is a sine qua non in engineering to employ systematic optimal design tools to any problem where the solution is not obvious from intuition alone. In contrast, scientific intuition has been the primary mode of guidance for choosing molecular scale manipulative optical fields. The long previous history of these laboratory studies amply provides evidence for the complexity of the task and the serious need for the introduction of systematic design techniques.

Although the field of molecular optimal control theory is in its infancy, it is clear that these tools present a rigorous foundation for lifting the subject beyond the Edisonian level of "cut and try" laboratory studies. Over the years many dreams have been put forth in the area of molecular manipulation with optical fields, and it may play out that many of these goals are ill-founded and will remain as dreams. However, a central component in this effort is the ability to shape and control optical pulses in the laboratory. This capability is improving rapidly and the current laboratory limitations in terms of optical pulse intensity, temporal feature size, spectral content, and pulse length are expected to evolve.[4,5] Although the present paper covers theoretical aspects of this research, the collaboration of experiment and theory is critical. The field of optically manipulated molecular structure and dynamics may provide a unique example of where both theory and experiment alone have limited utility but their cooperation could open up new capabilities.

In addressing questions of optical field manipulation of molecular motion, a host of questions arise including the desirability of using strong fields versus weak fields, how to treat Hamiltonian and optical coupling coefficient uncertainties, how to best utilize available laboratory techniques, etc? A thorough weighing of these issues surely will depend on the particular systems and only a beginning at answers is available. In general theoretical research has several roles in this topic. First, molecular optimal control theory needs to be employed in a practical design mode taking into account laboratory limitations to achieve practical field designs for laboratory implementation. Second, theory should address the fundamental question of whether there are any inherent limitations on the degree that molecular objectives may be reached. Thirdly, theory has the role of exploring the regime beyond currently available laboratory techniques to search for possibly interesting results or phenomena that might be achievable with the development of new optical field techniques. At this stage of development of molecular optimal control theory none of these tasks has seriously been broached, although insight has been obtained on all three. It is anticipated that the subject will evolve rapidly and some of the essential elements are outlined here.

The key step in molecular scale optimal control theory is the specification of what is called the control cost functional J[E] which depends on the time dependent optical electric field E(t). Regardless of the problem, the cost functional has three components.

$$J[E] = J_o + J_p + J_c \qquad (1)$$
$$= \text{objectives} + \text{penalties} + \text{constraints}$$

There is broad flexibility in the choice of these terms and it is the role of the molecular scale designer to wisely choose their form and relative weights. The complete mathematical analysis of this problem will not be presented here but a discussion of the physical significance of these terms will be given. First, the cost functional J[E] is constructed to be positive definite such that its minimization $\delta J/\delta E = 0$

provides an optimal solution for the sought after optical field E(t). First it must be recognized that there may be cases where no solution exists while other cases will arise involving multiple solutions (i.e., multiple minima of the cost functional). Experience to this point indicates that well posed physical problems most likely will fall into the latter category. The possibility of there being multiple good quality solutions to the design of optical fields meeting molecular objectives is a very important point allowing for broad flexibility in the choice of terms entering into eq (1).

The physical objective term J_o in eq(1) contains the explicit goals for achievement. For example, it may be desirable to attempt to steer a molecule such that its final state $|\psi(T)\rangle$ has maximum overlap with a desired product state $|p\rangle$. Thus J_o would be arranged to have a form to maximize the probability $|\langle \psi(T)|P\rangle|^2$. In general, physical objectives could include expectation values of any desired molecular operator or its classical analog including output channel fluxes if bond breaking is the objective. In a similar fashion, the penalty term J_p includes all physical processes that are deemed undesirable during the controlled evolution of the molecule. For example, if it is desirable to break some particular bond, then it is equally undesirable to break the remaining bonds. For similar or other purposes, it may be desirable to avoid access to certain molecular states or excitation in certain regions of a molecule. Penalties can also include restrictions on the nature of the optical field E(t) which might be expressed as a fluence term, frequency filtering or explicit introduction of limited forms for the field. Penalties entering J_p typically occur throughout the control interval $O \leq t \leq T$ and thus J_p is usually expressed as an integral cost to be minimized. An inherent degree of flexibility when combining the various terms to form $J_o + J_p$ is a choice of their relative weights through the coefficients ω_i i=1,2... The choice of these weights is a matter of judgement on the significance of the objective terms versus their penalty counterparts. For example, if optical field fluence is not a significant issue then its weight in the cost function of eq(1) would be chosen as relatively small. It is evident that the two terms in the sum $J_o + J_p$ may include many different components with the weights being chosen to turn on or turn off various contributions. Similarly, a detailed choice of these components can clearly control the ultimate form of the final optical field E(t) and indeed whether a viable field even exists.

There should be no surprise in this statement, recalling again that the problem is one of design. An over demanding criteria either in J_o or J_p could lead to unsatisfactory results including the prospect of there being no viable field in some instances. This is a matter that can only be assessed on a case by case basis.

Finally, the constraint term J_c is in a different category from J_o and J_p. The notion of constraints refers to demanded truisms such as the equations of motion being exactly satisfied. The equations of motion would be those of quantum mechanics or classical mechanics (or combinations of the two approaches if deemed appropriate). It is also possible to shift some objectives or penalties from J_o or J_p into the category of being exact demands in J_c. Clearly, exactly demanding that some aspect of the molecular evolution be precisely true may be a rather severe criteria. If a problem is well posed, such demands may be met although at possibly some additional cost such as in the form or nature of the optical field. In general, the constraints entering into J_c may be included through the introduction of Lagrange multipliers. Thus, by construction, the constraint term is strictly maintained as null $J_c = 0$. The constraint functions and Lagrange multipliers are determined upon minimization of the overall cost functional in eq(1).

After specification of the cost functional, the remaining task is minimization of eq(1), and a variety of means may be employed for this purpose. Conjugate gradient methods have become popular, although many other approaches may also be utilized. Much remains to be learned about functional minimization for molecular scale control problems, but successful numerical implementation has been achieved at this stage. Iteration is generally called for given the highly nonlinear nature of molecular optimal control theory, and it may be more than a curiosity that the optimal control of quantum mechanical systems described by the linear Shröedinger equation naturally leads to the need to solve a cubically nonlinear Shröedinger equation.[6] Problems described through classical mechanics also naturally lead to the need to solve nonlinear optimization equations. Molecular control problems reducible to an exact solution consist of those described by a linear forced set of equations of motion.

$$\frac{\partial z}{\partial t} = Az + \mu E(t) \tag{2}$$

This equation is exact for a coupled harmonic oscillator system where z is the dual vector of position and momentum expectation values.[7] In addition eq(2) would also arise within the framework of employing perturbation theory, and perturbation approaches may be valuable as a guide for more general problems. In the immediate future, the availability of only weak intensity fields may naturally lead to eq(2) as being the correct operative choice. These various broad theoretical issues and numerical matters will now be put aside and the following section will consider a few illustrations of the current methodology.

ILLUSTRATIONS

At this stage of development of molecular optimal control theory, numerous illustrations have been carried out with virtually all of them best being characterized as initial exercises of the control machinery. Continuing effort will be necessary to explore the capabilities of these new theoretical design tools at the molecular scale. However, simultaneously it is anticipated that realistic computations for actual laboratory studies might also be carried out keeping in mind the current state of the art of laboratory pulse shaping techniques. The illustrations below serve as a sampling of some recent work carried out at Princeton; parallel studies at other laboratories are also underway.

Optimal Control of Curve Crossing

The control over dynamics of crossing potential surfaces presents an interesting challenge. The surfaces are assumed to cross but also couple such that without control a wave packet would go through the crossing region to form some nominal distribution of final products. The goal is thus to employ control theory to steer the outgoing flux into the channel that would not normally be favored.[8] The cost functional is chosen to have a minimal field fluence and a bias away from strong DC or low frequency field structure which would be difficult to create in the laboratory. The potential curves involved are shown in figure 1. The initial state $|\psi_3(0)\rangle$ is taken as a Gaussian centered near $r \simeq 4$au on curve 3.

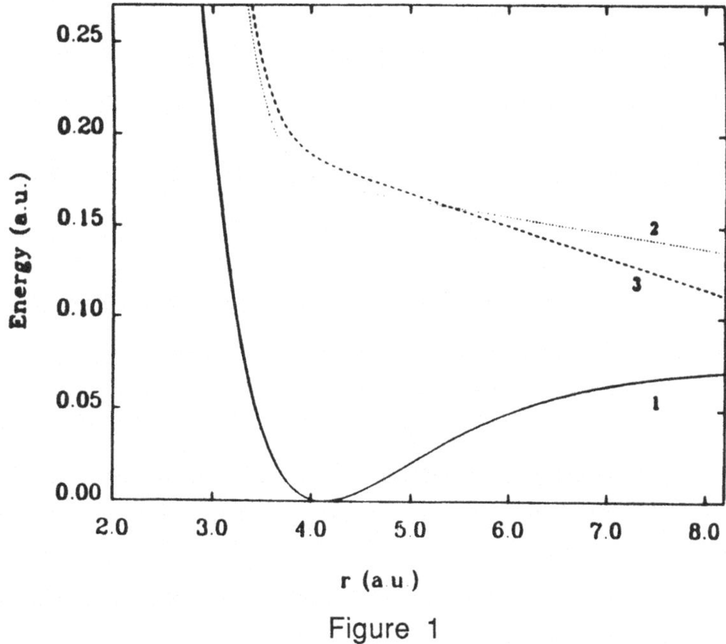

Figure 1

Electronic potential curves involved in the control of curve crossing. Curves 2 and 3 cross and couple with the initial state $|\psi_3(0)\rangle$ specified on curve 3 near $r \simeq 4$au. The optical control field acts between curves 1 and 3 except in the crossing region where additional mixing occurs. The goal is to steer the wave packet into the exit channel which would not otherwise be favored by the nominal Hamiltonian. In the present case the nominal Hamiltonian has strong coupling between curves 2 and 3 such that the initial state will naturally exit as flux j_2 on potential 2.

This state could possibly have been prepared by a transfer of amplitude from an even more highly excited electronic state. Various examples were considered but in the case shown here, the coupling between the crossing curves 2 and 3 was sufficiently strong such that without any controlled intervention, 87% of the flux j_m exited out channel m = 2 versus 13% out channel m = 3. The goal is to introduce an optimal optical field which couples curves 1 and 3 (except in the crossing region where strong mixing occurs between curves 2 and 3) and thereby predominantly steer the initial wave packet to remain on curve 3 such that we achieve the result $j_3 > j_2$ as best as possible.

The optimal field achieved (not shown here) through the control machinery was dominated by a strong carrier wave near frequency 0.15au with distinct lower amplitude side band structure apparently due to the vibrational states in potential well 1. The results of this effort are shown in figures 2 and 3. Figure 2 presents the integrated flux $\int_0^t j_m(t')dt'$ for channels $m = 2,3$. First the desired result $j_3 > j_2$ is quite well achieved with a complete reversal of the nominal uncontrolled case. Secondly, it is evident that two pulses of flux appear near times t~ 800au and 2200au. The underlying mechanism is illustrated in figure 3. The wave packet is effectively recycled by initially dumping most of it to lower surface 1 and utilizing its natural motion on that surface to recycle it once again to curve 3 and through the crossing region. The dynamics involved is quite subtle as even the portion of the wave packet remaining on curve 3 after the initial dump, successfully stays on that curve after passing through the crossing region. The important point to emphasize is that the mechanism depicted in figure 3 was not imposed in any fashion upon the control problem. Rather, only the cost functional for achieving the objective was specified and the computational machinery came up with the mechanism. More important than the example is the observation that this same generic design machinery will typically operate in the same fashion on other physical problems of more or less complexity.

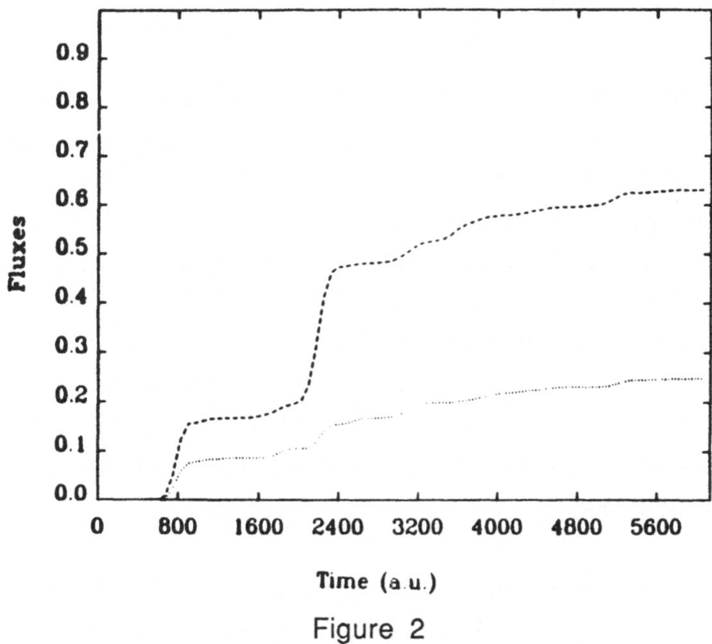

Figure 2

The integrated flux history $\int_0^t j_m(t')dt'$ is presented for control over the dynamics shown in Figure 1. The dashed course is the desired product m=3 and the dotted curve is m=2. The optimal field successfully achieves the result $j_3 > j_2$ with two dominant pulses corresponding to the mechanism in Figure 3.

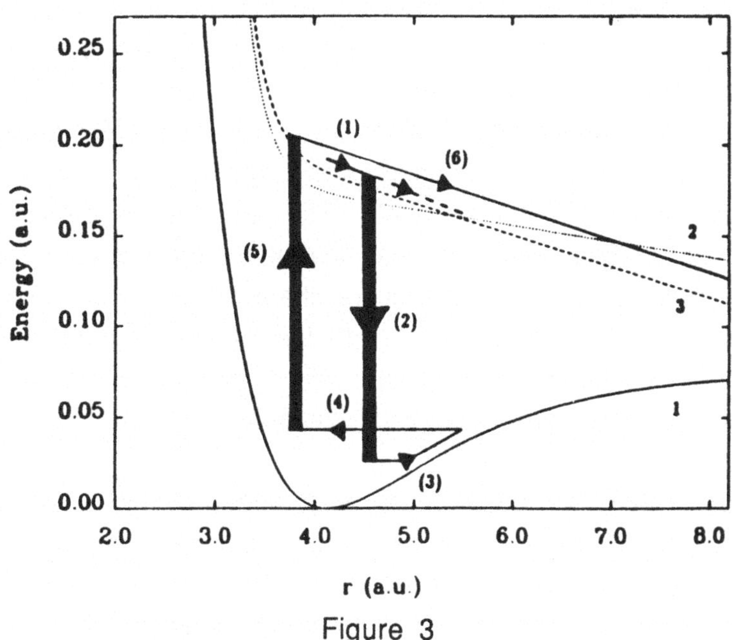

Figure 3

The control mechanism associated with the optimal design field. The latter field operates by a dump and pump mechanism to "recycle" the wave packet and thereby produce as much flux j_3 as possible exiting out of potential 3. The initial wave packet at point (1) is largely dumped (2) to the lower potential curve 1 where it carries its initial momentum forward (3) and then rebounds (4) to be pumped once again (5) to potential 3 to finally pass through the crossing region (6). The control involved is quite subtle as a detailed examination of the wave packets reveals. Even the portion of the wave packet remaining in the excited state after the initial dump still preferentially exits (indicated by the bold dashed curve after (1)) in the desired channel j_3 flux while its uncontrolled counterpart would naturally go out with dominate flux j_2. The key point is that this control mechanism was not imposed on the design process but rather fully determined on the computer by the optimal design machinery. Only the cost functional was specified as input.

Optimal Control Within Classical Molecular Dynamics

Naturally it would be desirable to always utilize quantum mechanics when treating the dynamics of molecules, but as is well known, this is not computationally feasible in general for molecules involving four or more atoms. Classical mechanics is routinely utilized in molecular dynamics studies, but in the present context, a number of special issues and questions arise. Most importantly there is the question of how good the design results will be? This question is in reference to the designs being reached using classical mechanics while real molecules in the laboratory will satisfy the laws of quantum mechanics. Therefore, will classically designed fields be accepted by real molecules to produce the desired results or will significant deviations arise? The well known prospect of finding chaotic dynamics with classical mechanics raises important issues as to whether the presence of chaos is significant and also whether it can be suppressed through the external control field. At this time, very little is known about these matters, although some interesting glimpses are becoming available.

Even within the obvious approximation of using classical mechanics the inherent molecular control mechanism still relies on delicate constructive and destructive interference processes. These latter processes are natural within quantum wave mechanics but similar interferences can arise when propagating a classical ensemble of trajectories; the wave-like collective motion character of classical particles is also well known.

An illustration of the classical optimal design machinery with the following quartic oscillator Hamiltonian was treated.[9]

$$H = 1/2\,(p_1^2 + p_2^2 + q_1^4 + q_2^4) - kq_1^2q_2^2 + q_1E(t) \qquad (3)$$

The last term proportional to the coordinate q_1 represents the control dipole interaction with field $E(t)$. The nominal Hamiltonian without the latter field is known to be fully chaotic for the coupling parameter $k>0.6$. This behavior is evident from propagating an initial ensemble of tightly bundled trajectories which rapidly disperse throughout the configuration space. The optimal control field aims to steer the initial ensemble as nearly as possible to a particular target value $<q_2(t)> = q_2^*$ regardless of the value of q_1. The average is in reference to an ensemble chosen from a Wigner distribution. Even without considering the chaotic aspects of the problem, this is a difficult task given that the control only depends on the coordinate q_1. Nevertheless, ignoring chaos and treating this problem with ordinary control machinery resulted in the objective being rather well achieved. However, the final state and intermediate trajectory variance $<(q_2(t)) - <q_2(t)>)^2>$ was quite large. Thus a variance cost was introduced into the term J_p of eq(1). This term had the effect of attempting to minimize the trajectory variance while simultaneously still attempting to achieve the target goal of $<q_2(t)> = q_2^*$. The variance was dramatically reduced but at the expense of increasing the field intensity. Although one should not conclude that increased field intensity will inevitably be the penalty for a variance reduction, typical extra demands such as these will have a price.

Concerning field intensity in general, an important point is that the overall action introduced by the optical field is likely the dominant physical variable. Thus increasing the pulse length with a concomitant increase in the fluence penalty can have the natural effect of deminishing the field intensity while spreading out the overall pulse. The attraction of such a control scenario will depend on a host of physical factors including the undesirability of certain types of electronic disturbances as well as the ability to create particular pulse shapes in the laboratory. These matter must await sorting out in future studies.

Robust Optimal Control: A Worst Case Analysis

In dealing with optimal field designs with the aim of taking them into the laboratory for implementation, it is not sufficient to merely take any "reasonable" model Hamiltonian and hope that it suffices. Control over molecular motion is a coherence phenomena and it is anticipated that there will be considerable sensitivity to errors in the Hamiltonian, optimal coupling coefficients and uncertainties arising in the laboratory generated fields. The level of tolerance to these uncertainties is still an open question and this problem looms as one of the most serious issues deserving attention in molecular control theory. This problem reduces to one of seeking robustness with respect to the various uncertainties or possible errors. Robustness can best be treated when there is specific knowledge about the statistics, errors bounds or other characteristics associated with the uncertainties. Model calculations along these lines have been performed.[10]

In contrast to the latter perspective, many realistic systems will have uncertainties that are not well characterized. In this situation, a conservative design approach can be attractive. This philosophy is the basis of the worst case molecular optimal control scenario recently formulated.[11] As usual, the goal is to steer a molecular system to some particular objective by means of an optimal field $E(t)$ which is designed through the minimization of a cost functional. However, the problem is now confounded by the Hamiltonian having the following form.

$$H = H_A + \mu E(t) + H_B \qquad (4)$$

Where H_A is the reference (known) portion of the molecular Hamiltonian and μ is the relevant molecular dipole moment. The new feature here is that H_B is a portion of the Hamiltonian which is not well known. The simplest characterization of H_B would be through an appropriately defined bound on its norm $||H_B||<E^*$ where E^* might be conservatively chosen or estimated on physical grounds. Thus, the optimal control problem reduces to seeking the best field $E(t)$ under the condition that H_B does the worst damage under the constraint $||H_B||<E^*$. This procedure can be expressed as a mini-max problem over the cost functional,

$$\min_{E(t)} \max_{H_B} J[E,H_B] \qquad (5)$$

subject to the norm constraint on the H_B portion of the Hamiltonian. The term H_B might arise due to various circumstances. In some cases the form of H_B may be known but detailed values of the coefficients involved are uncertain. A particularly attractive model arises when H_A represents a harmonic Hamiltonian and H_B contains all of the anharmonic terms. Regardless of the origin of H_B there will always remain some uncertainty in any Hamiltonian.

The mini-max procedure imbodied in eq(5) was recently examined[11] in a case where H_A was harmonic and H_B was characterized by a worst case equality $||H_B|| = E^*$. This latter criteria was built in with a Lagrange multiplier and the formulation led to H_B being proportional to the worst case unknown environmental temporal disturbance w(t). Illustrations were examined involving linear chain molecules seeking the best field E(t) and the worst disturbance w(t). The detailed behavior of these temporal functions is not as important as some general conclusions found in the process.

A) Nature of the worst case w(t). At first sight, one might expect that the worst environmental interaction w(t) would be stochastic in an attempt to undo the coherence generated by E(t). However, the worst disturbance is in fact also coherent and perhaps would best be refered to as anti-coherent. Thus a specific temporal history is spelled out in the function w(t) which gives the worst possible disturbance of bounded energy acting to counter the best control field E(t).

B) Conservative nature of the worst case analysis. It was found that the worst case field characterized by an energy of only ~10% of the average bond energy typically could have a dramatic effect in destroying the controlled coherence. The positive aspect of such worst case calculations is their conservative nature in that any temporal disturbance of the same energy is guaranteed to have a less diliterious effect. However, the results are probably more conservative than necessary and more experience with such problems will help weigh this matter.

C) Scaling of E(t) and w(t). Perhaps the most dramatic result from the computations was the approximate scaling $E(t) \simeq \alpha E_o(t)$ and $w(t) \simeq \beta w_o(t)$ where $\alpha \gg 1$ and $\beta \gg 1$. Here $E_o(t)$ is the optimal field determined for the same physical system with no disturbance $E^* = 0$ and $w_o(t)$ is a reference worst case disturbance determined asymtotically at a vanishingly small value of E^*. Thus it was found that as the energy E* of the environmental disturbance increased the best field E(t) and the worst disturbance w(t) approximately scaled up as multiples of their predetermined reference cases. Although this situation may not be general, it lends itself to an interesting suggestion: without any further knowledge a conservative control strategy might be to scale up the control field E(t) to be as high as practical in order to deal with any uncharacterized uncertainties.

Robust control techniques should play an increasing role at the molecular scale. It is anticipated that a variety of design tools besides worst case analysis will be helpful in this regard.

An Optimally Controlled Molecular Dynamics Simulator

Although optimal design work is underway on various molecular scale problems to bring them into the laboratory, such tasks are difficult and will likely remain so at least for the immediate future. With this in mind, an effort was proposed to develop a bench-top analog simulator of the molecular dynamics control problem.[12] This effort was initiated both as an educational tool as well as a means to get direct experience on the effect of laboratory noise or other uncertainties. Figure 4 presents a schematic diagram of the simulator designed to mimic a linear polyatomic molecule. The bonds were initially created by precision wound linear and nonlinear springs. However, recently these mechanical springs have been replaced by specially designed magnetic "springs" capable of fully simulating interatomic bonds including their dissociative properties.

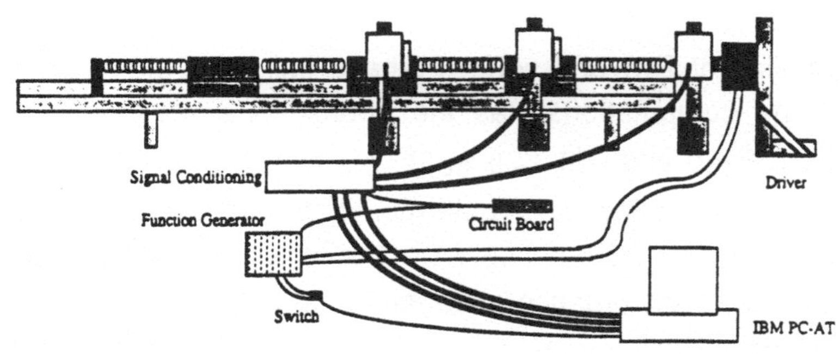

Figure 4

Schematic diagram of an optimally controlled molecular dynamics simulator. The bench-top simulator mimics a linear polyatomic molecule (in this case a triatomic) with the atoms replaced by carts on a nearly frictionless air track and the bonds created by precision mechanical or magnetic springs. The laser field is simulated by a linear motor. Although the motion of the "atoms" can be easily recognized by direct visual observation, optical encoders are utilized to precisely measure the position of the atoms for quantitative confirmation of the control design process. Exactly the same optimal control conceptual and numerical machinery employed at the molecular scale is utilized in the simulator. The simulator is capable of bringing in laboratory realism especially through the presence of inherent laboratory noise.

Clearly, the bench-top simulator follows the laws of classical mechanics but this circumstance may be quite advantageous as the work directly ties in with molecular scale classical optimal control theory.[9] Indeed, exactly the same optimal design software machinery employed at the molecular scale was implemented to create simulated field designs for the apparatus in Figure 4. Although the time scales are now boosted into the 1 sec ~ 1 min range, the optimal driving fields have the same essential structure as seen in their ultra-fast ~fsec molecular scale counterparts. Perhaps the most important result from this apparatus has been the rather surprising robustness of the results in the presence of laboratory noise. It is anticipated that this optimal control simulator or its evolutionary extension to two spatial dimensions should continue to prove to be a valuable and inexpensive means for obtaining hands on experience about what might actually happen with control at the true molecular scale.

CONCLUSION

At this stage of development of molecular optimal control theory perhaps the most important conclusion is that traditional engineering design tools may be carried over and implemented at the molecular scale. A key task at this point is to draw in laboratory studies by including explicit constraints from laboratory pulse shaping capabilities.[4,5] The goal is not merely one of demonstrating quantum coherence phenomena, as quantum beats have been observed for many years. Rather, an important goal will be to demonstrate the creation of special coherences by design (i.e., designed field $E_1(t)$ creates one particular observable coherence while designed field $E_2(t)$ creates another predictable coherence, etc.). Selective chemical bond breaking falls into this category as a dramatic goal but there are many interesting steps in between that may ultimately lead to equally valuable results or techniques based on optimal pump and probe experiments. The limited band width (~300 cm^{-1}) and visible carrier wave characteristics of current pulse shaping techniques suggest that electronic excitation or Raman experiments are reasonable goals with control over rotational or low frequency vibrational motion. Another interesting possibility is to side step much of the robustness problems and use various learning algorithms combined with the high duty cycle of pump-probe experiments as a means to "teach" lasers how to design their own pulse shapes.[13] A simulation of this suggestion has been performed and in the long run may be the best means to fully combine experiment and theory in the domain of optimal control over molecular motion. Putting aside the original dream of creating a molecular pair of scissors, the adoption of optimal design and control techniques at the molecular scale may have broad implications and applications for a variety of problems in the domain of molecular materials by design.

Acknowledgement

The author acknowledges support for aspects of this research from the Office of Naval Research, the Army Research office and the Air Force Office of Scientific Research.

References

(1) S. Shi and H. Rabitz, J. Chem. Phys., **92**, 364 (1990)
(2) S. Tersigni, T. Gaspard and S. Rice, J. Chem. Phys., **93**, 1670 (1990)
(3) A. Bryson and Y. Ho, Applied Optimal Control (Hemisphere, New York, 1975)
(4) A. Weiner, J. Heritage and S. Salehi, Opt. Lett., **13**, 300 (1988)
(5) M. Haner and W. Warren, Appl. Phys. Lett., **52**, 1459 (1988)

(6) E. Drazin and R. Johnson, Solitons: An Introduction, (Cambridge University Press, New York 1989)
(7) S. Shi, A. Woody and H. Rabitz, J. Chem. Phys., **88**, 6870 (1988)
(8) P. Gross, D. Neuhauser and H. Rabitz, "Optimal Control of Curve - Crossing Systems", J. Chem. Phys., Submitted
(9) C. Schwieters and H. Rabitz, "Optimal Control of Non-Linear Classical Systems: Application to Uni-Molecular Dissociation Reactions and Chaotic Potentials", Phys. Rev. **A**, (1991), in press
(10) M. Dahleh, A. Peirce and H. Rabitz, Phys. Rev. **A42**, 1065, (1990)
(11) H. Beumee and H. Rabitz, "Optimal Control of Uncertain Molecular Systems: A Worst Case Analysis", Phys. Rev. **A**, submitted
(12) M. Husman, C. Schwieters, M. Littman and H. Rabitz, "Molecular Dynamics Simulator for Optimal Control of Molecular Motion", Am. J. Phys., (1991), in press
(13) R. Judson and H. Rabitz, "Teaching Lasers to Control Molecules", Phys. Rev. Lett., submitted

CONTROL OF MOLECULAR VIBRATIONAL EXCITATION AND DISSOCIATION BY CHIRPED INTENSE INFRARED LASER PULSES

Szczepan Chelkowski and André D. Bandrauk

Département de Chimie, Faculté des Sciences, Université de Sherbrooke, Sherbrooke, Québec, J1K 2R1, Canada

INTRODUCTION

Multiphoton vibrational excitation of molecules is a subject of active research since many years[1-4]. One expects that this process will selectively excite specific molecules in a mixture or even a particular bond inside a polyatomic molecule without exciting other molecules or bonds. This would allow to control chemical reactions in a time scale shorter than various relaxation times in the medium. Besides the considerable research effort in the field only multiphoton dissociation of large molecules, such as SF_6 was demonstrated experimentally[2-4]. It is generally believed that a multiphoton dissociation of a diatomic or triatomic molecule by a monochromatic field is difficult to achieve[2,4-6]. The dissociation probability of an ionic diatomic molecule, such as HF, by a picosecond pulse approaches 0.1 for the intensity 2×10^{14} W/cm^2 and is smaller than 10^{-7} for laser intensities below 5×10^{13} W/cm^2 (see Fig. 1, for results of our simulations based on a model described below). At such high intensities dissociation cannot compete with atomic ionization since at an intensity 5×10^{13} W/cm^2 one is already above the ionization threshold[7]. Obviously, the anharmonicity of molecular vibrations is responsible for this difficulty: if the laser is tuned to the first vibrational transition the second transition is out of resonance, for HF molecule the respective detuning is 173 cm^{-1} and the subsequent transitions are detuned by multiples of this value. It is natural to expect that the vibrational excitation will increase dramatically if the laser frequency tunes itself to the subsequent transition frequencies when the inversion in the previous transition is already accomplished. This requires specific frequency sweeping (chirping) on a time scale shorter than relaxation times. The recently developed pulse shape techniques seem to

Coherence Phenomena in Atoms and Molecules in Laser Fields
Edited by A.D. Bandrauk and S.C. Wallace, Plenum Press, New York, 1992

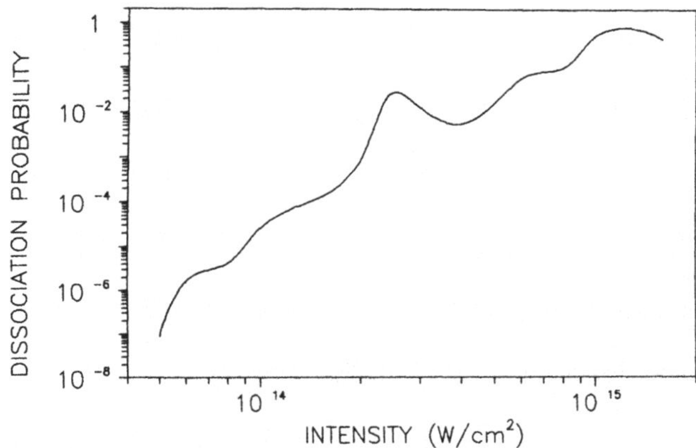

Fig. 1. Probability of dissociation of HF molecule at the end of a monochromatic laser pulse of duration ≅ 1 ps (pulse envelope is shown in Fig. 2a).

indicate that such chirped pulses can be prepared. Several authors have suggested[3,9-15], that the vibrational excitation rates can be increased if pulses with modulated amplitude[14] or phase[9-13] or several pulses[15] are used instead of one monochromatic pulse. In our previous papers[9,10] we guessed particular shapes of frequency sweeping and performed numerical simulations showing that indeed the appropriate chirping increases excitation and dissociation rates of a diatomic[9] and a triatomic molecule[10] by many orders of magnitude. We also showed that sufficiently fast chirping allows selective excition of a particular bond in a linear triatomic molecule[10]. Our results show that the chirped, intense mid infrared pulses may become an important tool in selective photochemistry by exciting faster and more efficiently than monochromatic pulses. In this paper we review and extend the results of these studies.

DESCRIPTION OF THE MODEL FOR A DIATOMIC MOLECULE

A. Dynamics of molecule-laser field interaction

Our calculation scheme describing the interaction of radiation with a Morse oscillator is based on the time dependent Schrödinger equation with the hamiltonian

$$H = H_0 + V(x,t) \quad , \qquad H_0 = \frac{-\hbar^2}{2\,m} \frac{\partial^2}{\partial x^2} \quad , \tag{1}$$

$$V(x,t) = D \left(1 - \exp(-ax) \right)^2 - \mu(x)\, E_M\, U(t) \cos(\, \Phi(t)\,),$$

where $x = r - r_0$, r_0 is the equilibrium separation of nuclei, m is the molecule's reduced mass, $\mu(x)$ is the molecular dipole function, E_M is the maximum value of the radiation electric field, U(t) is the pulse envelope chosen equal to 1 at the maximal value of the electric field and $\Phi(t)$ is the laser phase which will be specified in the next section. The energy eigenvalues E of the Morse oscillator are[6,16,17]

$$E = \hbar\omega \ (n + 1/2) \{1 - B \ (n + 1/2)/2\} \quad , \tag{2}$$

$$\text{where} \quad \hbar\omega_0 = 2 \ B \ D \quad , \quad B = \left(\frac{\hbar^2 a^2}{2m \ D} \right)^{1/2} , \quad n = 0,1,\ldots, < 1/B - 1/2 \quad . \tag{3}$$

The dipole moment function $\mu(x)$ was chosen in the form:

$$\mu(x) = d_1 \ [\ D(r) - D(r_0) \ ] \quad , \ D(r) = r \ \{ \ 1 + \exp \ [\alpha(r-r_M)]/(\alpha r_M - 1 \ ) \ \}^{-1} \ , \tag{4}$$

where d_1 is the dipole moment gradient. The function D(r) has a maximum at $r = r_M$ and is linear for $r << r_M$, as expected for ionic molecules. We have chosen r_M = 3 Å and $\alpha = 10$ Å^{-1} (at r = 3.5 Å D(r) is practically zero) in order to comply with spectroscopic results and ab initio[18] calculations for typical diatomics. In our numerical calculations we have used the following values of molecular parameters[6], corresponding to HF:

$B = 0.0419$, $D = 6.125$ eV , $a = 1.1741 \ a_0^{-1}$, $d_1 = 0.786$ Db/a_0 , $r_0 = 1.7329 \ a_0$. The molecule was assumed to be in its vibrational ground state (n=0) at t=0. The transition dipole moment $p_{0,1}$ for HF was calculated with the help of the expression:

$$p_{01} = d_1 < \psi_1 | \ x \ | \ \psi_0 > \cong d_1 (B/2)^{1/2} / \ a = 0.097 \text{ Db} \tag{5}$$

where ψ_n are Morse eigenfunctions[6,9,17]. For these parameters the Morse potential supports 24 bound states and its characteristic vibrational period is $t_0 = 2\pi \ / \ \omega_{0,1} = 8.41$ fs .

B. Construction of a frequency sweep adapted to molecular anharmonicity

In order to construct the instantaneous frequency sweep $\omega_{ins}(t) = d\Phi(t)/dt$ yielding the best excitation efficiency of the Morse oscillator, let us consider this system as a sequence of two-level systems. The function $\omega_{ins}(t)$ should decrease in such a way that it should be close to the transition frequency $\omega_{n,n+1} = (E_{n+1} - E_n)/\hbar$, where E_n are vibrational levels of the molecule, only until the complete inversion in levels n,n+1 occurs. We know from the theory of the two-level systems[19,20] that the transition amplitudes, in such systems, are functions of the pulse area defined by

$$\sigma(t) = (p / \hbar) \int_{-\infty}^{t} \varepsilon(t') \, dt' \tag{6}$$

where p is the transition dipole moment and $\varepsilon(t)$ is the slowly varying electric field envelope. The complete inversion of population occurs when the pulse area is equal to π. Thus one might expect that that a series of N pulses each of frequency $\omega_{n,n+1} = (E_{n+1} - E_n)/\hbar$, where E_n are vibrational levels of the molecule, and each having the area (defined by (1) in which $p = p_{n,n+1}$) equal to π should completely transfer the population from the ground state (n =0) to the N-th excited state.

Such pulse sequences can probably be constructed using emerging experimental techniques[8]. For low intensity radiation, dissociation would require many components to the pulse sequence, each satisfying exact resonance and constant area conditions. Since the transition frequencies and transition dipole moments between higher excited states are usually not well known, this may be impractical. Moreover, at low intensities the whole sequence would take longer than the various relaxation times in the molecular medium. Thus we concentrate on sufficiently high intensities that the dynamic Stark effects overwhelm these problems and a continuously chirped pulse can be used. By continuous chirping, even at lower intensities, we will certainly couple resonantly with all transitions and thus the method will be more robust than the method using the series of constant frequency pulses. Such chirped pulses can be comparatively easily generated using modern laser technology, especially in the high intensity regime which is necessary for rapid excitation.

Our task now is to find such a pulse shape $\varepsilon(t) = E_M U(t)$ and pulse frequency $\omega_{ins}(t)$ such that the resonance and area conditions mentioned in the introduction should be satisfied, i. e. we require that

$$(p_{0,1} /\hbar) \int_{0}^{t_1} \varepsilon(t) \, dt = S_0 \quad , \quad (p_{n,n+1}/ \hbar) \int_{t_n}^{t_{n+1}} \varepsilon(t) \, dt = S \quad , \tag{7}$$

$n = 1,\ldots,N-1$, where S_0, S are area values, N is the level number up to which the approximate resonance conditions:

$$\hbar \left(\omega_{ins}(t_n) + \omega_{ins}(t_{n+1}) \right) / 2 = E_{n+1} - E_n = \hbar\, \omega_0 \left\{ 1 - B\,(n+1) \right\} \qquad (8)$$

are imposed. Eqs. (7) fix the pulse area in the interval (t_n , t_{n+1}) and Eq. (8) ensures that in the center of this interval the pulse is resonant with the transition from the n-th to n+1 -th level. We consider S and S_0 as free control parameters and expect that the best efficiency of dissociation will be achieved for values close to π The following choice of pulse shape U(t)

$$U(t) = \begin{cases} \{ 1 - \mathrm{sech}(\alpha_0)\} \{ \mathrm{sech}\,[\alpha_0(t-t_0)/t_0] - \mathrm{sech}(\alpha_0) \} & \text{for } t < t_0 \\ 1 & \text{for } t_0 < t < t_c \ , \\ \{ 1 - \mathrm{sech}(\alpha_F)\}^{-1} \{ \mathrm{sech}\,[\alpha_F(t-t_c)/t_0] - \mathrm{sech}(\alpha_F) \} & \text{for } t < t_c \end{cases} \qquad (9)$$

allows us to integrate analytically conditions (7) and (8). In Eq. (9) α_0, α_F , t_c and t_0 are free parameters determining the pulse switching on and switching off. They are chosen here to be $\alpha_0 = 2.5$, $\alpha_F = 6.25$, $t_c = t_N + 0.2\ t_0$. We assumed that chirping starts at $t = t_0$ i. e. when the pulse achieved its maximal value. Thus

$$\omega_{ins}(t) = \rho\, \omega_{0,1} \qquad \text{for} \qquad t < t_0 \qquad (10)$$

where ρ is the coefficient determining the initial laser frequency, $\rho \cong 1$. Equations (7)-(10) determine $t_0, t_1, \ldots, t_N$ uniquely and the function $\omega_{ins}(t)$ can be found by interpolation . An analytical expression for $\omega_{ins}(t)$ can be found when the harmonic oscillator formula for dipole moments

$$p_{n,n+1} = (n + 1)^{1/2}\, p_{0,1} \qquad (11)$$

holds . For HF the Eq.(11) is a good approximation up to 12-th level. Using the eq. (11) the second part of (7) , with U(t) given by (9), becomes

$$t(n+1) - t(n) = Q\,(n + 1)^{-1/2} , \qquad (12)$$

where $n = 1,2,..,N$, $t(n) \equiv t_n$,

$$Q = S\,\hbar / (p_{0,1}\, E_M) . \qquad (13)$$

The function t(n) can be easily found from (12) if one uses the Taylor expansion $t(n+1) \cong t(n) + \frac{dt}{dn}$. Thus one gets

$$t(n) = 2\,Q\,\{ (n + 1)^{1/2} - 2^{1/2} \} + t_1 . \qquad (14)$$

and from (8) $\omega_{ins}(t_n) \equiv \omega(n) = \omega_0(1 - B\,n - B/2)$. Thus after calculating from Eq. (14) n as function of t one gets the following form of a chirp, which approximately satisfies requirements (7) and (8):

$$\omega_{ins}(t)/\omega_0 = \rho \left[-B(t - t_1)^2/(4Q^2) - 2^{1/2}B(t - t_1)/Q + 1 - 3B/2\right] . \quad (15)$$

The values of t_0 and t_1, calculated from the first part of Eq. (7) and from (10), are given by

$$t_0 = Q \frac{S_0/S - 2\{2^{1/2} - (3/2)^{1/2}\}}{A + 1} , \qquad t_1 = Q S_0/S - A t_0 \quad (16)$$

where $$A = \frac{2\{\pi/4 - \tan^{-1}[\exp(-\alpha_0)]\}/\alpha_0 - 1}{1 - \mathrm{sech}(\alpha_0)}$$

The pulse shape U(t) and the instantaneous frequency $\omega_{ins}(t)$ described by the above formulas, for the laser peak intensity $I = (c/8\pi) E_M^2$ (the incident radiation is assumed to be linearly polarized) 10^{13} and 10^{12} W/cm^2, are shown in Fig. 2a and 3a respectively.

C. Excitation via adiabatic following mechanism

The populations in two-level systems can be also inverted via an adiabatic following mechanism[19,21,22] if the pulse amplitude and frequency variation are sufficiently slow, i. e. if the following conditions hold:

$$|\Omega_R \, d\Omega_R/dt| << [\Omega_R^2 + \Delta^2]^{3/2} , \quad |d\Delta/dt| << \Omega_R^2 , \quad (17)$$

where Ω_R is the Rabi frequency ,

$$\Delta = \omega_{TR} - \omega_{ins}(t) = \omega_{TR} - d\Phi/dt \quad (18)$$

is the time dependent detuning and ω_{TR} is the transition frequency. The first inequality allows to prepare a specific dressed state and the second one requires the chirping rate be sufficiently slow to follow adiabatically this solution. The adiabatic following solution for the population inversion w in a two-level system is then (Ref. 19, pages 74 and 101):

$$w(t) = \pm \Delta(t) [\Omega_R^2 + \Delta(t)^2]^{-1/2} . \quad (19)$$

Note that special care must be taken when using the formulas from the literature for the case of time dependent frequencies; if the laser phase is written in the form $\Phi(t) = \omega(t)\, t$, then $\Delta(t)$ must be calculated from (18) and not from $\omega_{TR} - \omega(t)$; $\omega(t)$ and $\omega_{ins}(t)$ differ considerably in our case. The formula (19) tells us that 100% transfer of population will occur if after the frequency sweeping the Rabi frequency reduces to zero. This would require the amplitude modulation in addition to chirping, which we we preferred to avoid in our scheme. We see from (19) that efficient adiabatic transfer will also occur at intensities such that the total frequency sweep accompanying the inversion in one subsequent transition in our Morse oscillator is greater than Rabi frequency Ω_R . This means that we should have $\Omega_R << \omega_0 B = 173\ cm^{-1}$. In our model Ω_R calculated for 0→1 transition is equal to 140 cm^{-1} for I = 10^{13} W/cm^2 and equals 45 cm^{-1} for 10^{12} W/cm^2. Thus one expects that this mechanism prevails for 10^{12} W/cm^2 and lower intensities and the efficient adiabatic transfer of popultions will occur without additional amplitude modulation. . The resulting population transfer in the Morse oscillator will be for this case very robust since the chirping must be only sufficiently slow, according to (17). The transfer is then insensitive to the pulse area, unlike in the case when one transfers the populations using a series of monochromatic π pulses. This is indeed confirmed by the results for $I=10^{12}$ W/cm^2 presented here and in Ref. 9 . At intensity 10^{13} W/cm^2 the population transfer mechanism does not follow exactly the formula (19) since then the whole process does not occur as in a series of two-level systems since more than 2 levels are all the time coupled by the radiation because of the dynamic Stark shifts exceeding the detunings.

TECHNIQUE OF CALCULATION AND DISCUSSION OF RESULTS FOR A DIATOMIC MOLECULE

The calculations were performed with the help of the split operator technique [23], in which the evolution of the wave function over the time interval δt is given by

$$\psi(x,t+\delta t) = \exp\left(\frac{-iV(x,t')\delta t}{2\hbar}\right) \exp\left(\frac{-i H_0 \delta t}{\hbar}\right) \exp\left(\frac{-iV(x,t')\delta t}{2\hbar}\right) \psi(x,t) , \tag{20}$$

where $t' = t + \delta t/2$. The second exponential from this expression, containing the second derivative with respect to x was evaluated in momentum space, using fast Fourier transform techniques. The wave function was confined in a box of size L = 130 a. u. and 2048 space steps were used . The size of the box was chosen so that the wave function was rapidly decreasing well before the edges of the box. The time step δt was equal to $0.001\ t_0 = 8.41\times10^{-18}$ s. We calculated the populations $P_n(t)$ of all vibrational levels and the dissociation probability $P_D(t)$ using the formulas:

$$P_n(t) = | < \psi_n | \psi(t) > |^2 \quad , \quad P_D(t) = 1 - \sum_{n=0}^{23} P_n(t) \tag{21}$$

The time evolution of the populations $P_0(t)$, $P_1(t)$, $P_8(t)$, $P_{14}(t)$ and of the dissociation probability are shown in Fig.2b ($I = 10^{13}$ W/cm^2) and in Fig.3b ($I = 10^{12}$ W/cm^2) for the pulses displayed in Figs. 2a and 3a respectively. For comparison, the dissociation probability at the end of a pulse having no chirp, resonant with $\omega_{0,1}$, but otherwise similar to that in Fig. 2a is shown in Fig. 1 . We were unable to calculate it for the intensity 10^{13} W/cm^2 since its value falls below the numerical errors, we expect it to be much less than 10^{-10} by doing linear extrapolation in Fig.1. We observe therefore a very considerable increase of dissociation

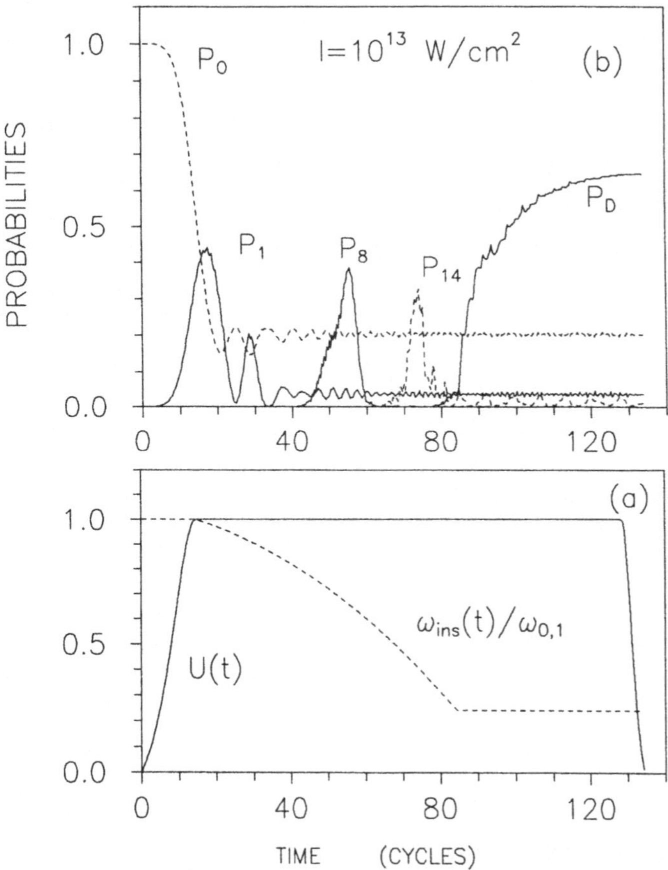

Fig. 2. Time dependence of (a) pulse shapes (——) $U(t) = \varepsilon(t) / E_M$, (----) chirps $\omega_{ins}(t) / \omega_{0,1}$, and (b) the resulting populations (----) $P_0(t)$, (——)$P_1(t)$, (——) $P_8(t)$, (----) $P_{14}(t)$ and dissociation probability (——)$P_D(t)$,at the intensity $I = 10^{13}$ W/cm^2 , $N = 18$, $S = S_0 = 0.8\times\pi$, $\rho = 1$, 1 cycle = 8.41 fm.

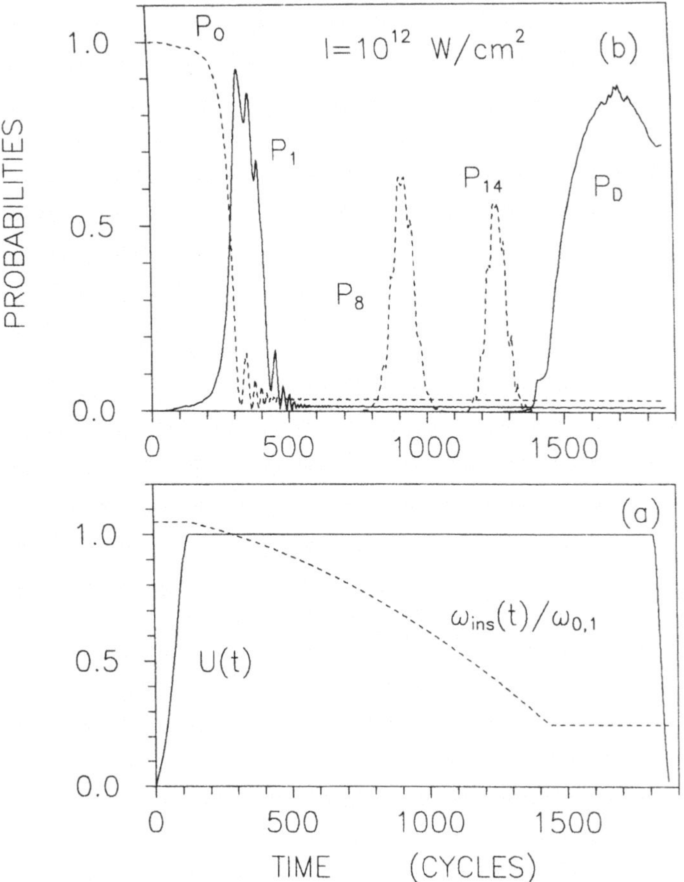

Fig. 3. The same as in Fig. 2 but for the intensity $I = 10^{12}$ W/cm^2 , $S = 4.5\times\pi$, $S_0 = 3\times\pi$, $\rho = 1.05$, 1 cycle = 8.01 fs.

probability due to chirping. One sees clearly that the frequency $\omega_{ins}(t)$, decreasing according to Eq. (15), constitutes the "force" driving populations upward beyond the N = 18 level (the chirp from Fig. 2a stops at the frequency corresponding to the transition from level 17 to 18) yielding 50% dissociation. The fact that such a chirp is sufficient to give such efficient dissociation results from dynamic Stark shifts[5,24] of higher vibrational levels. Already for the 1→2 transition, the Rabi frequency exceeds the detuning $\Delta = 2B^2D$ at an intensity of 10^{13} W/cm^2 [5]. This implies that for high intensities, our predictions are less sensitive to the asymptotic behavior of the molecular potential. The method appears to be fairly robust: Fig. 3 of Ref. 9 shows that greater than 25% dissociation is obtained when the area is varying by a factor of two (equivalent to a factor of 4 in pulse intensity or a factor of two in chirp rate) at an intensity 10^{13} W/cm^2. At lower intensities greater values of S are required but because of the fact that the adiabatic following conditions are satisfied the method appears to be even more robust (less sensitive to the chirping rate). We checked that similar efficiencies can be obtained also at $I=10^{11}$ W/cm^2 for 80% chirping in the interval 55 ps. The most serious difficulties for experimental preparation of such pulses are:

(i) to dispose enough power in order that the chirping be sufficiently fast to compete with relaxation times, e.g. when exciting a particular bond inside a polyatomic molecule (see the next section).

(ii) to achieve the dissociation probability greater than 0.5 in a picosecond time scale, Fig. 2, 76% total chirp of instantaneous frequency is required .This seems to be a very tough requirement, although for certain pulses a continuum having spectral width 6000 cm^{-1} has been already obtained[25]. Therefore we performed several simulations for lower total chirps but for pulses which are are monochromatic during an additional picosecond, after the end of chirping. We obtained at $I=10^{13}$ W/cm^2: P_D equal to 0.15, 0.01 and 10^{-5} for the corresponding values of total chirp 59% ,50% and 33% . All these results although more modest still show a many order effect compared to monochromatic pulses of same duration which yield $P_D < 10^{-10}$.

Note that in this paper when we are using the term chirp we are refering to the sweeping of $\omega_{ins}(t)$ while in the previous papers[9,10] we were interpreting our results in terms of $\omega(t)$.

SELECTIVE EXCITATION OF A TRIATOMIC MOLECULE

Many authors have studied the intra molecular vibrational redistribution of the initial excitation energy (IVR) resulting from some initially excited local bond [26,27]. One expects that such an excitation might demonstrate chemical reactivity radically different from a thermally excited state at the same energy or also from molecules electronically excited by lasers of visible or higher frequencies. However, because of the rapidity of IVR (picosecond time scale) it is difficult to prepare efficiently such bond-localized excitations using monochromatic nearly resonant pulses for bonds of small molecules exhibiting large anharmonicities. Another method of local bond excitation based on direct (one photon) overtone excitation[28,29] by monochromatic pulses is too slow to yield considerable populations of upper vibrational states[29]. The resulting transition probabilities to the states around v=4 for CH bonds are less than 10^{-7} for subpicosecond pulses of intensity 2×10^{11} W/cm^2 [29] . Both methods cannot yield efficient bond dissociation for intensities below ionization thresholds, as discussed in the Introduction. By contrast, we will present here results of numerical simulation showing that an appropriate chirped pulse can compete with I.V.R.

We present here results obtained from a model in which a triatomic molecule is represented as two coupled Morse oscillators. Our calculation scheme describing the molecule interaction with the radiation is based on the time dependent Schrödinger equation with the Hamiltonian

$$H = H_0 + V(x,y,t) \quad , \quad H_0 = P_y^2/\,2\mu_{CH} + P_x^2/\,2\,\mu_{CN} - P_x\,P_y\,/\,m_C \quad ,$$
$$V = D_{CH}\left(1 - \exp(-ay)\right)^2 + D_{CN}\left(1 - \exp(-bx)\right)^2 - \mu(x,y)\,E_M\,U(t)\,\cos(\Phi(t)) \quad , \tag{22}$$

where $x = r_{CN} - R_{CN}$, $y = r_{CH} - R_{CH}$, r_{CN} and r_{CH} are respective nuclear separations; R_{CN} , R_{CH} are their equilibrium separations; μ_{CH} and μ_{CN} are the respective reduced masses; m_C is the central atom mass; $\mu(x,y)$ is the molecular dipole moment function. For more details about this model we refer the reader to Ref. 10.

We have integrated numerically the time dependent Schrödinger equation with the Hamiltonian (22) for the laser peak intensity $I = 10^{13}$ W/cm^2 using the values of parameters present in (22) and (23) corresponding to the HCN molecule[10,30-32]. It was asssumed that the molecule was in its ground state[10]. The calculations were performed with the help of the technique described in the previous section. We calculated the populations $P_0(t)$ of the ground state and of the 8-th vibrational state of the CH bond $P_{8,0} = |\langle\psi_8^{HC}\psi_0^{CN}|\psi(t)\rangle|^2$, as well as, the probability of dissociation of the CH bond with the help of the formula:

$$P_D(t) = \int_{-R_{CN}}^{\infty} dx \int_{3/a}^{\infty} dy\; |\psi(x,y,t)|^2 \tag{23}$$

P_D is defined as probability of finding C and H atoms outside the range of attraction, which we assumed to correspond to inter atomic distance $r_{CH} > R_{CH} + 3/a$. We calculated also the average energies E_{CH}, E_{CN} present in the CH and CN bonds[10].

The pulse shape $U(t)=\varepsilon(t)/E_M$ and pulse chirp $\omega_{ins}(t)/\omega_{0,1}$ are shown in Fig. 4a and the molecular response to such a pulse is shown in Fig. 4b. This Figure shows the time dependence of the population of the ground state P_0 and of the 8-th vibrational state of the CH bond $P_{8,0}$ (CN being unexcited), the dissociation probability P_D of the CH bond, as well as, the average energy stored in each bond, E_{CH} and E_{CN}. We see that chirping by 20% gives us 10% population in 8-th vibrational state (E_{CH} = 1.6 eV) at t = 0.65 ps (65 cycles) and no excitation in the CN bond. Chirping by 70% yields a considerable dissociation of the CH bond, P_D = 0.5 and bond energy E_{CH} = 4 eV (the corresponding response of the HCN molecule to a similar but monochromatic pulse is: energy $E_{CH}(t)$ oscillates and never exceeds 0.8 eV and P_D is smaller than 10^{-10}). Thus we have two possibilities to control vibrational energy distribution inside HCN. <u>First</u>, we can stop our pulse at a specific moment and thus achieve a <u>specific</u> excitation of each bond. We checked that stopping the pulse at t=0.74 ps (73 cycles), Fig 2a, Ref.10, prepares the molecule in the state of average energy 1.7 eV (almost the energy of the fifth vibrational state of CH) and leaves CN unexcited. The

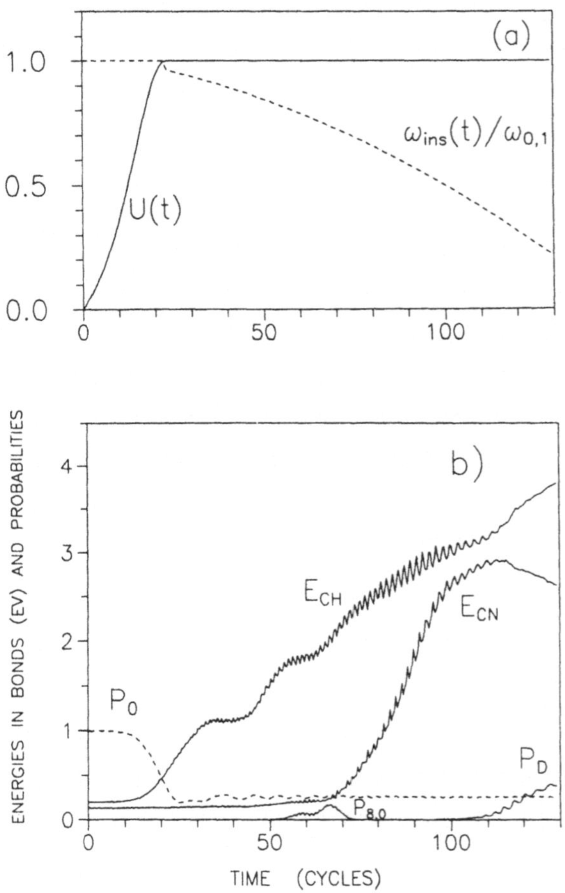

Fig. 4. Time dependence of (a): pulse shapes (——) $U(t) = \varepsilon(t)/E_M$; (----) chirps $\omega(t)_{ins} / \omega_{0,1}$; (b) the resulting populations: (---) $P_0(t)$; (——) $P_{0,8}(t)$; dissociation probability (——)$P_D(t)$; energy in CN bond E_{CN} E_{CN} and in CH bond E_{CH}. Intensity I = 10^{13} W/cm^2, 1 cycle = 10.2 fs.

state prepared by such a pulse does not relax to a the state of equal energy thermal redistribution predicted by statistical theories. The energy redistribution inside the molecule remains almost constant after the pulse is switched off. Thus our simulation supports the results of Ref.27 showing that the statistical theories do not apply for certain small molecules. Stopping the chirped pulse a little later, at t=0.97 ps (96 cycles) gives already 1.2 eV of energy in the CN bond and 2.8 eV in CH, again a rather unequal energy distribution. This result indicates that in HCN a considerable relaxation occurs when CH is excited around the v=8 level. The possibility of creating such unequal stable energy distributions inside HCN relies on the fact that the CH and CN frequencies differ considerably, being equal in our model to 3370 and 2180 cm^{-1} respectively. We performed similar simulations for DCN, for which the CD frequency is very close to the CN frequency , 2370 and 2180 cm^{-1} respectively. We obtained similar excitations and dissociation probabilities of CD as those of CH presented above. But because of fast IVR in the deuterated case the selectivity is less. Right from the beginning both modes are almost equally excited. <u>A second</u> possibility of control relies on the sensitivity of the excitation to the rate of chirping, which in our calculations is controlled by the area of the pulse on the interval (t_n, t_{n+1}) denoted by S (a larger interval Δt means slower chirping and a larger area S), as discussed in Ref. 10.

In order to achieve a good selectivity of excitation the chirping should be large and very very fast: in one picosecond the frequency should decrease by more than 50% . This is certainly a very tough requirement for experimentalists. However, the techniques developed in the visible region seem to suggest that our requirements are not completely unrealistic. Already six femtosecond pulses[8], which have about 30% frequency spread were obtained. Therefore, after the passage through an appropriate dispersive grating system[8] the desired chirp can be obtained. Also the chirp can be obtained directly by using self phase modulation techniques [8,25]. There is no physical reason to expect that in the mid infrared region such pulses could not be obtained.

CONCLUDING REMARKS

Summarizing, our study shows that chirped infrared pulses should become an important tool in photochemistry. They provide an efficient control of vibrational energy distribution inside a small molecule , such as HCN, and allows to dissociate efficiently a particular ionic bond via a multiphoton resonant process. Our simulations show that such pulses can thus prepare nonstatistical quantum vibrational states and control the reactivity of a molecule by varying the laser phase. The control implicit in chirped pulse dissociation is very strong if the pulse has a chirp sufficient to compensate the molecular anharmonicity and sufficiently intense if applied in polyatomics in order to compete with very fast I.V.R. If selectivity of excitation within a molecule is not required, less intense pulses, $I \cong 10^{11}$ W/cm^2 (or less but, but than pulse duration should be longer than 50 picosecond) but having a considerable chirp can be used. The results presented here were obtained using a model non-rotating molecule. We are currently performing the calculations for a rotating diatomic and will present the results in the near future.

Acknowledgements: We wish to thank the Natural Sciences and Engineering Research Council of Canada for the financial support and to P. B. Corkum (NRC-Ottawa) for valuable discussions. We thank also prof. J. Eberly for drawing our attention to the literature concerning adiabatic following mechanisms.

REFERENCES

1. J. Jortner and R. D. Levine, Adv. Chem. Phys. **47**, 1 (1981).
2. V. S. Letokhov, *Nonlinear Laser Chemistry*, (Springer-Verlag, Berlin, 1983).
3. N. Bloembergen and A. H. Zewail, J. Phys. Chem. **88**, 5459 (1984).
4. C. D. Cantrell, V. S. Lethokov and A. A. Makarov, in *Coherent Nonlinear Optics*, ed. by M. S. Feld and V. S. Lethokov, Springer Verlag, Berlin (1980); S. K. Gray, J. R. Stine and D. W. Noid, Laser Chem. **5**, 209 (1985).
5. S. Chelkowski and A. D. Bandrauk, Phys. Rev. **A 41**, 6480 (1990).
6. M. E. Goggin and P. W. Milonni, Phys. Rev. A, **37**, 796 (1988).
7. P. B. Corkum, N. H. Burnett and F. Brunel, Phys. Rev. Lett. **62**, 1259 (1989).
8. A. M. Weiner, J. P. Heritage and R. N. Thurston, Opt. Lett. **11**, 153 (1986); Opt. Lett. **10**, 609 (1985); P. B. Corkum, IEEE J. Quantum Electron. **21**, 216 (1985); P. B. Corkum and C. Rolland, Proc. SPIE **664**, 212 (1986); C. Rolland and P. Corkum, J. Opt. Soc. Am. B **5**, 641 (1988); P. B. Corkum, P. P. Ho, R. R. Alfano, and J. T. Manassah, Opt. Lett. 10, 624 (1985); R. L. Fork, C. H. Brito Cruz, P. C. Becker and C. V. Shank, Opt. Lett. **12**, 483 (1987).
9. S. Chelkowski, A. D. Bandrauk and P. B. Corkum , Phys. Rev. Lett. **65**, 2355 (1990).

10. S. Chelkowski and A. D. Bandrauk, Chem. Phys. Lett. in press.
11. W. S. Warren and A. H. Zewail, J. Chem. Phys. **78**, Part II, 3583 (1983).
12. J. C. Diels and S. Besnainou, J. Chem. Phys. **85**, 6347 (1986).
13. Xie Bo-Min and Ding Jian-Qiang; J. Chem. Phys. **84**, 3819 (1986); ibid **86**, 6579 (1987).
14. B. G. Dibble and R. B. Shirts, J. Chem. Phys. **94**, 3451 (1991).
15. J. R. Stine and D. W. Noid, Opt. Comm. **32**, 161 (1971); D. W. Noid and J. R. Stein, Chem. Phys. Lett. **65**, 153 (1979).
16. R. B. Walker and R. K. Preston, J. Chem. Phys. **67**, 2017 (1977).
17. F. V. Bunkin and I. I. Tugov, Phys. Rev. A **8**, 601 (1973).
18. G. Di Lonardo and A. E. Douglas, Can. J. Phys. **51**, 434, (1973); J. P. Ogilvie, W. R. Rodwell and R .H. Tipping, J. Chem. Phys. **73**, 5221 (1980).
19. J. Allen and J. H. Eberly, *Optical Resonance and Two-Level Atoms*, (Wiley, New York, 1975).
20. G. L. Lamb, Jr., Rev. Mod. Phys. **43**, 99 (1971).
21. C. Lindenbaum, S. Stolte and J. Reuss, Phys. Rep. **178**, 1 (1989) and references therein.
22. J. Oreg, F. T. Hioe and J. H. Eberly, Phys. Rev. A, **29**, 690 (1984); F. T. Hioe and C. E. Caroll.
23. M. D. Feit, J. A. Fleck, Jr., and A. Steiger, J.Comput. Phys. **47**, 412 (1982); M. D. Feit and J. A. Fleck, Jr., J. Chem. Phys. **78**, 2578 (1984); J. N. Bardsley, A. Szöke, and J. M. Comella, J. Phys. B **21**, 3899 (1988); H. Kono and S. H. Lin, J. Chem. Phys. **84**, 1071 (1986).
24. S. Chelkowski and A. D.Bandrauk, J. Chem. Phys. **89**, 3618 (1988).
25. R. R. Alfano, ed. *The Supercontinuum Laser Source* (Springer-Verlag, New York, 1989).
26. P. Brumer, Adv. Chem. Phys. **47**, 201 (1981); S. A. Rice, ibid. **47**, 117 (1981) and ref. therein; T. Uzer , Phys. Rep. **199**, 74 (1991); D. W. Noid, M. L. Koszykowski and R. A. Marcus, Ann. Rev. Phys. Chem. **32**, 267 (1981).
27. S. N. Rai and K. G. Kay, J. Chem. Phys. **80**, 4961 (1984).
28. M. S. Sage and J. Jortner, Adv. Chem. Phys. **47**, 293 (1981).
29. J. S. Hutchinson, J. Chem. Phys. **85**, 7087 (1987).
30. G. Herzberg, *Molecular Spectra and Molecular Structure, Vol. III. Electronic Spectra and Electronic Structure of Polyatomic Molecules* (Van Nostrand Reinhold Company, New York, 1966), p. 588.
31. G. Strey and I. M. Mills, Mol. Phys. **26**, 129 (1973).
32. R. L. DeLeon and J. S. Muenter, J. Chem. Phys. **80**, 3992 (1984).

TIME DEPENDENT QUANTUM CALCULATIONS ON THE FEMTOSECOND PUMP/PROBE IONIZATION SPECTROSCOPY OF MOLECULAR SODIUM

Volker Engel

Fakultät für Physik, Albert-Ludwigs-Universität
Hermann-Herder-Straße 3, D-7800 Freiburg i.Br., Germany
also Institute for Theoretical Physics, University of California
Santa Barbara, Ca 93106, USA

I. INTRODUCTION

The real time spectroscopy of small molecules in the gas phase has become an active field since ultrashort laserpulses are available[1]. The use of laser pulses of 50-100 fs duration allows the experimentalist to follow periodical bound state motion, to monitor molecular fragments while they dissociate[1] or to study the time dependence of the absorption properties of molecules[2]. In most experiments the system under consideration is electronically excited by a first laserpulse and a coherent superposition of states, i.e. a localized wave packet is created. The probe pulse prepares the system in yet another electronic state which afterwards decays via a radiative transition. The total fluorescence as a function of delay time between pump and probe characterizes the dynamics within the intermediate electronic state. Alternatively the second excitation step may lead to ionization if the probe wavelength is chosen properly. There are experimental[3] as well as theoretical[4] investigations on such ionization processes. Femtosecond pulses were used to study the competing processes of molecular autoionization and fragmentation[5]. More recently Gerber and coworkers carried out molecular beam experiments with sodium molecules[6]. The total amount of Na_2^+ ions produced by a multiphoton transition was detected as a function of the temporal separation between the ultrashort pump and probe pulses. Below we will use time dependent quantum mechanical methods to describe and analyze this process. The basic theory and the underlying model is outlined in Sec. II. The results are presented in Sec. III. A comparison with experiment and conclusions are given in the final section.

II. THE CALCULATION

The singly excited electronic states of the neutral[7] as well as the ground state of the ionic[8] sodium molecule are well known. A peak wavelength of about 625 nm which was used in the experiment accesses at least four states which have to be included in a theoretical model, describing the excitation process. Others are either forbidden by selection rules or have comparably small Franck-Condon factors. The present calculation is restricted to the electronic X $({}^1\Sigma_g^+)$, A$({}^1\Sigma_u^+)$, $\Pi({}^1\pi_g)$ states of the neutral molecule and the ground state of the ion. For convenience we assign them with $|0>, |1>, |2>$ and $|3>$, respectively. The coordinate dependence of the transition dipole moments is ignored, i.e. the Condon approximation for the electronic transitions is applied. It is not clear if this approximation is good[9] in the present case and an improved calculation will have to incorporate them. Since the experiment is performed in a molecular beam[6], the initial ro-vibrational state distribution is cold. Thus we assume the molecule to be in its rotational and vibrational ground state before the electronic excitation occures. No rotational excitation is considered in what follows.

Coherence Phenomena in Atoms and Molecules in Laser Fields
Edited by A.D. Bandrauk and S.C. Wallace, Plenum Press, New York, 1992

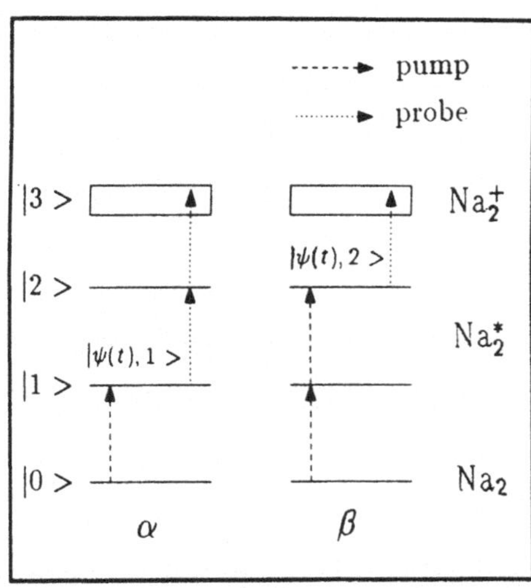

Fig. 1. Pump/probe excitation scheme for Na_2. Four electronic states are indicated. They correspond to the X-state ($|0>$), A-state ($|1>$), Π-state ($|2>$) and the ground state of the ion ($|3>$).

It was found numerically, that direct three photon processes have negligible intensity and thus this excitation step is not pursued further. The same result was found experimentally [10]. Within the present four state model two ionization pathways can be imagined (see Fig. 1):

(α):
The molecule absorbs one photon from the pump pulse and is excited from state $|0>$ to $|1>$. Another two photons from the probe pulse ionize the system via the intermediate state $|2>$.

(β):
A two photon transition $|0> \to |1> \to |2>$ is initiated by the pump pulse. One photon of the probe leads to ionization.

Since the experimental setup cannot distinguish between the two processes the ionic wave function is a coherent sum of the functions describing (α) and (β), respectively. Time dependent perturbation theory is used, to describe the ionization. In what follows we consider delay times when the two laserpulses have no temporal overlap. The quantum mechanical description of process (β) proceeds as follows: the two photon transition $|0> \to |1> \to |2>$ has to be calculated first. The second order formula for the nuclear wave function then writes as ($\hbar$=1)[11]:

$$|\psi(t),2> \sim \int_0^t dt'' U_2(t-t'')V_{21}(t'',\omega) \int_0^{t''} dt' U_1(t''-t')V_{10}(t',\omega)U_0(t')|\psi_i,0> . \quad (1)$$

This expression contains the initial wave function $|\psi_i, 0>$ in the electronic ground state. The time-dependent perturbations (for absorption) are given by

$$V_{nm}(t,\omega) = \vec{\mu}_{nm}\vec{E}_0 e^{-i\omega t} f(t), \quad (2)$$

where μ_{nm} (m=0,1;n=m+1) is the transition dipole moment between the states $|m>$ and $|n>$ and the electric field is characterized by the constant vector $\vec{E}_0$, frequency

ω and a function f(t) which characterizes the temporal shape of the laserpulse. The latter is chosen to be zero for negative times. Throughout the calculation we use Gaussians with a width (FWHM) of 80 fs for the envelopes f(t) and a frequency ω of 1.984 eV (~625 nm). $U_n(t)$ are the time evolution operators in the different electronic states. The absorption of another photon leads to ionization and within the Born-Oppenheimer approximation one obtains the ionic state vector

$$|\psi(t)>= \int_0^\infty dE'|E'>|3>|\psi_3(t)> . \qquad (3)$$

Here $|E'>$ denotes the state of the free electron, ejected with translational energy E'. In writing (3) we assumed the ionization cross section to be independent of E' [12]. The first order expression for the nuclear part of the wave function has the form [11]

$$|\psi(t),3>\sim \int_{t_3}^{t} dt' U_3(t-t',t_3)V_{32}(t',\omega-E')U_2(t',t_3)|\psi(t_3),2>, \qquad (4)$$

where t_3 is some intermediate time before the second pulse interacts with the molecules. Note that the field-molecule interaction V_{32} contains the difference between the field energy and the energy converted into the velocity of the electron. The dependence of the transition dipole moment on E' is ignored in what follows. After the probe pulse stops, the total ionic population settles to a constant and one obtains:

$$P^\beta = \lim_{t\to\infty} <\psi(t),3|\psi(t),3> . \qquad (5)$$

In the numerical calculation the continuum of translational energies E' is discretized (see below).
The alternative process (α) can be described as outlined above but with swapped indices for the electronic states in the respective expressions. This leads to the ionic population P^α. For fixed electric field parameters the populations depend exclusively on the delay time between pump and probe pulse, which is not explicitely retained in the equations above.
The calculation of the wave functions (Eq. 1,4) can be reduced to the time propagation of a wave packet[13]. This propagation is performed with the split operator formalism developed by Feit and Fleck[14].

III. RESULTS

In what follows the ionization pathways (α) and (β) (see Fig. 1.) will be discussed separately. A one photon absorption from the pump pulse prepares a coherent superposition of vibrational states in the bound electronic state $|1>$. The wave packet moves periodically between the classical turning points associated with its mean energy. A plot of the coordinate expectation value, calculated with the moving wave function exhibits a period of $\sim$ 310 fs. Because the wavelength of the pulse accesses the lower part of the bound state potential which is almost harmonic, it is not before 15 ps or so that dispersion becomes important. An analysis of the long time dynamics shows that after the packet has dispersed it takes about 48 ps ($\sim$ 157 vibrational periods) until it recovers its initial shape. This is known as "revival time"[15] and was discussed in connection with Rydberg wave packets [16] and also for the bound state motion of vibrational wave packets in single and double well potentials[17].
The periodic motion of the wave function is now probed by the time delayed second pulse. Molecular ions are built by a two photon transition via state $|2>$ and the initial state for this transition is the packet, localized in different areas of space when the probe pulse is switched on. In the middle panel of Fig. 2 we plot the ionic populations as a function of delay time between the pulses (i.e. the separation between the centers of the respective Gaussian shape functions f(t)) calculated with different values of $\omega - E'$.

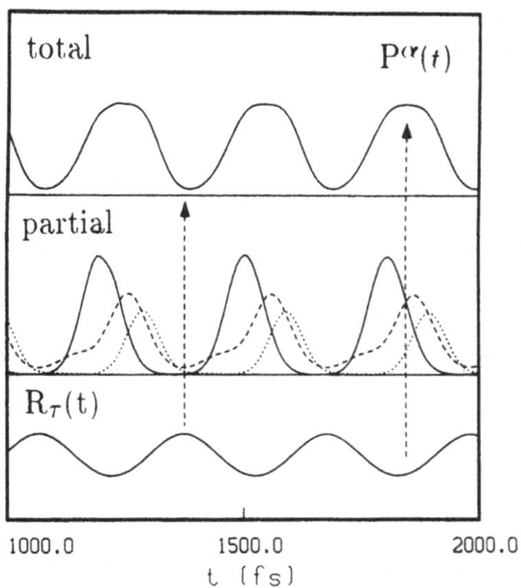

Fig. 2. Lower panel: the averaged coordinate expectation value (Eq. 6) in the electronic state |1 >. Middle panel: ionic population as a function of delay time between pump and probe pulse for fixed energies ω-E'of 1.06 eV (straight line), 1.10 eV (dashed line) and 1.14 eV (dotted line), respectively. Upper panel: total ionic population P^{α}.

The values of 1.06 eV (straight line), 1.10 eV (dashed line) and 1.14 eV (dotted line) are representative. For energies $\omega - E'$ outside of the interval 1.0 eV to 1.24 eV the calculated ionic population is negligible. The curves show a clearly oscillatory behavior with maxima separated by the period of the A-state motion ($\sim$ 310 fs). The extrema are shifted against each other. This is due to the different energies $\omega - E'$ which characterize the molecular ions within various vibrational states[19].
The upper panel of Fig. 2 contains the total ionic population. It is obtained by summing over all energies E' (see Eq. 3). The E' continuum was discretized with a grid step size of 0.02 eV which gives converged results. The summation broadens the peaks but the periodicity of the signal is maintained. To illustrate the connection between the signal and the wave packet dynamics we include the quantity

$$R_{\tau}(t) = \frac{1}{\tau} \int_{t-\tau/2}^{t+\tau/2} dt < \psi(t), 1|R|\psi(t), 1 > / < \psi(t), 1|\psi(t), 1 > \tag{6}$$

in Fig. 2. Here t is the center of the probe laserpulse andτ its width. R_{τ} is the coordinate expectation value in the state $|1>$ averaged over the length of the probe pulse and thus measures the average position of the packet during the excitation process. It can be seen in the figure that ions are built if the packet is localized near the inner turning point. Ions in lower vibrational states (low energy $\omega - E'$) stem from an excitation when $< R|\psi(t), 1 >$ moves inward and is close to the inner potential wall, while higher vibrational states of the ion are populated when the wave function moves outward and is somewhat right of the turning point. Note, that there is no one to one correspondence between a maximum in a curve for fixed energy $\omega - E'$ (Fig. 2, middle panel) and a certain value of $R_{\tau}(t)$. This is found in one photon transitions[18] as one would expect according to the Franck-Condon principle.

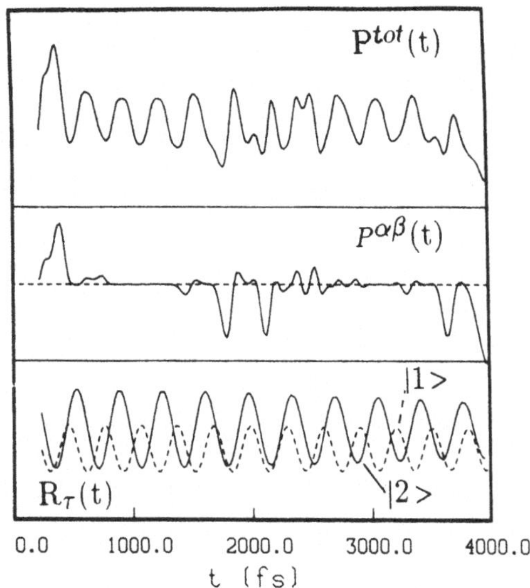

Fig. 3. Lower panel: averaged coordinate expectation value (Eq. 6) in the two electronic states as indicated. Middle panel: interference term $P^{\alpha\beta}$. Upper panel: total ionic population P^{tot} of Eq. 7. Arbitrary units are used and the dashed line corresponds to zero.

It does not hold for the present two photon excitation mechanism[19] where several electronic transitions occure. Essentially no ionic population is found when the packet is in the right hand region of the potential. The intermediate state $|2>$ which permits a favourable two photon transition at small distances but does not allow the same when the wave packet is located farther out, acts as a "filter" and the periodicity of the motion is carried over into the total ion signal. It is indeed the resonant transition via the intermediate state which is important here. The fraction of ions built by a nonresonant transition is negligible.
Another vibrational wave packet is built in state $|2>$, originating from the two photon transition as illustrated in Fig. 1 (process (β)). In this case a period of $\sim$ 360 fs can be extracted from the wave packet dynamics. The one photon ionization out of the Π-state ($|2>$) has been analyzed before[9] and the results can be summarized as follows: The ionic population for fixed energy $\omega - E'$ reflects the periodicity of the packet's motion in a similar way as found for process (α) (Fig. 2, middle panel). However, the sum over all final states (Eq. 3), yields an ion signal which no longer depends on the delay time between the pulses, i.e. the ionization probability is independent of the location of the wave packet during the ionization process. This, at the first glance, surprising result is due to the similarity of the potentials for the states $|2>$ and $|3>$, the chosen wavelength (in principle all vibrational ion states are accessible) and the Condon approximation [4,9].
Finally the coherent sum of the processes (α) and (β) has to be considered. The absolute square of the summed amplitudes yields

$$P^{tot} = P^{\alpha} + P^{\beta} + P^{\alpha\beta}. \tag{7}$$

Here P^{α} represents the oscillating signal produced by process (α), P^{β} the time independent signal corresponding to the alternative ionization pathway and $P^{\alpha\beta} = 2Re(<\psi^{(\alpha)}|\psi^{(\beta)}>)$ is an interference term. The latter is shown in Fig. 3 together with the total ion signal P^{tot} and the averaged positions (Eq. 6) in the states $|1>$ and $|2>$, respectively. If the interference term is zero the total population consists of the periodic signal P^{α} shifted by the constant background P^{β}. Strong interference is seen whenever the wave packets in the two states move in phase and are located in the same region of space. This can be taken from the expectation values in Fig. 3. The latter also reflect the different periods and spatial amplitudes of the bound state dynamics in the two electronic states. The total ion population P^{tot} has to be compared to the measured ion signal.

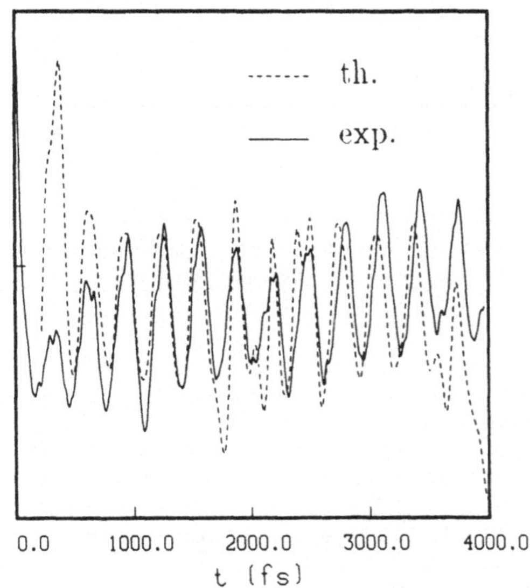

Fig. 4. Comparison between measured ion signal (straight line) and calculated total ionic population (dashed line). The curves are drawn as a function of delay time between pump and probe pulse and are normalized to each other at the fourth maximum of the theoretical curve.

4. COMPARISON WITH EXPERIMENT AND CONCLUSION

In Fig. 4 we compare the measured total ion signal[6] with the calculated total ion population for delay times up to 4 ps. The strong oscillations which stem from the wave packet motion in state $|1>$ are clearly seen in both curves. However, the interference pattern in the calculated population is by far too strong compared to the experimental curve. We note that the appearance of the interference features is strongly dependent on the pulse parameters used in the calculation. Instead of a large amplitude change the experiment shows only a soft modulation. The most probable explanation for the discrepancy is the excistence of another ionization process which is not included in the model presented here. It was proposed[6] that a doubly excited Rydberg state is involved in the ionization. There has been other evidence for the importance of such a state[5]. We have performed simulations and found the following: if contributions to the ionic population which stem from the one photon ionization out of state $|2>$ (i.e. process (β)) leading to lower vibrational states of the ion are neglected, a nearly perfect agreement between theory and experiment for delay times can be achieved shich points towards a competing path leading to ionization. The simulation, together with an extended analysis will be published in a forthcoming paper[19].

ACKNOWLEDGEMENTS

Stimulating discussions with T.Baumert and G.Gerber are acknowledged. The project was supported by the Deutsche Forschungsgemeinschaft through the SFB 276 "Korrelierte Dynamik hochangeregter atomarer und molekularer Systeme". This research was supported in part by the National Science Foundation under Grant No. PHY89-04035.

REFERENCES

1. For recent reviews see: L. R. Khundkar and A. H. Zewail, Ann. Rev. Phys. Chem. 41:15 (1990); H. Metiu and V. Engel, J. Opt. Soc. Am. B7:1709 (1990).
2. J. A. Misewich, J. H. Glownia, J. E. Rothenberg, and P. P. Sorokin, Chem. Phys. Lett. 150:374 (1988); J. H. Glownia, J. A. Misewich, and P. P. Sorokin, J. Chem. Phys. 92:3335 (1990); R. E. Walkup, J. A. Misewich, J. H. Glownia, and P. P. Sorokin, Phys. Rev. Lett. 65:2366 (1990).
3. J. M. Wiesenfeld and B. I. Greene, Phys. Rev. Lett. 51:1745 (1983); J. M. Smith, C. Lakshminarayan, and J. L. Knee, J. Chem. Phys. 93:4475 (1990).
4. V. Engel and H. Metiu, Chem. Phys. Lett. 155:77 (1989); M. Seel and W. Domcke, Chem. Phys. 1512:59 (1991).
5. T. Baumert, B. Bühler, R. Thalweiser and G. Gerber, Phys. Rev. Lett. 64:733 (1990).
6. T. Baumert, M. Grosser, V. Weiss, E. Wiedenmann and G. Gerber, Phys. Rev. Lett. (submitted); T. Baumert, V. Weiss and G. Gerber, J. Phys. Chem. (submitted).
7. A. J. Taylor, K. M. Jones, A. L. Schawlow, J. Opt. Soc. Am. 73:994 (1983); M. E. Kaminsky J. Chem. Phys. 66:4951 (1977).
8. C. Bordas, P. Labastie, J. Chevaleyre and M. Broyer, Chem. Phys. 129:21 (1989).
9. V. Engel; Chem. Phys. Lett. 178:130 (1991).
10. T. Baumert, G. Gerber private communication.
11. R. Loudon, "The quantum theory of light," Clarendon, Oxford (1983).
12. W. A. Chupka, in: "Ion-Molecule Reactions, Vol. 1," J. L. Franklin, ed., Butterworths, London, (1972).
13. V. Engel, Comput. Phys. Commun. 63:228 (1991).
14. J. A. Feit, J. R. Morris and M. D. Feit, Appl. Phys. 10:129 (1976).
15. J. Parker and C. R. Stroud Jr., Phys. Rev. Lett. 56:716 (1986); I. Sh. Averbukh and N. F. Perelman, Phys. Lett. A 139:449 (1989).
16. G. Alber and P. Zoller, Physics Reports,199:231 (1991).
17. W. Strunz, G. Alber and J. S. Briggs, J. Phys. B: At. Mol. Opt. Phys. 23:L697 (1990); ibid. (submitted).
18. V. Engel and H. Metiu,J. Chem. Phys. 93:5693 (1990).
19. T. Baumert, V. Engel and G. Gerber (to be published).

ANALYTIC MODELS OF PULSE EFFECTS IN TWO-PHOTON MOLECULAR SPECTROSCOPY

Eric E. Aubanel and André D. Bandrauk

Département de Chimie, Faculté des Sciences, Université de Sherbrooke, Sherbrooke, Québec, Canada, J1K 2R1

INTRODUCTION

Recent experiments with ultrafast lasers have demonstrated that bound and dissociating motion at small internuclear distances can be resolved[1]. This is accomplished using a pump-probe pulse sequence. The first pulse prepares a coherent superposition of excited state levels, the evolution of which can be followed by excitation to a final state using a probe pulse. The behaviour of the final signal, such as fluorescence from the final state, as a function of the time delay between the pulses, contains information on the nature of the intermediate potential, provided that the pulses are short enough to spacially resolve the motion.

The qualitative character of this behaviour can be examined using lowest-order perturbation theory. Radzewicz and Raymer[2] studied the evolution of wave packets in sodium dimer, using perturbation theory together with numerical stationary wave functions. Strunz *et al.*[3] derived a semi-classical expression for the final-state probability for two-photon excitation of diatomics, and characterized the motion of the vibrational wave-packet. Their work was based on the work of Alber *et al.* on atomic Rydberg wave packets[4]. We present analytic expressions for dynamics generated by femtosecond pump-probe pulses, with potentials approximated by harmonic oscillators and linear potentials. The potentials are illustrated in Fig. 1. In all three cases the probe pulse returns the wave packet to the ground state. In Fig. 1a the ground state is a harmonic oscillator (HO), and the excited state has a linear repulsive potential (model 1). The order of the

Coherence Phenomena in Atoms and Molecules in Laser Fields
Edited by A.D. Bandrauk and S.C. Wallace, Plenum Press, New York, 1992

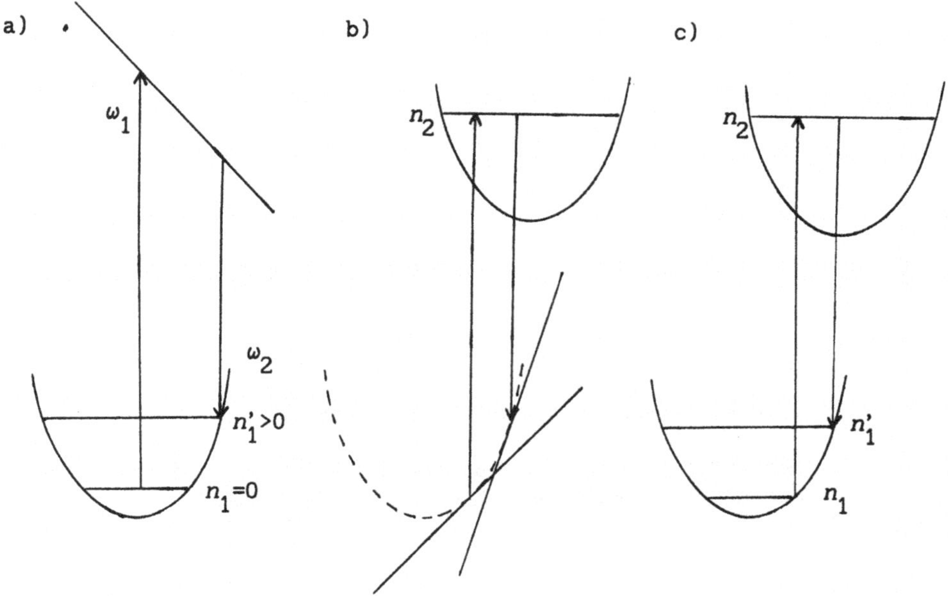

Fig. 1. a) model 1, b) model 2, c) model 3.

two potentials is reversed in Fig. 1b (model 2). In Fig. 1c the ground and excited states are represented by shifted HO's (model 3).

MODEL 1 : REPULSIVE EXCITED STATE

The two-photon transition probability is, from second-order perturbation theory :

$$P = \Bigg| \frac{1}{\hbar^2}\int_{-\infty}^{\infty} dE_c \int_{-\infty}^{\infty} \langle f|\vec{\mu}\cdot\vec{E}_2(t')|c\rangle\, e^{i(E_f-E_c)t'/\hbar} \cdot \int_{-\infty}^{t'} \langle c|\vec{\mu}\cdot\vec{E}_1(t'')|g\rangle\, e^{i(E_c-E_g)t''/\hbar}\, dt''dt' \Bigg|^2 \tag{1.1}$$

where |c> refers to the continuum state at energy E_c, |g> and |f> to the initial and final states respectively, and

$$\vec{E}_1(t) = \vec{\varepsilon}_1\mathcal{E}_1(t-t_1)e^{-i\omega_1 t} + \text{c.c.} \; ; \; \vec{E}_2(t) = \vec{\varepsilon}_2\mathcal{E}_2(t-t_2)e^{-i\omega_2 t} + \text{c.c.} \tag{1.2}$$

If the pulses do not overlap in time one can rewrite Eq. 1.1 as :

$$P = \left| \frac{1}{\hbar^2} \int_{-\infty}^{\infty} dE_c \, e^{-iE_c\tau/\hbar} \langle f|\vec{\mu}\cdot\vec{\varepsilon}_2|c\rangle\langle c|\vec{\mu}\cdot\vec{\varepsilon}_1|g\rangle \, \tilde{\mathcal{E}}_2(\Delta_2) \, \tilde{\mathcal{E}}_1(\Delta_1) \right|^2 \quad (1.3)$$

$$= \left| \frac{V^2}{\hbar^2} \int_{-\infty}^{\infty} dE_c \, e^{-iE_c\tau/\hbar} \langle f|c\rangle\langle c|g\rangle \, \tilde{\mathcal{E}}_2(\Delta_2) \, \tilde{\mathcal{E}}_1(\Delta_1) \right|^2 \quad (1.4)$$

where $\tau = t_2 - t_1$, $\tilde{\mathcal{E}}_i(\Delta_i) = \int_{-\infty}^{\infty} dt \, \mathcal{E}_i(t) \, e^{i\Delta_i t}$, $\Delta_1 = E_c - E_g - \hbar\omega_1$, $\Delta_2 = E_f - E_c + \hbar\omega_2$, and the usual Franck-Condon and rotating-wave approximations have been used (*cf* Eq. 1 in ref. 3).

We can rewrite this equation in terms of a Green's function, as follows

$$P = \left| \frac{V^2}{\hbar^2} \int_{-\infty}^{\infty} dt' \int_{-\infty}^{\infty} dt \, \exp\left[\left\{ (E_f + \hbar\omega_2)(t' - t_2) - (E_g + \hbar\omega_1)(t - t_1) \right\} i/\hbar \right] D_{fg} \right|^2$$

$$D_{fg} = \int_{-\infty}^{\infty} dE_c \, e^{-iE_c(t'-t)/\hbar} \langle f|c\rangle\langle c|g\rangle \, \mathcal{E}_2(t')\mathcal{E}_1(t) \quad (1.4a)$$

The above expression contains a time-dependent Green's function, which can be transformed to the energy representation :

$$G(t'-t) = \int_{-\infty}^{\infty} dE_c \, e^{-iE_c(t'-t)/\hbar} |c\rangle\langle c|$$

$$= \frac{i}{2\pi} \int_{-\infty}^{\infty} dE \, e^{-iE(t'-t)/\hbar} G(E)$$

$$G(E) = \int_{-\infty}^{\infty} dE_c \frac{|c\rangle\langle c|}{E - E_c + i\varepsilon} \quad (1.4b)$$

Therefore D_{fg} becomes :

$$D_{fg} = \frac{i\mathcal{E}_1(t)\mathcal{E}_2(t')}{2\pi} \int_{-\infty}^{\infty} dE \, e^{-iE(t'-t)/\hbar} \langle f|G(E)|g\rangle \quad (1.4c)$$

Inserting this into the first line of Eq. 1.4a yields :

$$P = \left| \frac{V^2}{\hbar^2} \int_{-\infty}^{\infty} dE \, \tilde{\mathcal{E}}_1(E_f - E + \hbar\omega_2) \, \tilde{\mathcal{E}}_2(E - E_g - \hbar\omega_1) \, \langle f|G(E)|g\rangle \, e^{-iE\tau/\hbar} \right|^2 \quad (1.4d)$$

This expression can easily be generalized to nth order (ie. for a succession of n pulses) :

$$P^{(n)} = \left| \frac{V^n}{\hbar^n} \int_{-\infty}^{\infty} dE_1 \ldots \int_{-\infty}^{\infty} dE_{n-1}\, \tilde{\mathcal{E}}_1(\Delta_1)\ldots\tilde{\mathcal{E}}_n(\Delta_n) \langle f|G(E_{n-1})\ldots G(E_1)|g\rangle \cdot e^{-i(E_1\tau_1+\ldots+E_{n-1}\tau_{n-1})/\hbar} \right|^2 \qquad (1.4e)$$

where $\Delta_n = E_f - E_1 \pm \hbar\omega_n$, $\Delta_i = E_{i-1} - E_i \pm \hbar\omega_i$, $\Delta_1 = E_{n-1} - E_g \pm \hbar\omega_1$, and $\tau_i = t_{i+1} - t_i$. Here we are concerned with two-photon processes, and will therefore concentrate on the second order expression (Eq. 1.4).

Let the pulse lineshapes be gaussian :

$$\mathcal{E}_i(t) = \exp\left(-\alpha(t-t_i)^2\right) \quad ; \quad t_2 - t_1 = \tau \qquad (1.5)$$

After some manipulation Eq. 1.4 becomes :

$$P = \left| \frac{\pi V^2}{\alpha \hbar^2} C \int_{-\infty}^{\infty} dE_c \langle f|c\rangle\langle c|g\rangle \exp\left[-\frac{\left(E_c - E_T/2\right)^2}{2\alpha\hbar^2} - \frac{i\tau E_c}{\hbar} \right] \right|^2$$

$$\text{where } C = \exp\left[-(E_f - E_g + \hbar\omega_2 - \hbar\omega_1)^2 / 8\alpha\hbar^2 \right]$$

$$\text{and } E_T = E_f + E_g + \hbar\omega_1 + \hbar\omega_2 \qquad (1.6)$$

Note that the factor of C ensures energy conservation. For the repulsive state we use a linear potential, $V_c = -Fx + E_0$, since the eigenfunction is known[5] :

$$\phi_c = N_c \,\mathrm{Ai}\left(-a_c(x+x_e)\right) \quad ; \; N_c = a_c F^{-1/2} \; ; \; a_c = (2\mu F/\hbar^2)^{1/3}$$

$$x_e = \frac{E_c - E_0}{F} \qquad (1.7)$$

We will take the initial state to be $n_1=0$, and the final state to be some $n_1' > 0$. To obtain an analytic expression, we make two approximations. For the first transition we employ the reflection method for the linear potential, where the Airy function is replaced by a delta function at the turning point[5]:

$$\phi_c^{rm} = F^{-1/2}\, \delta(x, x_c) \qquad ; \; x_c = -x_e \qquad (1.8)$$

The Franck-Condon overlap becomes :

$$\langle 0|c^{rm}\rangle = N_0 F^{-1/2} \exp\left[-0.5\beta x_c^{\,2}\right] \tag{1.9}$$

where $x_c=-x_e$ is the turning point of the linear potential, and $N_0=(\beta/\pi)^{1/4}$ and $\beta=\mu\omega_g/\hbar$ are the parameters of the harmonic oscillator (with frequency ω_g and reduced mass μ). For the second approximation we employ the reflection method for the ground state, where the region around the right turning point x_2 of the final state is represented by a linear potential (of slope $F_2=kx_2$, where k is the force contant of the ground state ; see Fig. 1b)[5] :

$$\phi_r = N_r \, \mathrm{Ai}\left(a_r(x-x_2)\right) \quad ; \; N_r = a_r F_2^{-1/2}(\omega_g/2)^{1/2} \quad ; \; a_r=(2\mu F_2/\hbar^2)^{1/2} \tag{1.10}$$

The second overlap integral is evaluated using the reflection method for the second linear potential :

$$\langle f|c\rangle = N_c(\omega_g/2F_2)^{1/2} \, \mathrm{Ai}\left(-a_c(x_2 + x_e)\right) \tag{1.11}$$

We could not employ the usual reflection method, i.e. for the upper linear potential, for the above overlap integral since the Franck-Condon region (around x_2) is to the right of the turning point ($x=x_c\cong0$) of the repulsive potential. Therefore the Airy function oscillates in the neighbourhood of x_2, and is not strongly peaked (this motivated the use of the stationary phase method ; see below). The reflection method for the ground state was found to give qualitative agreement of the two-photon transition probability with the exact expression for the case of two displaced harmonic oscillators (see next section).

After inserting Eqs. 1.9 and 1.11 into Eq. 1.6, and using the integral representation of the Airy function[6], one must evaluate the following integral

$$I = \int_{-\infty}^{\infty} dp \int_{-\infty}^{\infty} dE_c \exp\left[-\frac{\beta x_c^2}{2} + \frac{ip^3}{3} - ipa_c(x_2-x_c) - \frac{(E_c-E_T/2)}{2\alpha\hbar^2} - \frac{i\tau E_c}{\hbar}\right]. \tag{1.12}$$

After integrating over E_c, and rearranging, one is left with :

$$I = F\left(\frac{\pi}{a}\right)^{1/2} \exp\left[-\frac{E'^2}{2\alpha\hbar^2} - \frac{i\tau E_0}{\hbar} + \frac{b^2}{4a}\right] \int_{-\infty}^{\infty} dp \, \exp\left[-c_2^2p^2 - c_1p + ip^3/3 + ixp\right]$$

where

$$a = \frac{\beta}{2} + \frac{F^2}{2\alpha\hbar^2} \quad ; \; b = \frac{FE'}{\alpha\hbar^2} + i\tau F/\hbar \; ; \; c_1 = \frac{a_c \tau F}{2a\hbar} \quad ; \; c_2 = \frac{a_c^{\;2}}{4a}$$

$$x = \left[\frac{FE'}{2a\alpha\hbar^2} - x_2 \right] a_c \qquad E' = E_0 - E_T/2 \tag{1.13}$$

If x is negative and $|x|>>0$, then one can use the stationary phase method[7] to evaluate the integral. Rewriting the integral in Eq. 1.13 as :

$$I_2 = |x|^{1/3} \int_{-\infty}^{\infty} dy \, \exp\left[-c_2^2 |x|^{2/3} y^2 - c_1 |x|^{1/3} y \right] \exp\left[i|x| \left(y^3/3 - |x|^{1/3} y \right)\right] \tag{1.14}$$

one can see that the phase $\phi = y^3/3 - |x|^{1/3}y$ is stationary at $y_0 = \pm|x|^{1/6}$. Therefore, I_2 can be approximated by :

$$I_2 \simeq |x|^{1/3} \left[\frac{\pi}{|x|^{7/6}} \right]^{1/2} e^{-|x| c_2^2} \left\{ \exp\left[-c_1 |x|^{1/2} - \frac{2}{3} i |x|^{3/2} + i\frac{\pi}{4} \right] + \exp\left[c_1 |x|^{1/2} + \frac{2}{3} i |x|^{3/2} - i\frac{\pi}{4} \right] \right\}$$

$$= \left[\frac{\pi}{|x|^{1/2}} \right]^{1/2} \exp\left[-|x| c_2^2 + c_1 |x|^{1/2} + \frac{2}{3} i |x|^{3/2} - i\frac{\pi}{4} \right] \tag{1.15}$$

Assembling all of the terms into Eq. 1.6 yields :

$$P(\tau) = \left[\frac{CN_0 a_c^2 \pi^2 V^2}{\alpha\hbar^2} \right]^2 \frac{\omega_g}{2a|x|^{1/2} F_2} \exp\left[-E'^2 \left(\frac{1}{\alpha\hbar^2} - \frac{F^2}{2a\alpha^2\hbar^4} \right) - \frac{|x| a_c^2}{2a} \right]$$

$$\cdot \exp\left[-\frac{F^2\tau^2}{2a\hbar^2} + \frac{a_c F |x|^{1/2} \tau}{a\hbar} \right] \tag{1.16}$$

This is clearly a displaced gaussian, with maximum at :

$$\tau_m = a_c |x|^{1/2} \hbar/F$$

$$= \sqrt{2\mu(x_2 - FE'/2a\alpha\hbar^2)/F} \tag{1.17}$$

which is equal to the classical transit time from $x=0$ to $x=x_2$ in the limit that $x_2 >> FE'/2a\alpha\hbar^2$.

The expression for P is obtained by replacing the integral over the energy of the repulsive state in Eq. 1.6 by a sum over the bound states :

$$P = \left| \frac{\pi V^2 C}{\alpha \hbar^2} \sum_{n=0}^{\infty} \langle f|n\rangle\langle n|g\rangle \exp\left[- \frac{\left(E_n - E_T/2\right)^2}{2\alpha\hbar^2} - \frac{i\tau E_n}{\hbar} \right] \right|^2 \quad (1.13)$$

The oscillatory behaviour of P is evident when we rewrite Eq. 1.13 :

$$P = \left\{ \frac{\pi V^2}{\alpha \hbar^2} C \right\}^2 \left\{ \sum_{j=0}^{\infty} D_j^2 + 2 \sum_{j=1}^{\infty} \sum_{k=j}^{\infty} D_j D_k \cos\left[\tau(E_j - E_k)/\hbar\right] \right\} \quad (1.14)$$

$$\text{where } D_n = \langle f|n\rangle\langle n|g\rangle \exp\left[- \frac{\left(E_n - E_T/2\right)^2}{2\alpha\hbar^2} \right]$$

The overlap integral between an Airy function and harmonic oscillator wave function can be evaluated analytically using the reflection method. The overlap between two displaced harmonic oscillator wave functions is well known.

We will investigate model 2 as an approximation to model 3, ie. the Franck-Condon region of the ground state is approximated by a linear potential (see Eq. 1.10). This is expected to be reasonable for a highly excited ground state. The overlap with the harmonic oscillator wave function is, using the reflection method,

$$\langle n|c^{rm}\rangle = N_n \sqrt{\frac{\omega_g}{2F_2}} \exp\left[-0.5\beta_e x_2^2\right] H_n\left[\sqrt{\beta_e}\, x_2\right] \quad ; \beta_e = \mu\omega_e/\hbar \quad (1.16)$$

where ω_e is the frequency of the excited state. The two overlap integrals in Eq. 1.14 differ in the values of F_2 and x_2 if $n_1 \neq n_1$.

We used potential parameters corresponding to the transition between the ground state and the second bound excited (B) state of I_2 [8]: μ=63.45 amu, Δr_{eq}=0.3 Å, ω_g=210 cm^{-1}, ω_e=120 cm^{-1}, ΔE=1.6x10^4 cm^{-1} (energy difference between classical minima) ; we set V = 100 cm^{-1}.

In Fig. 2 we compare the transition probability versus the time delay between the pulses for models 2 and 3, for pulse widths ($t_a = 1/\sqrt{\alpha}$) of 10 (Fig. 2a) and 50 fs (Figs. 2b-f), and for selected initial and final states

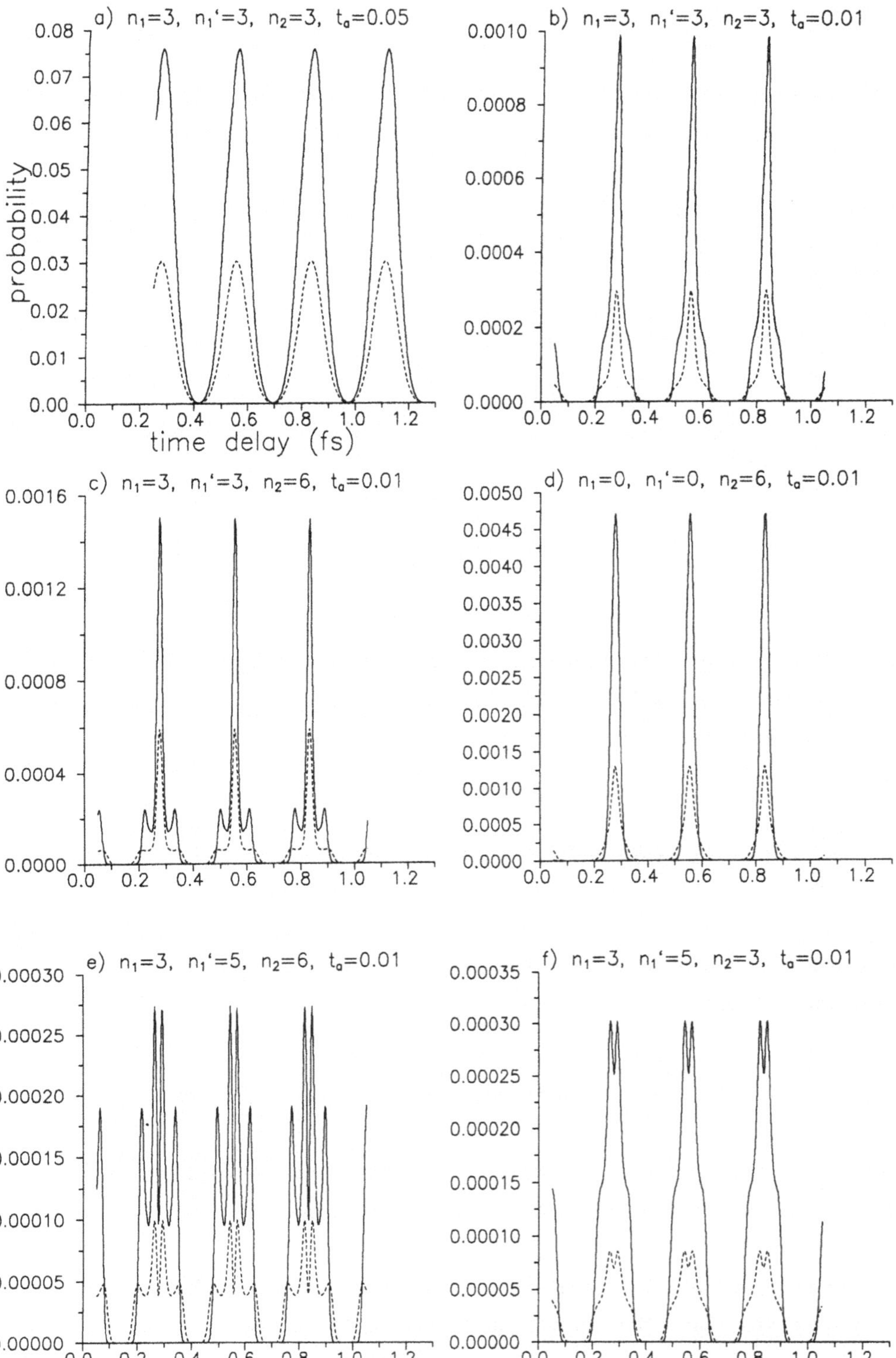

Fig. 2. Final state probability for pump-probe process $n_1 \rightarrow n_2 \rightarrow n_1'$ for displaced harmonic oscillators : ω_g= 210 cm^{-1}, ω_e=120 cm^{-1}, Δr_{eq}=0.3Å. Solid line : exact ; dashed lines : reflection method for ground state.

(n_1 and n_1') and frequencies ω_1,ω_2 (resonant with intermediate state n_2). It is evident that the reflection method reproduces qualitatively, but under-estimates the magnitude of, the shape of the transition probability curves. These exhibit oscillations with the period of the upper harmonic oscillator. In Fig. 2a $n_1=n_1'=3$, and $n_2=3$. Fig 2b corresponds to Fig. 2a, but with a shorter pulse width (10 fs). The wave packet produced by the shorter pulse is more spacially constrained, which results in sharper peaks in the final-state probability. There is more structure as well, which is a function of n_1,n_1' and n_2 : In Fig. 2c, ω_1 and ω_2 were increased to resonance with $n_2=6$; In Fig. 2d, $n_1=n_1'=0$, $n_2=6$. Even more structure is evident in Figs. 2e-f, where $n_1=3$ and $n_1'=5$; $n_2=6,3$ for Figs. 2e,f respectively. The results are identical for Figs. 2e,f when (n_1,ω_1) are exchanged with (n_2,ω_2).

DISCUSSION

The analytic expression for the pump-probe transition probability for model 1 is consistent with what one would expect classically, as long as the right turning point of the final state (x_2) is large enough. The magnitude of x_2 required is determined by the pulse duration : for a delta function pulse, the classical transit time on the repulsive surface τ_{cl} is reproduced for any $x_2>0$; for longer pulses the transition probability will peak at shorter times than τ_{cl} if x_2 is too small. The agreement with the classical result is due to the stationary phase approximation, which in turn was applicable because the Frank-Condon region of the final state overlapped the oscillating region of the excited state. In situations where the final state is not as localized non-classical results can be expected. Such a situation is found in a diatomic model of ICN, where the final state is a second repulsive potential[9] ; unfortunately the associated overlap integral cannot be evaluated analytically.

In the case of models 2 and 3, the probability depends not only on the pulse width, but also on the initial and final states, and on the radiation frequencies. The shape of the wave packets depends on the Franck-Condon factors (FCF) and on the pulse width. With a wide enough pulse (50 fs) there is no structure and one observes what one would expect for the classical motion of a gaussian wave packet. However, for narrower pulses and for excited initial states, the wave packets produced are no longer gaussian, and exhibit structure which depends on the above mentioned parameters. Excitation to and from the ground state $n_1=0$ yields recurring gaussians ; as the initial/final states are increased in energy, the different FCF result in more structure (compare Figs. 2c,d,e). The level of excitation of the upper state affects the structure of the probability as well (compare Figs. 2b-c, and Figs. 2e-f). Thus, as discussed by Bersohn et al.[9], quantum mechanics does not allow for a preparation-free discussion of the dynamics, nor does it permit the assumption of localized wave packets.

The extent of agreement between models 2-3 is surprising. The agreement does not seem to depend on the energy of the initial and final states. This suggests that it is mainly the turning point region of the initial/final wavefunctions which influence the final-state probability.

CONCLUSION

We have presented analytic models for the two-pulsed-photon transition probability. Using the stationary phase and reflection approximations, we have derived a semi-classical expression for a repulsive excited state, and bound initial/final states. For the case of a bound excited state, we found that the shape of the wave packet and transition probability depended on factors such as the pulse width, initial/final state energies, and laser frequencies. We intend to extend these models to two degrees-of-freedom, to model diatomics. Other developments could include extension to n-pulse processes (Eq. 1.4e), and to Morse oscillator bound states.

REFERENCES

1. M. Gruebele and A.H. Zewail, Phys. Today **43**, 24 (1990).
2. C. Radzewicz amd M.G. Raymer, in 'Atomic and Molecular Processes with Short Intense Laser Pulses', NATO ASI Series B vol. 171, A.D. Bandrauk ed., Plenum New York, 1988.
3. W. Strunz, G. Alber, and J.S. Briggs, J. Phys. B **23**, L697 (1990).
4. G. Alber, H. Ritsch, and P. Zoller, Phys. Rev. A **34**, 1058 (1986).
5. A.D. Bandrauk and J.-P. Laplante, Can. J. Chem. **55**, 1333 (1977).
6. J. Spanier and K.B. Oldham, 'An Atlas of Functions', Hemisphere, Washington, 1987.
7. G.F. Carrier, M. Krook, and C.E. Pearson, 'Functions of a Complex Variable', Mcgraw-Hill, New-York, 1966.
8. R,M, Bowman, M. Dantus, and A.H. Zewail, Chem. Phys. Lett. **161**, 297 (1989).
9. J.L. Krause, M. Shapiro, and R. Bersohn, J. Chem. Phys. **94**, 5499 (1991).

MOLECULAR EXCITATIONS CREATED BY AN ARBITRARILY SHAPED LASER PULSE: AN ADIABATIC WAVEPACKET PROPAGATION STUDY

T. Tung Nguyen-Dang, H. Abou-Rachid and D. Tanguay

Département de Chimie
Université Laval, Québec
Québec, G1K 7P4, Canada

I. INTRODUCTION

In pump-probe experiments, femtosecond laser pulses permit excited states to be accessed on a time-scale shorter than typical times for molecular motions, and are being used to investigate the dynamics of transitory species[1]. The theoretical descriptions of these experiments using wavepacket propagation techniques are currently limited to a first-order perturbative treatment of the time-dependent Schrödinger equation and rely on the rotating-wave approximation (RWA)[2]. A non-perturbative, non-RWA wavepacket description of molecular excitations created by ultra-short and intense laser pulses is proposed in the present communication, where a time-dependent electronic representation is introduced by solving the local, nuclear-coordinate parameterized Schrödinger equation describing the laser-driven, multistate electronic system. As shown in previous works[3], these solutions can be identified with the eigenstates of a time-dependent effective hamiltonian with respect to which the exact time evolution of the N-state electronic system is adiabatic. For a two-channel system, the time-dependent adiabatic representation depends on an effective area of the laser pulse and geometrical phases which are also functionals of the laser pulse-shape. Due to this explicit dependence on the characteristics of the laser pulse, the representation is appropriate for studying pulse-shape effects in the short-time dynamics of laser-driven molecular systems. The application of this representation to study the dynamics of the pre-dissociating NaI molecule by wavepacket propagation is given in detail below as an illustration of the methodology. Sample results of calculations on the dynamics of the collisional Na + Ar system excited by short pulses are also given to illustrate dramatic pulse shape effects on the excitation dynamics of this system.

Coherence Phenomena in Atoms and Molecules in Laser Fields
Edited by A.D. Bandrauk and S.C. Wallace, Plenum Press, New York, 1992

II. TIME-DEPENDENT ADIABATIC REPRESENTATION

The hamiltonian describing an N-channel molecular system interacting with an e.m. field through an electric dipole interaction is

$$\hat{H}\ (t)= \hat{T}_N + \int d^{3N}\vec{R}_\alpha\ |\vec{R}_\alpha>\hat{\mathcal{H}}_{el}(\vec{R}_\alpha, t)\ <\vec{R}_\alpha|, \tag{1}$$

where $\hat{T}_N$ is the nuclear kinetic energy operator, $\vec{R}_\alpha$ denotes collectively the nuclear coordinates and $\hat{\mathcal{H}}_{el}$ is the N-state electronic hamiltonian

$$\hat{\mathcal{H}}_{el}(\vec{R}_\alpha, t) = \sum_{i=1}^{N} \varepsilon_i(\vec{R}_\alpha)|i><i| + \left(\sum_{i\neq j} V_{ij}(\vec{R}_\alpha, t)|i><j| + h.c. \right), \tag{2}$$

$$V_{ij}(\vec{R}_\alpha, t)= \vec{\mu}_{ij}.\vec{E}(t), \tag{3}$$

$\vec{\mu}_{ij}$ being a matrix element of the total molecular dipole operator in the electronic adiabatic basis $\{\ |i> \equiv |i;\ R_\alpha>\ \}$.

We are interested in the dynamics of this N-channel system under the action of an ultrashort laser pulse. The short duration of the pulse suggests that laser-induced excitations occur almost vertically, i.e. at nearly frozen nuclear positions. The problem of strong laser-induced interactions between the N channels should then resemble a N-level problem defined at each nuclear location. This is the central physical idea underlying the introduction of a time-dependent electronic representation whose orthonormal basis states are obtained by solving the electronic Schrödinger equation

$$i\hbar\partial_t|\varphi_i;\ \vec{R}_\alpha, t> = \hat{\mathcal{H}}_{el}(t)|\varphi_i;\ \vec{R}_\alpha, t>, \tag{4}$$

with initial conditions

$$|\varphi_i;\ \vec{R}_\alpha, t_o> = |i;\ \vec{R}_\alpha>\ . \tag{5}$$

Alternatively, this representation may be defined as the span of the eigenvectors of a hermitian operator $\hat{h}_{el}$ defined by[3]

$$\left[\left[\hat{\mathcal{H}}_{el}\ , \hat{h}_{el} \right] -i\hbar\ \partial_t\hat{h}_{el}\ ,\ \hat{h}_{el} \right] = 0. \tag{6}$$

If the spectrum $\{\tilde{\varepsilon}_i(\vec{R}_\alpha, t)\ \}$ of $\hat{h}_{el}$ is non-degenerate for all $\vec{R}_\alpha$ and all t then the condition expressed by eq.(6) ensures that the electronic Schrödinger equation (4) is decoupled in the representation spanned by the eigenstates of $\hat{h}_{el}$. In particular, the state $|\varphi_i;\ \vec{R}_\alpha, t>$ defined by eqs.(4) and (5) can be identified simply as

$$|\varphi_i;\ \vec{R}_\alpha, t> = \exp\{-\frac{i}{\hbar}\int_{t_0}^{t} dt'\lambda_i(\vec{R}_\alpha, t')\}\ |\tilde{\varphi}_i;\ \vec{R}_\alpha, t>, \tag{7a}$$

where

$$\hat{h}_{el}(\vec{R}_\alpha, t)|\tilde{\varphi}_i; \vec{R}_\alpha, t> = \tilde{\varepsilon}_i(\vec{R}_\alpha, t)|\tilde{\varphi}_i; \vec{R}_\alpha, t> \tag{7b}$$

and

$$\lambda_i(\vec{R}_\alpha, t) \equiv <\tilde{\varphi}_i;\ \vec{R}_\alpha, t|(\ \hat{\mathcal{H}}_{el} - i\hbar\partial_t\ )|\tilde{\varphi}_i;\ \vec{R}_\alpha, t>. \tag{7c}$$

Let $D_i(\vec{R}_\alpha, t)$ denotes nuclear amplitudes associated with the electronic basis state $|\varphi_i;\ \vec{R}_\alpha, t>$, i.e.

$$|\Psi, t> = \sum_{i=1}^{N} \int dV_N\ D_i(\vec{R}_\alpha, t).|\varphi_i;\ \vec{R}_\alpha, t>|\vec{R}_\alpha>. \tag{8}$$

The total time-dependent Schrödinger equation for the laser-driven molecular system then gives

$$i\hbar\partial_t D_k = \hat{T}_N D_k + \sum_{i\neq k}^{N} \left(\sum_\alpha <\varphi_k|\hat{P}_\alpha|\varphi_i>(\hat{P}_\alpha/M_\alpha)D_i + <\varphi_k|\hat{T}_N|\varphi_i>D_i \right), \tag{9}$$

The coupled equations (9) involve time-dependent, laser-induced non-adiabatic coupling terms which arise from the parametric dependence of the new electronic basis states $|\tilde{\varphi}_i>$ on the nuclear coordinates $\vec{R}_\alpha$. These residual non-adiabatic couplings have been shown to be negligible if $(t-t_0)$ is sufficiently small and the dipole transition moments $\vec{\mu}_{ij}$ are sufficiently slowly varying with respect to the nuclear coordinates[4]. When these conditions are met, the nuclear amplitudes D_i become decoupled. Using the explicit expressions of the new basis states $|\varphi_i>$ in terms of the field-free states $|k>$, the nuclear amplitudes D_i can be converted to amplitudes or wavepackets χ_k in the field-free B.O. representation.

For a two-channel system, N = 2, the effective hamiltonian $\hat{h}_{el}$ and its eigenstates depend on an effective area $\mathcal{A}$ and phase angle δ defined by[3,4]

$$\mathcal{A}(\vec{R}_\alpha, t) \equiv \frac{2}{\hbar}\int_{t_0}^{t} dt'\ V_{12}(\vec{R}_\alpha, t')\sin\delta(\vec{R}_\alpha, t'), \tag{10a}$$

$$(\ \delta + \omega_{12}\ )\tan\mathcal{A}(\vec{R}_\alpha, t) = \frac{2}{\hbar} V_{12}(\vec{R}_\alpha, t)\cos\delta(\vec{R}_\alpha, t), \tag{10b}$$

with

$$\omega_{12}(\vec{R}_\alpha, t) \equiv (\ \varepsilon_1(\vec{R}_\alpha) - \varepsilon_2(\vec{R}_\alpha)\ )/\hbar. \tag{11}$$

Explicitly, defining

$$\hat{\mathcal{U}}_0(\vec{R}_\alpha, t) \equiv \exp\{ -\frac{i}{\hbar} \sum \varepsilon_k(\vec{R}_\alpha)|k\rangle\langle k| \}, \tag{12}$$

and

$$\hat{\tilde{h}}_{el}(\vec{R}_\alpha, t) = \hat{\mathcal{U}}_0^\dagger(\vec{R}_\alpha, t)\, \hat{h}_{el}(\vec{R}_\alpha, t)\hat{\mathcal{U}}_0(\vec{R}_\alpha, t), \tag{13}$$

we have

$$\hat{\tilde{h}}_{el}(\vec{R}_\alpha, t) \propto \Big\{ \cos\mathcal{A}(\vec{R}_\alpha, t)\ (\ |1\rangle\langle 1| - |2\rangle\langle 2|\ ) \tag{14}$$
$$+ (\ \sin\mathcal{A}(\vec{R}_\alpha, t)\ e^{i[\omega_{12}(t-t_0) + \delta(\vec{R}_\alpha, t)]}|1\rangle\langle 2| + h.c.) \Big\},$$

where $\propto$ indicates a proportionality relation modulo an arbitrary multiple of the identity operator.

The operator $\hat{\tilde{h}}_{el}$ admits two distinct eigenvalues equal to $\pm$ 1, and the respective orthonormal eigenvectors correspond to

$$|\tilde{\varphi}_1;\vec{R}_\alpha, t\rangle = e^{-i/\hbar\ \varepsilon_1(t - t_0)}(\ \cos(\mathcal{A}/2)|1\rangle + e^{-i\delta} \sin(\mathcal{A}/2)|2\rangle\ ), \tag{15a}$$

$$|\tilde{\varphi}_2;\vec{R}_\alpha, t\rangle = e^{-i/\hbar\ \varepsilon_2(t - t_0)}(\ \cos(\mathcal{A}/2)|2\rangle - e^{i\delta} \sin(\mathcal{A}/2)|1\rangle\ ). \tag{15b}$$

Note that the effective area defined in eq. (10a) vanishes at $t = t_0$ and the states $|\tilde{\varphi}_i;\vec{R}_\alpha, t\rangle$ reduce to the field-free states $|i\rangle$, $i = 1,2$, at the initial time $t = t_0$. They represent the exact time evolution from these initial states in the form of adiabatic transports of the associated eigenstates of $\hat{h}_{el}$. The Berry phases[5] associated with this adiabatic evolution are time-integrals of (cf. eq. (7a))

$$\lambda_1 = V_{12}\sin\mathcal{A}\,\sin\delta - \hbar(\ \dot{\delta} + \omega_{12}\ )(\ 1 - \cos\mathcal{A}\ )/2 = -\lambda_2 \tag{16}$$
$$= V_{12}\cos\delta(\ \frac{1 - \cos\mathcal{A}}{\sin\mathcal{A}}\ )$$

Substitution of eqs. (15) into eq. (8) gives an explicit expression of the total state $|\Psi, t\rangle$ in the field-free B.O. representation, in which the nuclear amplitudes are

$$\chi_1 = e^{-i/\hbar\varepsilon_1(t - t_0)}\cos(\mathcal{A}/2)\tilde{D}_1(R_\alpha, t) - e^{-i/\hbar\varepsilon_2(t - t_0)}\ e^{i\delta}\ \sin(\mathcal{A}/2)\tilde{D}_2(R_\alpha, t) \tag{17a}$$

$$\chi_2 = e^{-i/\hbar\varepsilon_2(t-t_0)}\cos(\mathcal{A}/2)\tilde{D}_2(R_\alpha,t) + e^{-i/\hbar\varepsilon_1(t-t_0)}e^{-i\delta}\sin(\mathcal{A}/2)\tilde{D}_1(R_\alpha,t) \tag{17b}$$

where the $\tilde{D}_i$'s are related to the D_i's in the same manner as the states $|\tilde{\varphi}_i\rangle$ are related to the new basis vectors $|\varphi_i\rangle$, eq. (7a).

Due to the appearance of the effective area $\mathcal{A}$ on the left-hand side, eq. (10b) is a non-linear integro-differential equation for the phase angle δ and represents the main difficulty in the present approach. One numerical integration method for solving eqs. (10) on a short time scale- which also ensures the decoupling of the amplitudes D_i in eq. (9)- is suggested by the observation that for $\omega_{12} = 0$, eqs. (10) admit the simple solution $\delta_0 = \pi/2$, $\mathcal{A}_0(t) = (2/\hbar)\int^t V_{12}(t')dt'$. Thus, writing

$$\delta(t) = \delta_0 + \Delta\delta(t), \tag{18}$$

and assuming that $\Delta\delta$ remains small in the interval $[t_0, t]$, eq. (10b) is linearised

$$\dot{\Delta\delta} \cong -\frac{d}{dt}[\ln(\sin\mathcal{A}_0)]\Delta\delta - \omega_{12}, \tag{19}$$

and yields

$$\Delta\delta \cong -\omega_{12}\int_{t_0}^{t}\sin\mathcal{A}_0(t')dt'/\sin\mathcal{A}_0(t). \tag{20}$$

This short-time approximation can be viewed as the starting point of a procedure to solve eq. (10b) iteratively. However, this has not been attempted in the numerical calculations reported below which employ only the first-order approximation for δ and $\mathcal{A}$ defined above. Note that eqs. (10) define the effective area and phase angle for a completely general time-dependent field, and furnish a formally exact solution to the laser-driven two-level problem beyond perturbation theory and the RWA.

A simple and efficient algorithm results from the above constructions, and permits the non-perturbative study of wavepackets in laser-coupled multichannel molecular systems within or without the RWA. This algorithm is expressed by eqs. (17) which give the amplitudes χ_i, i = 1,2, within a time interval short enough to justify the neglect of the non-adiabatic couplings in eq. (9). In this algorithm, the pulse duration is first divided into slices of equal widths δt. Within each slice, eqs. (17) holds with the amplitudes $\tilde{D}_i$ propagating freely under the action of the nuclear kinetic energy operator. Their initial values at the beginning of the slice are identified with the amplitudes χ_i generated at the end of the preceding slice. The phase angle δ is given by eqs.(18) and (20) and is used to generate the effective area $\mathcal{A}$ and Berry phase derivatives λ_i through eqs.(10a) and (16), respectively.

III. APPLICATIONS

1. PHOTODISSOCIATION OF NaI: EARLY DYNAMICS

The wavepacket dynamics during the pump phase of the pump-probe experiment on the NaI system has been studied as a test-case of this methodology. The crossing potential energy surfaces associated with the ground ionic state and the excited covalent state of NaI are taken from ref.[6]. The natural adiabatic coupling between these field-free diabatic states is also taken from this work and is operative only in a small neighbourhood of the crossing point at R = 6.93 Å. Pointwise diagonalization of the diabatic matrix give the adiabatic potential energy curves which represent the two coupled channels in the present application, neglecting the ensuing field-free non-adiabatic couplings. An inverse Smith transformation converts the propagated wavepackets from the (field-free) adiabatic to diabatic representation. Only a single nuclear degree of freedom associated with radial motions is considered and rotations are completely ignored in the calculations reported below. The field-induced interaction between the two field-free adiabatic channels is expressed as

$$V_{12}(t) = V_{12}^{max} f(t)\cos\omega_0 t, \tag{21}$$

where ω_0 corresponds to a wavelength of 310 nm. $V_{12}^{max}/\hbar$ is the peak Rabi frequency and is proportional to the square-root of the field-intensity. It is made to vary from 10^{-3} to 10^{-1} a.u. corresponding to a range of field-intensity varying from 10^{10} W/cm^2 to 10^{14} W/cm^2 for a typical value of the transition dipole moment (ca. 2.5 a.u.). The pulse shape function f(t) denotes a gaussian pulse, and is given by

$$f(t) = \exp[-\beta_f(t-t_f)^2], \tag{22}$$

with

$$\beta_f = \beta_0/\xi^2, \qquad t_f = \xi t_0, \tag{23}$$

where t_0 = 80 fs and β_0 = 1.109x10^{-3} fs^{-2} and the scaling factor ξ permits continuous, coherent variations of the pulse center t_f and width parameter β_f. These pulses are constructed such that the laser is switched on at t = 0 and is practically switched off at t = 2t_f. The values of t_0 and β_0 correspond to a pulse of 50 fs FWHM, a total width of 2t_f = 2t_0 = 160 fs.

In implementing the adiabatic wavepacket propagation algorithm, three levels of integrals arise and correspond to a) the definition of the Berry phases, b) the first-order representations of the effective area $\mathcal{A}$ and phase angle δ and c) the definition of the zeroth-order area $\mathcal{A}_0$. Accordingly, the numerical calculations of these integrals by Simpson's rule make use of three embedded grids covering the current time-slice and consisting of n_k subintervals, k = 1-3. A fast-Fourier transform algorithm is used to go from the coordinate representation, in which the dynamical and Berry phases are diagonal, to the momentum representation in which the free-particle propagator involving the nuclear kinetic energy operator is diagonal. Wavefunctions in coordinate and momentum representations are discretized on direct and reciprocal grids of N_{pos} points with stepsizes δR and $\delta k = 2\pi/(N_{pos}\delta R)$ respectively. The values of the various parameters introduced above are given in Table I for the present application. The initial state is represented by a gaussian wavepacket describing the ground vibrational state supported by the lower potential $\varepsilon_1(R)$ while the upper channel $\varepsilon_2(R)$ is assumed to be initially unexcited.

Three values of the width of the time-slices are given in table I. They are used in a convergence study of the adiabatic propagation algorithm. The results of this study are illustrated in fig. 1a, where the covalent-state's diabatic amplitude generated at t = 128 fs by a 50 fs FWHM pulse with $V_{12}^{max} = 10^{-2}$ a.u. is shown for these three values of δt. The results indicate a convergent representation for $\delta t \leq 10$ a.u. This estimate agrees with the conclusions of a previous study[4] which employed simpler model systems and established the convergence radius of the adiabatic approximation to be $\delta t \sim 8$ a.u. As noted above, the methodology is applicable both within and without the RWA. Fig. 1b compares the converged amplitude χ_2 taken at t = 96 fs for a 50 fs FWHM pulse of varying intensities treated within the RWA with corresponding results obtained beyond this approximation. Even at a relatively low intensity corresponding to $V_{12}^{max} = 3\text{x}10^{-3}$ a.u. (not shown), marked

Table I. parameters used in the propagation of coupled wavepackets for the early dynamics of the photodissociation of. NaI

parameter	value(s)
δt	5, 10, 20 a.u.
(n_1, n_2, n_3)	(100, 10, 10)
$(N_{pos}, \delta R, R_{beg}^{\dagger})$	(1024, .0075 a.u., 4.0 a.u.)

† $R_{beg.}$ is the first point of the spatial grid

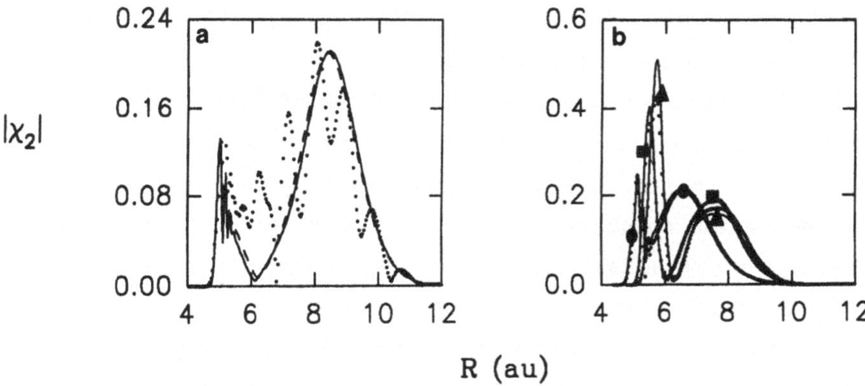

Fig. 1. (a) Diabatic amplitudes χ_2 at t = 128 fs under a 50 fs FWHM pulse with $V_{12}^{max} = 10^{-2}$a.u. calculated using δt = 5 a.u (——), 10 a.u. (- - -) and 20a.u (....). (b) RWA (....) and non-RWA (——) amplitudes χ_2 at t = 96 fs under a 50 fs FWHM pulse with $V_{12}^{max} = 10^{-2}$ a.u. (●), 5×10^{-2} a.u. (■) and 10^{-1} a.u (▲).

differences exist between the RWA and non-RWA results. The differences increase slightly as the field-intensity increases, but not as much as expected and, qualitatively, the RWA does reproduce the variation of the wavepacket dynamics with respect to field-intensity.

The actual dynamics of the diabatic wavepackets and their dependence on the pulse duration are illustrated in Fig.2, where the wavepackets supported by the upper diabatic channel are shown at three different times t within the duration of a 50 fs FWHM pulse (Fig.2a) and within a 100 fs FWHM pulse (Fig.2b). In both case the peak field-intensity corresponds to $V_{12}^{max} = 10^{-2}$ a.u.

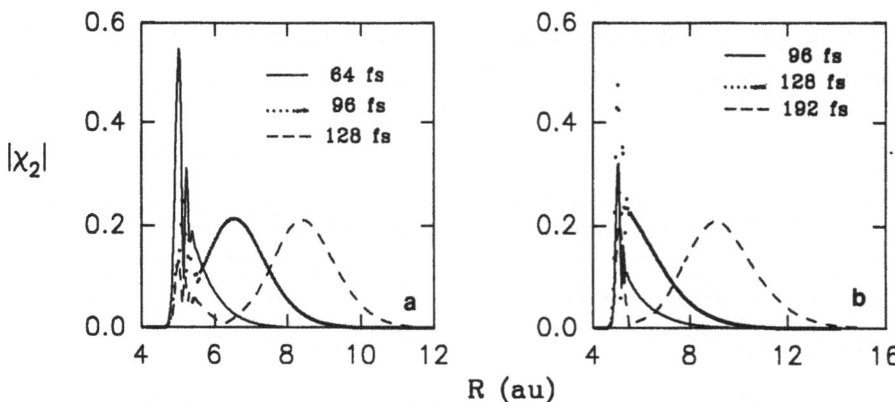

Fig. 2 Time evolution of the diabatic amplitudes χ_2 of NaI under (a) a 50 fs FWHM gaussian pulse and (b) a 100 fs FWHM pulse. For both pulses $V_{12}^{max} = 10^{-2}$a.u.

2. LASER-MODIFIED DYNAMICS OF Na - Ar COLLISIONS

The above system illustrates the application of the present methodology to a half-collision problem. A more interesting test-case is offered by the laser-modified dynamics of the collisions between ground state Na and Ar atoms[7,8]. Full details of the calculations on this system using the present methodology and the potential energy surfaces defined by Lee and George[8] have been reported elsewhere[9]. Only the observations pertaining to the dependence of the population dynamics on the laser pulse shape are summarized here. Starting with an initial gaussian amplitude which represents a narrow wavepacket entering the ground-state channel with a relative kinetic energy of 4.177×10^{-4} a.u., the laser-induced excitation dynamics were studied for two types of laser pulses of varying length and intensity: pseudo-rectangular pulses and smooth gaussian pulses of the same type as described above. Fig.3 gives sample results which illustrate both pulse-length and pulse-shape effects on the channel population dynamics. Two values of the field-intensity are represented in this figure.

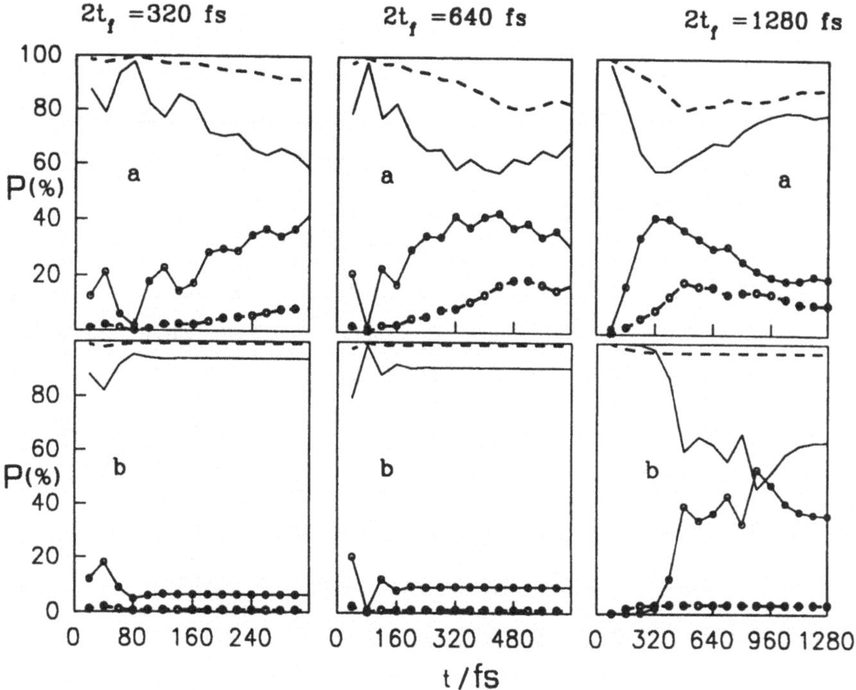

Fig. 3 Time evolution of the populations P_1 (unlabeled lines), P_2 (o-labeled lines) of the $X^2\Sigma$ and $A^2\Pi$ states of Na-Ar under irradiation by (a) rectangular pulses and (b) gaussian pulses of varying total length $2t_f$ and delivering peak Rabi-frequency V_{12}^{max} = 3×10^{-4} a.u. (....) and $V_{12}^{max} = 10^{-3}$ a.u. (——).

Previously, the collision time, the time required for a full reflection of the excited-state amplitude upon the upper channel repulsive wall, has been estimated to $\gtrsim$ 0.6 ps. The 320 fs and 640 fs pulses illustrated in Fig. 3 are short relative to this collision time whereas the 1280 fs is a long pulse. In the case of the rectangular pulses, the excited-state population exhibits a damped oscillatory pattern around a mean "trajectory" which denotes a gradual population built-up at short times, attains a maximum at t ~ 400 fs before declining at longer times. In contrast, under excitations by ultra-short (sub-picosecond) gaussian pulses, the excited state population exhibits an oscillatory pattern which is damped out much more quickly, within the rise-time of the pulses, before stabilizing to a final value which is relatively insensitive to variations of the pulse length and amounts to about 10% of final excitation, for a peak Rabi frequency of 10^{-3} a.u. The passage from this short-pulse regime to the long pulse regime is abrupt and gives way to a population dynamics that is reminiscent of the dynamics under a rectangular pulse.

ACKNOWLEDGMENTS: The financial support of NSERC of Canada is gratefully acknowledged.

REFERENCES

[1] T. S. Rose, M. J. Rosker and A. H. Zewail, J. Chem. Phys. 88:6672 (1988); M. J. Rosker, T. S. Rose and A. H. Zewail, Chem. Phys. Lett. 146:175 (1988); A. H. Zewail, Science 242:1645 (1988); J. Misewich, J. H. Glownia, J. E. Rothenberg and P. P. Sorokin, Chem. Phys. Lett. 150:374 (1988).

[2] V. Engel, H. Metiu, R. Almeida, R. A. Marcus and A. H. Zewail, Chem. Phys. Lett. 152:1 (1988); H. Metiu and V. Engel, J. Chem. Phys. 93:5693 (1990); S. O. Williams and D. G. Imre, J. Phys. Chem. 92:6648 (1988).

[3] T. T. Nguyen-Dang, S. Manoli and H. Abou-Rachid, Phys. Rev. A, 43:5012(1991); T. T. Nguyen-Dang, J. Chem. Phys. 90:2657 (1989).

[4] T. T. Nguyen-Dang and H. Abou-Rachid, Phys. Rev. A, submitted.

[5] M. V. Berry, Proc. R. Soc. London, Ser. A 392:45 (1984).

[6] V. Engel and H. Metiu, J. Chem. Phys. 90:6116 (1989).

[7] T. Sizer II and M. G. Raymer, Phys. Rev. Lett., 56:123 (1986).

[8] H. W. Lee and T. F. George, Phys. Rev. A, 35:4977 (1987).

[9] T. T. Nguyen-Dang and H. Abou-Rachid, J. Chem. Phys., submitted.

FEMTOSECOND PHOTON ECHO FORMATION IN LIQUIDS

E.T.J. Nibbering, D.A. Wiersma and K. Duppen

Department of Physical Chemistry
University of Groningen
The Netherlands

INTRODUCTION

When a quantum-mechanical system is isolated from its surroundings, the time evolution is completely determined by the Hamiltonian of the system alone. The physics of an ensemble of systems that are prepared in a certain state is then conveniently described by the time evolution of the wavefunctions that obey Schrödinger's equation. The propagation of molecular wavepackets on unperturbed molecular potential surfaces is an example of a possible theoretical treatment in these circumstances [1].

When a system is in contact with some sort of heat bath, the Hamiltonian that governs the dynamics of the states of the system has to take the system-bath interaction into account. In principle, the bath comprises all those degrees of freedom that do not form the system. For instance, the electronic states in a crystalline solid may interact mostly with the phonons as the most important bath states, through electron-phonon coupling. For electronic states of isolated gasphase atoms the only relevant bath states may be the vacuum fluctuations of the electromagnetic field. The discrimination between system and bath is often to a large extent arbitrary [2], and depends on detailed knowledge of all degrees of freedom that may affect the outcome of the experiment that is performed. With a macroscopic sample the number of degrees of freedom is staggering, and an exact evaluation is beyond all hope. It is often impossible to define relatively simple system states that form an ensemble, isolated from the environment and from each other.

Usually, the influence of a heat bath on a system has to be considered explicitly when a large number of degrees of freedom influences the motions of the few degrees of freedom that we choose to call the system. Since the details of the motions of the heat bath are unknown for a sample of macroscopic size, statistics has to be invoked to describe its influence on the ensemble of systems. It is then impossible to use wavefunctions in a description of the dynamical properties of the system, since the concept of an ensemble averaged wave function is meaningless. Instead, the dynamics can be described in terms of the density matrix, which evolves according to the Liouville-Von Neumann equation:

$$i\hbar\frac{\partial \rho}{\partial t} = [\mathcal{H}_0,\rho] + [\mathcal{H}_I,\rho] + i\hbar\left(\frac{\partial \rho}{\partial t}\right)_{system-bath} \qquad (1)$$

Here $\mathcal{H}_0$ is the Hamiltonian of the unperturbed system and $\mathcal{H}_I$ describes the coupling of the system with the electromagnetic field(s) that are applied in the experiment.

The random interactions that form the system-bath coupling have to be averaged over the macroscopic ensemble. It is often assumed that this always gives rise to time independent parameters in the equation of motion Eq. (1). An example of this approach are the two-level optical Bloch equations [3] with parameters called T_1, which affects the diagonal matrix elements of the density matrix, and T_2, which governs the relaxation of the off-diagonal elements. These two parameters are related by the pure dephasing constant T_2^*:

$$(T_2)^{-1} = (2T_1)^{-1} + (T_2^*)^{-1} \tag{2}$$

Although the Maxwell-Bloch approach has a wide range of applicability, there are limitations to the description of the relaxation behaviour in terms of time independent parameters. A prerequisite is that a distinct time scale separation exists between the system and the bath [4]. When the bath moves very rapidly on the time scale of the dynamics of the system, and is relatively weakly coupled to the system, it is always at thermodynamic equilibrium and its effect on the system can be time averaged to constant parameters. This is the so-called Markovian approximation of statistical physics. In this *fast modulation* or *Markovian limit*, the effects of the bath on the dynamics of the system of Eq. (1) are completely described, except for very short times, by phenomenological relaxation parameters such as T_1 and T_2 for the two-level case. As a consequence the optical transitions of the system are broadened to what is called the homogeneous line width.

In the opposite case, when the dynamics of the bath is very slow compared to that of the system, equation of motion (1) can be solved for a given configuration of the bath, i.e. for a given system energy, and subsequently an average can be calculated over the relevant distribution of bath states. This is the *slow modulation* or *static limit* of system-bath interactions, which gives rise to the inhomogeneous line width in a linear absorption spectrum.

In many cases both limits apply at the same time, and a well-defined separation occurs between very fast bath degrees of freedom and very slow bath degrees of freedom that both influence the levels of the system. For instance, the translational degrees of freedom provide a bath in the static limit (Doppler broadening) for optical transitions of atoms or molecules in dilute gasses, while collisions among the molecules may be treated in the Markovian (impact) limit [5]. Similarly, in solid materials at low temperature a static component is often present due to a distribution of crystal field strength at the various sites, while electron-phonon interactions give rise to Markovian behaviour of the system [6].

For molecules in liquids however, the validity of the time scale separation approach is questionable since neither the Markovian nor the static limit may apply [7,8]. A true static component is not expected since the velocities and surroundings of the molecules continuously change. It may be that (part of) these changes are sufficiently slow to give rise to static-like interactions, though. On the other end, optical dephasing of molecules in liquids is so extremely rapid, at the most tens of femtoseconds [9], that it is doubtfull whether the Markovian limit can be used. The assumption that such ultrafast system behaviour can be adequately explained by coupling to much faster bath movements seems far-fetched.

In this paper optical dephasing of dye molecules in liquids is studied by time-domain nonlinear optics at time scales of about 10 femtoseconds. This kind of resolution is necessary to characterize the fluctuations that occur in the electronic states of molecules. The experiments are designed to induce rephasing of optical coherences, i.e. are meant to generate photon echoes. In this way the possible effects of faster and slower fluctuations can be separated. In the analysis of the data the Markovian assumption of infinitely fast moving bath degrees of freedom is avoided. To substantiate this approach, the (linear) optical absorption spectrum is simulated within the same model with the same parameters. The consistency of a particular kind of non-Markovian approach is thus demonstrated.

The relevance of a study of the optical dynamics of molecules in liquids lies partly in the general implications for a theoretical description of optical phenomena. As discussed above, it can be expected that the usual approach which relies on Maxwell-Bloch type of relaxation parameters fails, since the Markov approximation probably does not apply. This necessitates an assessment of the application of alternative, more general approaches to optical dynamics. Another motivation for a study of this kind comes from chemistry. The fluctuations in the electronic states of the molecules that are studied not only give rise to optical effects, but are also responsible for most kinds of chemistry. In only a few practical situations can the reactants propagate on unperturbed potential surfaces. As a rule, however, chemistry occurs through the fluctuations that the reactive systems experience from the thermally activated surroundings. The vast majority of reactions is thus best performed in a liquid environment where fluctuations are fast and violent.

A STOCHASTIC MODEL OF OPTICAL DYNAMICS

When the Markov approximation fails because system and bath move at similar time scales, the obvious approach would be to include the most relevant bath states in the system Hamiltonian. The relaxation of the now larger system can be treated in the conventional way when the remaining, smaller bath does satisfy the Markov limitations. When only a few bath states matter, this can be done in a fully quantum mechanical context. However, for dye molecules dissolved in some liquid many bath states will be important. All molecules contain many atoms and all particles interact more or less randomly through many-particle interactions. When very many bath degrees of freedom are important for the dynamics, a fully quantum mechanical treatment of the relaxation behaviour becomes unfeasible. Instead, a classical treatment can be given which may be argued by the fact that the bath states are so closely spaced that they form a continuum.

The theory of the dynamics of a quantized system in contact with a classical, non-Markovian bath was originally introduced in Nuclear Magnetic Resonance to explain motional narrowing of solution phase linear absorption spectra [4,10-14]. The approach that was taken there is directly applicable to optical dynamics, and can be straightforwardly extended to include nonlinear response as well [15,16]. The Hamiltonian that determines the dynamics of what we have defined as the system is, in the absence of optical fields, of the form:

$$\mathcal{H} = \mathcal{H}_{system} + \mathcal{H}_{system\text{-}bath} \tag{3}$$

where the system and interaction Hamiltonians are:

$$\mathcal{H}_{system} = \sum_{\nu} |\nu\rangle \, [\varepsilon_{\nu} - (\tfrac{i}{2})\gamma_{\nu}] \, \langle\nu| \tag{4a}$$

$$\mathcal{H}_{system\text{-}bath} = \sum_{\nu} |\nu\rangle \, H^{\nu}_{SB}(Q_B) \, \langle\nu| \tag{4b}$$

Here ε_{ν} is the energy of the system state $|\nu\rangle$, γ_{ν} is the lifetime of that state, the system-bath Hamiltonian is taken to be diagonal in the system states, and Q_B represents the bath coordinates. When the bath degrees of freedom are assumed to be classical, and the bath is so large that its motions are independent of the system, the Hamiltonian of Eq. (3) can be written as [16]:

$$\mathcal{H} = \sum_{\nu} |\nu\rangle \, [\varepsilon_{\nu} - (\tfrac{i}{2})\gamma_{\nu} + \delta\varepsilon_{\nu}(t)] \, \langle\nu| \tag{5}$$

$\delta\varepsilon_\nu(t)$ describes the fluctuations in the energies of the system states due to the interaction with the bath. Thus, $\delta\varepsilon_\nu(t)$ contains all those bath coordinates that are relevant to the modulations. Without loss of generality it may be assumed that $\delta\varepsilon_\nu(t)=0$ for the ground state of the system. When a two-level system is considered, it then follows that $\delta\varepsilon_\nu(t)=\hbar\delta\omega(t)$ for the excited state. This means that the only fluctuation that matters experimentally is that of the two levels with respect to each other; $\delta\omega(t)$ describes this random fluctuation in the transition frequency of the system.

In case of a two-level system, the summation in the Hamiltonians of Eqs. (3-5) runs only over the two system states which may be called |a> and |b>. When this system is perturbed by an optical field of very short duration at time t=0, the time evolution of, for instance, the off diagonal matrix element ρ_{ba} can be written as:

$$\rho_{ba}(t) = \rho_{ba}(0) \langle \exp\{-\frac{i}{\hbar} \int_0^t \mathcal{H}_{ba}(\tau)d\tau\} \rangle$$

$$= \rho_{ba}(0) \exp\{-i\omega_{ba}t-\tfrac{1}{2}(\gamma_b+\gamma_a)t\} \langle \exp\{-i \int_0^t \delta\omega(\tau)d\tau\} \rangle \tag{6}$$

Here $\omega_{ba}=(\varepsilon_b-\varepsilon_a)/\hbar$ is the (unperturbed) transition frequency of the two-level system, and the brackets < > indicate averaging over the bath degrees of freedom. The evaluation of the relaxation that is due to the frequency fluctuation $\delta\omega(t)$ is not at all trivial. Details on the time dependence of the fluctuations are lacking since an entire ensemble of systems has to be considered. It is only possible to discuss its effect on the matrix element ρ_{ba} in statistical terms. To be able to work this out for a given statistical model the averaging has to be performed on the fluctuation itself, and not on some exponential form like in Eq. (6).

One way of treating this problem is an expansion of the expression of Eq. (6) in cumulants [12]. In this way it is found that the relaxation function can be written as:

$$R_1(t) \equiv \langle \exp\{-i \int_0^t \delta\omega(\tau)d\tau\} \rangle$$

$$\approx \exp\{-\tfrac{1}{2} \int_0^t d\tau_1 \int_0^t d\tau_2 \langle \delta\omega(\tau_1)\delta\omega(\tau_2) \rangle\} \tag{7}$$

To reach this result, $R_1(t)$ was expanded and subsequently all terms of the expansion but the quadratic one were neglected. Then the new exponential form was constructed with the averaging contained in the argument. This procedure is correct if Gaussian statistics is assumed for the stochastic modulations of the system energy levels in the liquid [12]. The Gaussian nature may be argued from the central limit theorem, which holds when many relatively weak random contributions add up to the stochastic effect under consideration. For a stationary Gaussian process the correlation function of frequency fluctuations of Eq. (7) is described by the following form [17]:

$$\langle \delta\omega(\tau_1)\delta\omega(\tau_2) \rangle = \langle \delta\omega(\tau_1-\tau_2)\delta\omega(0) \rangle = \Delta^2 \exp\{-\Lambda|\tau_1-\tau_2|\} \tag{8}$$

Here, Δ is the root mean square amplitude of the frequency excursions, and Λ is the inverse correlation time $(t_c)^{-1}$ of the interaction between the system quantum states and the levels of the bath. In the Markov limit the correlation time is infinitely short on the relevant experimental time scale ($t \gg t_c$), and in the static limit it is infinitely long ($t \ll t_c$). The equations above hold for any value for the ratio t/t_c.

The stochastic relaxation function $R_1(t)$ can now be evaluated by calculating the two-dimensional integral of Eq. (7) with the correlation function of Eq. (8). The result is [14]:

$$R_1(t) = \exp\{-g(t)\} \tag{9}$$

with

$$g(t) = \frac{\Delta^2}{\Lambda^2}\left[\exp(-\Lambda t)+\Lambda t-1\right] \tag{10}$$

In the Markovian and static limits the decays described by Eqs. (9) and (10) become exponential and Gaussian, respectively.

The macroscopic optical polarization that is induced by a very short coherent pulse in the two-level system decays proportional to the matrix element ρ_{ba} of Eq. (6). The spectral lineshape $I(\omega)$ that is observed in a linear absorption spectrum of this system is the Fourier-Laplace transform of the impulsive response [14], so from Eqs. (6)-(10) it follows that:

$$I(\omega-\omega_{ba}) \sim 2\,\mathrm{Re}\int_0^\infty \exp\{i(\omega-\omega_{ba})t-\tfrac{1}{2}(\gamma_b+\gamma_a)t-g(t)\}dt \tag{11}$$

In the Markovian and static limits the lineshapes are Lorentzian and Gaussian, respectively. In intermediate cases lineshapes will be observed that cannot be represented by simple analytical forms.

STOCHASTIC OPTICAL DYNAMICS IN NONLINEAR EXPERIMENTS

The generation of photon echoes is an excellent method to characterize modulations of energy levels. In these experiments two very short optical pulses excite a sample in such a way that a signal is generated separated in time from the externally applied optical fields. This is shown schematically in Fig. 1a. The experimental configuration can be chosen such that the signals propagate spatially separated as well, to facilitate the detection. This is shown in Fig. 1b. Photon echoes will be formed if rephasing of coherences is possible in the system. This will happen if phase information is lost in a well-defined manner and subsequently restored at a later time. For instance when the fluctuations are of a static character, dephasing of a macroscopic superposition state will be due to the apparent distribution in

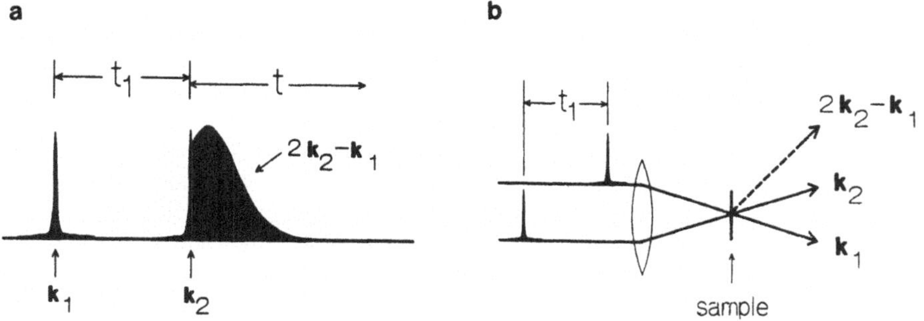

Fig. 1 Schematic representation of a two-pulse photon echo experiment. In (a) is shown how two pulses with wavevectors $\mathbf{k}_1$ *and* $\mathbf{k}_2$ *generate a signal which peaks after a certain time interval. In (b) the geometry of the experiment is displayed:* $\mathbf{k}_{signal} = 2\mathbf{k}_2 - \mathbf{k}_1$.

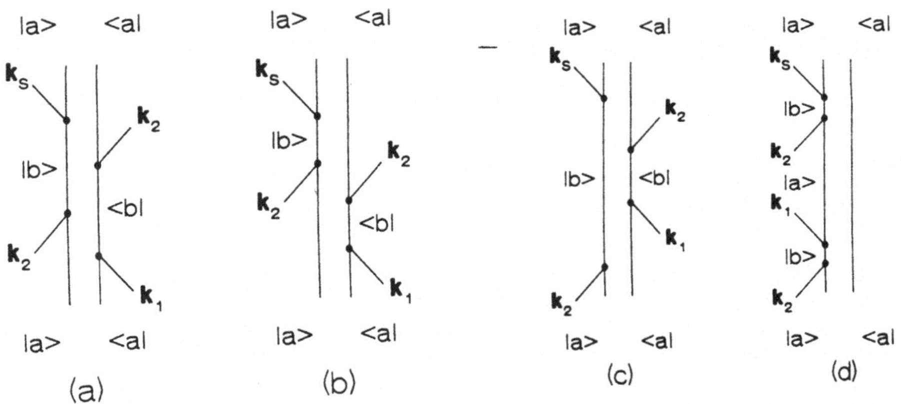

Fig. 2 *Double-sided Feynman diagrams for resonance four-wave-mixing in a two-level system with* $\mathbf{k}_{signal}=2\mathbf{k}_2-\mathbf{k}_1$. *In diagrams* (*a*) *and* (*b*) *an inversion of a coherent superposition state occurs; they contribute to a photon echo signal. Diagrams* (*c*) *and* (*d*) *are negligible when ultrashort pulses are used.*

transition frequencies (inhomogeneous broadening). When the superposition state of the system is inverted at a certain time, all phase factors acquired during the dephasing will be inverted and rephasing will start which then yields an echo.

In the Markovian limit, i.e. at times much larger than the correlation time t_c, all phase information is irretrievably lost. In those circumstances only free induction decays occur and rephasing is impossible (homogeneous broadening). Echo formation will not happen then. In the introduction a few situations were mentioned where both Markovian and static limits applied at the same time. In those cases the static component will allow for the generation of photon echoes, while the Markovian component will cause the echo signals to decay as a function of the time separation between the pulses. The echo experiment then provides a sensitive probe to the relative importance and properties of both components of the system-bath dynamics [18].

When neither of the limits of time scale separation applies, which we expect for molecules in liquids, of course the situation is somewhat harder to describe. However, the non-Markovian treatment that was sketched above for linear response theory can be extended in a straightforward manner to include nonlinear optical experiments such as photon echoes [19-23]. The simplest way of accomplishing this is through standard time domain perturbation theory. In the context of a perturbative treatment of nonlinear optical interactions the photon echo experiment is an example of time domain four wave mixing, i.e. a third order nonlinear effect.

The sequence of system coherences and populations that play a role in the photon echo experiment is conveniently depicted in the form of double sided Feynman diagrams. The scattered signal in the direction $2\mathbf{k}_2-\mathbf{k}_1$ that is due to two optical fields with wavevectors $\mathbf{k}_1$ and $\mathbf{k}_2$ is fully described by the four diagrams in Fig. (2). For a photon echo experiment in which very short pulses are used, the contribution of the terms (c) and (d) becomes negligible since the time ordering of the fields does not comply with the experimental conditions.

Two time periods can be distinguished in the experiment: time t_1 between the optical pulses $\mathbf{k}_1$ and $\mathbf{k}_2$, and time t after the application of the second pulse when the signal is formed. To evaluate the relaxation function that governs the experiment one simply has to follow the sequence of superpositions of the diagrams and write down the appropriate Hamiltonian analogous to the procedure that led to Eq. (6) in the previous paragraph. When population relaxation terms are neglected, the result of this recipe is a third order relaxation function [19–23] that is identical for diagrams (a) and (b):

$$R_3(t,t_1) = \langle \exp\{-i \left(\int_{t_1}^{t_1+t} \delta\omega(\tau)d\tau - \int_0^{t_1} \delta\omega(\tau)d\tau \right)\} \rangle \tag{12}$$

As before, the exponent can be expanded, only the second order terms are retained, and a new exponent is constructed with the averaging contained in the argument. The result is:

$$R_3(t,t_1) \approx \exp\{-\tfrac{1}{2} \left(\int_0^{t_1} d\tau_1 \int_0^{t_1} d\tau_2 \langle \delta\omega(\tau_1)\delta\omega(\tau_2) \rangle \right.$$

$$+ \int_{t_1}^{t_1+t} d\tau_1 \int_{t_1}^{t_1+t} d\tau_2 \langle \delta\omega(\tau_1)\delta\omega(\tau_2) \rangle$$

$$\left. - \; 2 \int_{t_1}^{t_1+t} d\tau_1 \int_0^{t_1} d\tau_2 \langle \delta\omega(\tau_1)\delta\omega(\tau_2) \rangle \right)\} \tag{13}$$

This stochastic relaxation function can be evaluated in the case of Gaussian statistics by inserting the correlation function of Eq. (8) in the two–dimensional integrals. This calculation then yields the following expression:

$$R_3(t,t_1) = \exp\{-2g(t)-2g(t_1)+g(t+t_1)\} \tag{14}$$

where the function g(t) is given by Eq. (10). Explicitly written down in terms of the stochastic amplitude Δ and the inverse correlation time Λ the result is:

$$R_3(t,t_1)=\exp\{- \frac{\Delta^2}{\Lambda^2}\left[2\exp(-\Lambda t)+2\exp(-\Lambda t_1)-\exp(-\Lambda[t+t_1])+\Lambda(t+t_1)-3\right]\} \tag{15}$$

In the Markovian limit ($t \gg t_c = 1/\Lambda$) this relaxation function yields a simple exponential decay. Rephasing of coherences is impossible. In the static limit ($t \ll t_c$) the signal peaks at $t=t_1$, i.e. a conventional photon echo will be observed that is due to complete rephasing of a macroscopic polarization. In intermediate cases rephasing is also possible, due to interference in the fluctuations during times t_1 and t. Thus photon echo type phenomena should be observable, although the standard Maxwell–Bloch type of description does not apply. Consequently, when the dynamics of the energy levels of molecules in liquids displays non–Markovian characteristics, the photon echoes can be expected to behave differently from those that are formed due to a real statical component in the interactions. We will return to this point in the discussion on the experimental results.

EXPERIMENTAL TECHNIQUES

Absorption spectra of molecules in liquids are relatively broad and structureless. This reflects the ultrafast dynamics that the electronic states of the molecules go through in a liquid environment. When these fluctuations are the subject of experimental research, the most direct information will be obtained by techniques which probe the dynamics in real time. Of course then an extremely high experimental time resolution is required. For the photon echo experiments of Fig. (1) optical pulses were employed with a duration of less than 10 femtoseconds.

The experimental set-up that was used to generate these short pulses is shown schematically in Fig. 3. The basic laser oscillator is a colliding pulse modelocked (CPM-)laser [24] (home-built), pumped by an argon ion laser (Coherent 90-6). The pulses from the CPM-laser, with a duration of about 40 fs, were amplified by a copper-vapor laser [25] (MetaLaser 2051) at a repetition rate of 8.3 kHz to a pulse energy of 2 μJ. A small fraction of this energy (ca. 10 nJ) was injected in an optical fiber to increase the spectral bandwidth by self-phase modulation [26]. The resulting spectral width of the pulses leaving the fiber was about 2600 cm^{-1}, enough to carry a 6 fs pulse. The pulse duration was stretched to over a picosecond, an effect which is mainly due to the linear dispersion in the fiber. In this way a more or less linear chirp is induced in the pulse, i.e. a continuous shift of instantaneous frequency from low to high [27].

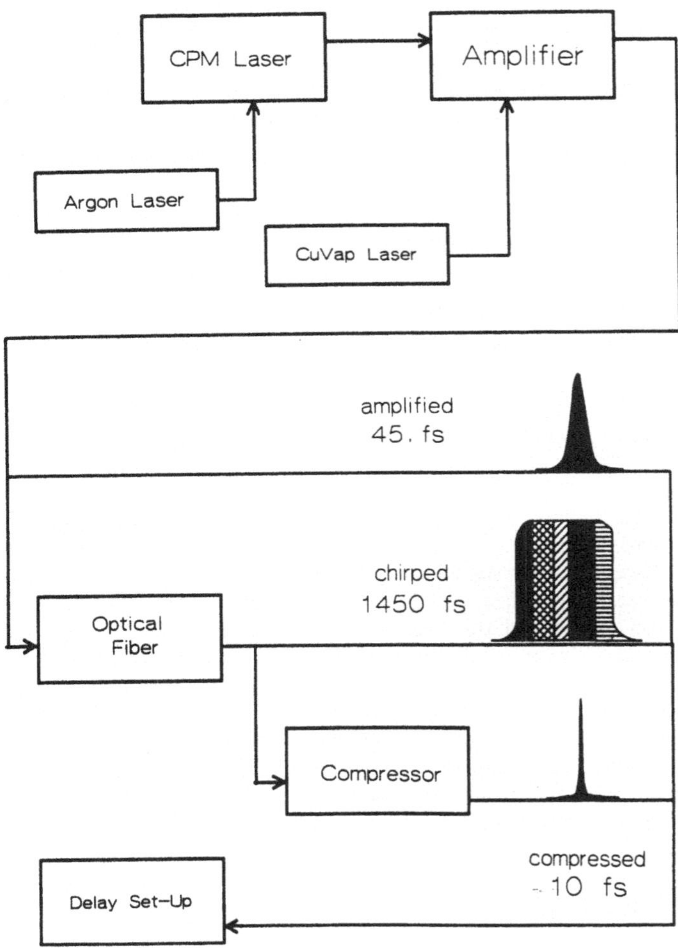

Fig. 3 Schematic flow diagram of the experimental femtosecond dye laser set up. Pulses of less than 10 *femtoseconds can be generated routinely with such an arrangement.*

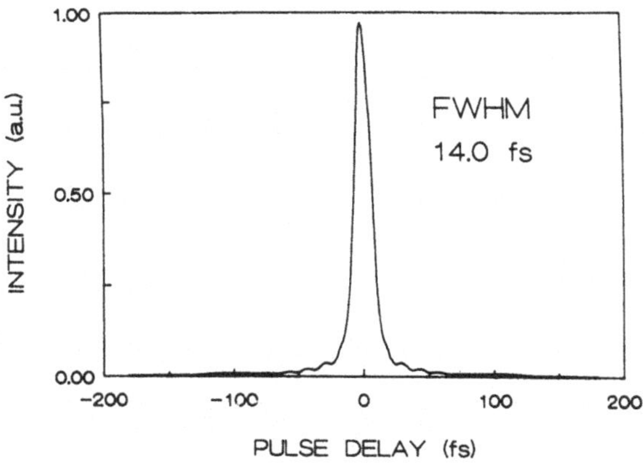

Fig. 4 Intensity autocorrelation trace which serves as an indication of experimental time resolution.

To realize pulse shortening from picosecond to less than 10 femtoseconds, the pulses were send through a compressor consisting of a set of two gratings and four prisms [28]. The temporal width of the optical pulse was measured in an autocorrelation experiment in a 150 μm thick KDP-crystal. The pulses crossed at an angle of 4.5° by focusing two parallel beams with a 75 mm lens. This is the same geometry that was used in the photon echo experiments. The observed autocorrelation trace is shown in Fig. (4). The FWHM of 14 femtoseconds means that the FWHM of the pulse width is about 9 femtoseconds, if a hyperbolic- secant or Gaussian shaped envelope is assumed.

For the impulsive photon echo experiments experiments, the KDP-crystal was simply replaced by a flowing jet stream with a thickness of 60 μm, containing the liquid that was studied. The geometry of the experiment was identical to that in the autocorrelation measurement, only the detector was moved to observe the signal in the direction $2\mathbf{k}_2-\mathbf{k}_1$. This detector was a photomultiplier (EMI 9816) at a considerable distance from the dye jet, to reduce stray light. The current from the photomultiplier was processed by a lock-in amplifier (EG&G 5209) operating at 1.2 kHz.

The samples consisted of the dyes Resorufin dissolved in dimethyl-sulfoxide (DMSO), and Rhodamine 700, dissolved in ethylene-glycol (EG). The blue side of the spectrum of the ultrashort pulse, centred at 620 nm, overlapped quite well with the absorption spectrum of Resorufin. Below 570 nm the optical amplitude decreased rapidly, so the molecular states in that region were coupled weaker to the radiation field. The entire absorption spectrum of Rhodamine 700 could be excited by the short optical pulses.

RESULTS AND DISCUSSION

The result of the femtosecond photon echo experiment on room-temperature Rhodamine in ethylene glycol is shown in Fig. 5. It is clear that dephasing of the electronic coherence occurs at a very short time scale. The fact that a small signal is observed when the pulses are not time-coincident indicates that rephasing processes play some role here. From the fast initial decay and the beating type of behaviour that is observed, it can be concluded that multiple transitions are excited by the optical fields. When many two-level coherences contribute to the macroscopic polarization, the analysis of the data becomes cumbersome, since all these coherences will beat against each other [29]. Unfortunately there is hardly any detailed information available

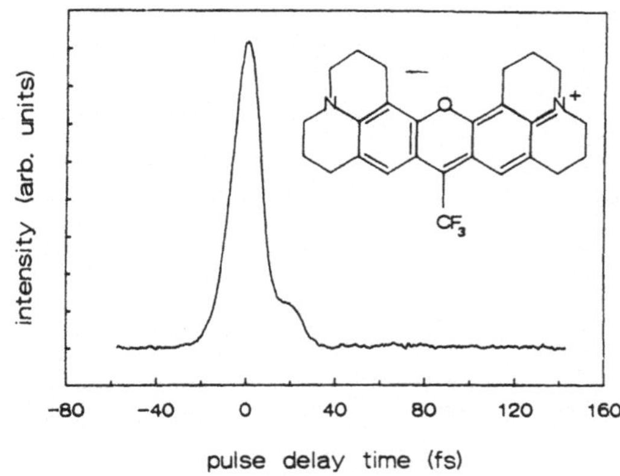

Fig. 5 *Photon echo signal for the dye Rhodamine* 700 *dissolved in ethylene glycol. The molecular structure of this molecule is shown as well. No fitting of the decay was attempted since the level structure is unknown.*

on the level structure of most dyes. Before such a decay as shown in Fig. (5) can be interpreted with any level of confidence, it is absolutely necessary to have some independent information on the level structure.

The photon echo signals that were observed for Resorufin in DMSO are shown in Fig. 6 as the dotted curve. In this case indeed information is available on the relevant level structure [30]. The solid line is a fit to the data that is based upon the stochastic theory described before. In principle many vibronic transitions are present in this system, but only two of these contribute significantly to the signals depicted here: the 0–0 transition at 16795 cm^{-1} and a vibronic transition at 17340 cm^{-1}. The neglect of all other vibronic transitions is permitted in consequence of the nonlinear character of the molecular response in the photon echo experiment. The signal amplitude scales as the fourth power of the transition moment and as the third power of the pulse spectral amplitude. The contribution from all

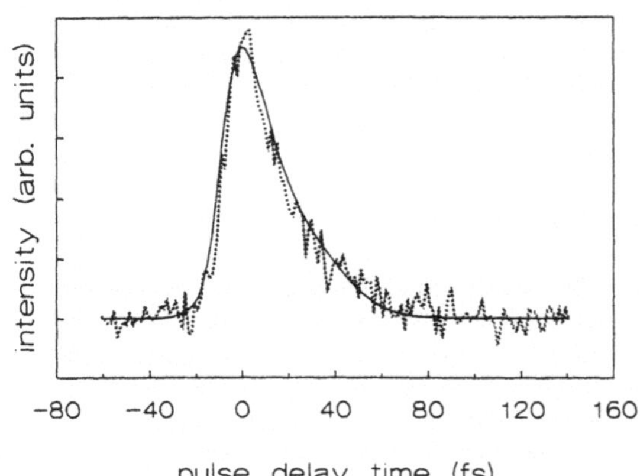

Fig. 6 *Photon echo signal for the dye Resorufin dissolved in dimethylsulfoxide. The solid line is a fit based upon a Gaussian stochastic model for optical dephasing with parameters* $\Delta = 41$ *THz (rad/sec) and* $\Lambda = 27$ *THz (1/sec).*

other vibronic transitions than the two mentioned is negligible on any part of the decay curve of fig. 6, due both to the smaller Franck–Condon factors and the spectral distribution of the short pulses. A detailed account of the level structure will be presented below in connection with simulations on the optical absorption spectrum.

The decay then consists of two components which in principle exhibit quantum beats, due to the interference of the coherences. In the simulation of the data the non-Markovian relaxation function displayed in Eq. (15) was taken as the basic description of the stochastic dephasing process. The solid line of Fig. (6) was constructed by (a) summing the two spectral components at frequencies ω_i with a ratio derived from the transition moments, (b) taking the square of that expression since an intensity is observed instead of an amplitude, (c) integrating the intensity over all times t since the signal is observed at all times, and (d) convoluting (designated by $\otimes$) the result with the experimental time resolution derived from the autocorrelation of Fig. (4). The full expression of the signal as a function of the pulse delay t_1 thus is:

$$I(t_1) = \left\{ \int_0^\infty \left| \sum_{i=1,2} \mu_i^4 \, e^{i\omega_i t} R_3(t,t_1) \right|^2 dt \right\} \otimes \left\{ T(t_1) \right\} \tag{16}$$

With this equation the stochastic parameters Δ and Λ of $R_3(t,t_1)$, that are defined by the correlation function of frequency fluctuations Eq. (8), were optimized to the decay of Fig. (6). This yielded the values Δ = 41 THz and Λ = 27 THz (t_c = 37 fs). Note that in this analysis just a single Gauss–Markov type modulation of the electronic ground and excited state manifolds was considered. Other processes such as vibrational dephasing or energy relaxation were assumed to be not effective on the ultrafast time scale of the electronic fluctuations. Also, the possible existence of slow fluctuations that might be treated in the static (inhomogeneous) limit was not included in the simulation.

What is the character of the signals that were generated by the two pulses? The values that were found for the parameters Δ and Λ indicate that the dynamics of system and bath occur on comparable time scales. Thus, the effect of a finite correlation time has to be explicitly considered. The general picture is that of every conventional photon echo experiment: the first pulse induces coherence which decays during time t_1; the second pulse starts rephasing by inverting the coherent superposition state. The rephasing is not due however to an independent static component in the bath degrees of freedom, but a consequence of interfering fluctuations during t_1 and t, which is made possible by the finite correlation time.

The shape of the signal in time can be calculated through the relaxation function $R_3(t,t_1)$ of Eq. (15). In Fig. (7) is shown how the calculated normalized signal shape varies as a function of pulse delay t_1. Clearly a distinct maximum developes well separated from the excitation pulses. The signals can therefore be classified as photon echoes. In Fig. (8) a three-dimensional plot is shown of the full relaxation function $R_3(t,t_1)$, which now also contains the overall decay of the signal. It turns out that the integrated (over all t) signal amplitude decays faster than exponential as a function of pulse delay t_1. This is a consequence of the fact that the echo cannot occur for a time delay greater than the correlation time after the second pulse. After that time all components of the fluctuations have lost all memory; there is no "inhomogeneous broadening" any more at those time scales.

The crucial role of the correlation time illustrates the most important aspect of photon echo formation in non-Markovian circumstances. The rephasing of the macroscopic coherence is a consequence of interfering fluctuations during times t_1 and t, and is not due to some sort of independent inhomogeneous broadening. Therefore, the correlation time also limits the

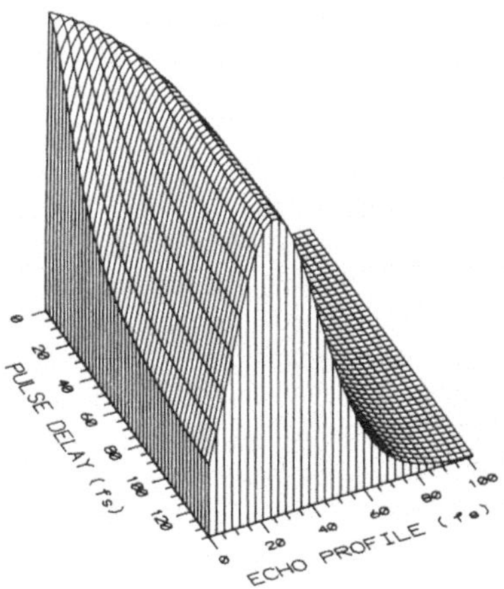

Fig. 7 Normalized photon echo profile for the parameter values of Fig. (6).

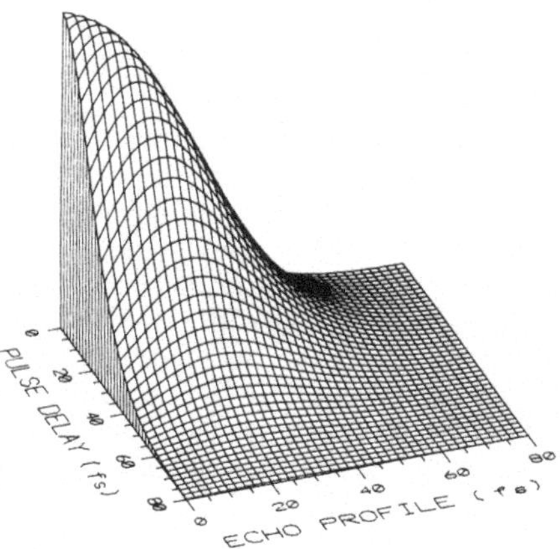

Fig. 8 The two dimensional relaxation function $R(t,t_1)$ for the parameter values of Fig. (6).

time interval between the second pulse and the maximum of the echo signal. Taking the derivative of $R(t,t_1)$ in Eq. (15) with respect to t shows that polarization amplitude has a maximum at:

$$t_{maximum} = \Lambda^{-1} \ln\{2-\exp(-\Lambda t_1)\} \tag{17}$$

Thus, the absence of a true static distribution of transition frequencies limits the time at which the echo can be emitted to less than the correlation time after the second pulse. In Fig. (9) the maximum of a conventional photon echo is compared to the maximum of the non-Markovian photon echo discussed here. In case of the conventional echo a very large inhomogeneous broadening was assumed to be present. With limited inhomogeneous broadening the line that indicates the maximum will be shifted to the right, but the slope remains the same in all situations. This is different when a finite correlation time exists between system and bath, without additional inhomogeneous broadening. In that case the maximum approaches a limit for

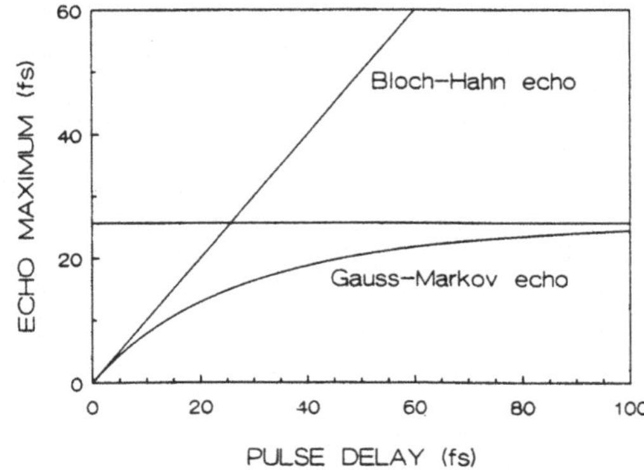

Fig. 9 *Time at which the maximum of the signal amplitude occurs for a conventional echo (Bloch–Hahn), and an echo with stochastic system modulations (Gauss–Markov).*

large pulse delays, as described by Eq. (17). This limit is indicated by the horizontal line in Fig. (9).

In fig. 10, the absorption spectrum of resorufin in DMSO is shown. It is obvious that, next to a pure electronic transition, a manifold of vibronic transitions contributes to this spectrum. As discussed before, in the femtosecond photon echo experiment only the strongest transitions participate due to the nonlinear character of the system-field interaction. While detailed knowledge about the vibrational frequencies and Franck-Condon factors is lacking for resorufin in solution, these parameters have been studied for resorufin in an organic glass [30].

Using the reported vibrational frequencies, an attempt was made to simulate the absorption spectrum with the same stochastic model that was used to explain the photon echo data. The envelope of the absorption spectrum $A(\omega)$ consists of many overlapping vibronic transitions that are all broadened by the coupling to the bath:

$$A(\omega) = \sum_{i} \mu_i^2 \, I(\omega-\omega_i) \qquad (18)$$

where $I(\omega-\omega_i)$ is given by Eq. (11). In the fit the values for the stochastic amplitude Δ and inverse correlation time Λ were taken from the echo decay data. The relative intensities of the vibronic transitions were adjusted to the observed spectrum. The dotted trace in Fig. (10) results from this procedure. The line positions and relative intensities are indicated by the sharp structure underneath the spectrum.

The vibronic intensities thus found agree well with those reported for resorufin in a glass [30]. The hot-band structure around λ = 620 nm was not included in the fit. Also, reliable information about high frequency vibrations was not available (λ<540 nm) and hence no attempt was made to fit this part of the spectrum. Altogether, however, we consider the simulated spectrum in excellent agreement with the measured one, indicating that the stochastic model for ultrafast modulations describes the optical dynamics of resorufin in DMSO remarkably well. Apparently slow fluctuations are unimportant in this system.

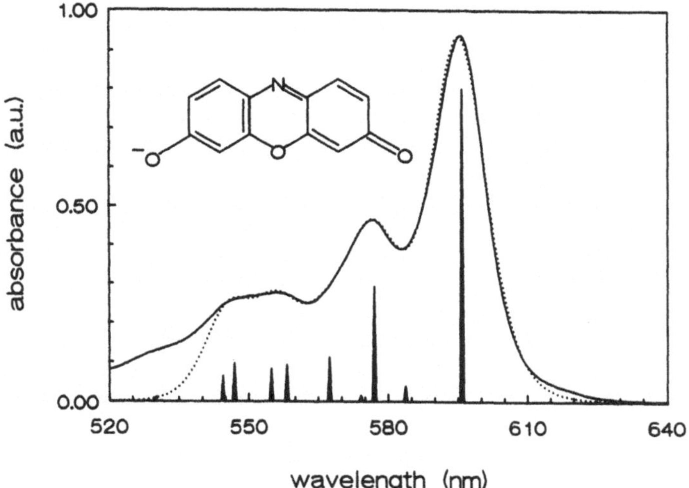

Fig 10 *Absorption spectrum of Resorufin dissolved in DMSO* (*solid line*). *The molecular structure of Resorufin is also shown. The dotted line is a fit based on the level structure and transition moments shown underneat. The parameter values are those of Fig.* (6).

SUMMARY

The optical dephasing of dye molecules in liquids occurs at ultrafast time scales. For a proper evaluation of experimental results information must be available on the level structure of the system. The dynamics of Resorufin in DMSO can be described very well by a stochastic modulation model. With Gaussian statistics both the femtosecond photon echo experiment and the steady state absorption spectrum can be adequately simulated with the same values for the stochastic parameters. In the first case the time scale of the experiment is that of the fluctuations themselves, which allows for rephasing of coherences; in the second case the effect of all possible fluctuations is sampled in the frequency domain.

The dynamics is shown to fall in the intermediate modulation regime, in which neither the Markovian nor the static approximations hold. Rephasing processes are possible as long as the correlation time of the system and bath movements is not exceeded. Photon echoes can therefore be generated which show unusual profiles and peak at times that are not expected in the conventional picture. Generally, the decay of these signals will be faster than exponential when the pulse separation is increased.

The fact that the absorption spectrum and the femtosecond photon echo decay of Resorufin in DMSO can be described with the same model and the same parameter values demonstrates the consistency of the stochastic non-Markovian approach. It also proves that additional fluctuations, on longer time scales, are negligible compared to the ultrafast dynamics that is probed by the femtosecond photon echo. However, at the moment this cannot be a general conclusion. It is quite possible that in other systems slow fluctuations do have considerable effect on the optical dynamics [31].

In the stochastic model the influence of a bath on system dynamics is treated in a non-Markovian way. The bath does not respond at all to the state of the system. It can be expected, however, that upon excitation of the system the bath relaxes towards a new equilibium state. In liquids that is known as solvent relaxation. Whether this behaviour will affect the outcome of the ultrafast dephasing experiments described in this paper depends on the magnitude and the speed of this bath relaxation behaviour. In this respect it should be noted that the experimental absorption profile of the 0–0 transition of Resorufin is slightly asymmetric, which does not show up in the stochastic line shape calculation. Microscopic models, for example based upon semiclassical stochastic Langevin equations, are capable of taking this kind of system–bath coupling into account, thus linking ultrafast dephasing to spectral solvent relaxation [32].

The microscopic physical origin of the amplitude Δ and spectral width Λ of the fluctuations, i.e. the nature of the relevant bath degrees of freedom, is as yet unclear. An explanation can possibly be found in the high-frequency tails of rotational/rovibrational correlation functions. The coupling to the electronic degrees of freedom may then occur through fluctuations of induced dipole moments in the surrounding solvent molecules, due to various motions of these molecules. However, in earlier studies on the system of azulene in liquid alkanes by absorption line shape analysis and resonance Raman it was found that the averaged molecular motions could not be linked directly to either Δ or Λ [8]. It therefore seems necessary to acquire much more data on a multitude of solvents and/or molecular probes to eludicate in any detail the microscopic origins of the ultrafast electronic behaviour. Molecular dynamics simulations will probably be very helpful in the evaluation of experimental results.

These investigations were supported by the Netherlands Foundations for Chemical Research (SON) and Physical Research (FOM) with financial aid from the Netherlands Organization for the Advancement of Science (NWO).

REFERENCES

1. E.J. Heller, Acc. Chem. Res. **14**, 368 (1981).
2. S. Mukamel, Chem. Phys. **31**, 327 (1978).
3. F. Bloch, Phys. Rev. **70**, 460 (1946).
4. P.W. Anderson and P.R. Weiss, Rev. Mod. Phys. **25**, 269 (1953).
5. P.R. Berman, J.M. Levy and R.G. Brewer, Phys. Rev. A **11**, 1668 (1975).
6. W.H. Hesselink and D.A. Wiersma, J. Chem. Phys. **73**, 648 (1980).
7. S. Mukamel, Adv. Chem. Phys. **70**, 165 (1988).
8. E.T.J. Nibbering, K. Duppen and D.A. Wiersma, J. Chem. Phys. **93**, 5477 (1990).
9. C.H. Brito-Cruz, R.L. Fork, W.H. Knox and C.V. Shank, Chem. Phys. Lett. **132**, 341 (1986).
10. P.W. Anderson, J. Phys. Soc. Jap. **9**, 316 (1954).
11. R. Kubo, J. Phys. Soc. Jap. **9**, 935 (1954).
12. R. Kubo, in: *Fluctuation, Relaxation and Resonance in Magnetic Systems*, D. ter Haar, ed., Oliver & Boyd, Edinburgh 1962, pg. 23.
13. J.R. Klauder and P.W. Anderson, Phys. Rev. **125**, 912 (1962).
14. R. Kubo, Adv. Chem. Phys. **15**, 101 (1969).
15. Z. Deng and S. Mukamel, J. Chem. Phys. **85**, 1738 (1986).
16. S. Mukamel, Phys. Rev. A **28**, 3480 (1983).
17. M.C. Wang and G.E. Uhlenbeck, Rev. Mod. Phys. **17**, 323 (1945).
18. N.A. Kurnit, I.D. Abella and S.R. Hartmann, Phys. Rev. Lett. **13**, 567 (1964).
19. M. Aihara, Phys. Rev. B **25**, 53 (1982).
20. B.D. Fainberg, Opt. Spectrosc. **55**, 669 (1983).
21. R.F. Loring and S. Mukamel, Chem. Phys. Lett. **114**, 426 (1985).
22. K.-E. Süsse, W. Vogel and D.-G. Welsch, Chem. Phys. Lett **162**, 287 (1989).
23. E.T.J. Nibbering, D.A. Wiersma and K. Duppen, Phys. Rev. Lett. **66**, 2464 (1991).
24. R.L. Fork, B.I. Greene and C.V. Shank, Appl. Phys. Lett. **38**, 671 (1981).
25. W.H. Knox, M.C. Downer, R.L. Fork and C.V. Shank, Opp. Lett. **9**, 552 (1984).
26. H. Nakatsuka, D. Grischkowsky and A.C. Balant, Phys. Rev. Lett. **47**, 910 (1981).
27. D. Grischkowsky and A.C. Balant, Appl. Phys. Lett. **41**, 1 (1982).
28. R.L. Fork, C.H. Brito-Cruz, P.C. Becker and C.V. Shank, Opt. Lett. **12**, 483 (1987).
29. P.C. Becker, H.L. Fragnito, J.-Y. Bigot, C.H. Brito-Cruz, R.L. Fork and C.V. Shank, Phys. Rev. Lett. **63**, 505 (1989).
30. R. van den Berg and S. Völker, Chem. Phys. **128**, 257 (1988).
31. J.-Y. Bigot, M.T. Portella, R.W. Schoenlein, C.J. Bardeen, A. Migus and C.V. Shank, Phys. Rev. Lett. **66**, 1138, (1991).
32. Y.J. Yan and S. Mukamel, J. Chem. Phys. **89**, 5160 (1988).

DESIGNING COHERENT ACOUSTIC WAVES BY OPTIMAL CONTROL THEORY

Young Sik Kim[1], Herschel Rabitz[2], Attila Askar[3*] and John B. McManus[4]

[1] Dept of Applied Science, Hong-Ik University, Seul 121, Korea
[2] Dept. of Chemistry, Princeton University, Princeton NJ 08544,USA
[3] Mathematics Dept. Boğaziçi University,80815 Bebek, Istanbul,Turkey
[4] Aerodyne Research, Inc., Billerica, Massachussetts 01821-3976, USA
* Presenting author

INTRODUCTION

There exists an intensive effort to design specific wave patterns in classical fields as well as in quantum chemistry [1-9]. In particular, there are a number of physical situations such as internal material diagnostics and modifications[10] where it would be useful to produce a specified acoustic wave structure within a solid by applying a pattern of forces on the solid's surface. The surface loads are created by using lasers, electron beams as well as through transducer arrays [11-15]. Waves in a solid are of two types as compressional and shear waves with respectively longitudinal and transverse propagation character. This added complexity offers in fact an additional flexibility for achieving a desired output. The design of a surface load pattern in both space and time for the coherent focusing of waves at a prescribed target volume at a prescribed time is studied in this paper. Posed in this manner, such a design is an inverse problem. In general, the guessing of the input surface load to achieve a prescribed wave pattern as output is very complicated and relies heavily on experience and intuition in the absence of a rational design procedure. The proposed scheme provides a rational design procedure to substitute intuition and to achieve constructions where intuition would fail. This is particularly true for generating coherent waves that interfere constructively as well as destructively in specific regions of space. The methodology is not confined to this specific case and the work presented here illustrates also the capabilities of the formulation. The theory has two main components: optimal control theory[16] and mechanics of solids[17] that are outlined in the subsequent two sections. The final section presents two sets of results and attempts to derive some general conclusions. A more detailed presentation and additional results are to be found in Ref.18.

OPTIMAL CONTROL THEORY

All control theory problems consist of the same elements: an "objective" to be met subject to some "costs" and "constraints". The objective is achieved by means of "controls". Mathematically, the objective, cost and constraint may be of various forms such as an algebraic, differential or integral expression. The constraints are usually the basic laws governing the process and the costs are choices made by the designer. The objective and the costs, which are introduced through some weights indicating their importance, form the functional to optimize. The "optimal control theory" does not try to achieve the goal exactly, which is not possible in most cases anyway, but rather tries to meet the objective as best as possible. It is only natural and common practice to introduce the constraints through Lagrange multipliers in the usual way. In the present formulation, the objective consists of the desire of depositing a prescribed total amount of energy at a target volume at a target time. The geometry is chosen to have a free surface, since this is the only experimentally feasible way to exert a force to a solid by a laser beam. In achieving this goal in the most efficient manner, the energy deposition at other locations in space but the target and the hitting of the surface too vigorously are discouraged. Thus, the physical objective functional may be written as

$$J(u_i, \tau_i) = \Phi(T) + L \tag{1}$$

with

$$\Phi(T) = \left| \int_{V_c} dV\, e(\mathbf{x},T) - E_p \right|$$

$$L_1 = w_1 \int_{V-V_c} dV \int_0^T dt\, e(\mathbf{x},t) \tag{2}$$

$$L_2 = \frac{1}{2} w_2 \int_S dS \int_0^T dt\, \tau_i^2$$

Above, the error function $\Phi(T)$ measures the degree of satisfaction of the objective of localizing the energy at the specified value E_p in the local target volume V_c at the target time $t = T$ and e is the energy density per unit volume. In competition with this goal are the penalty cost functions $L = L_1 + L_2$ respectively corresponding to the desire of minimizing the total energy of the system except in the volume V_c at $t = T$ and the total surface force during the energy focusing process.

The constraint L_2 above allows for the optimal surface loads to take on both positive and negative values. However, in the laboratory, it is quite difficult to generate negative surface loads with optical means (other techniques could achieve such negative forces by first uniformly preloading the surface and selectively relaxing the force). Thus, we may desire to constrain the surface load to assume only positive values. For this constraint, we define the surface traction as:

$$\tau_i = -p_i^2 \tag{3}$$

The optimal problem thus constrained can be reformulated with new cost function L_2 as

$$L_2 = \frac{1}{2} w_2 \int_S dS \int_0^T dt\, p_i^2 \tag{4}$$

The choice of the weight factors w_i allows for flexibility in balancing the role of the Φ and L contributions to J.

The dynamic optimization of the surface loading for the purpose of efficient focusing of acoustic energy within the solid interior may be stated as the problem of finding the optimal control function τ_i from a set of admissible controls such that the objective functional J is minimized under the constraint that the system obeys its evolution equation. For instance for a purely mechanical field, the evolution equation consists of the momentum conservation. For this case, the energy conservation is automatically satisfied as a consequence of the momentum conservation equation. Similarly, for a thermomechanical process, the field equations include both of the momentum and energy conservations. In this case, the energy conservation provides an independent equation which describes the heat flow. In the present formulation, let us consider the purely mechanical case where the equation of motion is:

$$t_{ij,j} - \rho \ddot{u}_i = 0 \tag{5}$$

Above t_{ij} is the stress and u_i is the displacement with the dot indicating the time derivative. Similarly, ρ is the mass density per unit volume. The summation convention on the repeated indices is adopted throughout the text. The constitutive law expressing the stress in terms of the displacement and related derivatives as well as the form of the accompanying strain energy distinguish between the various models of solid behaviour, with the most basic case being linear elasticity.

The constrained minimization problem can be treated as an unconstrained one by introducing Lagrange multiplier functions called as the costate function. Hence, instead of minimizing J with the constraints such as those in Eq.(3), one can minimize the modified objective functional

$$\bar{J}(u_i, \tau_i, \psi_i) = J - \int_V dV \int_0^T dt\, \psi_i(t_{ij,j} - \rho\, \ddot{u}_i) \tag{6}$$

The relevant equations are derived by following the usual prescription of variational calculus for arbitrary variations in τ_i ,u_i and ψ_i once a particular constitutive law for the solid is chosen.

ELASTIC SOLIDS

For the surface force deposited through a laser beam, it is only natural to generate heat on the surface. Similarly, a successful focusing of acoustic energy is accompanied by both large stresses and the release of these through heat generation. Again large stresses may result on the surface while the laser beam shines on the surface as well as the target when the waves achieve a high degree of focusing. In this context a natural hierarchy of models for the solid behaviour is suggested for accounting for various aspects of solid behaviour. The linear isotropic elasticity model for solid behaviour is the most fundamental and is treated in this work. A detailed treatment based on this procedure is presented in Ref(18). For this basic solid behaviour, the energy density and the stress-strain relations are:

$$e = \frac{\rho}{2}\dot{u}_i^2 + \frac{\lambda}{2}u_{k,k}\, u_{l,l} + \frac{\mu}{2}u_{i,j}(u_{i,j} + u_{j,i})$$

$$t_{ij} = \lambda\, u_{k,k}\, \delta_{ij} + \mu\, (u_{i,j} + u_{j,i}) \tag{7}$$

The corresponding equations of motion are obtained by the substitution of t_{ij} in (5) along with the boundary and initial conditions as:

$$\mu\, u_{i,jj} + (\lambda+\mu)\, u_{j,ji} = \rho\, \ddot{u}_i$$

$$\lambda\, u_{j,j}\, n_i + \mu\, (u_{i,j} + u_{j,i})\, n_j = \tau_i \qquad \mathbf{x} \in S \tag{8}$$

$$u_i(\mathbf{x},0) = \dot{u}_i\, (\mathbf{x},0) = 0$$

Linear problems exploit the advantage of a Green's function which eliminates the need to solve for the elasticity field in the whole domain and that the boundary conditions are satisfied automatically. The problem is hence formulated in terms of the surface traction alone. With the above expressions, integrations by parts and the use of the Green's function satisfying the free surface boundary conditions yield the solution for the displacement at any point in space as:

$$u_i(\mathbf{x},t) = \rho \int_V dV\left[G^+_{ij} \frac{\delta}{\delta t'} u_j(\mathbf{x}',t') - u_j(\mathbf{x}',t') \frac{d}{\delta t'} G^+_{ij} \right]_{t'=0} + \int_0^t dt' \int_S dS\, G^+_{ij}\, \tau_j(\mathbf{x}',t') \tag{9}$$

The equations for the optimized solution are now obtained by substituting the above expressions into the functional in (6). Certainly, for the functional derivative with respect to ψ_i , one recovers the equations of motion, i.e. the constraint as in all variational problems. The functional derivatives with respect to u_i and τ_i yield:

$$\mu\, \psi_{i,jj} + (\lambda+\mu)\psi_{j,ji} = \rho\ddot{\psi}_i$$

$$\lambda\psi_{j,j}\, n_i + \mu(\psi_{i,j} + \psi_{j,i})\, n_j = \tau^c_i \qquad \mathbf{x} \in S$$

$$\tau^c_i = w_1\ [(T-t)\, \dot{\tau}_i - \tau_i] \tag{10}$$

$$\psi_i(\mathbf{x},T) = (w_1 \pm 1)\ u_i(\mathbf{x},T) \qquad \dot{\psi}_i(\mathbf{x},T) = (w_1 \pm 1)\dot{u}_i(\mathbf{x},T) \qquad \mathbf{x} \in Vc$$

$$= 0 \qquad\qquad = 0 \qquad \mathbf{x} \notin Vc$$

It should be noted that the equation for the costate function ψ_i is the same as for the displacement field. This is a consequence of the self-adjointness of the wave operator. It is also seen that the initial and boundary conditions for ψ_i are different from those for u_i . The costate function may be solved in terms of the half space Green's funtion as:

$$\psi_i(\mathbf{x},t) = \rho \int_V dV\left[G^-_{ij} \frac{\delta}{\delta t'} \psi_j(\mathbf{x}',t') - \psi_j(\mathbf{x}',t') \frac{\delta}{\delta t'} G^-_{ij} \right]_{t'=T} + \int_0^T dt' \int_S dS\, G^-_{ij}\, \tau^c_j(\mathbf{x}',t') \tag{11}$$

The advanced dyadic Green's function G^-_{ij} is related to the retarded form G^+_{ij} used in (9) for solving u_i as:

$$G^-_{ij}\ (\mathbf{x},t;\mathbf{x}',t') = G^+_{ij}\ (\mathbf{x},-t;\mathbf{x}',-t') = G^+_{ij}\ (\mathbf{x},t';\mathbf{x}',t) \tag{12}$$

The continuous space-time optimal control problem is transformed to a mathematical programming problem through the discretization of the space and time. This leads to the task of minimizing a functional over a set of coefficients, subject to algebraic constraints. However, as a consequence of the use of the half-space dyadic Green's function, the

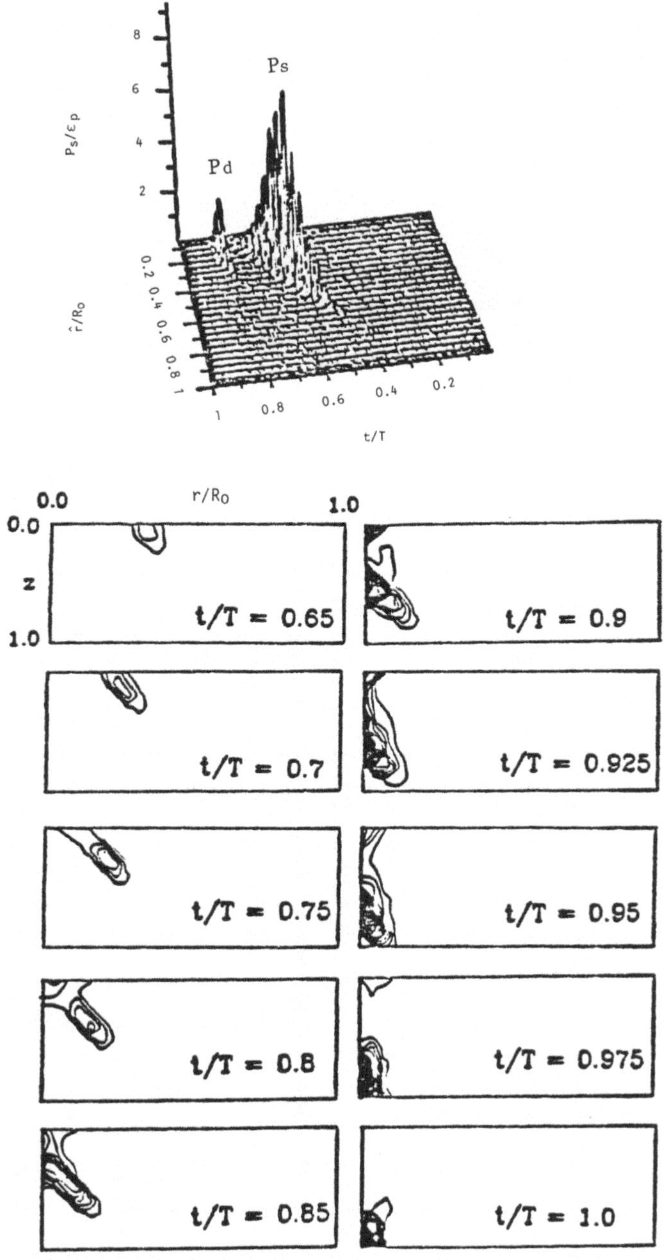

Figure 1. The optimal surface load and the corresponding time sequence of optimally controlled energy density contour maps for the weights $w_1 = 0.0$, $w_2 = 1.0$. The contour interval is $\Delta\varepsilon/\varepsilon P = 0.5$, where the target energy is defined as $\varepsilon P = EP/V_C$ and the time is nondimensional with the target time being $t/T = 1$. Note that by only applying a minimal surface force, most of the focused acoustic energy is generated by the shear motions and the maximal energy concentration occurring before the target time.

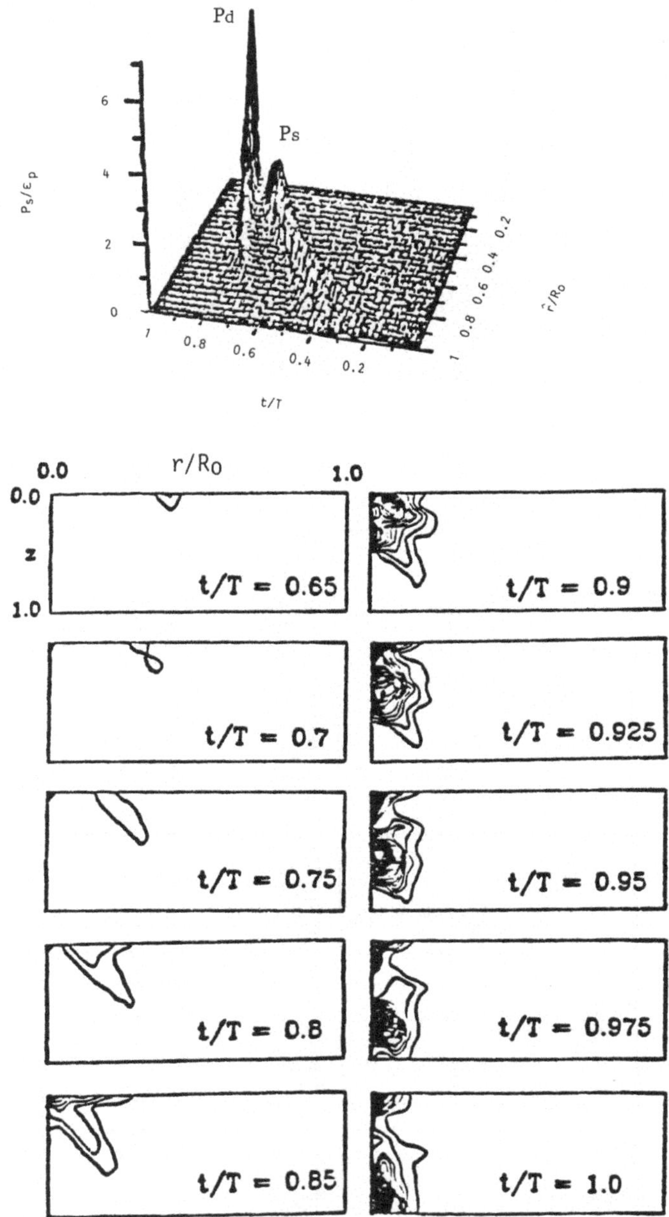

Figure 2. The optimal surface load and the corresponding time sequence of optimally controlled energy density contour maps for the weights $w_1 = 0.32$, $w_2 = 0.0$. The contour interval is $\Delta\varepsilon/\varepsilon p = 0.1$, where the target energy is defined as $\varepsilon p = Ep/V_c$ and the time is nondimensional with the target time being $t/T = 1$. Note that with only a minimum system disturbance, most of the focused acoustic energy is generated by the longitudinal motions due to the highly concentrated surface load near the center of the surface.

optimal control theory avoids the discretization of the whole space and the discretization of the boundary surface and target volume is sufficient. Special care is needed for the choice of time intervals and boundary discretization in order to avoid violating the causality property of the Green's function; that is, in a time step, the spatial nodes should not communicate and should be localized. This condition is met by having $c_d\Delta t/\Delta r<1$, where Δr and Δt are respectively the grid spacing and time increment and c_d is the longitudinal wave velocity. For the full treatment of the theoretical formulation and its numerical implementation for the linear elasticity the reader is referred to Ref. (18).

RESULTS

To illustrate the capabilities of the theory, Figures 1 and 2 show two typical sets of results for the surface load and the equal stress contours within the solid. Typically, the surface loads are made of two pulses travelling at the speeds of the compressional and shear waves. The optimal design consists of the constructive interference of the waves generated by these pulses within the target volume as well as their destructive interference outside it. The material parameters are those for Aluminum. The two cases correspond respectively to the choice of the weight parameters $w_1=0$, $w_2=1$ and $w_1=0.32$, $w_2=0$. Physically, the prescribed weigths correspond respectively to the cases where a cost is enforced on the surface load in the first case which amounts to keeping it from taking too large values. In the second case, the waves are discouraged to be found outside the target volume Vc. The comparison of the two figures indicates the effectiveness of the cost functionals in altering the design. These costs are reflected quantitatively in the observed load patterns where in the first case the level of surface force is less than that in the second case. Similarly, the calculated efficiency of energy deposition, i.e. the ratio of the energy at the target at time T to the work done by the applied surface load is respectively 13% and 65%. These yield ratios also reflect the assigned costs, since there was no cost in the first case for the waves to be outside the target volume. Especially this last yield is extremely high and it appears unlikely to guess using intuition or experience the force pattern that would create the necessary wave field.

REFERENCES

1. J. N. Brittingham, J. Appl. Phys. 54, 1179 (1983).
2. R. W. Ziolkowski, Phys. Rev. A39, 2005(1989).
3. R. W. Ziolkowski, D. K. Lewis, B. D. Cook, Phys.Rev. Lett. 62,147 (1989).
4. P.Hillion, J. Appl.Phys. 60, 2981 (1986); J. Math.Phys.28,1743 (1987).
5. T. T. Wu,J. Appl.Phys. 57, 2370 (1985); T.T. Wu, R.W.P. King and H.-M. Shen, J. Appl. Phys. 62, 4036 (1987).
6. J. Durnin, J. Opt. Soc. Am. A4, 651 (1987); J. Durnin, J. J. Miceli, Jr. and J. H. Eberly, Phys. Rev. Lett. 58, 1499 (1987).
7. H. E. Moses, J. Math. Phys. 25, 1905 (1984); H. E. Moses and R.T. Prosser, IEEE Trans. Antenna Propag. AP-34, 188 (1988).
8. E. Heyman and L.P. Felsen, IEEE Trans. Antenna Propag. AP-34, 1602 (1986); E. Heyman and B. Z. Steinberg, J. Opt. Soc.Am.A4, 473 (1987); E. Heyman, B. Z. Steinberg and L. P. Felsen, J. Opt. Soc. Am. A4, 2801 (1987).
9. S S. Shi, A. Woody, H. Rabitz, J. Chem. Phys. 88, 6870 (1988).
10. F. R. Breckenbridge, C. E. Tschiegg and M. Greenspan, J. Acoust. Soc. Am. 57, 626 (1975); N. N. Hsu and S. C. Hardy, in Elastic Waves and Non-Destructive Testing, Edited by Y. H. Pao (ASME, New York, 1978), pp. 85-106; For rewiev, see, e.g., R. B. Thompson, ASME J. Appl. Mech. 50, 1191 (1983).
11. H. W. Jones and H. W. Kwan, Ultrasonics 23, 63 (1985).
12. C. B. Scruby, R. J. Dewhurst, D. A. Hutchins and S. B. Palmer, J. Appl.Phys. 51, 6210 (1980); R. J. Dewhurst, D. A. Hutchins, S. B.

Palmer and C. B. Scruby, J. Appl. Phys. Lett. 38, 677 (1981); J. Appl. Phys. 53, 4064 (1982).
13. R. M. White, J. Appl. Phys. 34, 3559 (1963); J. E. Sinclair, J. Phys.D: Appl. Phys. 12, 1309 (1979); L. R. F. Rose, J. ,Acoust. Soc. Am. 75,723 (1984).
14. R. R. Boade and O. L. Burchett, J. Appl. Phys. 47, 3412 (1976); L. J. Balk, Can. J. Phys. 64, 1238 (1986).
15. A. J. A. Bruinsma, J. A. Vogel, Appl. Opt. 27, 4690 (1988).
16. D. G. Luenberger, "Introduction to Dynamical Systems: Theory, Models and Applications," [Wiley, New York, 1979]; A. Bryson and Y. Ho, Applied Optimal Control, Blaisdell, Waltham, Mass. (1969)
17. A. C. Eringen, Mechanics of Continua John Wiley and Sons (1967)
18. Y. S. Kim, H. Rabitz, A. Askar, J. B. McManus, "Optimal control of acoustic waves in solids,"Accepted for publication,Phys.Rev.B,1991.

AUTHOR INDEX

SUBJECT INDEX

The manufacturer's authorised representative in the EU is Springer Nature Customer Service Centre GmbH, Europaplatz 3, 69115 Heidelberg, Germany. If you have any concerns regarding our products, please contact ProductSafety@springernature.com

Printed and bound by CPI Group (UK) Ltd, Croydon, CR0 4YY
15/07/2026
02167619-0012